DATE DUE

| | |
|---|---|
| | |

# Progress in Nonlinear Differential Equations and Their Applications
Volume 21

# Partial Differential Equations and Mathematical Physics

## The Danish-Swedish Analysis Seminar, 1995

Lars Hörmander
Anders Melin
Editors

Birkhäuser
Boston · Basel · Berlin

Lars Hörmander
Anders Melin
Department of Mathematics
University of Lund
S-221 00 Lund
Sweden

**Library of Congress Cataloging-in-Publication Data**

Danish-Swedish Analysis Seminar (1995 : Copenhagen, Denmark, etc.)
Partial differential equations and mathematical physics : the Danish-Swedish Analysis Seminar, 1995 / Lars Hörmander and Anders Melin, editors.
p. cm. -- (Progress in nonlinear differential equations and their applications ; v. 21)
Includes bibliographical references.
ISBN 0-8176-3906-3 (h : alk. paper). -- ISBN 3-7643-3906-3 (h : alk. paper)
1. Differential equations, Partial--Congresses. 2. Mathematical physics--Congresses. I. Hörmander, Lars. II. Melin, Anders, 1943- . III. Title. IV. Series.
QC20.7.D5D36 1995 95-53771
515'.353--dc20 CIP

Printed on acid-free paper


*Birkhäuser*



ISBN 0-8176-3906-3
ISBN 3-7643-3906-3
Typeset in $\mathcal{AMS}$-TEX.
Printed and bound by Quinn-Woodbine, Woodbine, NJ.
Printed in the U.S.A.

9 8 7 6 5 4 3 2 1

## CONTENTS

# PREFACE

On March 17–19 and May 19–21, 1995, analysis seminars were organized jointly at the universities of Copenhagen and Lund, under the heading "Danish-Swedish Analysis Seminar". The main topic was partial differential equations and related problems of mathematical physics. The lectures given are presented in this volume, some as short abstracts and some as quite complete expositions or survey papers. They span over a large variety of topics. The most frequently occurring theme is the use of microlocal analysis which is now important also in the study of non-linear differential equations although it originated entirely within the linear theory. Perhaps it is less surprising that microlocal analysis has proved to be useful in the study of mathematical problems of classical quantum mechanics, for it received a substantial input of ideas from that field.

The scientific committee for the invitation of speakers consisted of Gerd Grubb in Copenhagen, Lars Hörmander and Anders Melin in Lund, and Johannes Sjöstrand in Paris. Lars Hörmander and Anders Melin have edited the proceedings. They were hosts of the seminar days in Lund while Gerd Grubb was the host in Copenhagen.

Financial support was obtained from the mathematics departments in Copenhagen and Lund, CNRS in France, the Danish and Swedish National Research Councils, Gustaf Sigurd Magnuson's foundation at the Royal Swedish Academy of Sciences, and the Wenner-Gren foundation in Stockholm. We want to thank all these organisations for their support.

We also wish to thank all the invited speakers for their inspiring lectures and contributions to this volume.

Copenhagen, Lund and Paris in October 1995

Gerd Grubb Lars Hörmander Anders Melin Johannes Sjöstrand

## Program for the seminars

**March 17, 1995 in Copenhagen**

L. Nirenberg
H. Brézis
S. Alinhac
J. Råde

**March 18, 1995 in Lund**

R. B. Melrose
B. Helffer
D. Robert
J. J. Duistermaat
G. Grubb

**March 19, 1995 in Lund**

J. M. Bony
J. Y. Chemin
G. Lebeau
G. Métivier

**May 19, 1995 in Copenhagen**

S. Agmon
J. Sjöstrand
E. B. Davies
G. Vodev
A. Melin

**May 20, 1995 in Lund**

F. Treves
J.-M. Trépreau
L. Boutet de Monvel
V. Guillemin
E. Skibsted
J. P. Solovej

**May 21, 1995 in Lund**

J. Brüning
M. Beals
C. Sogge
P. Gérard

# ON PERTURBATION OF EMBEDDED EIGENVALUES

SHMUEL AGMON

**1. Introduction.** It is well known that in a proper setup eigenvalues which belong to the discrete spectrum are stable. This property is the basis of the perturbation theory for such eigenvalues. On the other hand, the behavior of eigenvalues which are embedded in the continuous spectrum is completely different. Such eigenvalues may be very unstable under perturbations. A striking example of instability of embedded eigenvalues was given by Colin de Verdière [4]. Consider the Laplace operator $\Delta_g$ on a non-compact hyperbolic surface $M$ having a finite area, $g$ the metric. It is well known that the self-adjoint realization of $\Delta_g$ (denoted by the same symbol) has a continuous spectrum which is the half-line: $\left(-\infty, -\frac{1}{4}\right]$. The discrete spectrum of $\Delta_g$ is a finite set. However, there are many interesting examples where $\Delta_g$ has infinitely many eigenvalues embedded in the continuous spectrum. That these examples are exceptional in some sense is shown by the following theorem due to Colin de Verdière.

**Theorem 1.1.** *Given any open set $\Omega \Subset M$, there exists a metric $h$ conformal to $g$ on $M$, satisfying $h = g$ on $M \setminus \Omega$, $h$ arbitrarily close to $g$ in $\Omega$, such that $\Delta_h$ (the Laplacian in the metric $h$) has no eigenvalues embedded in the continuous spectrum.*

The phenomenon of instantaneous dissolution of embedded eigenvalues under arbitrary small perturbations was first observed and studied by physicists in the case of the many body Schrödinger operators. These investigations led to an interesting theory of resonances by which it is claimed that in certain situations when embedded eigenvalues are dissolved (under a perturbation) they immediately reappear as resonances. The mysterious objects called resonances had various definitions. It is now commonly accepted to identify resonances with certain poles of the resolvent operator function (suitably modified). These poles "live" in a Riemann surface obtained by analytic continuation of the modified resolvent through the continuous spectrum.

---

Partially supported by the Edmund Landau Center for research in Mathematical Analysis, sponsored by the Minerva Foundation (Germany).

That embedded eigenvalues which dissolve under an infinitesimal perturbation may turn into resonances was established rigorously in some interesting cases by various authors. We mention here in particular the works of Simon [8] and Phillips and Sarnak [7] who considered, respectively, the case of the $N$-body Schrödinger operator and the case of the automorphic Laplacian.

In this paper we show among other things that results of instability of embedded eigenvalues of the type of Theorem 1.1 hold for a general class of elliptic operators on Riemannian manifolds. The main assumption on the operators is that they possess resonances in the sense mentioned above. Our approach is based on a perturbation theory for resonances. A good part of this paper is devoted to the presentation of this theory. The theory has an abstract formulation but we discuss it here in the special framework of elliptic differential operators. (For other theories of perturbation of resonances see Howland [6], Albeverio and Høegh-Krohn [3], Gesztesy [5] and references given there).

Finally we note that this paper contains only few proofs. A more complete version of the results presented in this paper will be published elsewhere.

**2. The main results of instability of embedded eigenvalues.** Let $M$ be a $C^\infty$ Riemannian manifold (possibly with a boundary). We denote by $g$ and $d\mu$ the metric and the induced measure on $M$, respectively. Let $P$ be a $C^\infty$ elliptic differential operator on $M$. We assume that $P$ is formally self-adjoint in the sense that

$$\langle Pu, v\rangle = \langle u, Pu\rangle,$$

$\forall u, v \in C_0^\infty(\operatorname{int} M)$ ($\langle\cdot,\cdot\rangle$ denotes the inner product in $L^2(M) := L^2(M; d\mu)$).

We consider a self-adjoint realization of $P$ in $L^2(M)$, denoted by $\mathcal{P}$. We assume that

$$\operatorname{Dom}(\mathcal{P}) \supset C_0^\infty(\operatorname{int} M). \tag{2.1}$$

We also assume that $\sigma_c(\mathcal{P})$, the continuous spectrum of $\mathcal{P}$, has a non-empty interior.

Let $R(\lambda) = (\mathcal{P} - \lambda)^{-1}$ be the resolvent operator function with values in $\mathcal{L}(L^2(M))$. We denote by $\tilde{R}(\lambda)$ the resolvent $(\mathcal{P} - \lambda)^{-1}$ considered as an operator function with values in $\mathcal{L}(L^2_{\text{comp}}(M), L^2_{\text{loc}}(M))$ (space of cont. linear operators: $L^2_{\text{comp}}(M) \to L^2_{\text{loc}}(M)$). We refer to $\tilde{R}(\lambda)$ as the **modified** resolvent. Let $J$ be an open subset of $\operatorname{int}\sigma_c(\mathcal{P})$ ($J$ to be fixed throughout). We assume that the following basic condition holds.

**Hypothesis** $\mathcal{M}(J)$. *The operator function* $\tilde{R}(\lambda)$ *admits a meromorphic continuation with finite rank poles from* $C_+ = \{\lambda \in \mathbb{C} : \operatorname{Im}\lambda > 0\}$ *into a domain* $D_+ \supset C_+ \cup J$. *We denote this meromorphic continuation of* $\tilde{R}(\lambda)$ *to* $D_+$ *by* $R^+(\lambda)$.

It is readily seen (by consideration of adjoints) that Hypothesis $\mathcal{M}(J)$ implies that there also exists a meromorphic continuation with finite rank poles of $\tilde{R}(\lambda)$ from $\mathbb{C}_-$ to a domain $D_-$ where $D_- = \{\lambda \in \mathbb{C} : \bar{\lambda} \in D_+\}$. We denote this meromorphic continuation by $R^-(\lambda)$.

*Remark.* It can be shown that Hypothesis $\mathcal{M}(J)$ is equivalent to the assumption that the resolvent kernel of $\mathcal{P}$ (the Green's function) has a meromorphic continuation with finite rank kernel poles from $\mathbb{C}_+$ to $D_+$.

**Definition.** Poles of $R^+(\lambda)$ in $D_+ \cap \mathbb{C}_-$ and poles of $R^-(\lambda)$ in $D_- \cap \mathbb{C}_+$ are called **resonances.**

Observe that $\lambda_0$ is a pole of $R^+(\lambda)$ iff $\bar{\lambda}_0$ is a pole of $R^-(\lambda)$. Also note that $\lambda_0 \in J$ is an (embedded) eigenvalue of $\mathcal{P}$ iff $\lambda_0$ is a pole of $R^+(\lambda)$ and thus also a pole of $R^-(\lambda)$.

We are in a position to state our main results on instability of embedded eigenvalues.

**Theorem 2.1.** *Under the preceding conditions and notation the following holds. Suppose that* $P$ *is a second order elliptic operator. Then for any open set* $\Omega \Subset \operatorname{int} M$ *and any* $k \in \mathbb{N}$, *there exists a real* $C^\infty$ *function* $V$ *on* $M$ *with* $\operatorname{supp} V \subset \Omega$, $V$ *with arbitrary small* $C^k$ *norm, such that the operator* $\mathcal{P} + V$ *has no eigenvalues in* $J$.

**Theorem 2.2.** *Suppose that* $P = -\Delta_g$ *where* $\Delta_g$ *is the Laplacian in the metric* $g$. *Suppose also that* $\dim M = 2$. *Then, given any open set* $\Omega \Subset \operatorname{int} M$, *there exists a metric* $h$ *conformal to* $g$ *on* $M$ *such that* $h = g$ *on* $M \setminus \Omega$ *and* $h$ *is arbitrarily close to* $g$ *in* $\Omega$, *such that the self-adjoint realization* $\mathcal{P}'$ *of* $-\Delta_h$ *in* $L^2(M; d\mu')$ *has no eigenvalues in* $J$ *except perhaps* $\lambda = 0$. *Here* $d\mu'$ *is the measure induced on* $M$ *by* $h$ *and* $\mathcal{P}'$ *is defined as follows:* $\operatorname{Dom}(\mathcal{P}') = \operatorname{Dom}(\mathcal{P})$,

$$\mathcal{P}'u = -\Delta_h u \text{ for } u \in \operatorname{Dom}(\mathcal{P}').$$

Observe that Theorem 1.1 is essentially a special case of Theorem 2.2 since it is well known that the modified resolvent of $\Delta_g$ on a non-compact hyperbolic surface with a finite area verifies Hypothesis $\mathcal{M}(J)$ with $J = \operatorname{int}\sigma_c(\Delta_g)$ (cf. [1]).

The proofs of Theorem 2.1 and Theorem 2.2 are quite long. The proofs rely heavily on a perturbation theory for resonances. This theory has an independent interest. In the following sections we describe this theory for the special class of operators $P$ introduced above ($P$ not necessarily of the second order), and then indicate how the theory is applied to prove the theorems.

**3. On the definition of the multiplicity of a resonance.** In this section we introduce some notions associated with resonances. The basic among these is the notion of the algebraic multiplicity of a resonance. There are two definitions of this notion. We first give a definition which is not easy to apply but which is very similar to the definition of the multiplicity of an eigenvalue in the classical setup.

In the following and throughout most of the paper we shall consider without loss of generality only resonances located in $D^+$ (i.e. poles of $R^+(\lambda)$).

Let $\lambda_0$ be a pole of $R^+(\lambda)$ in $D_+$. Introduce the operator:

$$\Pi = -\frac{1}{2\pi i}\int_\gamma R^+(\lambda)d\lambda \tag{3.1}$$

where $\gamma$ is a contour in $D_+$, containing $\lambda_0$ and no other pole in its interior. Clearly, $\Pi : L^2_{\mathrm{comp}}(M) \to L^2_{\mathrm{loc}}(M)$ is a finite rank operator. It is not difficult to see (using (2.1) and elliptic theory) that functions $u$ in $\operatorname{Ran}\Pi$ are $C^\infty$ solutions of the differential equation:

$$(P-\lambda_0)^r u = 0 \quad \text{on } M, \tag{3.2}$$

where $r$ is the **order** of the pole $\lambda_0$ of $R^+(\lambda)$ in the function theoretic sense. More precisely, one can show that $P-\lambda_0$ is an invariant nilpotent linear operator on $\operatorname{Ran}\Pi$.

We shall say that a function $u \in L^2_{\mathrm{loc}}(M)$, $u \neq 0$, is a **resonance function** of $\mathcal{P}$ associated with the pole $\lambda_0$ iff $u \in \operatorname{Ran}\Pi$, and $u$ satisfies the equation:

$$(P-\lambda_0)u = 0 \quad \text{on } M. \tag{3.3}$$

From the preceding remark it follows that any pole $\lambda_0 \in R^+(\lambda)$ has an associated resonance function.

**Definition.** (i) The integer:

$$m(\lambda_0) := \dim \operatorname{Ran}\Pi \tag{3.4}$$

is called the multiplicity (or the **algebraic** multiplicity) of the pole $\lambda_0$.

(ii) The dimension of the linear set of resonance functions associated with $\lambda_0$ is called the geometric multiplicity of the pole $\lambda_0$.

It is clear that the geometric multiplicity of $\lambda_0$ never exceeds its algebraic multiplicity and that there is an equality between the geometric and algebraic multiplicities iff $\lambda_0$ is a pole of the first order.

If $\lambda_0$ is a real pole of $R^+(\lambda)$, then as was already remarked $\lambda_0$ is an embedded eigenvalue of $\mathcal{P}$. It is a pole of the first order and it is easy to see that $\Pi$ is the restriction to $L^2_{\text{comp}}(M)$ of the orthogonal projection on the $\lambda_0$-eigenspace of $\mathcal{P}$ in $L^2(M)$. Thus in this case the set of resonance functions of $\mathcal{P}$ at $\lambda_0$ coincides with the set of eigenfunctions at $\lambda_0$, and the multiplicity of the pole $\lambda_0$ equals the dimension of the corresponding eigenspace.

A pole (resonance) $\lambda_0$ is said to be simple if its multiplicity is one. It follows from the preceding remarks that a pole $\lambda_0$ is simple iff $\lambda_0$ is a pole of the first order and $\dim \operatorname{Ran} \Pi = 1$.

The notion of the multiplicity of a resonance as defined above does not lend itself easily to applications. We shall give another definition of this notion which is more useful. To this end consider the operator

$$\mathcal{P}_W = \mathcal{P} + W \tag{3.5}$$

where $W$ is a finite rank self-adjoint operator in $L^2(M)$ of the form

$$W = \sum_{i=1}^{N} \langle \cdot, g_i \rangle g_i, \tag{3.6}$$

where the $g_i$ are some real functions in $C_0^\infty(M)$. It is not difficult to see (invoking the analytic Fredholm theory) that the resolvent $(\mathcal{P}_W - \lambda)^{-1}$, when considered as an operator function on $\mathbb{C}_+$ with values in $\mathcal{L}(L^2_{\text{comp}}(M), L^2_{\text{loc}}(M))$, admits a meromorphic continuation (with finite rank poles) to $D_+$. We denote this operator function by $R_W^+(\lambda)$. Next, introduce the function:

$$f_W(\lambda) := \det(I - W R_W^+(\lambda) W) \tag{3.7}$$

where $W R_W^+(\lambda) W$ is considered as a meromorphic operator function in $D_+$ with values in $\mathcal{L}(L^2(M))$ and $I$ is the identity in $\mathcal{L}(L^2(M))$. Since $W R_W^+(\lambda) W$ is a finite rank operator for each $\lambda$ it follows that the determinant $f_W(\lambda)$ is well defined. It is a meromorphic function in $D_+$ with poles contained in the set of poles of $R_W^+(\lambda)$. It is easy to see that $f_W(\lambda) \not\equiv 0$. One can prove the following result.

**Theorem 3.1.** *Let $\lambda_0 \in D_+$, $\lambda_0$ not a pole of $R_W^+(\lambda)$. Then, $\lambda_0$ is a pole of $R^+(\lambda)$ iff $\lambda_0$ is a zero of $f_W(\lambda)$. Denoting by $\nu(\lambda_0; f_W)$ the order (multiplicity) of the zero $\lambda_0$ of $f_W$, we have*

$$\nu(\lambda_0; f_W) = m(\lambda_0) \tag{3.8}$$

*where $m(\lambda_0)$ is the multiplicity of the pole $\lambda_0$ of $R^+(\lambda)$ as defined above.*

We also have the following

**Proposition 3.2.** *Let $\lambda_0 \in D_+$ be a pole of $R^+(\lambda)$. There exists a finite rank operator $W$ of the form (3.6) such that $R^+_W(\lambda)$ is regular at $\lambda_0$.*

In view of Theorem 3.1 and Proposition 3.2 the integer $\nu(\lambda_0; f_W)$ gives an alternative useful definition of the multiplicity of a resonance.

**4. A perturbation theory for resonances.** We shall consider a family of elliptic operators $P_t$ on $M$ of order $\ell$ and of class $C^\infty$ depending on a real parameter $t$ in an interval $[a, b]$ where $a < 0 < b$. For simplicity we shall assume that the $P_t$ are real analytic in $t$ on $[a, b]$.

We assume that the differential operator $P_0$ is the operator $P$ studied in the preceding sections. Thus, $P_0$ is formally self-adjoint in $L^2(M) = L^2(M; d\mu)$ and it has a self-adjoint realization denoted here by $\mathcal{P}_0$. We denote by $R_0(\lambda)$ the $L^2$ resolvent of $\mathcal{P}_0$ and denote by $R_0^+(\lambda)$ the meromorphic continuation of the modified resolvent $\tilde{R}_0(\lambda)$ from $\mathbb{C}_+$ to $D_+ \supset \mathbb{C}_+ \cup J$ (Hypothesis $\mathcal{M}(J)$).

We assume that $P_t$ is a perturbation of $P_0$ on some fixed compact set, i.e. we assume that there exists an open set $\Omega \Subset \operatorname{int} M$ such that $P_t = P_0$ on $M \setminus \Omega$ for all $t \in [a, b]$. We also assume that $P_t$ is formally self-adjoint in $L^2(M; d\mu_t)$ where $d\mu_t$ is a measure of the form:

$$d\mu_t = e^{2f_t} d\mu, \quad t \in [a, b]. \tag{4.1}$$

Here $f_t$ is a real $C^\infty$ function on $M$ with $\operatorname{supp} f_t \subset \Omega$, which depends analytically on $t$ for $t \in [a, b]$, and $f_0 \equiv 0$ on $M$.

Next, we introduce the operators $\mathcal{P}_t$ defined by

$$\begin{aligned} \operatorname{Dom}(\mathcal{P}_t) &= \operatorname{Dom}(\mathcal{P}_0), \\ \mathcal{P}_t u &= P_t u \text{ for } u \in \operatorname{Dom}(\mathcal{P}_t). \end{aligned} \tag{4.2}$$

Using our assumption (2.1) it is not difficult to see that $\mathcal{P}_t$ is a self-adjoint operator in $L^2(M; d\mu_t)$.

The operator $\mathcal{P}_t$ has properties similar to that of $\mathcal{P}_0$. In particular, it can be shown that the modified resolvent $\tilde{R}_t(\lambda)$ of $\mathcal{P}_t$ admits a meromorphic continuation with finite rank poles from $\mathbb{C}_+$ to $D_+$. We denote this operator function by $R_t^+(\lambda)$.

It can further be shown that the poles of $R_t^+(\lambda)$ in $D_+$ depend analytically on the parameter $t$. More precisely, we have the following theorem.

**Theorem 4.1.** *Let $\mathcal{O}$ be any simply connected domain in the $\lambda$-plane such that $\mathcal{O} \Subset D_+$. There exists a function $F(\lambda, t)$, defined and analytic in $(\lambda, t)$ in a neighborhood of $\mathcal{O} \times [a, b]$ in $\mathbb{C}^2$, such that the following holds:*

(i) *$F(\lambda, t) \not\equiv 0$ in $\mathcal{O}$ for each fixed $t \in [a, b]$.*

(ii) *The set of poles of $R_t^+(\lambda)$ in $\mathcal{O}$, for any fixed $t \in [a, b]$, coincides with the set $\{\lambda \in \mathcal{O} : F(\lambda, t) = 0\}$.*

(iii) *If* $\lambda_0 \in O$ *is a pole of* $R_t^+(\lambda)$ *then its algebraic multiplicity coincides with the order of the zero* $\lambda_0$ *of the function* $F(\lambda, t)$.

We note that the proof of the theorem is achieved in two steps. In the first step one proves a **local** version of the theorem where $O$ is a small neighborhood of some point $\lambda_0 \in D_+$ and $t$ varies in a small interval. The global version of the theorem follows from the local one by complex analysis arguments. The local version of Theorem 4.1. follows easily from Theorem 3.1 and Proposition 3.2.

Suppose now that $\lambda_0$ is a simple pole of $R_0^+(\lambda)$ in $D_+$ and let $\omega$ be a closed disc centered at $\lambda_0$ such that $R_0^+(\lambda)$ is everywhere analytic in $\omega \backslash \{\lambda_0\}$. It follows from Theorem 4.1 that $R_t^+(\lambda)$ has a unique simple pole $\lambda(t)$ in $\omega$ for $t \in (-\varepsilon, \varepsilon)$ if $\varepsilon$ is sufficiently small, and that $\lambda(t)$ is an analytic function of $t$ in the interval $|t| < \varepsilon$ (we set $\lambda(0) = \lambda_0$).

One can derive formulas for the derivatives of $\lambda(t)$ at the origin. These formulas are similar to classical perturbation formulas for eigenvalues of nonself-adjoint operators. Before presenting such formulas we digress to introduce a notion which we shall need. To this end write the Laurent expansion of $R_0^+(\lambda)$ about $\lambda_0$:

$$R_0^+(\lambda) = \sum_{i=-1}^{\infty} (\lambda - \lambda_0)^i S_i \tag{4.3}$$

and observe that $S_{-1}$ is of rank 1 and that $S_i \in \mathcal{L}(L^2_{\text{comp}}(M), L^2_{\text{loc}}(M))$. Let $\psi$ be a resonance function associated with the pole $\lambda_0$. By definition:

$$\psi = \Pi f \tag{4.4}$$

for some $f \in L^2_{\text{comp}}(M)$ where $\Pi = -S_{-1}$. Next, observe that $\bar{\lambda}_0$ is a simple pole of $R_0^-(\lambda)$ (the meromorphic continuation of $\tilde{R}_0(\lambda)$ from $\mathbb{C}_-$ to $D_-$). Let $\psi^*$ be a resonance function associated with the pole $\bar{\lambda}_0$. Define

$$[\psi, \psi^*] := \langle f, \psi^* \rangle \tag{4.5}$$

where $f$ is any function in $L^2_{\text{comp}}(M)$ verifying (4.4). Here and throughout the remainder of the paper $\langle \cdot, \cdot \rangle$ denotes the inner product in $L^2(M, d\mu)$ ($d\mu = d\mu_0$, see (4.1)). One can show that $[\psi, \psi^*]$ depends only on $\psi$ and $\psi^*$ (and not on $f$) and that $[\psi, \psi^*] \neq 0$. We shall say that $\psi$. $\psi^*$ is a **normalized** pair of resonance functions if

$$[\psi, \psi^*] = 1.$$

We are now in a position to describe the perturbation formulas for resonances and embedded eigenvalues mentioned above.

**Theorem 4.2.** *The first two derivatives of $\lambda(t)$ at the origin ($\lambda(t)$ defined above, $\lambda(0) = \lambda_0$), are given by the following formulas:*

$$\begin{aligned} \dot{\lambda}(0) &= \langle \dot{P}_0 \psi, \psi^* \rangle, \\ \ddot{\lambda}(0) &= \langle \ddot{P}_0 \psi, \psi^* \rangle - 2\langle \dot{P}_0 S_0 \dot{P}_0 \psi, \psi^* \rangle, \end{aligned} \tag{4.7}$$

*where $\psi$, $\psi^*$ is a normalized pair of resonance functions ($\psi$ associated with the pole $\lambda_0$ of $R_0^+$ and $\psi^*$ is associated with the pole $\bar{\lambda}_0$ of $R_0^-$), and where $S_0$ is the zero'th coefficient in the Laurent expansion (4.3).*

In the special case when $\lambda_0$ is real (i.e. $\lambda_0$ is an embedded eigenvalue), we have:

$$\begin{aligned} \dot{\lambda}(0) &= \langle \dot{P}_0 \psi, \psi \rangle, \\ \ddot{\lambda}(0) &= \langle \ddot{P}_0 \psi, \psi \rangle - 2\langle \dot{P}_0 S_0 \dot{P}_0 \psi, \psi \rangle, \end{aligned} \tag{4.8}$$

where $\psi$ is an eigenfunction of $\mathcal{P}_0$ at $\lambda_0$ such that $\langle \psi, \psi \rangle = 1$.

We use here the notation:

$$\dot{\lambda}(0) := \frac{d}{dt}\lambda(t)|_{t=0}, \quad \dot{P}_0 := \frac{d}{dt}P_t|_{t=0}, \quad \text{etc.}$$

*Remark.* Note that the r.h. sides of (4.7)–(4.8) are well defined since the differential operators $\dot{P}_0$ and $\ddot{P}_0$ have compact supports. Note also that when $\lambda_0$ is real and $\psi$ is as above then $[\psi, \psi] = \langle \psi, \psi \rangle$.

**5. Generalized eigenfunctions.** We continue with our study of the family of self-adjoint operators $\mathcal{P}_t$ in $L^2(M; d\mu_t)$ introduced in the previous section. Let $R_t^{\pm}(\lambda)$ denote the meromorphic continuation of the modified resolvent $\tilde{R}_t(\lambda)$ of $\mathcal{P}_t$ from $\mathbb{C}_\pm$ to $D_\pm$, respectively. We set:

$$E_t(\lambda) = -i(R_t^+(\lambda) - R_t^-(\lambda)). \tag{5.1}$$

For each $t \in (a, b)$ $E_t(\lambda)$ is a well defined meromorphic operator function of $\lambda$ (with values in $\mathcal{L}(L^2_{\text{comp}}(M), L^2_{\text{loc}}(M))$) in the open set: $D_0 = D_+ \cap D_-$. Note that $D_0$ is symmetric with respect to the real axis and that $D_0 \cap \{\lambda : \operatorname{Im} \lambda = 0\} = J$, where $J$ is the open set in $\mathbb{R}$ which appears in Hypothesis $\mathcal{M}(J)$ for the operator $\mathcal{P}_0$.

It is clear that the non-real poles of $E_t(\lambda)$ in $D_0$ consist of the non-real poles of $R_t^+(\lambda)$ in $D_+$ and their complex conjugates. On the other hand it is not difficult to see that all real poles of $R_t^{\pm}(\lambda)$ (i.e. embedded eigenvalues of $\mathcal{P}_t$ in $J$) are removable singularities for $E_t(\lambda)$. Thus, the operator function $E_t(\lambda)$ is everywhere analytic on $J$.

Another observation is that $E_t(\lambda) \not\equiv 0$ in every connected component $J_i$ of $J$. This is clearly true for $t = 0$ since if $E_0(\lambda) \equiv 0$ in $J_i$ it would follow by standard arguments from complex function theory that $J_i$ does not belong to the continuous spectrum of $P_0$ which is in contradiction to Hypothesis $\mathcal{M}(J)$. That $E_t(\lambda) \not\equiv 0$ for any $t$ one proves in the same way by noting that $J \subset \sigma_c(P_t)$ for any $t$ (this can be proved by combining the result for $t = 0$ with Theorem 5.1).

From now on we shall assume that $\lambda$ is real. Functions in Ran $E_t(\lambda)$ for $\lambda \in J$ are referred to as the generalized eigenfunctions of $\mathcal{P}_t$ which correspond to the eigenvalue $\lambda$. It follows readily that these functions are $C^\infty$ solutions of the differential equation:

$$P_t u = \lambda u \text{ in } M. \tag{5.2}$$

The integer: dim Ran $E_t(\lambda)$ is called the multiplicity of the continuous spectrum of $\mathcal{P}_t$ at the point $\lambda$. This multiplicity is stable under perturbation. More precisely, one can prove the following result.

**Theorem 5.1.** *For any $\lambda \in J$, we have:*

$$\dim \operatorname{Ran} E_t(\lambda) = \dim \operatorname{Ran} E_0(\lambda) \tag{5.3}$$

*for all $t \in (a, b)$.*

Since $E_0(\lambda)$ is analytic and $\not\equiv 0$ in every component of $J$, it follows that $E_0(\lambda) \neq 0$ at all points $\lambda \in J$ except possibly for a discrete set of points. Hence, Theorem 5.1 yields the following

**Corollary 5.2.** *There exists a discrete (possibly empty) set of points $\{k_i\}$ in $J$ such that* dim Ran $E_t(\lambda) \geq 1$ *for all $\lambda \in J \setminus \{k_i\}$ and all $t \in (a, b)$.*

Next, consider the bilinear form

$$Q_\lambda(f, g) := \langle E_0(\lambda) f, g \rangle \tag{5.4}$$

defined on $L^2_{\text{comp}}(M)$ for any $\lambda \in J$. Noting that $Q_\lambda(f, f)$ is analytic in $\lambda$ on $J$ and that

$$Q_\lambda(f, f) = -\lim_{\varepsilon \to 0} i\langle (R_0(\lambda + i\varepsilon) - R_0(\lambda - i\varepsilon))f, f\rangle$$

for any $\lambda \in J$, $\lambda$ not a pole of $R_0^\pm$, it follows by classical results that $Q_\lambda$ is a positive Hermitian bilinear form. Let $\mathcal{K}$ be the subspace of functions $f \in L^2_{\text{comp}}(M)$ such that $Q_\lambda(f, f) = 0$. As is well known the form $Q_\lambda$ induces an inner product $(\cdot, \cdot)_\lambda$ on $L^2_{\text{comp}}(M)/\mathcal{K}$ defined by

$$(\xi, \eta)_\lambda = Q_\lambda(f, g) \quad \text{for } f \in \xi,\ g \in \eta \tag{5.5}$$

where $\xi, \eta \in L^2_{\text{comp}}(M)/\mathcal{K}$.

Next observe that the map:

$$L^2_{\text{comp}}(M)/\mathcal{K} \ni \xi \to E_0(\lambda)f, \ f \in \xi$$

establishes an isomorphism between $L^2_{\text{comp}}(M)/\mathcal{K}$ and $\operatorname{Ran} E_0(\lambda)$. We transport the inner product (5.5) onto $\operatorname{Ran} E_0(\lambda)$ so that we have for $u, v \in \operatorname{Ran} E_0(\lambda)$:

$$(u, v)_\lambda = \langle E_0(\lambda)f, g\rangle \tag{5.6}$$

for any $f, g$ in $L^2_{\text{comp}}(M)$ such that $u = E_0(\lambda)f$, $v = E_0(\lambda)g$.

Denote by $\{\varphi_i\}$, $i = 1, 2, \ldots$, a complete orthonormal sequence of functions in $\operatorname{Ran} E_0(\lambda)$ in the inner product (5.6). Then "Parseval's formula" for $\varphi = E_0(\lambda)f$ yields the following identity:

$$\langle E_0(\lambda)f, f\rangle = \sum_i |\langle f, \varphi_i\rangle|^2. \tag{5.7}$$

Formula (5.7) is of course also valid for $E_t(\lambda)$ if one replaces $\langle\cdot,\cdot\rangle$ by $\langle\cdot,\cdot\rangle_t$ (the inner product in $L^2(M; d\mu_t)$), etc.

**6. Perturbation of embedded eigenvalues.** We resume our study of the behavior of resonances of the family of operators $\mathcal{P}_t$ introduced in Section 4. Under the general assumptions made in that section, using the same notation, we now consider the case where $\mathcal{P}_0$ has a simple eigenvalue $\lambda_0 \in J$. As before we denote by $\lambda(t)$, $|t| < \varepsilon$, the (unique) simple pole of $R_t^+(\lambda)$ in $D_+$ located in some small neighborhood of $\lambda_0$ ($\lambda(0) = \lambda_0$). It is an analytic function of $t$ for $|t| < \varepsilon$, satisfying: $\operatorname{Im}\lambda(t) \leq 0$, $\operatorname{Im}\lambda(0) = 0$. In view of the analyticity of $\operatorname{Im}\lambda(t)$ it follows that either: (i) $\operatorname{Im}\lambda(t) \equiv 0$ for $|t| < \varepsilon$, or (ii) $\operatorname{Im}\lambda(t) < 0$ for $0 < |t| < \varepsilon'$ for some $\varepsilon' > 0$. In the first case $\lambda(t)$ is an embedded eigenvalue of $\mathcal{P}_t$ for all $t \in (-\varepsilon, \varepsilon)$, while in the second case $\lambda(t)$ is a resonance of $\mathcal{P}_t$ for $0 < |t| < \varepsilon'$. There are non-trivial examples showing that both cases may occur. In the following we give sufficient conditions for (ii) to hold.

Noting that $\operatorname{Im}\dot{\lambda}(0) = 0$, a sufficient condition for the alternative (ii) to hold is that $\operatorname{Im}\ddot{\lambda}(0) < 0$. Now, by the perturbation formula (4.8) we have:

$$\operatorname{Im}\ddot{\lambda}(0) = \operatorname{Im}\langle \ddot{P}_0\psi, \psi\rangle - 2\operatorname{Im}\langle \dot{P}_0 S_0 \dot{P}_0\psi, \psi\rangle \tag{6.1}$$

where $\psi$ is a normalized eigenfunction of $\mathcal{P}_0$ at the eigenvalue $\lambda_0$.

Assume first that the operators $\mathcal{P}_t$ are self-adjoint in the fixed space $L^2(M; d\mu)$ (i.e. $d\mu_t = d\mu$, see (4.1)). In this case the forms: $\langle \dot{P}_0 u, v\rangle$ and $\langle \ddot{P}_0 u, v\rangle$ are Hermitian forms on $C^\infty(M)$ (note again that $\dot{P}_0$ and $\ddot{P}_0$

have compact supports). Hence it follows that $\operatorname{Im}\langle \ddot{P}_0\psi, \psi\rangle = 0$ and we can rewrite (6.1) in the form:

$$\begin{aligned}\operatorname{Im}\ddot{\lambda}(0) &= -2\operatorname{Im}\langle \dot{P}_0 S_0 \dot{P}_0\psi, \psi\rangle \\ &= -2\operatorname{Im}\langle S_0\dot{P}_0\psi, \dot{P}_0\psi\rangle \\ &= -\langle E_0(\lambda_0)(\dot{P}_0\psi), \dot{P}_0\psi\rangle,\end{aligned} \tag{6.2}$$

where $E_0(\lambda_0)$ is defined by (5.1) and where the last equality in (6.2) follows by noting that for any $f \in L^2_{\text{comp}}(M)$:

$$\begin{aligned}\operatorname{Im}\langle S_0 f, f\rangle &= \lim_{\varepsilon\to 0}\operatorname{Im}\langle R_0^+(\lambda_0+\varepsilon)f, f\rangle \\ &= \frac{1}{2}\lim_{\varepsilon\to 0}\langle E_0(\lambda_0+\varepsilon)f, f\rangle = \frac{1}{2}\langle E_0(\lambda_0)f, f\rangle.\end{aligned}$$

Applying the identity (5.7) to the last term in (6.2), we can rewrite (6.2) in the form:

$$\operatorname{Im}\ddot{\lambda}(0) = -\sum_i |\langle \dot{P}_0\psi, \varphi_i\rangle|^2 \tag{6.3}$$

where $\{\varphi_i\}$ is a complete orthonormal system of functions in $\operatorname{Ran} E_0(\lambda_0)$ in the inner product $(\cdot,\cdot)_{\lambda_0}$ defined by (5.6).

We refer to formula (6.2), or to formula (6.3), as "Fermi's Golden Rule" (see [8;7]). We shall now show that Fermi's Golden Rule holds in the general case when $\mathcal{P}_t$ is a family of self-adjoint operators in $L^2(M; d\mu_t)$ with measure $d\mu_t$ depending on $t$: $d\mu_t = e^{2f_t}d\mu$, $f_t \in C_0^\infty(\Omega)$ etc. (see (4.1)). To this end we introduce a family of operators:

$$\mathcal{A}_t = e^{f_t}\mathcal{P}_t e^{-f_t}, \quad a < t < b \tag{6.4}$$

with $\operatorname{Dom}(\mathcal{A}_t) = \operatorname{Dom}(\mathcal{P}_0)$. From (6.4) and our assumptions on $\mathcal{P}_t$ it follows that $\mathcal{A}_t$ is a self-adjoint realization in $L^2(M; d\mu)$ of the differential operator

$$A_t = e^{f_t}P_t e^{-f_t}.$$

We also have: $\mathcal{A}_0 = \mathcal{P}_0$. From the similarity of $\mathcal{A}_t$ and $\mathcal{P}_t$ it follows that the two operators possess the same spectrum and the same eigenvalues and resonances. Hence applying formula (6.3) to $\mathcal{A}_t$ at $t = 0$ with $\lambda(t)$ and $\psi$ as above we find that

$$\operatorname{Im}\ddot{\lambda}(0) = -\sum_i |\langle (\dot{P}_0 + [\dot{f}_0, P_0])\psi, \varphi_i\rangle|^2 \tag{6.5}$$

where $[\dot{f}_0, P_0] = \dot{f}_0 P_0 - P_0\dot{f}_0$. Noting that $\psi$ and $\varphi_i$ are solutions of the differential equation: $P_0 u = \lambda_0 u$, it follows readily that

$$\langle [\dot{f}_0, P_0]\psi, \varphi_i\rangle = 0. \tag{6.6}$$

Combining (6.5) with (6.6) we establish (6.3) in the general case. An immediate corollary of Fermi's Golden Rule is the following

**Proposition 6.1.** *With the same notation as above, a necessary and sufficient condition that* $\operatorname{Im} \ddot{\lambda}(0) < 0$ *is that*

$$\langle \dot{P}_0 \psi, \varphi \rangle \neq 0 \tag{6.7}$$

*for some* $\varphi \in \operatorname{Ran} E_0(\lambda_0)$.

We conclude this study by giving some details of the proofs of Theorem 2.1 and Theorem 2.2.

We consider first the case where $\mathcal{P}$ is a self-adjoint realization in $L^2(M; d\mu)$ of a second order elliptic operator $P$. We assume that $\mathcal{P}$ satisfies all the conditions imposed at the beginning of Section 2. Let $\Omega$ be some open set with compact closure in $M$. We shall consider self-adjoint perturbations of $\mathcal{P}$ of the form:

$$\mathcal{P}_t^W := \mathcal{P} + tW, \quad t \in \mathbb{R}, \tag{6.8}$$

$(\operatorname{Dom}(\mathcal{P}_t^W) = \operatorname{Dom}(\mathcal{P}))$ where $W$ is a multiplication operator in $L^2(M; d\mu)$ by a real function $W \in C_{0R}^\infty(\Omega)$. Here $C_{0R}^\infty(\Omega)$ denotes the class of real $C^\infty$ functions on $M$ with support in $\Omega$. It is readily checked that $\mathcal{P}_t^W$ is a family of self-adjoint operators in $L^2(M; d\mu)$ possessing all the properties of the family $\mathcal{P}_t$ discussed in the previous sections. We set

$$iE_t^W(\lambda) := R_t^{W+}(\lambda) - R_t^{W-}(\lambda), \quad \lambda \in J,$$

where $R_t^{W\pm}(\lambda)$ are the generalized resolvents of $\mathcal{P}_t^W$ defined in $D_\pm$. Note that $R_0^{W\pm}(\lambda) = R^\pm(\lambda)$ where $R^\pm(\lambda)$ are the generalized resolvents of $\mathcal{P}$. We set:

$$E(\lambda) = -i(R^+(\lambda) - R^-(\lambda)), \quad \text{for } \lambda \in J,$$

noting again that $E(\lambda) = E_0^W(\lambda)$. We recall that by Corollary 5.2 it follows that

$$\dim \operatorname{Ran} E_t^W(\lambda) \geq 1 \tag{6.9}$$

for $\lambda \in J \setminus \{k_j\}$ for all $t \in (a, b)$ and $W \in C_{0R}^\infty(\Omega)$ where $\{k_j\}$ is some discrete set of points in $J$ independent of $t$ or $W$.

Let $\Lambda$ be a union of a finite number of compact intervals contained in $J$ such that $\partial\Lambda$ does not contain any eigenvalue of $\mathcal{P}$. We denote by $\ell_1, \ldots, \ell_m$ the eigenvalues of $\mathcal{P}$ contained in $\Lambda$ and assume that all these eigenvalues are simple. We also assume that $\ell_i \neq k_j$ for $i = 1, \ldots, m$ and all $j$. The last assumption and (6.9) imply that there exist non-zero functions $\varphi_i \in \operatorname{Ran} E(\ell_i)$, $i = 1, \ldots, m$. We denote by $\psi_i$ a normalized eigenfunction of $\mathcal{P}$ with the eigenvalue $\ell_i$. Since $\varphi_i$ and $\psi_i$ are solutions of second order elliptic equations with real principal parts it follows from the unique continuation theorem that the zeros of the function: $\prod_i \psi_i \varphi_i$ are nowhere dense. Hence, it is easy to see that there exists a function $W \in C^{\infty}_{0R}(\Omega)$ such that

$$\langle W\psi_i, \varphi_i \rangle \neq 0 \tag{6.10}$$

for $i = 1, \ldots, m$. We have:

**Proposition 6.2.** *With $W$ satisfying (6.10), there exists $\varepsilon > 0$ such that for any $t \in (-\varepsilon, \varepsilon) \setminus \{0\}$ the operator $\mathcal{P}_t^W$ has no eigenvalues in $\Lambda$.*

*Proof.* Let $I_i$, $i = 1, \ldots, m$, be compact intervals in $J$ such that $\ell_i \in \operatorname{int} I_i$ and such that no other eigenvalue of $P$ is contained in $I_i$. Taking note of the remarks made at the beginning of this section, using the analytic dependence in $t$ of poles of $R_t^{W+}$ (see Theorem 4.1), one arrives at the following conclusion. There exists an $\varepsilon > 0$ such that

(a) All eigenvalues of $\mathcal{P}_t^W$ in $\Lambda$ for any $t \in (-\varepsilon, \varepsilon)$ are contained in $\bigcup_{i=1}^m I_i$.
(b) In each interval $I_i$ the following alternative holds. Either (i) $I_i$ does note contain any eigenvalue of $\mathcal{P}_t^W$ for any $t \in (-\varepsilon, \varepsilon) \setminus \{0\}$, or (ii) $I_i$ contains a single simple eigenvalue $\lambda_i(t)$ for any $t \in (-\varepsilon, \varepsilon)$ ($\lambda_i(t)$ analytic in $t$, $\lambda_i(0) = \ell_i$).

However, it is readily seen that in view of (6.10) the alternative (ii) in (b) cannot hold, since assuming (ii) for some interval $I_i$ yields a contradiction to Proposition 6.1. This proves Proposition 6.2.

We also have the following result which we state without a proof.

**Proposition 6.3.** *Let $\Lambda$ be any compact subset of $J$. Given $\varepsilon > 0$ and $j \in \mathbb{N}$, there exists a function $W_1 \in C^{\infty}_{0R}(\Omega)$, with a $C^j$ norm less than $\varepsilon$, such that $\mathcal{P} + W_1$ has only simple eigenvalues in $\Lambda$, none of which is contained in the set $\{k_i\}$ introduced above.*

Propositions 6.2 and 6.3 hold, of course, when $\mathcal{P}$ is replaced by $\mathcal{P} + W_0$, any $W_0 \in C^{\infty}_{0R}(\Omega)$. Combining the two propositions we obtain:

**Theorem 6.4.** *Let $\Lambda$ be any compact subset of $J$. Given $\varepsilon > 0$ and $j \in \mathbb{N}$ there exists a function $V \in C^{\infty}_{0R}(\Omega)$, $V$ with $C^j$ norm less than $\varepsilon$, such that $\mathcal{P} + V$ has no eigenvalues in $\Lambda$.*

Theorem 2.1 is the statement that Theorem 6.4 holds with $\Lambda$ replaced by $J$. We omit the additional arguments, based on Baire category theorem, which deduce Theorem 2.1 from Theorem 6.4 (compare [2]).

The proof of Theorem 2.2 with some modifications is the same as that of Theorem 2.1. Here $\mathcal{P}$ is a self-adjoint realization of $-\Delta_g$ in $L^2(M; d\mu)$.

Given $\Omega \Subset M$ and $W \in C^{\infty}_{0R}(\Omega)$, we introduce the family of metrics $g_t$ on $M$ defined by

$$g_t = e^{tW} g, \quad t \in \mathbb{R}.$$

We denote by $\mathcal{P}_t^W$ the self-adjoint realization of $-\Delta_{g_t}$ in $L^2(M; d\mu_t)$ ($d\mu_t$ the measure induced by $g_t$) defined as follows:

$$\begin{aligned} &\operatorname{Dom}(\mathcal{P}_t^W) = \operatorname{Dom}(\mathcal{P}), \\ &\mathcal{P}_t^W u = -\Delta_{g_t} u \quad \text{for} \quad u \in \operatorname{Dom}(\mathcal{P}_t^W). \end{aligned} \tag{6.11}$$

Noting that

$$\frac{d}{dt} \Delta_{g_t}|_{t=0} = -2W\Delta_{g_0}$$

one finds that Proposition 6.2 holds for the family $\mathcal{P}_t^W$ just defined, except that in this case one has to replace $\Lambda$ by $\Lambda \setminus \{0\}$. (The proof of the Proposition is exactly the same as above, it relies on Fermi's Golden Rule for the family $\mathcal{P}_t^W$). The remainder of the proof of Theorem 2.2. parallels that of Theorem 2.1.

## References


1. S. Agmon, *Topics in spectral theory of Schrödinger operators on non-compact Riemannian manifolds with cusps*, Mathematical quantum theory II: Schrödinger operators (J. Feldman, R. Froese and L.M. Rosen, eds.), CRM Proceedings and Lecture Notes **8**(1995), 1–33.
2. S. Agmon, I. Herbst and E. Skibsted, *Perturbation of embedded eigenvalues in the generalized N-body problem*, Comm. Math. Phys. **122** (1989), 411-438.
3. S. Albeverio and R. Høegh-Krohn, *Perturbation of resonances in quantum mechanics*, J. Math. Anal. Appl. **101** (1984), 491–513.
4. Y. Colin de Verdière, *Pseudo-Laplacians II*, Ann. Inst. Fourier **33** (1983), 87–113.
5. F. Gesztesy, *Perturbation theory for resonances in terms of Fredholm determinants*, Resonances - Models and Phenomena (S. Albeverio, L.S. Ferreira and L. Streit, eds.), Springer-Verlag, Berlin, Lecture Notes in Physics **211** (1984), 78–104.
6. J.S. Howland, *Puiseux series for resonances at an embedded eigenvalue*, Pac. J. Math. **55** (1974), 157–176.
7. R. Phillips and P. Sarnak, *Perturbation theory for the Laplacian on automorphic functions*, J. Amer. Math. Soc. **5** (1992), 1–32.
8. B. Simon, *Resonances in N-body quantum systems with dilation analytic potentials and the foundation of time dependent perturbation theory*, Ann. of Math. **97** (1973), 247–274.



Institute of Mathematics, The Hebrew University of Jerusalem, Givat Ram, 91904 Jerusalem, Israel.

*E-mail address*: agmon@math.huji.ac.il

# BLOWUP OF CLASSICAL SOLUTIONS OF NONLINEAR HYPERBOLIC EQUATIONS: A SURVEY OF RECENT RESULTS

SERGE ALINHAC

**Abstract.** Some recent results on blowup of classical solutions to hyperbolic systems or equations are presented. Emphasis is put on the description of the mechanism, which can be given for some semilinear cases or for cases in one space dimension. For higher dimensions, only asymptotic results have been obtained so far.

**Introduction.** We will consider here, in $\mathbf{R}_x^n \times \mathbf{R}_t$, quasilinear systems

$$\partial_t u + \sum_{j=1}^{n} A_j(u)\partial_j u + B(u) = 0 \tag{1}$$

or quasilinear wave equations

$$\sum_{0\leq i,j\leq n} g_{ij}(\nabla u)\partial_{ij}^2 u + F(u, \nabla u) = 0, \quad x_0 = t. \tag{2}$$

The systems are assumed to be strictly hyperbolic (or symmetric hyperbolic), and the wave equations to be either semilinear with principal part equal to the wave operator

$$\partial_t^2 - \Delta_x$$

or to have principal part close to the wave operator.

In both cases, the Cauchy problem with data in $C_0^\infty$ has a smooth (classical) solution in a strip $\{0 \leq t \leq T\}$ for some $T > 0$ (see for instance Majda [19]). We denote by $\overline{T}$ the supremum of all such $T$ and call it the "lifespan" of the solution.

For large classes of nonlinear equations, it is well known that such classical solutions do not remain smooth for all times (that is, $\overline{T} < \infty$): this phenomenon is called "blowup".

The aim of this survey is to present some recent results about blowup.

Of course, there is a huge literature about this problem; let us mention here only general references such as Hörmander [8], Alinhac [4] and the survey papers by John [9], Strauss [22] and Zuily [23].

One can roughly say that most of this literature is devoted to determine whether the lifespan is finite or not. In general, little attention is paid to what actually happens when $t \to \overline{T}$. Though some general "functional analysis" principles tell us that either $u$ or $\nabla u$ have to be unbounded if blowup is to occur in case (1) (in case (2), this corresponds to $u$, $\nabla u$ or $\nabla^2 u$ unbounded), precise information about the rate, the locus, in one word the mechanism of the blowup is seldom available.

Let us explain by a very simple example a typical device proving $\overline{T} < \infty$ without giving any information on what actually takes place (see [4] for details).

Consider Burgers' equation

$$\partial_t u + u \partial_x u = 0 \tag{3}$$

with an initial value $u_0 \in C_0^\infty([a,b])$. For $t < \overline{T}$, $u(.,t) \in C_0^\infty$ is supported in $[a,b]$, because the characteristics starting outside $[a,b]$ are vertical lines on which $u = 0$. Choose $\phi \in C^\infty([a,b])$, $\phi' > 0$, and set

$$F(t) = \int \phi(x) u(x,t) dx.$$

Using the equation and the Cauchy-Schwarz inequality, one easily obtains

$$2CF'(t) \geq F^2(t), \quad C = \int_a^b \frac{\phi^2(x)}{\phi'(x)} dx.$$

Thus, if $F(0) = \int \phi(x) u_0(x) dx > 0$, $F$ cannot remain smooth beyond $T_0 = 2C/F(0)$, hence $\overline{T} \leq T_0$.

We see that blowup of $u$ is proved by blowup of $F$; but the method of characteristics shows that, for $t < \overline{T}$, $|u(x,t)| \leq \max |u_0|$: blowup of $F$ in fact never occurs! What actually happens when $t \to \overline{T}$ is $\max |\partial_x u(.,t)| \to \infty$.

It turns out that, for semilinear wave equations (2), one can define a priori a "maximal influence domain" $\Omega$ where the classical solution exists; in this case, the lifespan is simply

$$\overline{T} = \inf\{t, (x,t) \in \partial\Omega\}.$$

Hence a richer concept is available in the semilinear case, and this is why we distinguish in what follows the semilinear case (part I) from the general (quasilinear) case (part II). However, in some special quasilinear cases with at least two propagation speeds, a concept of maximal domain seems also to be available as the result of a mathematical analysis (see for instance [7]).

Finally, it is convenient to consider, in contrast with the global blowup situations discussed so far, local blowup situations. By this we mean that a smooth solution $u$ of (1) or (2) is given in an open domain $D$, $u$ is singular at $x_0 \in \overline{D}$, but this singularity cannot be traced back along characteristics of the linearized operator to singularities of data determining $u$. In other words, the singularity is created, not propagated.

Hence the content of this paper is organized as follows: in each case (semilinear, quasilinear), we display first patterns of local blowup situations; then we check whether classical solutions in global blowup situations behave according to these local patterns.

## I. Semilinear problems for the wave equation.

*I.1 Local blowup patterns.* A space-like surface

$$S = \{(x,t), t = \psi(x)\}, \quad \gamma = 1 - |\nabla\psi|^2 > 0$$

being given, it is possible to obtain solutions singular on $S$ by inserting in the equation an appropriate Ansatz. This approach, well-known from the formal point of view, has been recently analytically clarified by Kichenassamy and Littman (cf. [12], [13]).

For instance, consider the equation

$$\partial_t^2 u - \Delta u = \exp u. \tag{4}$$

Define a new unknown $w$ by

$$u = \ln\frac{2}{T^2} + \ln\gamma - (\Delta\psi)\frac{T}{\gamma} - \frac{2R}{3\gamma}T^2\ln T + T^2 w, \tag{5}$$

where $T = t - \psi(x)$ and $R$ is a nonlinear funtion of $\nabla\psi, \nabla^2\psi$ (the "curvature" of $S$). Substituting into (4), we obtain an equation on $w$ which can be reduced to a symmetric first order system of the form

$$Q(NW + AW) = t_0 \sum A_j\partial_j W + f(t_0, t_1, x, W),\ t_0 = T,\ t_1 = T\ln T. \tag{6}$$

Here $Q$ and the $A_j$ are symmetric matrices ($Q \gg 0$), $N = t_0\partial_{t_0} + (t_0 + t_1)\partial_{t_1}$, $A$ is constant and $f$ vanishes for $t_0 = t_1 = 0$. It turns out that systems like (6), called "generalized Fuchsian", can be solved in Sobolev spaces if the eigenvalues of $A$ have non-negative real parts. Returning to equation (4), we obtain for any $\psi$ and $w_0$ a smooth function $w(T, T\ln T, x)$, $w(0,0,x) = w_0(x)$ yielding, via (5), a singular solution of (4).

A completely similar method can be applied to other types of nonlinearities. For instance, consider the equation

$$\partial_t^2 u - \Delta u = \frac{2(m+1)}{(m-1)^2} u^m + \sum_{-\infty}^{m-1} a_j(x,t) u^j, \tag{7}$$

where the sum is finite and the $a_j$'s smooth. Then, for any space-like $S$, there exists a solution $u$ of (7) for $t > \psi(x)$ which satisfies

$$\lim_{t \to \psi(x)} (t - \psi(x))^{\frac{2}{m-1}} u(x,t) = \gamma^{\frac{1}{m-1}}.$$

The structure of $u$ can be described as follows: with $\tau = T^{\frac{1}{m-1}}$, we have

$$u = \tau^{-2} v, \quad v = \sum_0^{2m+1} v_j(x)\tau^j + \tau^{2m+2}[v_{2m+2}(x)\ln\tau + w], \tag{8}$$

where the functions $v_j$ $(j = 0, \ldots, 2m+2)$ are determined by $\psi$ and the $a_j$'s $(v_0 = \gamma^{\frac{1}{m-1}})$. The function $w = w(\tau, \tau\ln\tau, x)$ is as before a smooth function, solution of an appropriate generalized Fuchsian equation ($w$ is determined by its initial value $w(0,0,x)$).

In the preceding examples, the blowup surface is a priori smooth and non-characteristic. There are some simple examples where the blowup surface is characteristic at some point $0 \in S$. For instance, in the plane with coordinates $(x,t)$,

$$u(x,t) = h(t) + \int_{-1}^{x} \frac{ds}{s^3 - t} \tag{9}$$

is a solution of the semilinear wave equation

$$\partial_{xt}^2 u = u_x^2 \tag{10}$$

for $(x,t)$ close to the origin, $t > x^3$. This solution is (nonuniformly) blowing up on $S = \{t = x^3\}$, which is characteristic only at the origin.

We do not know in general which structure of nonlinearities $F$ allows such solutions, and what could be a general Ansatz to obtain them.

*I.2. Global situations.* For general semilinear wave equations, it is easy to define, using local existence and uniqueness of smooth solutions, a maximal influence domain $\Omega$ of existence for a solution with smooth compactly supported data; in fact, we can take the union of all open influence domains containing $\{t \geq 0\}$ where the solution exists (see [18] or [4] for details). The boundary of $\Omega$ is always space-like or characteristic, and we expect the solution or its gradient to become infinite near it. More precisely, at non-characteristic points of $\partial\Omega$, we expect this boundary to be smooth enough and the solution to behave like the local models displayed in I.1.

The first case where this has been proved is the case of the equation

$$\partial_t^2 u - \Delta u = F(u) = u^p \tag{11}$$

handled by Caffarelli and Friedman. For $n = 1$, they prove ([6]) the differentiability of the blowup curve $S$ and the "conormal" behavior of the solution, that is

$$u \sim d^{\frac{-2}{p-1}}, \ \nabla u \sim d^{\frac{-2}{p-1}-1}, \ \ldots, \quad d = t - \psi(x).$$

Moreover, they show that $S$ cannot be characteristic, thus approaching very closely the local model. In [5], they achieve the same goal at the price of some strong assumptions on the initial data.

The second case is a perturbation situation for (4), that we now describe. Equation (4) admits the particular solution

$$u = \ln \frac{2}{t^2},$$

corresponding to the Ansatz (5) with $\psi = w = 0$, and Cauchy data $(\ln 2, -2)$ on $\{t = 1\}$. In [12], Kichenassamy proves the following theorem.

**Theorem.** *Let $u$ be a solution of (4) with*

$$(u(x,1) - \ln 2, \partial_t u(x,1) + 2) \in H^s \times H^{s-1}, \quad s > 167 + 10\left[\tfrac{n}{2}\right].$$

*If the data are sufficiently close to $(\ln 2, -2)$, there is a function $\psi \in H^r$ $(r = s - 146 - 9[\frac{n}{2}])$ such that $u - \ln \frac{2}{(t-\psi(x))^2}$ is bounded near $S$.*

In fact, the proof establishes that $u$ is given by (5) for some $\psi, w$ close to zero; hence we have a complete control of the differentiability of $S$ and of the behavior of $u$ in this case.

In the general case, many questions remain open: Is the boundary of $\Omega$ always differentiable? In which case can it be characteristic? How can we describe the behavior of $u$? This last point has been discussed in general situations by Lindblad [18] for equation (11).

## II. Quasilinear problems.

*II.1 Local blowup patterns.* For general quasilinear equations or systems (1), (2), the blowup mechanism described in Section I, due to self-increase of $u$ through a "source" term $F(u, \nabla u)$, can of course still be observed. However, in "truly" quasilinear cases, a new mechanism appears, due to "focusing of characteristics".

Let us explain how we can construct local examples of this phenomenon.

Denote by $\Phi : (X,T) \to (x,t)$ a smooth map near $M_0 = (X_0, T_0)$, with $\Phi'(M_0)$ noninvertible; assume that there exists an open domain $D$ ($\Phi(M_0) = m_0 \in \overline{D}$) and a map $\Psi$ defined and continuous on $\overline{D}$ with

$$\Psi(m_0) = M_0, \ (x,t) \in D \Rightarrow \Phi(\Psi(x,t)) = (x,t), \ \det \Phi'(\Psi(x,t)) \neq 0.$$

Assume now that we can find such a map $\Phi$ and a smooth $v(X,T)$ so that

$$u(x,t) = v(\Psi(x,t))$$

is a solution of (1): this solution $u$ (or more precisely its gradient) in general blows up at $m_0$; we call this "geometric blowup".

It turns out that the construction of such solutions can be carried out for large classes of quasilinear systems in any number of space dimensions (see [1]). More precisely, looking for $\Phi$ of corank one in the form

$$\Phi(X,T) = (\phi(X,T), X_2, \dots, X_n, T),$$

one can write down a new system for $v, \phi$ (the "blowup system") such that to any smooth solution of this system and to any couple $D, \Psi$ corresponds a blowup solution $u$ of the original system.

For instance, consider the system

$$\partial_t u + A(u)\partial_x u = 0, \tag{12}$$

where the matrix $A$ has real distinct eigenvalues $\lambda_1(u) < \dots < \lambda_N(u)$ with left and right eigenvectors $l_j(u), r_j(u)$. To each $j$ can be associated the blowup system

$$\partial_T \phi = \lambda_j(v), {}^t l_j(v)\partial_T v = 0,$$

$${}^t l_k(v)[\partial_X \phi \partial_T v + (\lambda_k - \lambda_j)(v)\partial_X v] = 0, \quad k \neq j.$$

In this case, we just have straightened out the $j$-characteristics.

Remark that all the quantities ${}^t l_k(u)\partial_x u$ ($k \neq j$) remain bounded, while ${}^t l_j(u)\partial_x u$ blows up provided ${}^t l_j(v)\partial_X v \neq 0$.

The two simplest cases for $\Phi$ are the fold and the cusp.

If $\partial_X^2\Phi(M_0) \neq 0$, $\Phi$ has a fold singularity, and we can take for $D$ one side of some blowup surface $S$; the gradient of $u$ becomes infinite on $S$ like $d^{-\frac{1}{2}}$ ($d$ being the distance to $S$).

If

$$\partial_X^2\phi(M_0) = 0, \quad \partial_X^3\phi(M_0) \neq 0,$$

$\Phi$ has a cusp singularity, and we can take for $D$, for instance, the exterior of the cusp; the gradient of $u$ becomes then infinite only at $m_0$.

In this framework, the distinction hinted at in the Introduction between singular and blowup solutions can be made more precise. The question: "How nonlinear has the system to be to possess a true blowup solution?" has the simple answer: the eigenvalue $\lambda_j$ has to be "genuinely nonlinear" in the sense of Lax, that is $r_j\nabla_u\lambda_j \neq 0$ (see [1] for precise statements).

*II.2 Global situations.* It is clear that for scalar first order equations

$$\partial_t u + \sum a_j(u)\partial_j u = 0$$

the blowup of any solution is of the "geometric" type described in section II.1. When a source term is present, one may consult Natalini [20].

Consider again systems of the form (12) in one space dimension. John [10] has proved that, if some eigenvalue is genuinely nonlinear at the origin and the Cauchy data are small enough, blowup occurs in finite time. Since some "asymptotic decoupling" of the various modes of propagation has already taken place then, one may suspect that the blowup mechanism in John's case is like in the scalar case, i.e. of geometric nature. This is actually the case, as proved in [3] for $N = 3$.

If the Cauchy data are not small, blowup may well occur before any perceptible decoupling has taken place. However, in the case $N = 2$ where the eigenvalues are genuinely nonlinear, we formulate the following conjecture:

For generic data $u_0$, blowup occurs at time $\overline{T}$ at a single point $m_0$, and the solution $u$ coincides for $t < \overline{T}$ near $m_0$ with a blowup solution associated with some mode $\lambda_j$ and a map $\Phi$ with a cusp singularity.

In fact, for $2 \times 2$-systems which can be diagonalized using Riemann invariants, a slightly weaker result is proved in [3]. If $N \geq 3$, the situation is not clear; results in [11] give a hint that maybe a different kind of interaction between the various modes is taking place.

For several space dimensions, only the case of equation (2) with small data of size $\epsilon$ has been studied (see John [9], Hörmander [8]). Even in this special case, there is no result giving a relevant upper bound for the lifespan, and a fortiori no description of the blowup mechanism. Only results of asymptotic nature have been obtained so far. Let us explain briefly some approach using geometric blowup for the case $n = 2$.

Recall some notations. We assume in (2) $F \equiv 0$ and

$$g_{ij} = g_{ji},\ g_{00} = 1,\ i \geq 1 \Rightarrow g_{i0}(0) = 0,\ i,j \geq 1 \Rightarrow g_{ij}(0) = -\delta_{ij}.$$

The Cauchy data $f(x,\epsilon)$ and $g(x,\epsilon)$ on $t = 0$ are smooth for $(x,\epsilon) \in \mathbf{R}^n \times [0,\epsilon_0]$, and compactly supported in $|x| \leq M$. Setting

$$g_{ij}(\eta) = g_{ij}(0) + \sum g_{ij}^k \eta_k + O(\eta^2),$$

we define, for $\omega \in S^1$,

$$g(\omega) = \sum_{0 \leq i,j,k \leq n} g_{ij}^k \omega_i \omega_j \omega_k, \quad \omega_0 = -1,\ \omega_1 = \cos\omega,\ \omega_2 = \sin\omega.$$

Finally, set

$$F_0(\sigma,\omega) = \frac{1}{2}(2\pi)^{-\frac{1}{2}(n-1)}\chi_-^{-\frac{1}{2}(n-1)} * [R(s,\omega,g) - \partial_s R(s,\omega,f)],$$

where the star denotes convolution in $s$, $\chi_-^\lambda = \frac{|x|^\lambda}{\Gamma(\lambda+1)}\mathbf{1}_{x<0}$ and

$$R(s,\omega,f) = \int_{x\omega = s} f(x)dS(x)$$

is the Radon transform of $f$.

First, asymptotic techniques (known as "nonlinear geometrical optics" in the literature, see [4]) allow a normalization of the variables; we introduce

$$\sigma = |x| - t, \quad \omega = \frac{x}{|x|}, \quad \tau = \epsilon\sqrt{t},$$

and represent the solution $u$ in the form

$$u(x,t) = \epsilon|x|^{-\frac{1}{2}} w(\sigma,\omega,\tau).$$

We expect second order derivatives of $w$ to blowup at a finite time $\bar{\tau}$ (thus, the lifespan of $u$ would be $\overline{T} = \epsilon^{-2}\bar{\tau}^2$).

Translated into a problem on $w$ in the $(\sigma,\omega,\tau)$ variables, the question of blowup becomes local (in space-time $(\sigma,\tau)$); an oversimplified model problem would be the following.

Given $w_0(\sigma,\omega,\epsilon)$, vanishing for $\sigma \geq 0$, find the lifespan and the blowup mechanism for the solution $w$ of the Goursat problem

(13) $\partial_{\sigma\tau}^2 w + \partial_\sigma w \partial_\sigma^2 w + \epsilon\partial_\omega^2 w = 0,\ w(\sigma,\omega,0) = w_0(\sigma,\omega),\ \sigma \geq 0 \Rightarrow w = 0.$

We see that this is a multidimensional perturbation of Burgers' equation in the $(\sigma,\tau)$ variables.

The blowup system corresponding to equation (13) can be written

$$(14)\quad \partial_T\phi = v+\epsilon(\partial_\omega\phi)^2,\ \partial^2_{XT}v-2\epsilon\partial_\omega\phi\partial^2_{X\omega}v+\epsilon\partial_X\phi\partial^2_\omega v-\epsilon\partial^2_\omega\phi\partial_X v=0,$$

where $\Phi:(X,T,\omega)\to(\sigma=\phi(X,T,\omega),\tau=T,\omega=\omega)$, and $\partial_\sigma w(\Phi)=v$. If it were possible to solve (14) with smooth data, one would prove a statement close to the one obtained for John's case above: generically, $\partial_\sigma w$ would be a cusp solution.

Unfortunately, we do not know how to solve (14) with smooth data, and the problem of blowup remains open.

The following information of asymptotic nature can however be obtained by this approach.

**Theorem.** *Assume that the function $-g(\omega)\partial^2_\sigma F_0(\sigma,\omega)$ has a unique negative minimum at $(\sigma_0,\omega_0)$ with positive definite Hessian. Then there exists a function $\overline{T}^a=\overline{T}^a(\varepsilon)$ with the two following properties:*

(i) *For all $N$, $\overline{T}\geq\overline{T}^a-\epsilon^N$ for $0<\epsilon\leq\epsilon_N$.*

(ii) *There exists $C>0$ such that, for $(C\epsilon)^{-2}\leq t\leq\overline{T}^a-\epsilon^N$ and $0<\epsilon\leq\epsilon_N$,*

$$(C(\overline{T}^a-t))^{-1}\leq||\nabla^2u(.,t)||_{L^\infty}\leq C(\bar{T}^a-t)^{-1}.$$

*Moreover, the function $\overline{T}^a$ is of the form*

$$\overline{T}^a=\epsilon^{-2}\bar{\tau}^a(\epsilon,\epsilon^2\ln\epsilon)^2,$$

*where $\bar{\tau}^a$ is a $C^\infty$ function of its arguments. In particular, $\bar{\tau}^a=A_0+A_1\epsilon+O(\epsilon^2\ln\epsilon)$, with*

$$A_0^{-1}=\max\sum g^k_{ij}\omega_i\omega_j\omega_k\partial^2_\sigma F_0(\sigma,\omega).$$

We call the function $\overline{T}^a$ "asymptotic lifespan".

## References

1. Alinhac S., *Explosion géométrique pour des systèmes quasilinéaires*, Séminaire d'Équations aux dérivées partielles, Exp. no 1, École Polytechnique, Paris, (1993), and article to appear in Amer. J. Math. (1995).
2. ———, *Temps de vie et comportement explosif des solutions d'équations d'ondes quasilinéaires en dimension deux II*, Duke Math. J. **73** no 3 (1994), 543–560.
3. ———, *Temps de vie précisé et explosion géométrique pour des systèmes hyperboliques quasi-linéaires en dimension un d'espace*, Séminaire d'Équations aux

dérivées partielles, Exp. no 6, École Polytechnique, Paris, (1995), and article to appear in Annali di Pisa (1995).

4. ———, *Blowup for nonlinear hyperbolic equations*, Birkhäuser, 1995, Progress in Nonlinear Differential Equations and their Applications.
5. Caffarelli L. and Friedman A., *The blowup boundary for nonlinear wave equations*, Trans. Amer. Math. Soc. **297** no 1 (1986), 233–241.
6. ———, *Differentiability of the blowup curve for one dimensional nonlinear wave equations*, Arch. Rat. Mech. Anal. **91** no 1 (1985), 83–98.
7. Christodoulou D., *Bounded variation solutions of the spherically symmetric Einstein scalar field equations*, Comm. Pure Appl. Math. **46** (1993), 1131–1220.
8. Hörmander L., *The lifespan of classical solutions of nonlinear hyperbolic equations*, Springer Lecture Notes in Math. **1256** (1986), 214–280.
9. John F., *Nonlinear wave equations. Formation of singularities*, Amer. Math. Soc., Providence, R. I., 1990, University Lecture Series 2, Lehigh University.
10. ———, *Formation of singularities in one-dimensional nonlinear wave propagation*, Comm. Pure Appl. Math. **27** (1974), 377–405.
11. Joly J. L., Metivier G., Rauch J., *A nonlinear instability for* $3 \times 3$ *systems of conservation laws*, Preprint, Université de Bordeaux (1993).
12. Kichenassamy S. and Littman W., *Blowup surfaces for nonlinear wave equations I and II*, Comm. PDE **18** (1993), 431–452 and 1869–1899.
13. Kichenassamy S., *The blow-up problem for exponential nonlinearities*, Preprint (1995).
14. Klainerman S. and Majda A., *Formation of singularities for wave equations including the nonlinear vibrating string*, Comm. Pure Appl. Math. **33** (1980), 241–263.
15. Lax P. D., *Development of singularities of solutions of nonlinear hyperbolic partial differential equations*, J. Math. Phys. **5** no 5 (1964), 611–613.
16. ———, *Hyperbolic systems of conservation laws II*, Comm. Pure Appl. Math. **10** (1957), 537-566.
17. Lebaud M. P., *Description de la formation d'un choc dans le p-système*, J. Math. Pure Appl. **73** (1994), 523–565.
18. Lindblad H., *Blowup of solutions of* $\partial_t^2 u - \Delta u = |u|^p$ *with small initial data*, Comm. PDE **15** no 6 (1990), 757–821.
19. Majda A., *Compressible fluid flow and systems of conservation laws*, Springer Verlag, 1984, Appl. Math. Sci. 53.
20. Natalini R., *Unbounded solutions for conservation laws with sources*, Nonlinear Anal. **21** no 5 (1993), 349–362.
21. Smoller J., *Shock waves and reaction diffusion equations*, Springer Verlag, New York, 1983, Grundlehr. d. Math. Wiss. 258.
22. Strauss W., *Nonlinear wave equations*, Conf. Board Math. Sc. **73** (1989).
23. Zuily C., *Solutions en grand temps d'équations d'ondes non linéaires*, Séminaire Bourbaki 779, Paris, (1993/1994).

Département de Mathématiques, Université Paris-Sud, F-91405 Orsay, France.
*E-mail address*: salinhac@matups.matups.fr

# GLOBAL TIME DECAY OF THE AMPLITUDE OF A REFLECTED WAVE

Michael Beals

**1. Introduction and statement of the main results.** We are interested in global time estimates for the $L^p$ norms of solutions to the wave equation on the exterior of a smooth compact strictly convex obstacle in $\mathbf{R}^n$, $n \geq 2$, with vanishing Dirichlet data on the boundary:

$$\square u = 0 \text{ on } \Omega \times \mathbf{R};\ u(0) = f_0,\ u_t(0) = f_1,\ u|_{\partial\Omega\times\mathbf{R}} = 0. \tag{1.1}$$

The behavior of solutions to the classical wave equation in Lebesgue spaces other than $L^2$ has been studied both for its intrinsic interest and for its applications in linear and nonlinear theory. Estimates have been used to study small time existence, long time existence, and scattering properties of solutions to hyperbolic wave equations; see for example Ginibre-Velo [7], Strauss [22], Shatah-Struwe [18], Lindblad-Sogge [11]. An example of the extra information contained in other $L^p$ spaces is the long time behavior — unlike the energy norm, certain other norms of solutions can decay as the time $t$ approaches infinity. For instance, if $u$ satisfies the free-space problem

$$\square u = 0 \text{ on } \mathbf{R}^n \times \mathbf{R};\ u(0) = f_0,\ u_t(0) = f_1, \tag{1.2}$$

then the following time decay holds:

$$\|u(t)\|_\infty \leq Ct^{-(n-1)/2} \quad \text{as } t \to \infty, \tag{1.3}$$

as long as the initial data $f_0$ and $f_1$ and sufficiently many of their derivatives are in $L^1(\mathbf{R}^n)$. As shown in Marshall-Strauss-Wainger [12], the estimate (1.3) holds with minimal smoothness hypotheses on the data. The usual energy estimates then yield the interpolated results on the norms as maps from $L^p(\mathbf{R}^n)$ to $L^{p'}(\mathbf{R}^n)$, for $p'$ the dual index to $p$, of the operators $E(t)f$ and $E_0(t)f$ given by

$$E(t)f(x) = \int e^{ix\cdot\xi}[(\sin(t|\xi|))/|\xi|]\hat{f}(\xi)\, d\xi, \tag{1.4}$$

Research supported in part by NSF grant DMS-9401819

and

$$E_0(t)f(x) = \int e^{ix\cdot\xi}[(\cos(t|\xi|))/|\xi|]\hat{f}(\xi)\,d\xi. \tag{1.5}$$

For example, if $n \geq 2$, $1/p = 1/2 + 1/(n+1)$, $1/p' = 1/2 - 1/(n+1)$, then as established in Strichartz [23],

$$\|E(t)f\|_{p'} \leq Ct^{-(n-1)/(n+1)}\|f\|_p. \tag{1.6}$$

The methods in Strauss [21] can then be applied to the long-time existence and scattering for certain nonlinear equations $\Box u = g(u)$ on $\mathbf{R}^n \times \mathbf{R}$, since

$$u(t) = \int E(t-s)g(u(s))\,ds \text{ satisfies} \tag{1.7}$$

$$\Box u = g \text{ on } \mathbf{R}^n \times \mathbf{R};\ u(0) = 0,\ u_t(0) = 0.$$

Here we consider the analogues of (1.3) and (1.6) for the solution of the exterior problem (1.1). We will treat data of support in a fixed compact subset $F \subset \Omega$. After finite time, it is well known that, since $\Omega$ is non-trapping, the norms of $u(t)$ on any compact subset of $\Omega$ decay as $t \to \infty$, exponentially in $t$ for $n$ odd and as a negative power of $t$ for $n$ even. (See e.g. Ralston [17].) On the other hand, the primary contribution to the size of $u$ is near the union of the light rays in $\Omega$ over $F$ (including reflected rays). Even for smooth data there seem to be no previous estimates analogous to (1.3) or (1.6) for the spatially global long time behavior, since unlike the free space case, the rate of decay in $t$ cannot be deduced from a simple scaling argument once the boundedness of the corresponding operators is known. Our main result is the following estimate, which is best possible in terms of decay and regularity as seen from the free space results.

**Theorem 1.1.** *Let $\Omega \subset \mathbf{R}^n$ denote a domain which is the complement of a smooth convex compact obstacle, and such that the boundary $\partial\Omega$ has nonvanishing total curvature at each point. Let $F \subset \Omega$ denote a fixed compact set. Suppose that $f_0, f_1$ have support in $F$, with $f_0 \in L^{1,\sigma}(\mathbf{R}^n)$, $f_1 \in L^{1,\sigma-1}(\mathbf{R}^n)$, $\sigma > (n+1)/2$. If $u$ is the solution to the initial-boundary problem (1.1), then there are constants $T = T_{F,\Omega}$ and $C = C_{F,\Omega}$ such that*

$$\|u(t)\|_\infty \leq C\langle t-T\rangle^{-(n-1)/2}(\|(1-\Delta)^\sigma f_0\|_1 + \|(1-\Delta)^{\sigma-1}f_1\|_1),\ t \geq T. \tag{1.8}$$

*If $\sigma = (n+1)/2$ and $f_0 \in \mathscr{H}^{1,\sigma}(\mathbf{R}^n)$, $f_1 \in \mathscr{H}^{1,\sigma-1}(\mathbf{R}^n)$, then the estimate (1.8) holds with $\|\ \|_1$ denoting the norm in the local Hardy space $\mathscr{H}^1(\mathbf{R}^n)$.*

As a direct consequence of (1.8), the standard energy estimate and analytic interpolation of the corresponding family of operators, we can deduce the optimal decay of the solutions to (1.1) for $f_0 = 0$, $f_1 \in L^p(\Omega)$, as treated in the case $\Omega = \mathbf{R}^n$ in Marshall-Strauss-Wainger [12]. As in (1.4), we define $E_\Omega(t)f$ as follows:

(1.9)

$$v(t) = E_\Omega(t)f \text{ satisfies } \Box v = 0 \text{ on } \Omega \times \mathbf{R};\ v(0) = 0, v_t(0) = f,\ v|_{\partial\Omega\times\mathbf{R}} = 0.$$

Then for the inhomogeneous problem

$$(1.10)\qquad u(t) = \int_0^t E_\Omega(t-s)g(s)\,ds \quad \text{satisfies}$$
$$\Box u = g \text{ on } \Omega \times \mathbf{R};\ u(0) = 0,\ u_t(0) = 0,\ u|_{\partial\Omega\times\mathbf{R}} = 0.$$

The boundedness of $E_\Omega(t)$ as an operator $L^p(\Omega) \to L^p(\Omega)$ for the optimal range of $p$ has been established in Smith-Sogge [19], with uniform control of the norm for $t$ in a bounded set. (The range of $p$ is as in the free-space case, Peral [16], Beals [1]; an analogous estimate for the Dirichlet problem with data on $\partial\Omega$ was treated in Beals [2] for solutions near $\partial\Omega$.) This result is applied in Smith-Sogge [20] to a nonlinear problem. The boundedness of $E_\Omega(t)$ as a map $L^p(\Omega) \to L^{p'}(\Omega)$ can be established similarly, so it is really the time decay in the following estimate which is new.

**Theorem 1.2.** *Let $\Omega, F$ be as in Theorem 1.1. Suppose $f$ has support in $F$ and $f \in L^p(\Omega)$, $1/p = 1/2 + 1/(n+1)$, $1/p' = 1/2 - 1/(n+1)$. If $E_\Omega(t)f$ is given by (1.9), then there are constants $T = T_{F,\Omega}$ and $C = C_{F,\Omega}$ such that*

$$(1.11)\qquad \|E_\Omega(t)f\|_{p'} \le C\langle t - T\rangle^{-(n-1)/(n+1)}\|f\|_p.$$

The analogue of this estimate has been proved previously for one other case in which a simple scaling argument cannot be used: for appropriate potentials $V$, the problem $(\Box + V)u = 0$ on $\mathbf{R}^n \times \mathbf{R}$; $u(0) = 0$, $u_t(0) = f$, is treated in Beals-Strauss [4] (for compactly supported data) and [5] (for general data). See also Yajima [26] and Zhong [27] for related results.

The main ideas in the proof of Theorem 1.1 are as follows. When localized in space, the solution to (1.1) decays more rapidly than the rate in (1.8). Thus it suffices to treat $\beta u$ for $\beta \in C^\infty(\mathbf{R}^n)$ identically zero near the obstacle and identically one further away from $\partial\Omega$. In that case, $\beta u$ satisfies an inhomogeneous equation of the form $\Box\beta u = g$ in $\mathbf{R}^n \times \mathbf{R}$, for $g$ with compact $x$-support near $\partial\Omega$. Thus from (1.7) the primary term to be estimated is

$$\int_0^t E(t-s)g(s)\,ds,$$

with the time dependence contained in $E$ given explicitly by (1.4). If $s$ is sufficiently large, say $s \ge T$ as determined by the support of the data and $\Omega$, then $g(s)$ is smooth and sufficiently rapidly decaying, so that the estimate corresponding to (1.3) can be applied to $E(t-s)g(s)$. Thus it remains to treat $E(t-s)g(s)$ on the compact set $0 \leq s \leq T$. The details of this reduction to the action of the explicit operator $E(t-s)$ on the compactly supported term $g(s)$ derived from $u$ are given in Section 2.

The inhomogeneous term $g(s)$ above is obtained from the initial data $(f_0, f_1)$ by the action of compositions of Fourier integral operators as in Hörmander [9] and Fourier-Airy operators as in Taylor [24]. Since zeroth order operators of this type are not bounded on $L^p(\mathbf{R}^n)$ for $p \neq 2$ (except when $n = 1$), the estimate (1.6) cannot be used to treat $E(t-s)g(s)$ to obtain a result like (1.11), let alone (1.8), since $E(t-s)$ only maps into $L^{p'}(\mathbf{R}^n)$ with a loss of derivatives. On the other hand, as shown in Beals [3], the composition of $E(t-s)$ with certain other Fourier integral operators of appropriate order again yields an operator with the desired decay estimate. The argument in [3] is in the context of the problem $(\Box + V)u = 0$ on $\mathbf{R}^n \times \mathbf{R}$; in Section 3 the corresponding kernels for the compositions of Fourier and Fourier-Airy operators in the problem (1.1) are derived. The integral representations are explicit because $g(s)$ can be taken to have support as near to $\partial\Omega$ as necessary for the use of the Taylor-Melrose parametrix [24]. In Section 4 the estimates on the kernels are sketched, with arguments similar to those given in [3]. Full details will appear elsewhere.

The author would like to thank Walter Strauss for many conversations about the problem (1.1) over the years. Our joint work on the problem $(\Box + V)u = 0$ was an attempt to gain insight into decay estimates in the absence of scaling arguments; that insight has been invaluable in leading to the results here.

**Notation.** $\| \ \|_p$ denotes the norm in the Lebesgue space $L^p(\mathbf{R}^n)$ or $L^p(\Omega)$, $1 \leq p \leq \infty$; $\| \ \|_1$ will also denote the norm in the local Hardy space $\mathscr{H}^1(\mathbf{R}^n)$ when so indicated. This space is defined as follows: $f \in \mathscr{H}^1(\mathbf{R}^n)$ is equivalent to $a(D)f \in L^1(\mathbf{R}^n)$ for all pseudodifferential operators with symbols $a(\xi) \in S^0$. Equivalently, fix $\beta \in C^\infty(\mathbf{R}^+)$ with $0 \leq \beta \leq 1$, $\beta(\varrho) \equiv 1$ for $\varrho \geq 2$, $\beta(\varrho) \equiv 0$ for $\varrho \leq 1$; let $R_j$ denote the Riesz transforms, $1 \leq j \leq n$, (i.e. $R_j = r_j(D)$, with $r_j(\xi) = \xi_j/|\xi|$), and require $f$, $\beta(D)f$, $R_j(D)\beta(D)f \in L^1(\mathbf{R}^n)$. See Goldberg [8].

$\Delta$ denotes the Laplacian on $\mathbf{R}^n$, $\Box$ denotes the d'Alembertian $\partial_t^2 - \Delta$, and $\Lambda = (1-\Delta)^{\frac{1}{2}}$. We write $L^{p,\sigma}(\mathbf{R}^n) = \{f : \Lambda^\sigma f \in L^p(\mathbf{R}^n)\}$ and $\mathscr{H}^{1,\sigma}(\mathbf{R}^n) = \{f : \Lambda^\sigma f \in \mathscr{H}^1(\mathbf{R}^n)\}$.

$A \approx B$ means there are constants $C > c > 0$ with $cA \leq B \leq CA$. $\langle a \rangle = \sqrt{1+a^2}$.

**2. Reduction to an inhomogeneous problem.** Let $\Omega \subset \mathbf{R}^n$ denote the complement of a smooth compact convex obstacle as in Section 1, and let $F \subset \Omega$ denote a fixed compact set. Let $\alpha \in C^\infty(\mathbf{R}^n)$ have compact support. Since $\Omega$ is nontrapping, there is a time $T$ determined by $F$ and $\Omega$ such that whenever $u$ satisfies the initial-boundary problem (1.1) for data $(f_0, f_1)$ with support in $F$, $\alpha(x)u(t,x)$ is smooth for time $t \geq T$. (See e.g. Melrose-Sjöstrand [14].) Moreover, from Lax-Phillips [10] (for $n$ odd), Morawetz-Ralston-Strauss [15] (for $n = 2$), Melrose [13], Vainberg [25] and Ralston [17] (for $n$ even), there is local decay of the norms in all Sobolev spaces:

(2.1)

$$\|\Lambda^M \alpha u(t)\|_2 \leq C_M e^{-c(t-T)}(\|\nabla f_0\|_2 + \|f_1\|_2) \text{ for } t \geq T, \ n \text{ odd};$$

$$\|\Lambda^M \alpha u(t)\|_2 \leq C_M \langle t - T\rangle^{-(n-1)}(\|\nabla f_0\|_2 + \|f_1\|_2) \text{ for } t \geq T, \ n \text{ even}.$$

The hypotheses on the data given in Theorem 1.1 are not quite sufficient to yield $(f_0, f_1) \in L^{2,1}(\mathbf{R}^n) \times L^2(\mathbf{R}^n)$, since the Sobolev embedding theorem only implies that $L^{1,\sigma-1}(\mathbf{R}^n) \subset L^2(\mathbf{R}^n)$ for $\sigma \geq (n+2)/2$. On the other hand, it follows easily from the decay estimate (2.1) that for $t \geq T$

$$\|\Lambda^M \alpha u(t)\|_2 \leq C_{M+1} e^{-c(t-T)}(\|\Lambda^{-1}\nabla f_0\|_2 + \|\Lambda^{-1} f_1\|_2) \ n \text{ odd};$$

$$\|\Lambda^M \alpha u(t)\|_2 \leq C_{M+1} \langle t - T\rangle^{-(n-1)}(\|\Lambda^{-1}\nabla f_0\|_2 + \|\Lambda^{-1} f_1\|_2) \ n \text{ even}.$$

Thus the analogous local decay estimate in the norms under consideration for this problem follows:

(2.2)

$$\|\Lambda^M \alpha u(t)\|_\infty \leq C e^{-c(t-T)}(\|\Lambda^\sigma f_0\|_1 + \|\Lambda^{\sigma-1} f_1\|_1) \text{ for } t \geq T, \ n \text{ odd};$$

$$\|\Lambda^M \alpha u(t)\|_\infty \leq C \langle t - T\rangle^{-(n-1)}(\|\Lambda^\sigma f_0\|_1 - \|\Lambda^{\sigma-1} f_1\|_1) \text{ for } t \geq T, \ n \text{ even}.$$

Let $\beta \in C^\infty(\mathbf{R}^n)$ be smooth, identically one outside a small neighborhood of the complement of $\Omega$, identically zero on a smaller neighborhood of the complement of $\Omega$. If we write

$$u = (1 - \beta)u + \beta u, \tag{2.3}$$

then (2.2) shows that $(1 - \beta)u$ decays at a faster rate than we are trying to establish in Theorem 1.1. On the other hand, $\beta u$ is the solution to the following free-space inhomogeneous problem:

$$\Box \beta u = -2\nabla\beta \cdot \nabla u - (\Delta\beta)u \text{ on } \mathbf{R}^n \times \mathbf{R}; \ \beta u(0) = f_0, \ \beta u_t(0) = f_1. \tag{2.4}$$

Thus from the expressions for the free-space problem as in (1.4), (1.5), and (1.7),

$$\Box u = E_0(t) f_0 + E(t) f_1 + \int_0^t E(t-s) U(s)\, ds; \ U = -2\nabla\beta \cdot \nabla u - (\Delta\beta)u. \tag{2.5}$$

Since $v = E_0(t)f_0 + E(t)f_1$ satisfies the homogeneous free-space problem, the desired decay estimate for $v$ holds (Marshall-Strauss-Wainger [12]; Beals [3]):

$$\|E_0(t)f_0 + E(t)f_1\|_\infty \le C|t|^{-(n-1)/2}(\|\Lambda^\sigma f_0\|_1 + \|\Lambda^{\sigma-1}f_1\|_1). \tag{2.6}$$

On the other hand, for $t \ge T$ it follows that $U$ is smooth and decays with $t$ on the support of the derivatives of $\beta$, as long as $\beta$ is identically one except on a sufficiently small neighborhood of $\partial\Omega$. Thus from (2.5) and (2.2), for $t \ge T$,

$$\begin{gathered}\|\Lambda^M U(t)\|_\infty \le C\delta(t-T)(\|\Lambda^\sigma f_0\|_1 + \|\Lambda^{\sigma-1}f_1\|_1);\\ \delta(t-T) = e^{-c(t-T)},\ n \text{ odd};\ \delta(t-T) = \langle t-T\rangle^{-(n-1)},\ n \text{ even}.\end{gathered} \tag{2.7}$$

As long as $M > (n-1)/2$, the known free-space estimate yields

$$\|E(t-s)\Lambda^{-M}g\|_\infty \le C\langle t-s\rangle^{-(n-1)/2}(\|g\|_1 + \|g\|_\infty). \tag{2.8}$$

(It is convenient to map out of the space $L^1(\mathbf{R}^n)\cap L^\infty(\mathbf{R}^n)$ rather than $L^1(\mathbf{R}^n)$, so that we can use the bound $\langle t-s\rangle^{-(n-1)/2}$ rather than the bound $|t-s|^{-(n-1)/2}$.) Then from (2.7) and (2.8) the following estimate holds.

**Lemma 2.1.** *Let $U(s)$ be given by (2.5). For $t \ge T$,*

$$\left\| \int_T^t E(t-s)U(s)\,ds \right\|_\infty \le C\langle t-T\rangle^{-(n-1)/2}(\|\Lambda^\sigma f_0\|_1 + \|\Lambda^{\sigma-1}f_1\|_1).$$

*Proof.* Write

$$\begin{aligned}\int_T^t E(t-s)U(s)\,ds &= \int_T^t E(t-s)\Lambda^{-M}\{\Lambda^M U(s)\}\,ds\\ = \int_T^{(t+T)/2} E(t-s)\Lambda^{-M}\{\Lambda^M U(s)\}\,ds &+ \int_{(t+T)/2}^t E(t-s)\Lambda^{-M}\{\Lambda^M U(s)\}\,ds.\end{aligned}$$

From (2.7) and (2.8), the second term on the right hand side is bounded by

$$\begin{aligned}C\delta(t-T)\Big(\int_{(t+T)/2}^t \langle t-s\rangle^{-(n-1)/2}\,ds\Big)&(\|\Lambda^\sigma f_0\|_1 + \|\Lambda^{\sigma-1}f_1\|_1)\\ \le C\delta(t-T)\langle t-T\rangle^{\frac12}&(\|\Lambda^\sigma f_0\|_1 + \|\Lambda^{\sigma-1}f_1\|_1) \text{ as long as } n \ge 2,\\ &\le C\langle t-T\rangle^{-(n-1)/2}(\|\Lambda^\sigma f_0\|_1 + \|\Lambda^{\sigma-1}f_1\|_1).\end{aligned}$$

Similarly, the first term on the right hand side is bounded by

$$C\langle t-T\rangle^{-(n-1)/2}\Big(\int_T^{(t+T)/2}\delta(t-s)\,ds\Big)(\|\Lambda^\sigma f_0\|_1+\|\Lambda^{\sigma-1}f_1\|_1)$$
$$\le C\langle t-T\rangle^{-(n-1)/2}(\|\Lambda^\sigma f_0\|_1+\|\Lambda^{\sigma-1}f_1\|_1)\text{ as long as } n\ge 3.$$

If $n=2$, instead of (2.8) we use the refined estimate on the kernel $K(t-s,x)$ corresponding to the operator $E(t-s)\Lambda^{-M}$, $M>(n-1)/2$, (see e.g. Beals [3]):

$$|K(t-s,x)|\le C\langle t-s\rangle^{-(n-1)/2}\langle t-s-|x|\rangle^{-(n-1)/2}.$$

Then (with $G$ denoting a compact set in $\Omega$ containing $\operatorname{supp} U$), the term

$$\Big\|\int_T^{(t+T)/2}E(t-s)\Lambda^{-M}\{\Lambda^M U(s)\}\,ds\Big\|_\infty$$

can be estimated by

$$C\sup_x\int_T^{(t+T)/2}\langle t-s\rangle^{-\frac12}\int_G\langle t-s-|x-y|\rangle^{-\frac12}|\Lambda^M U(s,y)|\,dy\,ds$$
$$\le C\Big(\sup_x\int_G\int_T^{(t+T)/2}\langle t-s\rangle^{-\frac12}\langle t-s-|x-y|\rangle^{-\frac12}\,ds\,dy\Big)(\|\Lambda^\sigma f_0\|_1+\|\Lambda^{\sigma-1}f_1\|_1)$$
$$\le C\langle t-T\rangle^{-\frac12}(\|\Lambda^\sigma f_0\|_1+\|\Lambda^{\sigma-1}f_1\|_1),$$

because for $r=|x-y|$ and $s=t-r$

$$\int_T^{(t+T)/2}\langle t-s\rangle^{-\frac12}\langle t-s-|x-y|\rangle^{-\frac12}\langle s\rangle^{-1}\,ds$$
$$\le C\langle t-T\rangle^{-\frac12}\int_T^{(t+T)/2}\langle t-s-r\rangle^{-\frac12}\langle s\rangle^{-1}\,ds$$
$$\le C\langle t-T\rangle^{-\frac12}\int_{-\infty}^{\infty}\langle S-s\rangle^{-\frac12}\langle s\rangle^{-1}\,ds\le C\langle t-T\rangle^{-\frac12}.$$

This completes the proof of Lemma 2.1.

In order to complete the analysis of (2.5), we are therefore reduced to studying

$$(2.9)\quad \int_0^T E(t-s)U(s)\,ds,\quad U=-2\nabla\beta\cdot\nabla u-(\Delta\beta)u,\ T>0\text{ fixed},\ t\ge T.$$

The argument used in the proof of Lemma 2.1 depended on having $U$ be smooth (with appropriate norms bounded by norms of the data), as well as on having decay estimates for large $s$ where the integration took place. In (2.9), any local (or microlocal) portion of $U$ which is again smooth with appropriate norm bounds can be be handled in the same fashion. (Decay is no longer an issue since the integration in $s$ is now over a fixed interval.) This will allow us easily to handle all of the contributions to $u$ (the solution of the initial-boundary problem (1.1)) which are microlocally rapidly decreasing away from the null bicharacteristics over the singular support of the initial data, as well as remainder terms of sufficiently rapid decay (in the Fourier variables) in asymptotic expansions for the solution near those null bicharacteristics. This argument is formulated precisely as follows.

**Lemma 2.2.** *Let $R(s)$, $0 \le s \le T$, have compact support and satisfy*

$$\|\Lambda^{M+1}R(s)\|_\infty \le C(\|\Lambda^\sigma f_0\|_1 + \|\Lambda^{\sigma-1}f_1\|_1) \tag{2.10}$$

*for some $M > (n-1)/2$. Then for $t \ge T$,*

$$\begin{aligned}\left\|\int_0^T E(t-s)[-2\nabla\beta\cdot\nabla R(s) - (\Delta\beta)R(s)]\,ds\right\|_\infty \\ \le C\langle t-T\rangle^{-(n-1)/2}(\|\Lambda^\sigma f_0\|_1 + \|\Lambda^{\sigma-1}f_1\|_1).\end{aligned}$$

*Proof.* Write

$$\begin{aligned}&\int_0^T E(t-s)[-2\nabla\beta\cdot\nabla R(s) - (\Delta\beta)R(s)]\,ds \\ &\qquad = \int_0^T E(t-s)\Lambda^{-M}\{\Lambda^M[-2\nabla\beta\cdot\nabla R(s) - (\Delta\beta)R(s)]\}\,ds\end{aligned}$$

and use (2.8) and (2.10).

**3. Kernels near the singularities.** In order to study the remaining term (2.9) in the solution $u$ to the initial-boundary problem (1.1), we know from Lemma 2.2 that we can neglect the contributions due to smooth perturbations of the data. Thus instead of the free-space solutions of the initial value problem for the wave equation with data $(f_0, f_1)$, for convenience we consider the solutions $u_\pm$ given by

$$u_\pm(r,w) = \iint e^{i[(w-y)\cdot\eta \pm r|\eta|]} f(y)\,d\eta\,dy, \tag{3.1}$$

for $f \in L^{1,\sigma}(\mathbf{R}^n)$, $\sigma > (n+1)/2$, or $f \in \mathscr{H}^{1,\sigma}(\mathbf{R}^n)$, $\sigma = (n+1)/2$. If we set

$$(3.2) \qquad g = \Lambda^\sigma f, \quad g \in L^1(\mathbf{R}^n) \text{ or } \mathscr{H}^1(\mathbf{R}^n),$$

then this equation can be rewritten as

$$(3.3) \qquad u_\pm(r,w) = \iint e^{i[(w-y)\cdot\eta \pm r|\eta|]} \langle\eta\rangle^{-\sigma} g(y)\, d\eta\, dy.$$

The function $g$ will not (in general) have compact support, but if $\gamma \in C^\infty(\mathbf{R}^n)$ is identically one on the fixed set $F$ (which contains the support of $f$) and has compact support near $F$, then $f = \Lambda^{-\sigma}\gamma g$ modulo a smooth remainder (rapidly decreasing away from $F$). It follows from Lemma 2.2 that the term corresponding to $(1-\gamma)g$ satisfies the desired estimates. Thus we replace (3.3) with

$$(3.4) \qquad u_\pm(r,w) = \iint e^{i[(w-y)\cdot\eta \pm r|\eta|]} \langle\eta\rangle^{-\sigma} \gamma(y) g(y)\, d\eta\, dy.$$

In order to correct the free-space solution to account for the Dirichlet condition on the boundary, we will localize near a fixed point of $\partial\Omega$, say $(w', w_n) = (0,0)$, with coordinates chosen so that locally

$$(3.5) \qquad \Omega = \{w_n > \varphi(w')\}, \quad \varphi(w') = -Q(w') + \mathscr{O}\{|w'|^3\},$$

for $Q$ a positive definite quadratic form. Let $\alpha' \in C^\infty(\mathbf{R}^{n-1})$ have compact support and be identically one near $\{w' = 0\}$ and let $\alpha_0 \in C^\infty(\mathbf{R})$ have compact support and be identically one near a fixed time $r_0$. The localized boundary value is then given by

$$(3.6) \qquad \alpha_0(r)\alpha'(w')u_\pm(r,w',\varphi(w')) = \alpha_0(r)\alpha'(w') \iint e^{i[(w',\varphi(w')-y)\cdot\eta \pm r|\eta|]} \langle\eta\rangle^{-\sigma}\gamma(y)g(y)\, d\eta\, dy.$$

The forward outgoing solution to the wave equation on $\Omega \times \mathbf{R}$ with this data on $\partial\Omega \times \mathbf{R}$ is (up to a smooth remainder term) given by decomposing the expression (3.6) microlocally in the variable $\zeta = (\zeta', \zeta_n)$ dual to $(w', r)$, depending on whether the directions correspond to null bicharacteristics intersecting the boundary transversally or near grazing (see Taylor [24]). The smooth zero-order pseudodifferential symbol $\beta(\zeta)$ will denote the microlocalizer in what follows.

For the transverse terms, we must analyze expressions of the form

$$(3.7) \qquad u_{\mathrm{tr}}(s,z) = \iiint e^{i[\psi_{\mathrm{tr}}(z,\zeta) + (s-r)\zeta_n - w'\cdot\zeta']} a_{\mathrm{tr}}(s,z,\zeta)\chi(\zeta)\, d\zeta \times \alpha_0(r)\alpha'(w')u_\pm(r,w',\varphi(w'))\, dw'\, dr.$$

Here the phase function $\psi_{\mathrm{tr}}(z,\zeta)$ satisfies the eiconal equation

$$(3.8)\qquad |\nabla_z\psi|^2=\zeta_n^2,\quad \psi(z',\varphi(z'),\zeta)=z'\cdot\zeta',$$

so that on $\partial\Omega\times\mathbf{R}$ it reduces to the usual phase function for the Fourier transform. The amplitude $a_{\mathrm{tr}}(s,z,\zeta)\in S^0$ has principal symbol identically one, and satisfies the usual transport equations; $\chi(\zeta)$ has conic support where $\zeta_n\neq 0$.

Near grazing, the Taylor-Melrose ansatz (Taylor [24]) provides the solution locally near the boundary. Using the notation in [24] we have the primary term

(3.9)

$$u_{\mathrm{gr}}(s,z)=\iiint e^{i[\psi_{\mathrm{gr}}(z,\zeta)+(s-r)|\zeta'|-\psi_0(w',\zeta)]}[A(\varrho/|\zeta'|^{\frac13})/A(-\zeta_n/|\zeta'|^{\frac13})]$$
$$\times\, a_{\mathrm{gr}}(s,z,\zeta)\chi(\zeta)\,d\zeta\,\alpha_0(r)\alpha'(w')u_{\pm}(r,w',\varphi(w'))\,dw'\,dr.$$

Here the pair $(\psi_{\mathrm{gr}},\varrho)$ satisfy the eiconal equations

$$(3.10)\qquad |\nabla_z\psi|^2+(\varrho/|\zeta'|)|\nabla_z\varrho|^2=|\zeta'|^2,\quad \nabla_z\psi\cdot\nabla_z\varrho=0.$$

Moreover,

$$(3.11)\qquad \psi_0(w',\zeta)=\psi_{\mathrm{gr}}(w',\varphi(w'),\zeta)+\mathcal{O}\{|\zeta'|(\zeta_n/|\zeta'|)^\infty\},$$

and $A$ is an Airy function. The amplitude $a_{\mathrm{gr}}(s,z,\zeta)\in S^0$ satisfies the corresponding analogue of the transport equations, and $\chi(\zeta)$ has conic support where $|\zeta'|\neq 0$. The primary contribution occurs where $\zeta_n<0$; in that case the Airy function has an oscillating factor which can be rewritten as part of the phase function. The result is then

$$(3.12)\quad u_{\mathrm{gr}}(s,z)=\iiint e^{i[\psi_{\mathrm{gr}}(z,\zeta)+(s-r)|\zeta'|-\psi_0(w',\zeta)+\frac23|\zeta'|(\varrho_0^{\frac32}(z,\zeta)-\theta^{\frac32}]}$$
$$\times\,\{\Phi(\varrho_0|\zeta'|^{\frac23})/\Phi(\theta|\zeta'|^{\frac23})\}a_{\mathrm{gr}}(s,z,\zeta)\chi(\zeta)\,d\zeta$$
$$\times\,\alpha_0(r)\alpha'(w')u_{\pm}(r,w',\varphi(w'))\,dw'\,dr.$$

Here

(3.13)

$$\varrho_0(z,\zeta)=\varrho(z,\zeta)/|\zeta'|,\ \theta=-\zeta_n/|\zeta'|,\ \text{and}\ \Phi(\mu)\sim\mu^{-\frac14}(c_0+c_1\mu^{-\frac32}+\dots).$$

As remarked previously, we are primarily interested in the action of the operator $E(t-s)$ given by (1.4) on localized versions of the functions $u_{\mathrm{tr}}$ and $u_{\mathrm{gr}}$ and their derivatives. Let $\alpha(z)$ be a smooth cutoff with compact support in the interior of $\Omega$ and for which supp $\alpha$ is sufficiently near to $\partial\Omega$ that the expressions (3.7) and (3.12) are valid. Since the Fourier multiplier $[\sin((t-s)|\xi|)]\xi/|\xi|$ corresponds to the operator $E(t-s)\nabla_x$, the main terms to be treated involve the action of the operators $E_\pm(t-s)$ corresponding to the multipliers $e^{\pm i(t-s)|\xi|}$ on $\alpha(z)u_{\mathrm{tr}}$ and $\alpha(z)u_{\mathrm{gr}}$. Then for the transverse microlocal portion,

$$E_\pm(t-s)\alpha(z)u_{\mathrm{tr}}(s,z) = \iint e^{i[(x-z)\cdot\xi \pm (t-s)|\xi|]}\alpha(z)u_{\mathrm{tr}}(s,z)\,d\xi\,dz. \tag{3.14}$$

From (3.6), (3.7), and (3.14), the action on the modified initial data $g = \Lambda^\sigma f$ is given by

$$E_{\mathrm{tr}}(t,s)g(x) = \int K_{\mathrm{tr}}(t,s;x,y)g(y)\,dy, \tag{3.15}$$

with

$$\begin{aligned} &K_{\mathrm{tr}}(t,s;x,y) \\ &= \int\cdots\int e^{i[(x-z)\cdot\xi \pm (t-s)|\xi| + \psi_{\mathrm{tr}}(z,\zeta) + (s-r)\zeta_n - w'\cdot\zeta' + ((w',\varphi(w'))-y)\cdot\eta \pm r|\eta|]} \\ &\qquad \times a_{\mathrm{tr}}(s,z,\zeta)\chi(\zeta)\langle\eta\rangle^{-\sigma}\alpha(z)\alpha_0(r)\alpha'(w')\gamma(y)\,d\xi\,d\zeta\,d\eta\,dz\,dr\,dw'. \end{aligned} \tag{3.16}$$

For the grazing microlocal piece,

$$E_\pm(t-s)\alpha(z)u_{\mathrm{gr}}(s,z) = \iint e^{i[(x-z)\cdot\xi \pm (t-s)|\xi|]}\alpha(z)u_{\mathrm{gr}}(s,z)\,d\xi\,dz. \tag{3.17}$$

The expression analogous to (3.15) then follows from (3.6), (3.12), and (3.14):

$$E_{\mathrm{gr}}(t,s)g(x) = \int K_{\mathrm{gr}}(t,s;x,y)g(y)\,dy, \tag{3.18}$$

with

$$\begin{aligned} &K_{\mathrm{gr}}(t,s;x,y) \\ &= \int\cdots\int \exp\big(i[(x-z)\cdot\xi \pm (t-s)|\xi| + \psi_{\mathrm{gr}}(z,\zeta) + (s-r)|\zeta'| - \psi_0(w',\zeta) \\ &+ \tfrac{2}{3}|\zeta'|(\varrho_0^{\frac{3}{2}}(z,\zeta) - \theta^{\frac{3}{2}}) + ((w',\varphi(w'))-y)\cdot\eta \pm r|\eta|]\big)\,\{\Phi(\varrho_0|\zeta|^{\frac{2}{3}})/\Phi(\theta|\zeta'|^{\frac{2}{3}})\} \\ &\qquad \times a_{\mathrm{gr}}(s,z,\zeta)\chi(\zeta)\langle\eta\rangle^{-\sigma}\alpha(z)\alpha_0(r)\alpha'(w')\gamma(y)\,d\xi\,d\zeta\,d\eta\,dz\,dr\,dw'. \end{aligned} \tag{3.19}$$

In Section 2 we have indicated that estimates of the following form are the key to proving our main theorem. The proofs will be sketched in Section 4.

**Theorem 3.1.** *Let $E_{\mathrm{tr}}(t,s)$ be the operator defined by (3.15), with kernel $K_{\mathrm{tr}}(t,s;x,y)$ given by (3.16). Let $T$ be determined by $\Omega$ and the compact set $F$ supporting the data as in Section 2. If $\sigma > (n+1)/2$, then*

$$|K_{\mathrm{tr}}(t,s;x,y)| \leq C\langle t-T\rangle^{-(n-1)/2} \quad \text{for } s \leq T,\ t \geq T.$$

*Thus $E_{\mathrm{tr}}(t,s): L^1(\mathbf{R}^n) \to L^\infty(\mathbf{R}^n)$ defines a bounded map for $0 \leq s \leq T$, of norm at most $C\langle t-T\rangle^{-(n-1)/2}$. If $\sigma = (n+1)/2$, then $E_{\mathrm{tr}}(t,s)$ : $\mathscr{H}^1(\mathbf{R}^n) \to L^\infty(\mathbf{R}^n)$ defines a bounded map, and the norm is at most $C[\langle t-T\rangle^{-(n-1)/2}$.*

**Theorem 3.2.** *Let $E_{\mathrm{gr}}(t,s)$ be the operator defined by (3.18), with kernel $K_{\mathrm{gr}}(t,s;x,y)$ given by (3.19). Let $T$ be determined by $\Omega$ and the compact set $F$ supporting the data as in Section 2. If $\sigma > (n+1)/2$, then*

$$|K_{\mathrm{gr}}(t,s;x,y)| \leq C\langle t-T\rangle^{-(n-1)/2} \quad \text{for } s \leq T,\ t \geq T.$$

*Thus $E_{\mathrm{gr}}(t,s): L^1(\mathbf{R}^n) \to L^\infty(\mathbf{R}^n)$ defines a bounded map for $0 \leq s \leq T$, of norm at most $C\langle t-T\rangle^{-(n-1)/2}$. If $\sigma = (n+1)/2$, then $E_{\mathrm{gr}}(t,s)$ : $\mathscr{H}^1(\mathbf{R}^n) \to L^\infty(\mathbf{R}^n)$ defines a bounded map, and the norm is at most $C\langle t-T\rangle^{-(n-1)/2}$.*

We can immediately deduce the desired estimate on the main portion of the integral (2.9).

**Corollary 3.3.** *Let $E(t,s)$ denote either of the operators $E_{\mathrm{tr}}(t,s)$ or $E_{\mathrm{gr}}(t,s)$ and let $T$ be as in Theorems 3.1 and 3.2. If $\sigma \geq (n+1)/2$, then for $t \geq T$,*

$$\left\| \int_0^T E(t,s)g\,ds \right\|_\infty \leq C\|g\|_1\langle t-T\rangle^{-(n-1)/2}.$$

*(If $\sigma = (n+1)/2$, then $\|\ \|_1$ denotes the norm in the local Hardy space $\mathscr{H}^1(\mathbf{R}^n)$.)*

The proofs of the main results now follow.

*Proof of Theorem 1.1.* Let $\beta \in C^\infty(\mathbf{R}^n)$ be identically zero on a neighborhood of the complement of $\Omega$ and identically one outside of a larger neighborhood, such that $\nabla\beta$ has support near enough to $\partial\Omega$ that the representations (3.7) and (3.9) are valid on $\mathrm{supp}(\nabla\beta)$. (It is assumed that the boundary values in (3.6) of the solution to (2.4) have been microlocalized via smooth partitions of unity on the compact sets $\partial\Omega$ and $S^{n-1}$ (the unit sphere in the dual variables), and by an affine change of coordinates the support is assumed to be near $0 \in \partial\Omega$.) From (2.2), (2.3), (2.6), and Lemma 2.1, it is enough to estimate the integral in (2.9). From Lemma 2.2, the solution to the initial-boundary problem $u$ in (2.9) can be replaced by $u_\pm$ as in (3.3). Moreover, let $\alpha(z)u_{\text{tr}}$ in (3.14) denote either $(\nabla\beta(z))\cdot\nabla_z\Lambda^{-1}u_{\text{tr}}$ or the smoother term $(\Box\beta(z))\Lambda^{-1}u_{\text{tr}}$, and let $\alpha(z)u_{\text{gr}}$ have the analogous connotation in (3.17). Then Lemma 2.2 allows the replacement of the integral (2.9) by

$$\int_0^T E_{\text{tr}}(t,s)g\,ds, \tag{3.20}$$

with $E_{\text{tr}}(t,s)g$ given by (3.15) and (3.16), and

$$\int_0^T E_{\text{gr}}(t,s)g\,ds, \tag{3.21}$$

with $E_{\text{gr}}(t,s)g$ given by (3.18) and (3.19). The estimate (1.8) now follows from (3.2) and Corollary 3.3.

*Proof of Theorem 1.2.* If $\beta \in C^\infty(\mathbf{R}^n)$ has compact support in a small neighborhood of $F$ and is identically one on $F$, it follows easily from conservation of energy for the solution to (1.1) that for $f \in \mathscr{S}(\mathbf{R}^n)$, $\|E_\Omega(t)\beta(-\Delta)^{\frac{1}{2}}f\|_2 \le \|f\|_2$. Since $(-\Delta)^{-\frac{1}{2}}\Lambda$ is bounded as a map from $L^2(F)$ to $L^2(\mathbf{R}^n)$ (at least for $n \ge 3$; if $n = 2$ a slightly different argument is necessary), it follows that for $f \in \mathscr{S}(\mathbf{R}^n)$,

$$\|E_\Omega(t)\beta\Lambda\beta f\|_2 \le \|f\|_2. \tag{3.22}$$

Moreover, if $\theta \in \mathbf{R}$, then $\|\Lambda^{i\theta}f\|_2 = \|f\|_2$ and $\|\Lambda^{i\theta}f\|_1 \le C_\theta\|f\|_1$ (where now the norms are in $\mathscr{H}^1(\mathbf{R}^n)$ and the bounds $C_\theta$ have admissible growth in $\theta$, as in Fefferman-Stein [6]). Thus from the estimates (1.8) and (3.22),

$$\|E_\Omega(t)\beta\Lambda^{\sigma_j}\beta f\|_{p_j'} \le C_\theta\langle t-T\rangle^{d_j}\|f\|_{p_j}$$

if $t \ge T$, for $\operatorname{Re}\sigma_0 = 1$, $p_0 = p_0' = 2$, $d_0 = 0$, and for $\operatorname{Re}\sigma_1 = -(n-1)/2$, $p_1 = 1$, $p_1' = \infty$, $d_1 = -(n-1)/2$. The result (1.11) now follows by Stein's interpolation theorem ([6]) applied to the family of operators $\{E_\Omega(t)\beta\Lambda^\sigma\beta\}$.

**4. Decay estimates on kernels.** It remains to analyze the kernels $K_{\text{tr}}$ and $K_{\text{gr}}$ given by (3.16) and (3.19). As usual, the terms without compact support (involving $\xi$, $\eta$, and $\zeta$) are treated as oscillatory integrals — implicitly it is assumed that smooth compactly supported cutoff functions have been inserted, and the estimates obtained (using integration by parts and the method of stationary phase) are independent of the supports. Although the analysis of $K_{\text{tr}}$ (a composition of three Fourier integral operators) essentially follows from that given in Beals [3], it is included here since part of the argument carries over in handling $K_{\rm gr}$. (Also, in [3] a much more technical analysis was necessary because the analogue of the integration in $s$ took place over an interval of length $t$, so that extra estimates were needed in order to recover the decay rate $t^{-(n-1)/2}$.)

*Proof of Theorem 3.1.* The phase function for $K_{\rm tr}$ is

(4.1)

$$(x-z)\cdot\xi \pm (t-s)|\xi| - \psi_{\rm tr}(z,\zeta) + (s-r)\zeta_n - w'\cdot\zeta' + ((w',\varphi(w')) - y)\cdot\eta \pm r|\eta|.$$

The (transverse) singularities correspond microlocally to a conic set where $\zeta_n \neq 0$, so by Lemma 2.2 we can assume that the amplitude contains a smooth homogeneous cutoff supported on such a region. From (3.8),

$$\psi_{\rm tr}(z,\zeta) = z'\cdot\zeta' + (z_n - \varphi(z'))\psi_1(z,\zeta),$$

and it is assumed that $a_{\rm tr}(s,z,\zeta)$ has support as close to $\partial\Omega = \{z_n = \varphi(z')\}$ as necessary. Moreover, the smooth partition of unity on the boundary allows the assumption that $z'$ is close to zero. Since $\varphi$ vanishes to second order at $z' = 0$ (see (3.5)), then $\nabla_z\psi_{\rm tr}(0,\zeta) = (\zeta', \psi_1(0,\zeta))$. Thus from (3.8), $\psi_1(0,\zeta) = \pm\sqrt{\zeta_n^2 - |\zeta'|^2}$. Hence

$$\psi_{\rm tr}(z,\zeta) = z'\cdot\zeta' \pm (z_n - \varphi(z'))\sqrt{\zeta_n^2 - |\zeta'|^2} + \psi_2(z,\zeta), \tag{4.2}$$

with

$$\psi_2(z,\zeta) = \mathcal{O}\{(z_n - \varphi(z'))(|z'| + |z_n - \varphi(z')|)|\zeta|\}. \tag{4.3}$$

Since singularities on the boundary propagate into $\Omega$ along null bicharacteristics where $\nabla_\zeta[\psi_{\rm tr}(z,\zeta) + (s-r)\zeta_n - w'\cdot\zeta'] = 0$, and since

$$\partial_{\zeta_n}[\psi_{\rm tr}(z,\zeta) + (s-r)\zeta_n - w'\cdot\zeta'] = \pm(z_n - \varphi(z'))\zeta_n/\sqrt{\zeta_n^2 - |\zeta'|^2} + \mathcal{O}\{|z|^2\} + s - r,$$

on the outgoing ray (where $s > r$, $z_n > \varphi(z')$), necessarily $\pm\zeta_n < 0$. Thus assuming without loss of generality that $\zeta_n > 0$, $|\zeta'| \ll \zeta_n$, and hence $|\zeta| \approx \zeta_n$ on the support of $a_{\rm tr}$, by Lemma 2.2 the expressions (4.2) and (4.3) can be replaced by

$$\psi_{\rm tr}(z,\zeta) = z'\cdot\zeta' - z_n\sqrt{\zeta_n^2 - |\zeta'|^2} + \mathcal{O}\{|z|^2|\zeta|\}. \tag{4.4}$$

Similarly, since singularities propagating from the initial data to the boundary travel along null bicharacteristics where $\nabla_\eta[((w',\varphi(w')) - y)\cdot\eta \pm r|\eta|] = 0$, the nongrazing hypothesis implies that by Lemma 2.2 it can be assumed that the amplitude contains a smooth cutoff on a conic neighborhood where $\eta_n \neq 0$. Since

$$\partial_{\eta_n}[((w',\varphi(w')) - y)\cdot\eta \pm r|\eta|] = \varphi(w') - y_n \pm r\eta_n/|\eta_n|$$

on the incoming ray (where $r > 0$ and $y_n > \varphi(w')$), necessarily $\pm\eta_n > 0$. Thus assuming without loss of generality that $\eta_n > 0$, and hence $|\eta| \approx \eta_n$ on the support of the amplitude, by Lemma 2.2 the expression (4.1) can be replaced by

$$\begin{aligned}(4.5)\quad \psi(x,t,z,s,w',r,y,\xi,\zeta,\eta) &= (x-z)\cdot\xi \pm (t-s)|\xi| + \psi_{\mathrm{tr}}(z,\zeta)\\ &\quad + (s-r)\zeta_n - w'\cdot\zeta' - ((w',\varphi(w')) - y)\cdot\eta + r|\eta|.\end{aligned}$$

From (4.4) and (4.5) it follows that

$$(4.6)\qquad \nabla_z\psi = (-\xi' + \zeta', -\xi_n - \sqrt{\zeta_n^2 - |\zeta'|^2}) + \mathcal{O}\{|z||\zeta|\},$$

and (with (3.5)),

$$(4.7)\qquad \nabla_{(w',r)}\psi = (-\zeta' + \eta', -\zeta_n + |\eta|) + \mathcal{O}\{|w'||\eta|\}.$$

Since the integrand in (3.16) has compact support in $(z,w',r)$, repeated integration by parts in those variables yields a sum of expressions of the following form

$$\begin{aligned}(4.8)\quad K_{\mathrm{tr}}(t,s;x,y) &= \int\cdots\int e^{i\psi}\tilde{a}(s,z,\zeta)\chi_1(\zeta)\chi_2(\eta)\langle\eta\rangle^{-\sigma}\mathcal{O}\{\langle\nabla_z\psi\rangle^{-M}\}\\ &\quad\times \mathcal{O}\{\langle\nabla_{(w',r)}\psi\rangle^{-M}\}\tilde{a}(z)\tilde{a}_0(r)\tilde{a}'(w')\gamma(y)\eta(y)\,d\xi\,d\zeta\,d\eta\,dz\,dr\,dw'.\end{aligned}$$

Here the zero order symbols $\chi_1$ and $\chi_2$ are conically supported where $|\zeta'| \ll \zeta_n$ and $|\eta| \approx \eta_n$.

Since it can be assumed that $t - s \geq T$, and since the singularities travel along the null bicharacteristics where $\nabla_\zeta\psi = x - z \pm (t-s)\xi/|\xi| = 0$, Lemma 2.2 allows the assumption that the $\xi$ integration occurs on a conic neighborhood of the stationary point $\omega^0 = \omega^0(t,x,s,z) = \pm(x-z)/(s-t)$. It follows from (4.6), (4.7), and the usual stationary phase argument in $\xi$ (see e.g. Beals [1], [3]) that (4.8) can be rewritten as

$$\begin{aligned}(4.9)\quad K_{\mathrm{tr}} &= \int\cdots\int\int_0^\infty e^{i[|x-z|\pm(t-s)|\xi|]}\mathcal{O}\{\langle t-s\rangle^{-(n-1)/2}\}\\ &\quad\times \mathcal{O}\{|\xi|^{-(n-1)/2}\}\mathcal{O}\{\langle\nabla_z\tilde{\psi}\rangle^{-M}\}|\xi|^{n-1}d|\xi|\tilde{a}(s,z,\zeta)\chi_1(\zeta)\chi_2(\eta)\langle\eta\rangle^{-\sigma}\\ &\quad\times \mathcal{O}\{\langle\nabla_{(w',r)}\psi\rangle^{-M}\}\tilde{a}(z)\tilde{a}_0(r)\tilde{a}'(w')\gamma(y)\,d\zeta\,d\eta\,dz\,dr\,dw'.\end{aligned}$$

(The asymptotic expansion in negative powers of $|\xi|$ obtained from the stationary phase method can be stopped after the first step to leave a remainder satisfying the same bound.) Here $\nabla_z\tilde{\psi} = \nabla_z\psi|_{\xi=|\xi|\omega^0}$.

Since $|z|$ and $|w'|$ can be taken as small as necessary, the integrand is rapidly decreasing in $(|\xi|,\zeta,\eta)$ away from any conic neighborhood of

$\{|\zeta| \approx \zeta_n \approx |\xi| \approx |\eta|\}$. Thus the term $\langle\eta\rangle^{-\sigma}$ can be replaced by $\mathcal{O}\{\langle\xi\rangle^{-\sigma}\}$. Moreover, on such a conic neighborhood the change of variables

$$\tilde{\zeta} = \nabla_z \tilde{\psi}, \quad \tilde{\eta} = \nabla_{(w',r)} \psi,$$

can be made. It is then easily shown that, for $|z|$ and $|w'|$ small enough,

$$\left|\det \frac{\partial(\tilde{\zeta}, \tilde{\eta})}{\partial(\zeta, \eta)}\right| \approx 1.$$

Thus from (4.9), for arbitrarily large $M$,

$$\begin{aligned} &K_{\mathrm{tr}}(t, s; x, y) \\ &= \mathcal{O}\{\langle t-s\rangle^{-(n-1)/2} \iint \int_0^\infty \langle|\xi|\rangle^{-\sigma+(n-1)/2}\, d|\xi| \langle\tilde{\zeta}\rangle^{-M} \langle\tilde{\eta}\rangle^{-M}\, d\tilde{\zeta}\, d\tilde{\eta}\}. \end{aligned}$$

If $\sigma > (n+1)/2$, the proof of Theorem 3.1 is then complete. If $\sigma = (n+1)/2$ a modification similar to that in [1] and [3] applies; after first acting on $f$ with a pseudodifferential operator of order zero (bounded on $\mathcal{H}^1(\mathbf{R}^n)$) the resulting composed kernel satisfies the same estimate.

Two parts of the grazing kernel $K_{\mathrm{gr}}$ given by (3.19) differ from the corresponding expressions in $K_{\mathrm{tr}}$. Instead of the smooth phase function $\psi_{\mathrm{tr}}(z, \zeta) + (s-r)\zeta_n$, the singular phase function

$$\psi_{\mathrm{gr}}(z, \zeta) + (s-r)|\zeta'| - \psi_0(w', \zeta) + \tfrac{2}{3}|\zeta'|(\varrho_0^{\frac{3}{2}}(z, \zeta) - \theta^{\frac{3}{2}})$$

occurs, with $\varrho_0$ and $\theta$ homogeneous of degree zero in $\zeta$ and given by (3.13). Also, the amplitude contains the term $\Phi(\varrho_0|\zeta'|^{\frac{2}{3}})/\Phi(\theta|\zeta'|^{\frac{2}{3}})$, which does not in general behave like a pseudodifferential symbol of type $S^0_{1,0}$. We are dealing with the primary term (see Taylor [24]), for which $|\zeta| \approx |\zeta'|$ and

$$c|\zeta'|^{\frac{1}{3}} \le -\zeta_n \ll |\zeta'|, \quad \text{i.e. } c|\zeta|^{-\frac{2}{3}} \le \theta \ll 1. \tag{4.10}$$

It can be assumed that smooth cutoffs to this set have been inserted into the amplitude in (3.19); these involve pseudodifferential symbols of type $S^0_{\frac{1}{3},0}$. The remaining terms are handled similarly, or else by Lemma 2.2.

*Proof of Theorem 3.2.* From (3.19), the phase function for $K_{\mathrm{gr}}$ is

$$\begin{aligned} (x-z)\cdot\xi \pm (t-s)|\xi| + \psi_{\mathrm{gr}}(z, \zeta) + (s-r)|\zeta'| - \psi_0(w', \zeta) \\ + \tfrac{2}{3}|\zeta'|(\varrho_0^{\frac{3}{2}}(z, \zeta) - \theta^{\frac{3}{2}}) + ((w', \varphi(w')) - y)\cdot\eta \pm r|\eta|. \end{aligned} \tag{4.11}$$

In addition to (4.10), it can be assumed without loss of generality (after a microlocalization and rotation of coordinates) that

$$\zeta'/|\zeta'| \equiv (\zeta_1, \zeta'')/|\zeta'| \approx (1,0), \quad \text{i.e. } \delta = |\zeta''|/|\zeta'| \ll 1. \tag{4.12}$$

An analysis similar to the derivation of (4.4), now using (3.10), (3.13), (4.10), (4.12), the assumption that $z$ is close to zero, and the properties of $\psi_{\text{gr}}(z,\zeta)$ and $\varrho_0(z,\zeta)$ as in Taylor [24], pp. 238–246, shows that there exists $\nu' = (\nu_1, \nu'') \in \mathbf{R}^{n-1}$, $\nu_1 > 0$, such that

$$\begin{aligned}\psi_{\text{gr}}(z,\zeta) = z' \cdot [\zeta' + \theta\nu'|\zeta'| + \mathcal{O}\{\theta(\delta+\theta)|\zeta'|\}] \\ + z_n \mathcal{O}\{\theta^2|\zeta'|\} + \mathcal{O}\{|z|^2|\zeta'|\},\end{aligned} \tag{4.13}$$

and

$$\begin{aligned}\varrho_0(z,\zeta) = \theta + (z_n - \varphi(z'))[\sqrt{2\nu_1} + \mathcal{O}\{(\delta+\theta)\}] \\ + \mathcal{O}\{(z_n - \varphi(z'))|z| + \theta^2\}.\end{aligned} \tag{4.14}$$

Then from (3.11) and (4.13), the expression (4.11) can be replaced by

$$\begin{aligned}\psi(x,t,z,s,w',r,y,\xi,\zeta,\eta) &= (x-z)\cdot\xi \pm (t-s)|\xi| \\ &+ (z'-w')\cdot(\zeta' + \theta\nu'|\zeta'|) + (z'-w')\cdot\mathcal{O}\{\theta(\delta+\theta)|\zeta'|\} + z_n\mathcal{O}\{\theta^2|\zeta'|\} \\ &+ \mathcal{O}\{(|z|^2 + |w'|^2 + \theta^\infty)|\zeta'|\} + \tfrac{2}{3}|\zeta'|(\varrho_0^{\frac{3}{2}}(z,\zeta) - \theta^{\frac{3}{2}}) \\ &+ (s-r)|\zeta'| + ((w',\varphi(w')) - y)\cdot\eta \pm r|\eta|.\end{aligned} \tag{4.15}$$

From (3.5), (4.14), and (4.15), it then follows that

$$\begin{aligned}\nabla_{z'}\psi &= -\xi' + \zeta' + \theta\nu'|\zeta'| + \mathcal{O}\{[|z| + \theta(\delta+\theta)]|\zeta'|\} + \varrho_0^{\frac{1}{2}}|\zeta'|\mathcal{O}\{|z| + \theta^2\}, \\ \partial_{z_n}\psi &= -\xi_n + \varrho_0^{\frac{1}{2}}|\zeta'|\sqrt{2\nu_1} + \mathcal{O}\{[|z| + \theta^2]|\zeta'|\} + \varrho_0^{\frac{1}{2}}|\zeta'|\mathcal{O}\{|z| + \delta + \theta\},\end{aligned} \tag{4.16}$$

and

$$\begin{aligned}\nabla_{w'}\psi &= -\zeta' - \theta\nu'|\zeta'| + \eta' + \eta_n\nabla\varphi(w') + \mathcal{O}\{[|w'| + \theta(\delta+\theta)]|\zeta'|\}, \\ \partial_r\psi &= -|\zeta'| \pm |\eta|.\end{aligned} \tag{4.17}$$

As in the earlier case, we can integrate by parts in $(z, w', r)$ in (3.19), and obtain a sum of terms of the following form

$$\begin{aligned}K_{\text{gr}}(t,s;x,y) = \int\cdots\int e^{i\psi}\tilde{a}(s,z,\zeta)\langle\eta\rangle^{-\sigma}\mathcal{O}\{\langle\nabla_z\psi\rangle^{-M}\} \\ \times\ \mathcal{O}\{\langle\nabla_{(w',r)}\psi\rangle^{-M}\}\tilde{a}(z)\tilde{a}_0(r)\tilde{a}'(w')\gamma(y)\,d\xi\,d\zeta\,d\eta\,dz\,dr\,dw'.\end{aligned} \tag{4.18}$$

Here the symbol $\tilde{a}(s,z,\zeta) = \nabla_z^M[\Phi(\varrho_0|\zeta'|^{\frac{2}{3}})/\Phi(\theta|\zeta'|^{\frac{2}{3}})a_{\text{gr}}(s,z,\zeta)]$ is supported where (4.10) and (4.12) hold. Moreover, since $z \notin \partial\Omega$ on its support, $z_n - \varphi(z') \geq c > 0$ so from (4.10) and (4.13), $\varrho_0 \geq \theta + c \geq c$. Then from (3.13) it follows that $|\tilde{a}(s,z,\zeta)|$ is uniformly bounded.

Exactly as in the derivation of (4.9), stationary phase arguments in $\xi$ yield

$$
\begin{aligned}
(4.19)\quad K_{\text{gr}} = \int \cdots \int \int_0^\infty & e^{i[|x-z|\pm(t-s)|\xi|]}\mathcal{O}\{\langle t-s\rangle^{-(n-1)/2}\} \\
& \times \mathcal{O}\{|\xi|^{(n-1)/2}\}\mathcal{O}\{\langle\nabla_z\tilde{\psi}\rangle^{-M}\}\, d|\xi|\tilde{a}(s,z,\zeta)\langle\eta\rangle^{-\sigma}\mathcal{O}\{\langle\nabla_{(w',r)}\psi\rangle^{-M}\} \\
& \times \tilde{\alpha}(z)\tilde{\alpha}_0(r)\tilde{\alpha}'(w')\gamma(y)\, d\zeta\, d\eta\, dz\, dr\, dw'.
\end{aligned}
$$

Since $|z|$ and $|w'|$ can be taken as small as necessary, the integrand in (4.19) is rapidly decreasing in $(|\xi|,\zeta,\eta)$ away from any conic neighborhood on which $|\zeta| \approx \zeta_1 \approx |\xi| \approx |\eta|$. Therefore we again have $\langle\eta\rangle^{-\sigma} = \mathcal{O}\{\langle\xi\rangle^{-\sigma}\}$. Moreover, on the support of $\tilde{\alpha}(z)$, $(z_n - \varphi(z')) \approx \varepsilon^2$ and $|z'| \approx \varepsilon > 0$, because the $z$-support of the integrand is near (but not on) $\partial\Omega$ and is microlocally near the grazing null bicharacteristics over $\{(w',\varphi(w')) = (0,0)\}$. For the change of variables

$$\tilde{\zeta} = \nabla_z\tilde{\psi},$$

it then follows from (4.10), (4.12), (4.14), and (4.16) that $\partial\tilde{\zeta}/\partial\zeta$ has the "columns"

$$
\begin{pmatrix}
I + \mathcal{O}\{\varepsilon+\theta\} + \varrho_0^{\frac{1}{2}}\mathcal{O}\{\varepsilon+\theta\} + \varrho_0^{-\frac{1}{2}}\mathcal{O}\{(\varepsilon+\theta)^2\} \\
\mathcal{O}\{\varepsilon+\theta^2\} + \varrho_0^{\frac{1}{2}}\mathcal{O}\{1\} + \varrho_0^{-\frac{1}{2}}\mathcal{O}\{\varepsilon+\theta\}
\end{pmatrix}
\text{ and }
\begin{pmatrix}
\nu' + \mathcal{O}\{\varepsilon+\delta+\theta\} + \varrho_0^{\frac{1}{2}}(1+\varrho_0^{-1})\mathcal{O}\{\varepsilon+\theta\} \\
\mathcal{O}\{\varepsilon+\theta\} + \varrho_0^{\frac{1}{2}}\mathcal{O}\{1\} + \varrho_0^{-\frac{1}{2}}[-\sqrt{\nu_1/2} + \mathcal{O}\{\varepsilon+\theta\}]
\end{pmatrix}.
$$

From (4.14), we can write $\varrho_0 \approx \theta + \varepsilon^2$ and $\varrho_0^{-\frac{1}{2}} \approx (\varepsilon + \theta^{\frac{1}{2}})^{-1}$. It follows that, for $\theta$, $\delta$, and $\varepsilon$ sufficiently small

$$
\frac{\partial\tilde{\zeta}}{\partial\zeta} = \begin{pmatrix} I + \mathcal{O}\{\varepsilon+\delta\} & \nu' + \mathcal{O}\{1\} \\ \mathcal{O}\{\varepsilon+\delta^{\frac{1}{2}}\} & -\varrho_0^{-\frac{1}{2}}\sqrt{\nu_1/2} + \mathcal{O}\{1\} \end{pmatrix};
$$

$$
\left|\det\frac{\partial\tilde{\zeta}}{\partial\zeta}\right| \approx (\varepsilon+\theta^{\frac{1}{2}})^{-1} + \mathcal{O}\{1\} \geq C > 0.
$$

Next, for the change of variables $\tilde{\eta} = \nabla_{(w',r)}\psi$, it follows from (4.17) that as long as $|\eta_n| \geq \kappa|\eta'|$ and $w'$ is sufficiently small,

$$
\frac{\partial\tilde{\eta}}{\partial\eta} = \begin{pmatrix} I & \nabla\varphi(w') \\ \pm\eta'/|\eta| & \pm\eta_n/|\eta| \end{pmatrix};
$$

$$
\left|\det\frac{\partial\tilde{\eta}}{\partial\eta}\right| = |\eta_n - \nabla\varphi(w')\cdot\eta'|/|\eta| \approx \kappa + \mathcal{O}\{|w'|\} \geq C > 0.
$$

In that case, it follows from (4.19) that

$$K_{\mathrm{gr}}(t,s;x,y) = \mathcal{O}\{\langle t-s\rangle^{-(n-1)/2}\iint\int_0^\infty \langle|\xi|\rangle^{-\sigma+(n-1)/2}\,d|\xi|\,\langle\tilde\zeta\rangle^{-M}\langle\tilde\eta\rangle^{-M}\,d\tilde\zeta\,d\tilde\eta\},$$

and the proof is completed as in the case of $K_{\mathrm{tr}}$.

On the other hand, since the analysis is taking place near a grazing null bicharacteristic, the microlocal support is near $\eta_n = 0$; the preceding paragraph merely allows the assumption that $|\eta_n| \le \kappa|\eta'|$ for $\kappa$ as small as necessary. Then after a microlocal partition of unity, without loss of generality, $\eta' = (\eta_1, \eta'')$ satisfies $|\eta''| \ll \eta_1$. Hence from (4.15),

$$\text{(4.20)}\quad \nabla_\eta\psi = \big(w_1 - y_1 \pm r\eta_1/|\eta|,\ w'' - y'' \pm r\eta''/|\eta|,\ \varphi(w') - y_n \pm r\eta_n/|\eta|\big).$$

Microlocally near the grazing null bicharacteristic, on which $\nabla_\eta\psi = 0$, it therefore follows that $r \approx \pm(y_1 - w_1)$.

For $\big|r \pm (w_1 - y_1)|\eta|/\eta_1\big| \le |\eta_n - \nabla\varphi(w')\cdot\eta'|/|\eta|$, the $r$-integration in the expression (4.19) yields a factor which is $\mathcal{O}\{\det(\partial\tilde\eta/\partial\eta)|\}$. This term is then of size

$$\mathcal{O}\{\langle t-s\rangle^{-(n-1)/2}\iint\int_0^\infty \langle|\xi|\rangle^{-\sigma+(n-1)/2}\,d|\xi|\,\langle\tilde\zeta\rangle^{-M}\langle\tilde\eta\rangle^{-M}\,d\tilde\zeta\,d\tilde\eta\},$$

and the proof is completed as before.

For $\big|r \pm (w_1 - y_1)|\eta|/\eta_1\big| \ge |\eta_n - \nabla\varphi(w')\cdot\eta'|/|\eta|$, it follows from (4.20) that after repeated integrations by parts with respect to $\eta_1$, the integral in (4.19) can be assumed to have an additional factor which is estimated by $\mathcal{O}\{\big((\eta_n - \nabla\varphi(w')\cdot\eta')/|\eta|\big)^{-M}\}$. Thus if the change of variables is now given by

$$\bar\eta' = \nabla_{w'}\psi, \quad \bar\eta_n = \eta_n - \nabla\varphi(w'),$$

it follows that the remaining term is of size

$$\mathcal{O}\{\langle t-s\rangle^{-(n-1)/2}\iint\int_0^\infty \langle|\xi|\rangle^{-\sigma+(n-1)/2}\,d|\xi|\,\langle\tilde\zeta\rangle^{-M}\langle\bar\eta'\rangle^{-M}\langle\bar\eta_n\rangle^{-M}\,d\tilde\zeta\,d\bar\eta\},$$

which suffices to finish the proof.

## References


1. M. Beals, *$L^p$ boundedness of Fourier integral operators*, Mem. Amer. Math. Soc. **38** (No 264) (1982).
2. ———, *Local $L^p$ estimates for Fourier-Airy integral operators*, Comm. Pure Appl. Math. **35** (1982), 751–769.

3. ———, *Optimal $L^\infty$ decay for solutions to the wave equation with a potential*, Comm. in PDE **19** (1994), 1319–1369.
4. M. Beals and W. Strauss, *Time decay estimates for a perturbed wave equation*, Journées Équations aux dérivées partielles, St Jean de Monts (1992), Exp. no XIII.
5. ———, *$L^p$ estimates for the wave equation with a potential*, Comm. in PDE **18** (1993), 1365–1397.
6. C. Fefferman and E. M. Stein, *$H^p$ spaces of several variables*, Acta Math. **129** (1972), 137–193.
7. J. Ginibre and G. Velo, *The global Cauchy problem for the nonlinear Klein-Gordon equation*, Math. Z. **189** (1985), 487–505.
8. D. Goldberg, *A local version of real Hardy spaces*, Duke Math. J. **46** (1979), 27–42.
9. L. Hörmander, *Fourier integral operators I*, Acta Math. **127** (1971), 79–183.
10. P. D. Lax and R. S. Phillips, *Scattering Theory*, Academic Press, New York, London, 1967.
11. H. Lindblad and C. D. Sogge, *On existence and scattering with minimal regularity for semilinear wave equations*, J. Funct. Anal. (to appear).
12. B. Marshall, W. Strauss and S. Wainger, *$L^p$-$L^q$ estimates for the Klein-Gordon equations*, J. Math. Pures Appl. (A) **59** (1980), 417–440.
13. R. B. Melrose, *Singularities and energy decay in acoustical scattering*, Duke Math. J. **46** (1979), 43–59.
14. R. B. Melrose and J. Sjöstrand, *Singularities of boundary value problems I*, Comm. Pure Appl. Math. **31** (1978), 593–617.
15. C. S. Morawetz, J. V. Ralston and W. A. Strauss, *Decay of solutions of the wave equation outside nontrapping obstacles*, Comm. Pure Appl. Math. **30** (1977), 447–508.
16. J. Peral, *$L^p$ estimates for the wave equation*, J. Funct. Anal. **36** (1980), 114–145.
17. J. Ralston, *Note on the decay of acoustic waves*, Duke Math. J. **46** (1979), 799–804.
18. J. Shatah and M. Struwe, *Regularity results for nonlinear wave equations*, Ann. of Math. **138** (1993), 503–518.
19. H. F. Smith and C. D. Sogge, *$L^p$ regularity for the wave equation with strictly convex obstacles*, Duke Math. J **73** (1994), 97–155.
20. ———On the critical semilinear wave equation outside convex obstacles, Preprint.
21. W. A. Strauss, *Nonlinear scattering at low energy*, J. Funct. Anal. **41** (1981), 110–133.
22. ———, *Nonlinear Wave Equations*, CBMS Reg. Conf. Series in Math. no 73, Amer. Math. Soc., Providence (1989).
23. R. S. Strichartz, *A priori estimates for the wave equation and some applications*, J. Funct. Anal. **5** (1970), 218–235.
24. M. Taylor, *Pseudodifferential Operators*, Princeton University Press, Princeton, 1981.
25. B. R. Vainberg, *On the short wave asymptotic behavior of solutions of stationary problems.*, Russ. Math. Surveys **30:2** (1975), 1–58.
26. K. Yajima, *The $L^p$ continuity of wave operators for Schrödinger operators*, Preprint.
27. Jiaping Zhong, *The $L^p$-$L^q$ estimates for the wave equation with a nonnegative potential*, Comm. in PDE **20** (1995), 315–334.



DEPARTMENT OF MATHEMATICS, RUTGERS UNIVERSITY, NEW BRUNSWICK, NJ 08903, USA

*E-mail address*: beals@math.rutgers.edu

# WEYL QUANTIZATION AND FOURIER INTEGRAL OPERATORS

JEAN-MICHEL BONY

**1. Introduction.** The Weyl-Hörmander calculus [Hö1], [Hö2] provides a framework for the theory of pseudo-differential operators which is very general and, in spirit, quite simple. The data is just a riemannian metric on the phase space satisfying some conditions, and to this metric is associated a graded algebra of pseudo-differential operators having essentially the properties one can expect.

On the other hand, the theory of Fourier integral operators is far from being developed in such a generality. The classical theory, with homogenous phase functions, is linked to the pseudo-differential calculus associated to the standard metric $dx^2 + d\xi^2/\langle\xi\rangle^2$. A theory of Fourier integral operators linked to the euclidean metric $dx^2 + d\xi^2$ is also well known (see [A&F], [D&G]). Except for the case of metrics not too far from the standard metric, which is studied in [Hö1], very little is known in the general case.

We shall consider the subject in a rather abstract way, described by our Problem 1 in the next section. We are essentially looking for operators $P$ such that conjugates $P^*AP$ of pseudo-differential operators $A$ associated to a metric $g_1$ are pseudo-differential operators associated to another metric $g_2$.

The theory is particularly simple in the case of symplectic metrics. In Section 4, using the concept of confinement introduced in [B&L], one is naturally led to a definition where metaplectic operators play a central role: Fourier integral operators are just superpositions of operators $U \circ A$, where $A$ is confined in a ball of the metric and where $U$ is a metaplectic operator.

We just give some indications on the case of general metrics in Section 5. It relies heavily on the second microlocalization as developed in [B&L]. The main step is to develop the analogue of Section 4 when pseudo-differential operators are replaced by 2-microlocal operators associated to a couple of metrics $g \leq \gamma$ with $\gamma$ symplectic.

In Section 6, we reap the benefits of our abstract approach in the study of evolution operators $P_t = e^{itA}$, where $A$ is given by its (real) symbol. Trying to write $P_t$ as a Fourier integral operator in our sense can be done in an essentially unique way, and we get sharp conditions allowing to settle this point.

Parts of the results of this paper appear in [Bo], where the reader will find more details on proofs, as well as complements in the symplectic case.

**2. Weyl quantization and Hörmander classes.** We shall consider the phase space $\mathbf{R}^{2n}_X = \mathbf{R}^n_x \times \mathbf{R}^n_\xi$, equipped with its canonical symplectic form $[X, Y] = y{\cdot}\xi - x{\cdot}\eta$. To any temperate distribution $a$ on $\mathbf{R}^{2n}$, the Weyl quantization associates the operator $a^w$ defined by

$$a^w u(x) = \iint e^{i(x-y)\cdot\xi} a\big(\tfrac{x+y}{2}, \xi\big) u(y)\, dy\, d\xi / (2\pi)^n.$$

The map $a \mapsto a^w$ is an isomorphism of $\mathcal{S}'(\mathbf{R}^{2n})$ onto $\mathcal{L}(\mathcal{S}(\mathbf{R}^n), \mathcal{S}'(\mathbf{R}^n))$ and an isomorphism of $\mathcal{S}(\mathbf{R}^{2n})$ onto $\mathcal{L}(\mathcal{S}'(\mathbf{R}^n), \mathcal{S}(\mathbf{R}^n))$.

When the composition $a^w \circ b^w$ is defined as an operator from $\mathcal{S}$ into $\mathcal{S}'$, one can define the symbol $a\#b$ by $(a\#b)^w = a^w \circ b^w$. An explicit formula is given by

$$a\#b(X) = \pi^{-2n} \iint e^{-2i[X-Y_1, X-Y_2]} a(Y_1) b(Y_2)\, dY_1\, dY_2.$$

A *Hörmander metric* $g$ is a riemannian metric on $\mathbf{R}^{2n}$ satisfying the three conditions below. For each $Y \in \mathbf{R}^{2n}$, one denotes by $g^\sigma_Y$ the dual of the quadratic form $g_Y$, when the dual of $\mathbf{R}^{2n}$ is identified to $\mathbf{R}^{2n}$ itself via the symplectic form:

$$g^\sigma_Y(T) = \sup_{W \neq 0} \frac{[T, W]^2}{g_Y(W)}.$$

**Uncertainty Principle:** $g_X(\cdot) \le g^\sigma_X(\cdot)$.
**Slowness:** There exists $C > 0$ such that

$$g_X(X-Y) \le C^{-1} \implies \big(g_Y(\cdot)/g_X(\cdot)\big)^{\pm 1} \le C.$$

**Temperance:** There exist $C > 0$ and $N \in \mathbf{N}$ such that

$$\big(g_Y(\cdot)/g_X(\cdot)\big)^{\pm 1} \le C\big(1 + g^\sigma_Y(Y-X)\big)^N.$$

We shall say that $g$ is a *symplectic metric* if $g = g^\sigma$. The classical metrics of type $(\rho, \delta)$ defined by

$$g_Y(dx, d\xi) = \langle\eta\rangle^{2\delta} dx^2 + \langle\eta\rangle^{-2\rho} d\xi^2, \tag{1}$$

are Hörmander metrics when $0 \le \delta \le \rho$ and $\delta < 1$; they are symplectic if $\rho = \delta$.

A *weight* is a positive function on $\mathbf{R}^{2n}$ such that there exist $C$ and $N$ with

$$\big(M(Y)/M(X)\big)^{\pm 1} \leq C \quad \text{for} \quad g_X(X-Y) \leq 1/C,$$
$$\big(M(Y)/M(X)\big)^{\pm 1} \leq C\big(1 + g_Y^\sigma(Y-X)\big)^N.$$

The class of symbols $S(M,g)$ is the space of functions $a \in C^\infty(\mathbf{R}^{2n})$ such that the following semi-norms

$$\|a\|_{k;S(M,g)} = \sup_{\substack{l \leq k;\ X \in \mathbf{R}^{2n} \\ g_X(T_j) \leq 1}} |\partial_{T_1} \ldots \partial_{T_l} a(X)|/M(X),$$

are finite, where $\partial_T a$ means the directional derivative $\langle T, da\rangle$. We shall denote by $\Psi(M,g)$ the space of pseudo-differential operators of weight $M$, i.e. the set of operators $a^w$ with $a \in S(M,g)$.

If $A$ belongs to $\Psi(M,g)$, it maps $\mathcal{S}(\mathbf{R}^n)$ into itself, $S'(\mathbf{R}^n)$ into itself and, if $M = 1$, it maps $L^2(\mathbf{R}^n)$ into itself. The set of weight functions is a multiplicative group, and $\Psi(\cdot, g)$ is an algebra graded by this group, which means that

$$S(M_1,g) \# S(M_2,g) \subset S(M_1 M_2, g); \quad \Psi(M_1 . g) \circ \Psi(M_2, g) \subset \Psi(M_1 M_2, g).$$

Summing up, to a "simple" structure (the riemannian metric $g$), the Weyl-Hörmander calculus associates a "big" structure: an algebra $S(\cdot, g)$ of operators graded by the group of weight functions. We can now state the problem which we want to solve: given a transformation of the "simple" structure (i.e. a diffeomorphism $F$ of $\mathbf{R}^{2n}$), define isomorphisms of the "big" structure which are "over" $F$. More precisely:

**Problem 1.** Given two Hörmander metrics $g_1$ and $g_2$ on $\mathbf{R}^{2n}$, and a (conveniently related) diffeomorphism $F$ from $\mathbf{R}^{2n}$ onto itself, define a class $\mathrm{FIO}(F, g_1, g_2)$ of operators such that
(i) These operators are bounded on $L^2(\mathbf{R}^n)$.
(ii) $\mathrm{FIO}(\mathrm{Id}, g, g) = \Psi(1, g)$.
(iii) If $P \in \mathrm{FIO}(F, g_1, g_2)$, then $P^* \in \mathrm{FIO}(F^{-1}, g_2, g_1)$.
(iv) If $A \in \Psi(M, g_2)$ and $P \in \mathrm{FIO}(F, g_1, g_2)$ then $P^* A P \in \Psi(M \circ F, g_1)$.

This is a minimal set of requirements, and we shall say more on the composition of Fourier integral operators and their boundedness in "Sobolev spaces".

Our second problem is quite natural: up to what point are the evolution operators associated to a hamiltonian Fourier integral operators in our sense. A positive answer will give one parameter groups of such operators and in particular will prove the existence of invertible Fourier integral operators.

**Problem 2.** Given a real valued $C^\infty$ function $a$ on the phase space and a Hörmander metric $g_0$, under which conditions do the operators $e^{ita^w}$ belong to $\mathrm{FIO}(F, g_0, g_t)$ for convenient metrics $g_t$?

We shall give an essentially complete answer to these two problems when the metrics are symplectic. It is a particular case which is interesting by itself, and which is the first step for the study of general metrics. We shall give just a few indications on this latter case.

**3. Confinement.** Let $g$ be a Hörmander metric. We denote by $B_{Y,r}$ the $g_Y$-ball of radius $r$ centered at $Y$. If $A$ and $B$ are two subsets of $\mathbf{R}^{2n}$ then $g_Y(X-A)$ (resp. $g_Y(A-B)$) is the infimum of $g_Y(X-X')$ (resp. $g_Y(X'-X'')$) for $X' \in A$ (and $X'' \in B$).

**Definition 3.1.** For $r > 0$, the space $\mathrm{Conf}(g, Y, r)$ is the Schwartz space $\mathcal{S}(\mathbf{R}^{2n})$ equipped with the following sequence of semi-norms

$$\|a\|_{k;\mathrm{Conf}(g,Y,r)} = \sup_{\substack{l\le k;\ X\in\mathbf{R}^{2n}\\ g_Y(T_j)\le 1}} |\partial_{T_1}\dots\partial_{T_l} a(X)|\,(1+g_Y^\sigma(X-B_{Y,r}))^{k/2}\,.$$

A family $(a_Y)_{Y\in\mathbf{R}^{2n}}$ is *uniformly confined in* $B_{Y,r}$ if $\|a_Y\|_{k;\mathrm{Conf}(g,Y,r)}$ is bounded by a constant depending on $k$ but not on $Y$.

A (uniformly) equivalent sequence of semi-norms is given by

$$\begin{aligned}&\|a\|'_{k;\mathrm{Conf}(g,Y,r)}\\ &\quad= \sup_X |a(X)|\,(1+g_Y^\sigma(X-B_{Y,r}))^{k/2} + \sup_P |\widehat{a}(P)|\,(1+g_Y^\sigma(P))^{k/2}\,|g_Y|^{1/2},\end{aligned}$$

where the Fourier transform is defined by $\widehat{a}(P) = \int e^{-2i[X,P]}\,a(X)\,dX$, and $|g_Y|$ is the determinant of $g_Y$.

When the metric $g$ is symplectic, the semi-norms corresponding to a positive value of $r$ are uniformly equivalent to those corresponding to $r = 0$. We shall make this choice and just write $\mathrm{Conf}(g, Y)$.

The most important result about confinement is the following theorem [B&L] which uses the function

$$\Delta_r(Y,Z) = 1 + g_Y^\sigma(B_{Y,r}-B_{Z,r}) + g_Z^\sigma(B_{Y,r}-B_{Z,r}), \tag{2}$$

measuring the "distance" between $Y$ and $Z$. When the metric is symplectic, it suffices to consider the case $r = 0$, and the corresponding function will be denoted by $\Delta(Y, Z)$.

**Theoreme 3.2 (biconfinement).** *If* $(a_Y)$ *and* $(b_Y)$ *are uniformly confined in* $B_{Y,r}$ *then, for any* $N$*, the family* $\Delta_r(Y,Z)^N(a_Y\# b_Z)$ *is uniformly confined in* $B_{Y,r}$ *and in* $B_{Z,r}$.

We shall now recall the important relation between the Weyl quantization and the symplectic structure. Given an affine symplectic (or canonical) transformation $\chi$ of the phase space, there exists a unitary operator $U$, uniquely determined up to a multiplicative factor of modulus 1, called a *metaplectic operator* such that for any symbol $a$, one has $U^* a^w U = (a\circ\chi)^w$ (see, for instance, [Hö2], Theorem 18.5.9).

When the metric $g$ is symplectic, given any two points $Y$ and $Z$, there exists a (not unique) symplectic transformation $\chi_{YZ}$ mapping $B_{Y,r}$ onto $B_{Z,r}$. We shall denote by $U_{YZ}$ an associated metaplectic operator. It is clear from the definition that

$$\|a\circ\chi_{YZ}\|_{k;\mathrm{Conf}(g,Y)} = \|a\|_{k;\mathrm{Conf}(g,Z)}$$

for any $a\in\mathcal{S}(\mathbf{R}^{2n})$.

**Theorem 3.3.** *Let* $g$ *be a symplectic Hörmander metric,* $(a_Y)$ *and* $(b_Y)$ *be uniformly confined, and let* $(T_Y)$ *be a bounded family of operators in* $L^2$. *Define* $c_Y$ *by* $c_Y^w = a_Y^w T_Y b_Y^w$. *Then the family* $(c_Y)$ *is uniformly confined.*

For a fixed $Y$, it is clear that $c_Y^w$ maps $\mathcal{S}'(\mathbf{R}^n)$ into $\mathcal{S}(\mathbf{R}^n)$ and thus $c_Y\in\mathcal{S}(\mathbf{R}^{2n})$. So, for any $k$, there exist $C$ and $l$ such that, for $c^w = a^w T b^w$, one has

$$\|c\|_{k;\mathrm{Conf}(Y)} \le C\,\|a\|_{l;\mathrm{Conf}(Y)}\,\|T\|_{\mathcal{L}(L^2)}\,\|b\|_{l;\mathrm{Conf}(Y)}\,, \tag{3}$$

and what remains to prove is uniformity. But conjugation by $U_{ZY}$ preserves the norm in $\mathcal{L}(L^2)$ and exchanges the semi-norms of confinement at $Y$ and $Z$. The inequality (3) is thus valid at any point $Z$ with the same constants, which proves the theorem.

For a nonsymplectic metric, the result is true if $(T_Y)$ is a bounded family of pseudo-differential operators of weight 1, but not if it is just a bounded family in $\mathcal{L}(L^2)$. In some sense, the $g$-balls of a symplectic metric are so small that, after microlocalization in such a ball, one cannot see the difference between a pseudo-differential operator and an operator which is just bounded on $L^2$.

**4. Fourier integral operators associated to symplectic metrics.** We first give a characterization of pseudo-differential operators showing the role played by the diagonal of the phase space.

**Theorem 4.1.** *Let $g$ be a symplectic Hörmander metric. Let $a \in \mathcal{S}'(\mathbf{R}^{2n})$ and put $A = a^w$. The following four properties are equivalent.*
(a) $a \in S(1,g)$ *i.e.* $A \in \Psi(1,g)$.
(b) *There exists a uniformly confined family $(a_Y)$ such that*

$$a = \int a_Y \, dY. \tag{4}$$

(c) *There exist uniformly confined families $(a_Y)$ and $(b_Y)$, and operators $T_{YZ}$, such that*

$$A = \iint a_Z^w \, T_{YZ} \, b_Y^w \, dY \, dZ,$$

*with*

$$\forall N, \quad \|T_{YZ}\|_{\mathcal{L}(L^2)} \le C_N \Delta(Y,Z)^{-N}.$$

(d) *For a convenient choice of uniformly confined families $(\psi_Y)$ and $(\theta_Y)$, one has*

$$\forall N, \quad \left\| \theta_Z^w \, A \, \psi_Y^w \right\|_{\mathcal{L}(L^2)} \le C_N \Delta(Y,Z)^{-N}.$$

The implications (a) $\iff$ (b) $\Longrightarrow$ (c) $\iff$ (d) (which are still true for a general Hörmander metric once modified by introducing a small positive radius of confinement and replacing $dY$ by $|g_Y|^{1/2}\, dY$) are consequences of the existence of partitions of unity of the following type

$$1 = \int \varphi_Y \, dY = \int \psi_Y \# \theta_Y \, dY, \tag{5}$$

where $(\varphi_Y)$, $(\psi_Y)$ and $(\theta_Y)$ are uniformly confined.

This defines the "convenient choice" of $(\psi_Y)$ and $(\theta_Y)$. For any operator $A$, one has

$$A = \iint \psi_Z^w \left\{ \theta_Z^w A \psi_Y^w \right\} \theta_Y^w \, dY \, dZ,$$

and this gives the equivalence of (c) and (d), just playing with the theorem of biconfinement and the fact that $\sup_Y \int \Delta(Y,Z)^{-N} \, dZ$ is finite for $N$ large enough.

The implication (c) $\Longrightarrow$ (b) is specific to symplectic metrics. It comes from the fact that, putting $c_{YZ}^w = a_Z^w \, T_{YZ} \, b_Y^w$ and using Theorem 3.3, one has

$$\|c_{YZ}\|_{k;\mathrm{Conf}(g,Y)} \le C \, \|a_Z\|_{l;\mathrm{Conf}(g,Y)} \, \|T_{YZ}\|_{\mathcal{L}(L^2)} \, \|b_Y\|_{l;\mathrm{Conf}(g,Y)},$$

and from the estimate

$$\forall k, \; \exists C, \; \exists N, \qquad \|a\|_{k;\mathrm{Conf}(g,Y)} \le C \Delta(Y,Z)^N \, \|a\|_{k;\mathrm{Conf}(g,Z)}.$$

Integrating with respect to $Z$ gives a decomposition (4) for $A$.

Our definition of Fourier integral operators will be quite analogous, replacing the diagonal of the phase space by the graph of some function $F$. We shall need the following condition.

**Definition 4.2.** Given two symplectic Hörmander metrics $g_1$ and $g_2$, a diffeomorphism $F$ from $\mathbf{R}^{2n}$ onto itself is $(g_1, g_2)$-tempered if there exist $C$ and $N$ such that

$$\forall Y,\ \forall Z, \quad C^{-1}\Delta_1(Y,Z)^{1/N} \leq \Delta_2(F(Y), F(Z)) \leq C\Delta_1(Y,Z)^N,$$

where $\Delta_j$ is the function defined by (2) for $g_j$.

This condition implies that the image of the $g_1$-unit ball centered at $Y$ contains (resp. is contained in) the $g_2$-ball of fixed radius $C'$ (resp. $C'^{-1}$) centered at $F(Y)$. The volume is thus approximately preserved. However, the condition does not imply that the image of any symplectic ball centered at $Y$ is approximately symplectic, and $F$ can be very far from being a canonical transformation.

**Definition 4.3.** Let $g_1$ and $g_2$ be symplectic Hörmander metrics and $F$ be a $(g_1, g_2)$-tempered diffeomorphism. We shall say that an operator $P$ belongs to $\mathrm{FIO}(F, g_1, g_2)$ if the three equivalent conditions below are satisfied:

(b′) Let $U_{Y,F(Y)}$ be metaplectic operators associated to an affine symplectic transformation mapping the $g_1$-unit ball centered at $Y$ onto the $g_2$-unit ball centered at $F(Y)$. Then one has

$$P = \int U_{Y,F(Y)}\, a_Y^w\, dY,$$

where $(a_Y)$ is $g_1$-uniformly confined.

(c′) There exist $(a_Y)$ and $(b_Y)$ uniformly confined for $g_2$ and $g_1$ respectively, and operators $T_{YZ}$ such that

$$A = \iint a_Z^w\, T_{YZ}\, b_Y^w\, dY\, dZ,$$

with

$$\forall N, \quad \|T_{YZ}\|_{\mathcal{L}(L^2)} \leq C_N \Delta_2(F(Y), Z)^{-N}.$$

(d′) Introducing partitions of unity (5) for $g_1$ and $g_2$, one has

$$\forall N, \quad \|\theta_{2Z}^w\, P\, \psi_{1Y}^w\|_{\mathcal{L}(L^2)} \leq C_N \Delta_2(F(Y), Z)^{-N}.$$

The proof of (b′) $\Longrightarrow$ (c′) $\Longleftrightarrow$ (d′) is quite analogous to the corresponding proof for pseudo-differential operators. Assuming now (c′), we first remark that the semi-norms of $a_Z$ in $\mathrm{Conf}(g_2, F(Y))$ are controlled by powers of $\Delta_2(Z, F(Y))$. We can write

$$\begin{aligned} a_Z^w\, T_{YZ}\, b_Y^w &= U_{Y,F(Y)} U_{Y,F(Y)}^* a_Z^w U_{Y,F(Y)} U_{Y,F(Y)}^* T_{YZ} b_Y^w \\ &= U_{Y,F(Y)} \{c_{YZ}^w S_{YZ} b_Y^w\} \end{aligned}$$

where $c^w_{YZ} = U^*_{Y,F(Y)} a^w_Z U_{Y,F(Y)}$, which implies that the semi-norms of $c_{YZ}$ in $\mathrm{Conf}(g_1, Y)$ are controlled by powers of $\Delta_2(Z, F(Y))$, and where the norm of $S_{YZ} = U^*_{Y,F(Y)} T_{YZ}$ in $\mathcal{L}(L^2)$ is controlled by any negative power of $\Delta_2(Z, F(Y))$. Integrating with respect to $Z$, we obtain (b′).

Our Problem 1 is solved in the case of symplectic metrics. It is clear from Theorem 4.1 that $\mathrm{FIO}(\mathrm{Id}, g, g)$ is precisely $\Psi(1, g)$. It is also easy to see that if $F$ is $(g_1, g_2)$-tempered, then $F^{-1}$ is $(g_2, g_1)$-tempered. Moreover, for $P \in \mathrm{FIO}(F, g_1, g_2)$ the condition (c′) and the conservation of the volume implies immediately that $P^*$ satisfies the same condition for $\mathrm{FIO}(F^{-1}, g_2, g_1)$.

Now, if $G$ is $(g_2, g_3)$-tempered and $Q \in \mathrm{FIO}(G, g_2, g_3)$, then $G \circ F$ is $(g_1, g_3)$-tempered and, writing $P$ and $Q$ as in (b′) one gets easily $Q \circ P$ as in (c′) which shows that $Q \circ P \in \mathrm{FIO}(G \circ F, g_1, g_3)$. Using this property of composition for $P^*AP$, with $A \in \Psi(1, g_2)$, one gets $P^*AP \in \mathrm{FIO}(\mathrm{Id}, g_1, g_1) = \Psi(1, g_1)$.

More generally, if $M$ is a $g_2$-weight and $A \in \Psi(M, g_2)$, then $M \circ F$ is a $g_1$-weight and $P^*AP \in \Psi(M \circ F, g_1)$. A related result is that $P$ maps continuously the "Sobolev space" $H(M \circ F, g_1)$ defined in [B&C] into $H(M, g_2)$.

*Remarks 4.4.* Metaplectic operators are typical examples of Fourier integral operators, where $F$ is the underlying affine symplectic transformation and $g_2$ is the direct image $F_* g_1$ of any metric.

The fact that the definition is modelled on the properties of pseudo-differential operators (and coincide for $F = \mathrm{Id}$) is a good guarantee that the class is neither too large nor too small. Actually definition (d′) seems to be the weakest one can think of, while property (b′) looks quite strong.

The direct analogue of these definitions cannot work for a Hörmander metric which is not symplectic. A definition using (d′) would be too broad, even for $F = \mathrm{Id}$. A definition using (b′) is inadequate too: the operators $P^*P$ would not be pseudo-differential operators in general.

**5. Fourier integral operators for general metrics.** We shall just give a few indications on the general construction. In this section, $g$ or $g_1$ will be a Hörmander metric satisfying the additional condition

$$g^\sigma_Y \leq C \lambda(Y)^N g_Y \tag{6}$$

for convenient $C$ and $N$, where the function $\lambda$ is defined by

$$\lambda^2(Y) = \inf_{T \neq 0} \frac{g^\sigma_Y(T)}{g_Y(T)}.$$

This condition is satisfied by all the metrics $g_{\rho,\delta}$ defined by (1).

An important observation is that, if one considers all the symplectic metrics $\gamma$ with $g \leq \gamma$ which are *relatively slow:*

$$g_Y(Y - Z) \leq C^{-1} \implies (\gamma_Y(\cdot)/\gamma_Z(\cdot))^{\pm 1} \leq C,$$

then

$$S(1,g) = \bigcap_\gamma S(1,\gamma).$$

Actually it is sufficient to take a finite number of well chosen metrics $\gamma$. For instance, it is easy to see that

$$S(1,g_{1,0}) = S(1,g_{0,0}) \bigcap S(1,g_{1,1}).$$

The metric $g_{1,1}$ is a very interesting metric, but it is not tempered.

If it were true in general that it is possible to choose the metrics $\gamma$ tempered, defining Fourier integral operators would be easy. Taking for simplicity a symplectomorphism $F$ of $\mathbf{R}^{2n}$ onto itself, and metrics $g_1$ and $g_2$ such that $g_{2,F(Y)}$ is the image of $g_{1,Y}$ by the differential $F'(Y)$, one would consider the relatively slow metrics $\gamma_1$ which are larger than $g_1$ and their direct images $\gamma_2$. The condition on $F$ would be that it is $(\gamma_1,\gamma_2)$-tempered for any $\gamma_1$ in the sense of Definition 4.2. Then $\mathrm{FIO}(F,g_1,g_2)$ would be defined as the intersection of the classes $\mathrm{FIO}(F,\gamma_1,\gamma_2)$. The stability by composition, by adjoint, and the fact that $\mathrm{FIO}(\mathrm{Id},g,g) = \Psi(1,g)$ would be evident.

Actually, we shall almost exactly fulfil this program, but the lack of temperance forces us to use second microlocalizations, as introduced in [B&L].

Let us consider the situation where $g$ satisfies (6) and where $\gamma \geq g$ is symplectic and relatively slow. For defining the second microlocalization, the main point is to introduce new spaces $\mathrm{Conf}(g\gamma, Y, r)$, which are equal to $\mathcal{S}(\mathbf{R}^{2n})$ equipped with the semi-norms

$$\begin{aligned}\|a\|_{k;\mathrm{Conf}(g\gamma,Y,r)} = \sup_X |a(X)| \left(1+g_Y^\sigma(X-B_{Y,r})\right)^{k/2} \left(1+\gamma_Y(X-Y)\right)^{k/2} \\ + \sup_P |\widehat{a}(P)| \left(1+g_Y^\sigma(P-B_{0,Y,r})\right)^{k/2} \left(1+\gamma_Y(P)\right)^{k/2},\end{aligned}$$

where $B_{Y,r}$ and $B_{0,Y,r}$ are the $g_Y$-balls of radius $r$, centered at $Y$ and at 0 respectively.

A family $(a_Y)$ which is uniformly $(g\gamma,r)$-confined is uniformly $\gamma$-confined. On the other hand, if $(a_Y)$ and $(c_Y)$ are uniformly $(g,r)$-confined and if $(b_Y)$ is uniformly $\gamma$-confined, then $(a_Y \# b_Y \# c_Y)$ is uniformly $(g\gamma,r)$-confined.

The biconfinement theorem is valid with some loss on the radius of confinement (assumed to be small enough): there exists $C$ such that, if $(a_Y)$ and $(b_Y)$ are uniformly $(g\gamma,r)$-confined, then, for any $N$, the family $\Delta_{Cr}(Y,Z)^N a_Y \# b_Z$ is uniformly $(g\gamma, Cr)$-confined at $Y$ and at $Z$. Here, the function $\Delta_r(Y,Z)$ is the product $\Delta_{r,g}(Y,Z)\Delta_\gamma(Y,Z)$ of the functions $\Delta$ corresponding to $g$ and $\gamma$, which means that one cumulates the almost orthogonality due to these two metrics.

The analogue of Theorem 3.3 is still valid for the $(g\gamma, r)$-confinement.

We define $\Psi_r(M, g\gamma)$ as the class of operators

$$A = \int M(Y) a_Y^w \, dY$$

where $(a_Y)$ is uniformly $(g\gamma, r)$-confined. For $r$ small enough, these operators map $\mathcal{S}$ [resp. $\mathcal{S}'$] into itself and $L^2$ into itself for $M = 1$. The property of algebra becomes

$$\Psi_r(M_1, g\gamma) \circ \Psi_r(M_2, g\gamma) \subset \Psi_{Cr}(M_1 M_2, g\gamma).$$

The analogue of Theorem 4.1 characterizing pseudo-differential operators remains valid for $\Psi_r(M, g\gamma)$ with the following changes: $\gamma$-confinement should be replaced by $(g\gamma, r)$-confinement; one of the statements for a radius of confinement $r$ implies the other ones with radius $Cr$. This gives a characterization of $\Psi(M, g)$ thanks to the relation

$$\Psi(M, g) = \bigcap_{r,\gamma} \Psi_r(M, g\gamma),$$

the intersection being taken for all $\gamma$ symplectic, larger than $g$ and relatively slow (a finite number of them is actually sufficient), and all $r > 0$ small enough.

It is not difficult now to define first the spaces $\mathrm{FIO}(F, g_1\gamma_1, g_2\gamma_2)$ by the analogue of Definition 4.3, proving that any property (b′), (c′) or (d′) for a radius $r$ implies the other ones for $Cr$. The intersection, for all $(\gamma_1, r)$, of these classes will be the definition of $\mathrm{FIO}(F, g_1, g_2)$.

With this definition, one gets easily the stability by composition, and the fact that conjugates of pseudo-differential operators are pseudo-differential operators. What is less evident is the fact that $\mathrm{FIO}(F, g_1, g_2)$ contains nontrivial elements. This can be proved by two methods. The first one is to solve evolution equations $dP_t/dt = ia^w P_t$ and to prove, as we shall do in the next section for the first microlocalization, that $P_t \in \mathrm{FIO}(F_t, g_0\gamma_0, g_t\gamma_t)$. This works only for canonical transformations coming from a hamiltonian field $H_a$ but it allows one to prove the existence of invertible Fourier integral operators.

One can also consider operators of the following type

$$P = \int a_{F(Y)} T_Y b_Y |g_{1Y}|^{1/2} \, dY,$$

where $(a_Y)$ and $(b_Y)$ are uniformly $g_2$- and $g_1$-confined respectively, and where $T_Y$ is defined with a phase (corresponding to $F$ in the $g_{1Y}$-ball) and an amplitude. One has to prove that for any $\gamma_1$, and for $\gamma_2 = F_*\gamma_1$, the operator $P$ belongs to $\mathrm{FIO}(F, g_1\gamma_1, g_2\gamma_2)$. This leads to an asymptotic symbolic calculus.

**6. Evolution operators.** We shall just treat the case of symplectic Hörmander metrics. Let $g_0$ be such a metric and let $a \in C^\infty(\mathbf{R}^{2n})$ be a real valued function. For $t \in [0,T]$, we are looking for metrics $g_t$ and diffeomorphisms $F_t$ such that $P_t \in \mathrm{FIO}(F_t, g_0, g_t)$, where $P_t = e^{ita^w}$ is the solution of

$$\partial_t P_t = ia^w P_t, \qquad P_0 = \mathrm{Id}.$$

Using a partition of unity $1 = \int \varphi_Y \, dY$, we write $P_{tY} = P_t \varphi_Y^w$ and we should solve

$$\partial_t P_{tY} = ia^w P_{tY}, \qquad P_{0Y} = \varphi_Y^w.$$

According to Definition 4.3 (b′), we want to write

$$P_{tY} = U_{tY} \circ \alpha_{tY}^w$$

with $U_{tY}$ metaplectic, and $\alpha_{tY}$ uniformly confined for $g_0$.

The group of metaplectic operators is a Lie group, whose Lie algebra is identified with the space $\mathcal{P}_2$ of real polynomials of degree $\leq 2$, equipped with the Poisson bracket. This means that the family $U_{tY}$ should satisfy an equation of the following type

$$\partial_t U_{tY} = ib_{tY}^w U_{tY}, \qquad U_{0Y} = \mathrm{Id} \tag{7}$$

with $b_{tY} \in \mathcal{P}_2$ and that conversely any solution of such an equation is a family of metaplectic operators. Then the equation for $\alpha_{tY}$ is

$$\partial_t \alpha_{tY}^w = i\, U_{tY}^* (a - b_{tY})^w\, U_{tY} \alpha_{tY}^w, \qquad \alpha_{0Y} = \varphi_Y. \tag{8}$$

To have the best chances of getting uniform confinement for $(\alpha_{tY})$, we should certainly make $a - b_{tY}$ as small as possible near the point which is the image of $Y$ by the affine symplectic transformation associated to $U_{tY}$ and so choose $b_{tY}$ equal to the Taylor polynomial of $a$ at that point. Actually, it can be proved that if our problem admits a positive answer, then it is given by this choice. Now, everything is determined:

- The map $(t, Y) \mapsto F_t(Y)$ is the flow of the hamiltonian field $H_a$, i.e. the points $Y_t = F_t(Y)$ satisfy

$$dY_t/dt = H_a(Y_t), \qquad Y_0 = Y.$$

- The affine symplectic map $F_{tY}$ associated to $U_{tY}$ is tangent to $F_t$ at $Y$, i.e.

$$F_{tY}(X) = F_t(Y) + F_t'(Y) \cdot (X - Y).$$

- The metric $g_t$ is the direct image of $g_0$ by $F_t$: its unit ball at $F_t(Y)$ is the image by $F_{tY}$ of the $g_0$-unit ball at $Y$.
- The polynomial $b_{tY}$ is the second order Taylor polynomial of $a$ at $F_t(Y)$; $U_{tY}$ and $\alpha_{tY}$ are given by (7) and (8).

**Geometric Conditions 6.1.**
— The flow should be globally defined up to time $T$.
— The metrics $g_t$ (which are symplectic) should be tempered.
— The maps $F_t$ should be $(g_0, g_t)$-tempered (Definition 4.2).

If these conditions are not satisfied, then $P_t$ is not a Fourier integral operator (at least when $g_0$ is the initial metric). If they are satisfied, it remains to prove the uniform $g_0$-confinement of $\alpha_{tY}$ at $Y$. Proving the confinement at one point $Y$ is just solving (8) in the Schwartz space $\mathcal{S}(\mathbf{R}^{2n})$ and is essentially a matter of energy estimates (a controlled loss of regularity/decay is admissible, which explains the logarithm below). Any two balls of $g_0$ being isomorphic via a symplectic transformation, uniformity follows immediately.

**Theorem 6.2.** *Assume that the geometric conditions 6.1 are satisfied for $t \in [0,T]$ and that there exist constants $C_k$ such that*

$$|\partial_{T_1} \dots \partial_{T_k}(a - b_{tY})(X)| \le C_k(1 + g_{tY_t}(X - Y_t))^{1/2} \log(2 + g_{tY_t}(X - Y_t))$$

*for $k \ge 1$ and $g_{tY_t}(T_j) \le 1$. Then one has*

$$P_t \in \mathrm{FIO}(F_t, g_0, g_t)$$

*for $t \in [0,T]$.*

**Examples 6.3.** The conditions are obviously satisfied if $a$ is a polynomial of degree 2, for any choice of $g_0$, in which case $(P_t)$ is a one parameter group of metaplectic operators.

If $a$ is a $C^\infty$ function whose derivatives of order $\ge 2$ are bounded, and if $g_0$ is the euclidean metric (which corresponds to $\rho = \delta = 0$ in (1)), then the conditions are satisfied for any $t$ and one has $g_t = g_0$. For $|t|$ small enough, it is known [D&G] that one has $P_t u = \int e^{iS_t(x,\xi)} a_t(x,\xi)\widehat{u}(\xi)\,d\xi$ with conditions of type $(0,0)$ on $S_t$ and $a_t$. Our definition does not see the caustics and works for any $t$.

The following example shows that one can have distorted metrics $g_t$, even if $g_0$ is the euclidean metric. With $n = 1$ we take $a(x,\xi) = x\xi \log(2 + |\xi|^2)$. The conditions above are not satisfied uniformly in the whole of $\mathbf{R}^{2n}$, but they are satisfied once restricted to $K \times \mathbf{R}^n$ for any compact $K$ of $\mathbf{R}^n$, and one gets the following. For any $Q \in \Psi(1, g_0)$ which is compactly supported in $x$, one has $P_t \circ Q \in \mathrm{FIO}(F_t, g_0, g_t)$, where $g_t$ is the metric of type $(\rho, \delta)$ with $\rho = \delta = 1 - e^{-t}$.

## References

[A&F] K. Asada and D. Fujiwara, *On some oscillatory integral transformations in $L^2(\mathbf{R}^n)$*, Japan. J. Math. **4** (1978), 299–361.

[B&C] J.-M. Bony and J.-Y. Chemin, *Espaces fonctionnels associés au calcul de Weyl-Hörmander*, Bull. Soc. Math. France **122** (1994), 77–118.

[B&L] J.-M. Bony and N. Lerner, *Quantification asymptotique et microlocalisations d'ordre supérieur I*, Ann. Sci. École Norm. Sup. 4e série **22** (1989), 377–433.

[Bo] J.-M. Bony, *Opérateurs intégraux de Fourier et calcul de Weyl-Hörmander (cas d'une métrique symplectique)*, Actes Journées E.D.P. St. Jean de Monts (1994), exposé n° 9.

[D&G] J. Derezinski and C. Gérard, *Asymptotic completeness of N-particle systems*, Preprint, École Polytechnique (1995).

[Hö1] L. Hörmander, *The Weyl calculus of pseudo-differential operators*, Comm. Pure Appl. Math. **32** (1979), 359–443.

[Hö2] ———, *The analysis of linear partial differential operators*, Springer-Verlag, 1985.

CENTRE DE MATHÉMATIQUES, ÉCOLE POLYTECHNIQUE, F-91128 PALAISEAU CEDEX, FRANCE
*E-mail address*: bony@math.polytechnique.fr

# THE LOCAL INDEX THEOREM WITHOUT SMOOTHNESS

JOCHEN BRÜNING

**1. Introduction.** The local index theorem has attracted much interest of analysts, geometers, and physicists over the last two decades, rendering its proof more and more perspicuous. All this work pertains to the smooth case whereas it is known that in the special case of the signature operator much less regularity is required, at least for an "almost local" signature theorem cf. [MoWu,CSuT]. Motivated by this, we study here the differentiability requirements necessary to formulate and prove the local index theorem for general Dirac operators. We will present a version for $C^{1,1}$ structures which seems close to the minimum requirement. Our proof relies on abstract heat kernel estimates and is perhaps general enough to extend to other singular situations.

To formulate the results, we begin by recalling the setting of the local version of the Atiyah-Singer index theorem for twisted Dirac operators. Let $M$ be a smooth compact spin manifold of dimension $m = 2k$. We fix a spin structure, $P_{\mathrm{spin}}M$, and a smooth Riemannian metric, $g$, on $M$. Let $S$ be a spin bundle over $M$, equipped with the natural hermitian structure and unitary connection, $\nabla^S$, defined by $g$. The bundle $S$ is a left module over the bundle of complex Clifford algebras $\mathbb{C}\ell M$; the action of $\sigma \in \mathbb{C}\ell M$ will be denoted by $c(\sigma)$. Choose a hermitian bundle of coefficients, $E$, with unitary connection, $\nabla^E$, and form the tensor product $S \otimes E$ which inherits naturally a hermitian structure and a unitary connection, $\nabla^{S\otimes E}$. The Dirac operator with coefficients in $E, D^E$, is then a first order elliptic differential operator on $C^\infty(S \otimes E)$, defined by

$$D^E f = \sum_{i=1}^{m} c(e_i)\, \nabla^{S\otimes E}_{e_i} f, \qquad f \in C^\infty(S \otimes E), \tag{1.1}$$

where $(e_i)_{i=1}^m$ is any (oriented) local orthonormal frame for $TM$. $D^E$ is symmetric and essentially self-adjoint in $L^2(S \otimes E)$; the unique self-adjoint extension will also be denoted by $D^E$. To obtain a nontrivial index we bring in the involution on $S$ defined by

$$c(\omega_{\mathbb{C}}) := \sqrt{-1}^{\,k} c(e_1) \circ \ldots \circ c(e_m) \tag{1.2}$$

which splits $S = S_+ \otimes S_-$. Then $\varrho := c(\omega_{\mathbb{C}}) \otimes I$ splits $S \otimes E = S_+ \otimes E \oplus S_- \otimes E$, and we obtain a decomposition

$$D^E = \begin{pmatrix} 0 & D^E_- \\ D^E_+ & 0 \end{pmatrix}. \tag{1.3}$$

$D^E_+$ is a closed Fredholm operator with adjoint $D^E_-$, and its index is given by the celebrated formula of Atiyah and Singer [AS]:

**Theorem 1.1.**

$$\operatorname{ind} D^E_+ = \int_M \hat{A}(M) \wedge \operatorname{ch} E. \tag{1.4}$$

Here $\hat{A}(M)$ and $\operatorname{ch} E$ are certain characteristic differential forms which can be computed locally from the curvature of $\nabla^{TM}$, the Levi-Civita connection defined by $g$, and $\nabla^E$. This "locality" of the index can be made more precise if one brings in the heat kernel. As observed by McKean and Singer [McKS],

$$\operatorname{ind} D^E_+ = \operatorname{tr}_{L^2(S\otimes E)} \left[\varrho\, e^{-t(D^E)^2}\right], \tag{1.5}$$

for all $t > 0$. Since $(D^E)^2$ has discrete spectrum, $e^{-t(D^E)^2}$ is an operator with smooth kernel,

$$e^{-t(D^E)^2} f(p) = \int_M K_t(p,q)\, f(q)\, \operatorname{vol}_M(q), \tag{1.6}$$

where $\operatorname{vol}_M(q) = e_1^\flat \wedge \ldots \wedge e_m^\flat(q)$ (with $\flat : TM \to T^*M$ the 'musical' isomorphism defined by $g$) is the volume form and

$$K_t(p,q) \in \operatorname{End}((S \otimes E)_q, (S \otimes E)_p).$$

Now it follows from a fundamental idea of Hadamard [H], further developed by Minakshisundaram and Pleijel [MiPl], Seeley [Se], and Greiner [Gr] that we have an asymptotic expansion

$$K_t(p,p) \sim_{t\to 0+} \sum_{j\geq 0} t^{j-m/2} U_j(p), \quad p \in M. \tag{1.7}$$

In local frames and coordinates, the endomorphisms $U_j$ are recursively defined as polynomials in the derivatives of $g$ and the data on $E$ such that $U_j$ contains $2j$ derivatives of the metric. Using this in (1.5) one finds

$$\int_M \operatorname{tr}_{S\otimes E} [\varrho U_j(p)] \operatorname{vol}_M(p) = \begin{cases} 0, & j < m/2, \\ \int_M \hat{A}(M) \wedge \operatorname{ch} E, & j = m/2. \end{cases} \tag{1.8}$$

The local index theorem now asserts that the identities (1.8) hold even pointwise.

**Theorem 1.2.** *For all $p \in M$,*

$$\operatorname{tr}_{S\otimes E}[\varrho\, U_j(p)] = \begin{cases} 0, & j < m/2, \\ \hat{A}(M) \wedge \operatorname{ch}E(e_1, \dots, e_m)(p), & j = m/2. \end{cases} \tag{1.9}$$

Thus, 'massive cancellations' occur upon taking pointwise supertraces to the effect that the final answer contains only two derivatives of all coefficients involved. Theorem 1.2 has been proved by Kotake [Ko] for the case of Riemann surfaces and later by Patodi for the Gauß-Bonnet operator [P1] and the Hirzebruch signature operator [P2]. Building on methods of invariant theory introduced by Gilkey [Gi], Atiyah, Bott, and Patodi [ABP] proved Theorem 1.2 for twisted Dirac operators, thus yielding also another proof of the full Atiyah-Singer index theorem. Their method did not allow a direct identification of $\operatorname{tr}_{S\otimes E}[\varrho\, U_{m/2}(p)]$ but only up to certain universal constants which had to be computed from examples. Following ideas of the physicists Alvarez-Gaumé [Al] and Friedan and Windy [FW], Getzler [Ge, BeGeVe] designed a direct proof based on a scaling argument and Mehler's formula. His proof has been modified by many authors in order to clarify the subtleties of the analysis involved; we mention only the approach of B. Simon [Si] which is somewhat close to what we are going to present.

Our starting point in this paper is an apparent asymmetry in the statement of the local index theorem: we require $m$ derivatives to formulate it (i.e. to construct $U_{m/2}$) but the answer will contain only two. Thus it is natural to look for a (possibly weaker) statement which requires less regularity. Clearly, if we allow $C^1$-structures everywhere then $D^E$ is still perfectly well defined, with the same properties as listed before and the same index. But we also can present the index as an integral, starting from (1.5), as follows. Choose a family $\{\chi_{t,p} \mid t \in (0,1], p \in M\} \subset C(M)$ with the following properties:

$$\chi_{t,p} \geq 0 \text{ and } \operatorname{supp}\chi_{t,p} \subset B_{\sqrt{t}/3}(p), \tag{1.10a}$$

$$\text{the map } p \mapsto \chi_{t,p} \text{ is continuous,} \tag{1.10b}$$

$$\int_M \chi_{t,p}(q)\operatorname{vol}_M(p) = 1 \quad \text{for all } q \in M,\ t \in (0,1], \tag{1.10c}$$

$$\lim_{t\to 0+} \int_M \chi_{t,p}(q)\operatorname{vol}_M(q) = 1 \text{ for all } p \in M. \tag{1.10d}$$

Such a family is easy to construct, cf. (2.6) below. Now the map

$$M \ni p \mapsto \varrho\chi_{t,p}e^{-t(D^E)^2} \in C_1(L^2(S\otimes E)),$$

(where $C_q$ denotes the von Neumann-Schatten class of order $q > 0$), is continuous, hence from (1.5)

$$
\begin{aligned}
\operatorname{ind} D_+^E &= \int_M \operatorname{tr}_{L^2(S\otimes E)} \left[\varrho\chi_{t,p}\, e^{-t(D^E)^2}\right] \operatorname{vol}_M(p) \\
&=: \int_M F_t^E(p)\operatorname{vol}_M(p).
\end{aligned} \tag{1.11}
$$

In the smooth case, we have from Theorem 1.2 and the conditions (1.10), uniformly in $p \in M$,

$$\lim_{t\to 0+} F_t^E(p) = \hat{A}(M) \wedge \operatorname{ch} E(e_1, \dots, e_m)(p).$$

This leads us to the following definition: We say that the *weak local index theorem holds for* $D_+^E$ if there is $F^E \in L^1(M)$, independent of the choice of the family $(\chi_{t,p})$, such that for any such choice we have

$$\lim_{t\to 0+} F_t^E =: F^E \quad \text{in } L^1(M), \tag{1.12a}$$

and

$$\operatorname{ind} D_+^E = \int_M F^E(p)\operatorname{vol}_M(p). \tag{1.12b}$$

It is fairly obvious that no local index theorem can hold if we require just $C^1$-structures on the (compact smooth) manifold $M$. In particular, to identify $F^E$ with the Atiyah-Singer integrand we have to make sure that the latter exists and is integrable. Our main result thus reads as follows.

**Theorem 1.3.** *Let $M$ be a compact smooth spin manifold and $E$ a smooth complex vector bundle over $M$. If we equip $M$ with a $C^{1,1}$ metric, and $E$ with a $C^{1,1}$ hermitian structure and a unitary Lipschitz connection then the weak local index theorem holds for $D_+^E$. Moreover, for almost all $p$ in $M$,*

$$F^E(p) = \hat{A}(M) \wedge \operatorname{ch} E(e_1, \dots, e_m)(p). \tag{1.13}$$

The original motivation to investigate this problem was to design a proof of the local index theorem à la Getzler which is general enough to carry over to stratified situations like wedges. We will return to this question in a future publication.

The assumptions of Theorem 1.3 are perhaps not optimal. It would be interesting to find the precise minimal regularity condition under which the theorem remains true.

The paper is organized as follows. In Section 2 we give the outline of the proof. In Section 3 we compute the transformation of $(D^E)^2$ under the scaling map. Section 4 contains the main analytic facts needed in the proof of Theorem 1.3 which we present in a more general version than actually needed here since we could not find easy references in the literature. All assertions used in Sec. 2 are proved in Sec. 5 which deals with $C_q$-estimates for heat kernels of certain elliptic systems of second order. We avoid higher regularity essentially by employing throughout the von Neumann-Schatten scale of operator ideals, the crucial norm estimate being given in Theorem 5.2.

During the preparation of this work I have profited from conversations with Jean-Michel Bismut and Henri Moscovici, and I am grateful for the hospitality of the Ohio State University, the Université Paris-Sud, the Institut des Hautes Etudes Scientifiques, and the Université Paris-Nord where this paper was eventually completed. This work has also been supported by the EU under the GADGET project.

**2. Outline of the proof.** From now on, we are dealing with a compact smooth spin manifold, $M$, equipped with a $C^{1,1}$ metric, $g$. That is, the metric is a $C^1$ section of the bundle of symmetric two-tensors and we can find a finite $C^{2,1}$ atlas, $(U_i, \varphi_i)_{i=1}^N$, for $M$ with $D(\varphi_i \circ \varphi_j^{-1}) \in C^{1,1}$ for all $i, j$ and $(\varphi_i^{-1})^* g \in C^1$, with all partial derivatives of first order Lipschitz. Similarly, we consider a smooth complex coefficient bundle, $E$, over $M$ equipped with a $C^{1,1}$ hermitian structure and a Lipschitz unitary connection, $\nabla^E$. Under these conditions it is well known that the exponential map $\exp_p : \{s \in T_pM \mid |s| < i(M)\} \to B_{i(M)}(p)$ is a Lipschitz homeomorphism. We may assume that $M$ has injectivity radius larger than one, $i(M) > 1$.

Our convergence proof for $F_t$ in (1.11) will make use of the "Getzler scaling", defined in (2.10) below. This requires two technical adjustments: first, we replace $t$ by $\tilde{t} := \varepsilon^2 t$ with $\varepsilon, t \in (0,1]$, and second, we replace the coefficient bundle $E$ by

$$E(p) := (M \times S_p^*) \otimes E. \tag{2.1}$$

The product bundle $M \times S_p^*$ is equipped with the obvious hermitian structure and the flat connection, $E(p)$ carries the tensor product structures. Then, clearly,

$$\begin{aligned} F_t^E(p) &= 2^{-k} \mathrm{tr}_{L^2(S \otimes E(p))} \left[ \varrho \chi_{t,p} e^{-t(D^{E(p)})^2} \right] \\ &=: 2^{-k} \tilde{F}_t^E(p). \end{aligned} \tag{2.2}$$

Next we give an explicit description of a family $(\chi_{t,p})$ with the properties (1.10). Denote by $d_M$ the Riemannian distance and define

$$\exp_p^*(\mathrm{vol}_M)(s) =: \Theta_p(s)ds \qquad |s| < 1. \tag{2.3}$$

Then choose an even function $\tilde{\eta} \in C_0^\infty(-1/3, 1/3)$ with $\tilde{\eta} \geq 0$ and such that for $\eta(s) := \tilde{\eta}(|s|)$ we have

$$\int_{\mathbb{R}^m} \eta(s)\, ds = 1. \tag{2.4}$$

Then, with

$$\eta_t(s) := t^{-m/2}\ \eta(s/\sqrt{t}) = t^{-m/2}\tilde{\eta}(|s|/\sqrt{t}) =: \tilde{\eta}_t(|s|), \tag{2.5}$$

we define

$$\chi_{t,p}(q) := \tilde{\eta}_t(d_M(p,q))(\Theta_q(\exp_q^{-1} p))^{-1}. \tag{2.6}$$

The first step in the proof of Theorem 1.3 consists in a suitable transformation of $(D^{E(p)})^2$, following Getzler. Fix $p \in M$ and $\varepsilon \in (0,1]$, then we obtain the diagram:

(2.7)

$$\begin{array}{ccc} L^2(B_1(0), \Lambda_p^* \otimes E_p) & \xrightarrow{\Psi^1_{\varepsilon,p}} & L^2(B_\varepsilon(0), \mathbb{C}\ell_p \otimes E_p) \\ & & \big\downarrow \Psi^2_{\varepsilon,p} \\ L^2(B_\varepsilon(0), S_p \otimes E(p)_p) & \xrightarrow{\Psi^3_{\varepsilon,p}} & L^2(S \otimes E(p) | B_\varepsilon(p)). \end{array}$$

To describe the various maps in (2.7) we fix a local orthonormal frame, $(e_i)^m$, for $TM$ in $B_1(p)$ and write for any strictly ordered multiindex, $I = \{i_1, \ldots, i_l\} \subset \{1, \ldots, m\}$,

$$\mu(I) := l, \tag{2.8a}$$

$$e_I(q) := e_{i_1} \cdot \ldots \cdot e_{i_l}(q) \in \mathbb{C}\ell_q M, \tag{2.8b}$$

$$e_I^\flat(q) := e_{i_1}^\flat \wedge \ldots \wedge e_{i_l}^\flat(q) \in \Lambda_q^* M, \qquad q \in B_1(p). \tag{2.8c}$$

Here, again, $\flat : TM \to T^*M$ is the 'musical' isomorphism defined by the metric. Then

$$\Psi^1_{\varepsilon,p}(e_I^\flat(p) \otimes \omega)(s) := \varepsilon^{\mu(I)} e_I(p) \otimes \omega(s/\varepsilon), \tag{2.9a}$$

$$\omega \in L^2(B_1(0), E_p),\ |s| < \varepsilon.$$

For the definition of $\Psi^2_{\varepsilon,p}$ we recall the existence of a $\mathbb{C}\ell_p$-equivariant isomorphism

$$\alpha_p : \mathbb{C}\ell_p M \to S_p \otimes S_p^*.$$

Then

$$\text{(2.9b)} \quad \Psi^2_{\varepsilon,p} g(s) := (\alpha_p \otimes I_{E_p})(g(s)), \quad g \in L^2(B_\varepsilon(0), \mathbb{C}\ell_p \otimes E_p),\ |s| < \varepsilon.$$

Finally, we need the parallel transport along radial geodesics from $p$: if $F$ is any complex vector bundle over $M$ with $C^{1,1}$ hermitian structure and Lipschitz unitary connection, $\nabla^F$, then we get parallel transport

$$\mathrm{P}^F_p(s) : F_p \to F_{\exp_p(s)}, \quad |s| < 1,$$

with Lipschitz dependence on $s$. We define

$$\text{(2.9c)} \quad \Psi^3_{\varepsilon,p}\, g(\exp_p s) := \mathrm{P}^{S\otimes E(p)}_p(s)(g(s)),$$
$$g \in L^2(B_\varepsilon(0), S_p \otimes E(p)_p),\ |s| < \varepsilon.$$

Finally, let

$$\text{(2.10a)} \qquad \Phi_{\varepsilon,p} := \Psi^3_{\varepsilon,p}\Psi^2_{\varepsilon,p}\Psi^1_{\varepsilon,p}.$$

We note the explicit formula

$$\text{(2.10b)} \quad \Phi_{\varepsilon,p}(e^\flat_I(p) \otimes \omega)(\exp_p s)$$
$$= \varepsilon^{\mu(I)} \mathrm{P}^{S\otimes S^*_p}_p(s)(\alpha_p(e_I(p))) \otimes \mathrm{P}^E_p(s)(\omega(s/\varepsilon)).$$

Now, with (2.2) we arrive at

$$\text{(2.11)} \quad \tilde{F}^E_{\varepsilon^2 t}(p)$$
$$= \mathrm{tr}_{\Lambda^*_p \otimes L^2(B_1(0),E_p)} \left[\Phi^{-1}_{\varepsilon,p}\, \varrho\, \Phi_{\varepsilon,p}\, \Phi^{-1}_{\varepsilon,p}\, \chi_{\varepsilon^2 t,p}\, e^{-t\varepsilon^2 (D^{E(p)})^2} \Phi_{\varepsilon,p}\right].$$

We examine next the transformed heat kernel. Formally, $P_{\varepsilon,p}(t) := \Phi^{-1}_{\varepsilon,p}\, e^{-t\varepsilon^2(D^{E(p)})^2} \Phi_{\varepsilon,p}$ solves the equation

$$0 = \left[\frac{\partial}{\partial t} + \Phi^{-1}_{\varepsilon,p}\, \varepsilon^2 (D^{E(p)})^2\, \Phi_{\varepsilon,p}\right] P_{\varepsilon,p}(t) =: \left[\frac{\partial}{\partial t} + \tau_{\varepsilon,p}\right] P_{\varepsilon,p}(t).$$

Here $\tau_{\varepsilon,p}$ is a second order elliptic operator on $C^\infty_0(B_1(0), \Lambda^*_p \otimes E_p)$ but not symmetric in general. In Section 3 we will prove

**Lemma 2.1.** *We can write*

$$\tau_{\varepsilon,p} = -\sum_{|\beta|+|\gamma|\leq 2} M_\beta \partial_s^\gamma A_{\varepsilon,p}^{\beta\gamma}, \tag{2.12}$$

*where $M_\beta$ denotes multiplication by $s^\beta$. Moreover, if $|\gamma| = 2$, we have*

$$A_{\varepsilon,p}^{0\gamma} =: A_{\varepsilon,p}^{\gamma} = a_{\varepsilon,p}^{\gamma} I_{\Lambda_p^*\otimes E_p} \tag{2.13a$'$}$$

*and, for some $\lambda_1, \lambda_2 > 0$ and uniformly in $\varepsilon$ and $p$,*

$$\lambda_1|\xi|^2 \leq \sum_{\gamma_1,\gamma_2} a_{\varepsilon,p}^{\gamma}\xi_{\gamma_1}\xi_{\gamma_2} \leq \lambda_2|\xi|^2, \tag{2.13a$''$}$$

*and we have the regularity properties*

$$A_{\varepsilon,p}^{\beta\gamma} \in C^{|\gamma|}(B_1(0), \operatorname{End}\Lambda_p^*\otimes E_p) \quad (\textit{the } C^2 \textit{ case}) \tag{2.13b$'$}$$

*or, more generally,*

$$A_{\varepsilon,p}^{\beta\gamma} \in \begin{cases} C^{1,1} & \textit{if } |\gamma| = 2 \\ \text{Lip} & \textit{if } |\gamma| = 1, \\ L^\infty & \textit{if } |\gamma| = 0 \quad (\textit{the } C^{1,1} \textit{ case}), \end{cases} \tag{2.13b$''$}$$

*depending on whether or not the metric is actually $C^2$. If we decompose the coefficients with respect to the basis $(\epsilon_I^\flat)$ (as in (2.20) below) then*

$$\left[A_{\varepsilon,p}^{\beta\gamma}\right]_{IJ} = 0 \quad \textit{if} \quad |\beta| - |\gamma| - |\mu(I) - \mu(J)| < -2. \tag{2.13c}$$

*Finally, we have a limit in the sense that, for $f \in C_0^\infty(B_1(0), \Lambda_p^*\otimes E_p)$,*

$$\lim_{\varepsilon\to 0} \tau_{\varepsilon,p} f = \tau_{0,p} f,$$

*where $\tau_{0,p}$ is the "Getzler operator":*

$$\tau_{0,p} = -\sum_{i=1}^m \left[\frac{\partial}{\partial s_i} - \frac{1}{8}\sum_{j=1}^m s_j\, \Omega_{ij}(p)\otimes I_{E_p}\right]^2 + \Omega^E(p). \tag{2.13}$$

*If we denote by $R^M, R^E$ the respective curvature tensors and by 'w' the operation of wedge multiplication, then we have*

$$\Omega_{ij}(p) = \sum_{k,l=1}^m R_{klij}^M w(e_k^\flat \wedge e_l^\flat(p)),$$

$$\Omega^E(p) = \frac{1}{2}\sum_{i,j=1}^m w(e_i^\flat \wedge e_j^\flat(p)) \otimes R_{e_i,e_j}^E(p).$$

Next we have to introduce "boundary conditions". It seems most natural in this context to use the Friedrichs extension as defined by Kato [Ka, p. 325]. The details will be given in Section 4; let us thus define $T_{\varepsilon,p}$ as the Friedrichs extension of $\tau_{\varepsilon,p}$ in $\mathcal{H} := L^2(B_{2/3}(0), \Lambda_p^*\otimes E_p)$ with domain $C_0^\infty(B_{2/3}(0), \Lambda_p^*\otimes E_p)$. Then $T_{\varepsilon,p}$ generates an analytic semigroup, $e^{-tT_{\varepsilon,p}}$, which is a good replacement for $P_{\varepsilon,p}(t)$:

**Theorem 2.2.** *Denote by $C_q$ the von Neumann-Schatten class of order $q > 0$ (such that $C_1$ is the trace class). Then, for $\chi_1, \chi_2 \in C_0^\infty(B_{1/2}(0))$ and $N \in \mathbb{N}$,*

$$\|\chi_1(e^{-tT_{\varepsilon,p}} - P_{\varepsilon,p}(t))\chi_2\|_{C_1(\mathcal{H})} \leq C_N(\varepsilon, p, \chi_1, \chi_2)t^N.$$

This result will also be proved in Section 4. To analyze $e^{-tT_{\varepsilon,p}}$ it is important to have a good approximation. For this, somewhat surprisingly, the heat kernel of the Getzler operator does not seem convenient. Instead, we look at the operator which consists of the scalar principal part in (2.12),

$$\tau^0_{\varepsilon,p} := - \sum_{|\gamma|=2} \partial_s^\gamma A^\gamma_{\varepsilon,p}. \tag{2.14}$$

Its Friedrichs extension, $T^0_{\varepsilon,p}$, is defined as before and it turns out that the operators $T_{\varepsilon,p}$ and $T^0_{\varepsilon,p}$ all have the same domain. This is the basis for the following representation by a Neumann series, also to be proved in Section 4.

**Theorem 2.3.** *Introduce for $t \in (0, 1]$*

$$R^0_{\varepsilon,p}(t) := e^{-tT^0_{\varepsilon,p}}, \tag{2.15a}$$

(2.15b)

$$R^{\nu+1}_{\varepsilon,p}(t) := \int_0^t R^\nu_{\varepsilon,p}(t-u) \sum_{|\beta|+|\gamma|\leq 2, |\gamma|\leq 1} M_\beta \partial_s^\gamma A^{\beta\gamma}_{\varepsilon,p} e^{-uT^0_{\varepsilon,p}}\, du, \ \nu \in \mathbb{Z}_+.$$

*Then we have*

$$e^{-tT_{\varepsilon,p}} = \sum_{\nu\geq 0} R^\nu_{\varepsilon,p}(t), \tag{2.16}$$

*and the sum converges in operator norm, uniformly in $t, \varepsilon \in (0, 1]$ and $p \in M$.*

Next, a little calculation using (2.10b) shows that

$$\chi_{\varepsilon^2 t,p}\, \Phi_{\varepsilon,p} = \Phi_{\varepsilon,p} \varepsilon^{-m} \eta_t \zeta_{\varepsilon,p}, \quad \zeta_{\varepsilon,p} \in C^\infty(B_1(0)), \tag{2.17a}$$

where, uniformly in $s, \varepsilon, p$,

$$\lim_{\varepsilon \to 0} \zeta_{\varepsilon,p}(s) = 1. \tag{2.17b}$$

Thus we obtain from (2.11) and Theorem 2.2 (with $N = m/2 + 1$), setting $H := L^2(B_1(0), E_p)$,

$$\tilde{F}^E_{\varepsilon^2 t}(p) = \varepsilon^{-m} \mathrm{tr}_{\Lambda_p^* \otimes H} \left[ \Phi_{\varepsilon,p}^{-1} \varrho \Phi_{\varepsilon,p} \eta_t \zeta_{\varepsilon,p} e^{-tT_{\varepsilon,p}} \right] + O_\varepsilon(t). \tag{2.18}$$

It remains to deal with the involution $\varrho_{\varepsilon,p} := \Phi_{\varepsilon,p}^{-1}\, \varrho\, \Phi_{\varepsilon,p}$. In preparation of the relevant statement we introduce a matrix decomposition of $A \in \mathcal{L}(\Lambda_p^* \otimes K)$, $K$ a Hilbert space, via

$$A(e_I^\flat(p) \otimes x) =: \sum_J e_J^\flat(p) \otimes [A]_{JI}(x), \ [A]_{JI} \in \mathcal{L}(K). \tag{2.19a}$$

Hence, with $\psi_I : x \mapsto e_I^\flat(p) \otimes x$ we have $[A]_{JI} = \psi_J^* A \psi_I$. In particular, we write, with $\sim \emptyset := \{1, \dots, m\}$,

$$[A]_{\mathrm{top}} := [A]_{\sim\emptyset,\emptyset}. \tag{2.19b}$$

Then we have the following variant of the Berezin-Patodi lemma.

**Theorem 2.4.** *For $t, \varepsilon \in (0,1]$, $p \in M$, we have*

$$F^E_{\varepsilon^2 t}(p) = (-2\sqrt{-1})^k \mathrm{tr}_H \left[ \eta_t\, \zeta_{\varepsilon,p}\, e^{-tT_{\varepsilon,p}} \right]_{\mathrm{top}} + O_\varepsilon(t). \tag{2.20}$$

*Proof.* With the notation in (2.19a) we have

(2.21)
$$\begin{aligned}\varepsilon^{-m} \mathrm{tr}_{\Lambda_p^* \otimes H} \left[ \varrho_{\varepsilon,p} \eta_t \zeta_{\varepsilon,p} e^{-tT_{\varepsilon,p}} \right] &= \varepsilon^{-m} \sum_{I,J} \mathrm{tr}_H \Big[ \varrho_{\varepsilon,p} \Big]_{IJ} \left[ \eta_t \zeta_{\varepsilon,p} e^{-tT_{\varepsilon,p}} \right]_{JI} \\ &=: G_{\varepsilon,t}(p).\end{aligned}$$

For any multiindex $I \subset \{1, \dots, m\}$ we denote by $\sim I$ the (strictly ordered) complementary multiindex and by $\mathrm{sgn}(I, \sim I)$ the sign of the permutation $\{I, \sim I\}$ of $\{1, \dots, m\}$. Then we have

$$\begin{aligned}\omega_{\mathbb{C}} \cdot e_I(p) = (-1)^{\mu(I)} e_I \cdot \omega_{\mathbb{C}}(p) &= (-1)^{\mu(I)} \sqrt{-1}^k e_I \cdot e_I \cdot \mathrm{sgn}(I, \sim I) e_{\sim I}(p) \\ &= \sqrt{-1}^{k + \mu(I)(\mu(I)-1)} \mathrm{sgn}(I, \sim I) e_{\sim I}(p),\end{aligned}$$

hence (2.10b) and the equivariance of $\alpha_p$, the canonical isomorphism $\mathbb{C}\ell_p M \to S_p \otimes S_p^*$, imply (with $*$ the Hodge star operator)

(2.22)
$$\begin{aligned}\varrho_{\varepsilon,p}(e_I^\flat(p) \otimes \omega) &= \varepsilon^{2\mu(I)-m} \sqrt{-1}^{k + \mu(I)(\mu(I)-1)} * e_I^\flat(p) \otimes \omega \\ &=: \varepsilon^{2\mu(I)-m} (\tau_p \otimes I)(e_I^\flat(p) \otimes \omega).\end{aligned}$$

Here $\tau$ is the involution defining the signature operator on $\Omega(M)$. (2.21) and (2.22) yield

$$(2.23)\qquad G_{\varepsilon,t}(p) = \varepsilon^{-m}\sum_I \mathrm{tr}_H[\varrho_{\varepsilon,p}]_{I,\sim I}[\eta_t\,\zeta_{\varepsilon,p}\,e^{-tT_{\varepsilon,p}}]_{\sim I,I}$$

$$= \sum_I \varepsilon^{-2\mu(I)}(-1)^k\sqrt{-1}^{k+\mu(I)(\mu(I)+1)}\mathrm{sgn}(\sim I, I)\mathrm{tr}_H\left[\eta_t\zeta_{\varepsilon,p}e^{-tT_{\varepsilon,p}}\right]_{\sim I,I}$$

$$=: \sum_I G^I_{\varepsilon,t}(p).$$

Now observe that the theorem is proved if we can show that $G^I_{\varepsilon,t}(p) = G^{\emptyset}_{\varepsilon,t}(p)$ for all $I$. To see this, we infer from the construction of $\tau_{\varepsilon,p}$ (cf. Section 3) that the operators $A^{\beta\gamma}_{\varepsilon,p}$ in (2.12) act on $\Lambda^*_p$ only through operators of the form

$$\prod_{i\in I}[\varepsilon^{-1}w(e^\flat_i(p)) - \varepsilon\,\lrcorner(e_i(p))],\quad I\subset\{1,\ldots,m\},$$

where $\lrcorner$ denotes interior multiplication, i.e. they act only through scaled Clifford multiplication. In view of Theorem 2.3, the same is true for $e^{-tT_{\varepsilon,p}}$ and we can write

$$e^{-tT_{\varepsilon,p}} =: \sum_J\prod_{i\in J}[\varepsilon^{-1}w(e^\flat_i(p)) - \varepsilon\,\lrcorner(e_i(p))]\otimes A^J_{\varepsilon,p}(t).$$

A contribution to (2.23) can arise only if $J = \{1,\ldots,m\}$, and we obtain (with obvious notation)

$$\left[\eta_t\zeta_{\varepsilon,p}e^{-tT_{\varepsilon,p}}\right]_{\sim I,I}$$

$$= (-1)^{\mu(I)}\varepsilon^{2\mu(I)-m}\mathrm{sgn}(\sim I, I)\left[w(e^\flat_{\sim I}(p))\lrcorner(e_I(p))\right]_{\sim I,I}\eta_t\zeta_{\varepsilon,p}A^{\sim\emptyset}_{\varepsilon,p}(t)$$

$$= (-1)^{\mu(I)+\mu(I)(\mu(I)-1)/2}\mathrm{sgn}(\sim I, I)\varepsilon^{2\mu(I)-m}\eta_t\zeta_{\varepsilon,p}A^{\sim\emptyset}_{\varepsilon,p}(t).$$

The proof is complete. □

It is natural to expect convergence in (2.20) as $\varepsilon\to 0$. However, $\eta_t$ involves a factor $t^{-m/2}$ and we are taking traces. The estimate necessary to handle this problem is the analytic core of the paper and will be given in Section 5 below. As a simple consequence, also to be proved in Section 5, we obtain the following result.

**Theorem 2.5.** *There is a function $\phi$ with $\lim_{\varepsilon\to 0}\phi(\varepsilon) = 0$ such that*

$$(2.24)\qquad \|\eta_t\left[\zeta_{\varepsilon,p}\,e^{-tT_{\varepsilon,p}} - e^{-tT_{0,p}}\right]_{\mathrm{top}}\|_{C_1(\mathcal{H})} \le C(\phi(\varepsilon) + t^{1/2}),$$

*uniformly in $t\in(0,1]$ and $p\in M$.*

The final step in the argument uses again Getzler's calculations based on Mehler's formula; it will be given in Section 4.

**Lemma 2.6.**

$$|(-2\sqrt{-1})^k \operatorname{tr}_H \left[\eta_t e^{-tT_{0,p}}\right]_{\text{top}} - \hat{A}(M) \wedge \operatorname{ch}E(e_1, \ldots, e_m)(p)| = O(\sqrt{t}). \tag{2.25}$$

The preceding results now easily yield the

*Proof of Theorem 1.3.* Combining (2.20), (2.24), and (2.25) we find

$$|F^E_{\varepsilon^2 t}(p) - \hat{A}(M) \wedge \operatorname{ch}E(e_1, \ldots, e_m)(p)| \le C_\varepsilon t + C(\phi(\varepsilon) + \sqrt{t}).$$

□

**3. Scaling.** We now want to prove Lemma 2.1. This calculation is essentially well known but we have to redo it in our special context. So we fix $p \in M$ and a local orthonormal frame, $(e_i)_{i=1}^m$, for $TB_1(p)$ as before. We choose a canonically associated local orthonormal frame $(\sigma_\alpha)$ for $S|B_1(p)$ and a local orthonormal frame $(\tau_\beta)$ for $E|B_1(p)$; $(\sigma^*_\alpha)$ denotes the dual frame for $S^*|B_1(p)$. The first important fact we need is the Bochner-Lichnerowicz-Weitzenboeck formula [LM, p. 164]:

$$\left(D^{E(p)}\right)^2 = -\sum_{i=1}^m \left[\nabla^{S\otimes E(p)}_{e_i}\nabla^{S\otimes E(p)}_{e_i} - \nabla^{S\otimes E(p)}_{\nabla_{e_i} e_i}\right] + \tfrac{1}{4}\kappa + \tfrac{1}{2}\sum_{i,j=1}^m c(e_i)c(e_j) \otimes R^E_{e_i,e_j}, \tag{3.1}$$

where $\nabla := \nabla^{TM}$ and $\kappa$ is the scalar curvature of $M$. It is apparent from (3.1) that the main computation concerns $\nabla^{S\otimes E(p)}_{e_i}$. Thus we choose $\omega \in C^\infty(B_1(0), E_p)$ and observe that (2.10b) implies

$$\Phi_{\varepsilon,p}\left(e^\flat_I(p) \otimes \omega\right) =: \varepsilon^{\mu(I)} \widetilde{\alpha_p(e_I(p))} \otimes \widetilde{\omega^\varepsilon}, \tag{3.2}$$

where a tilde denotes parallel translates i.e.

$$\begin{aligned} \widetilde{\alpha_p(e_I(p))}(\exp_p s) &= \mathrm{P}^{S\otimes S^*_p}_p(s)(\alpha_p(e_I(p))), \\ \widetilde{\omega^\varepsilon}(\exp_p s) &= \mathrm{P}^E_p(s)(\omega(s/\varepsilon)), \end{aligned}$$

are parallel along radial geodesics from $p$. Hence we obtain

$$\begin{aligned} \nabla^{S\otimes E(p)}_{e_i} \Phi_{\varepsilon,p}(e^\flat_I \otimes \omega) &= \nabla^{S\otimes E(p)}_{e_i}(\varepsilon^{\mu(I)}[\alpha_p(e_I(p))]^\sim \otimes \widetilde{\omega^\varepsilon}) \\ &= \varepsilon^{\mu(I)}\left[(\nabla^{S\otimes S^*_p}_{e_i}[\alpha_p(e_I(p)))]^\sim \otimes \widetilde{\omega^\varepsilon} + [\alpha_p(e_I(p))]^\sim \otimes \nabla^E_{e_i}\widetilde{\omega^\varepsilon}\right] =: \mathrm{I} + \mathrm{II}. \end{aligned} \tag{3.3}$$

We evaluate I, noting that the spin connection is given by the following formula for *any* section $\sigma \in C^\infty(S|B_1(p))$ which is parallel along radial geodesics from $p$ [LM, p. 110]:

$$\nabla^S_e \sigma = \frac{1}{4} \sum_{j,k=1}^m \langle \nabla^{TM}_e e_j, e_k \rangle c(e_j \cdot e_k)\sigma, \; e \in C^\infty(TB_1(p)). \tag{3.4}$$

We will write

$$\Gamma^i_{jk} := \left\langle \nabla^{TM}_{e_i} e_j, e_k \right\rangle, \tag{3.5}$$

and find, with $\beta_p : \Lambda^*_p \to \mathbb{C}\ell_p$ the canonical isomorphism:

$$\begin{aligned}
\mathrm{I} &= \varepsilon^{\mu(I)/4} \sum_{j,k=1}^m \Gamma^i_{jk} c(e_j \cdot e_k)[\alpha_p(e_I(p))]^{\sim} \otimes \widetilde{\omega^\varepsilon} \\
&= \varepsilon^{\mu(I)/4} \sum_{j,k=1}^m [\alpha_p\Big(c(e_j \cdot e_k(p))e_I(p)\Big)]^{\sim} \otimes \Big[\Big((\Gamma^i_{jk} \circ \exp_p)^{1/\varepsilon}\omega\Big)^\varepsilon\Big]^{\sim} \\
&= \varepsilon^{\mu(I)/4} \sum_{j,k=1}^m \Big[\Big(\alpha_p\beta_p(w(e^\flat_j) - \lrcorner(e_j))(w(e^\flat_k) - \lrcorner(e_k))e^\flat_I(p)\Big)\Big]^{\sim} \otimes \\
&\qquad\qquad \otimes \Big[\Big((\Gamma^i_{jk} \circ \exp_p)^{1/\varepsilon}\omega\Big)^\varepsilon\Big]^{\sim} \\
&= \Phi_{\varepsilon,p}\Big(\frac{1}{4} \sum_{j,k=1}^m \Big(\Gamma^i_{jk} \circ \exp_p\Big)^{1/\varepsilon} \Big(\varepsilon^{-1} w(e^\flat_j) - \varepsilon \lrcorner(e_j)\Big) \cdot \\
&\qquad\qquad \cdot \Big(\varepsilon^{-1} w(e^\flat_k) - \varepsilon \lrcorner(e_k)\Big) e^\flat_I(p) \otimes \omega\Big).
\end{aligned} \tag{3.6}$$

Observe that the Lipschitz function $\Gamma^i_{jk} \circ \exp_p$ can be written as

$$\Gamma^i_{jk} \circ \exp_p(s) = \sum_{l=1}^m s_l g^{il}_{jk}(s),$$

with $g^{il}_{jk} \in L^\infty(B_1(0))$. Moreover, by Rademacher's theorem,

$$\frac{\partial}{\partial s_l}(\Gamma^i_{jk} \circ \exp_p) \in L^\infty(B_1(0)).$$

Next, we choose $p \in M$ such that all second order partial derivatives of the metric and the hermitian structure as well as all first order partial derivatives of the connection coefficients of $\nabla^E$ exist at $p$. So, we can further evaluate this using the well known Taylor expansion

$$
\begin{aligned}
\Gamma^i_{jk}(\exp_p \varepsilon s) &= -\varepsilon/2 \sum_{l=1}^{m} s_l \Big[ \langle R^M(e_i, e_l)e_j, e_k\rangle - \hat{\Gamma}^{il}_{jk}(p; \varepsilon s) \Big] \\
&=: -\varepsilon/2 \sum_{l=1}^{m} s_l \Big( R^M_{iljk}(p) + \hat{\Gamma}^{il}_{jk}(p; \varepsilon s) \Big),
\end{aligned} \tag{3.7}
$$

with some uniformly bounded functions $\hat{\Gamma}^{i\cdot}_{jk} \in C(M \times B_1(0))$ satisfying $\lim_{\varepsilon \to 0} \hat{\Gamma}^{il}_{jk}(p, \varepsilon s) = 0$, for all $s \in B_1(0)$. It follows that

$$
\begin{aligned}
\varepsilon\, \mathrm{I} = \Phi_{\varepsilon,p}\Big[ -\frac{1}{8} \sum_{l=1}^{m} s_l \big( R^M_{iljk}(p) + \hat{\Gamma}^{il}_{jk}(p; \varepsilon s)\big) \cdot \\
\cdot \big( w(e_j^\flat) - \varepsilon^2 \lrcorner(e_j)\big)\big( w(e_k^\flat) - \varepsilon^2 \lrcorner(e_k)\big) \otimes \omega \Big].
\end{aligned} \tag{3.8}
$$

To evaluate II we observe that

$$
\begin{aligned}
\nabla^E_{e_i} \widetilde{\omega^\varepsilon}(q) &= \nabla^E_{e_i}\big( \mathrm{P}^E_p(\omega^\varepsilon \circ \exp_p^{-1})\big)(q) \\
&= \mathrm{P}^E_p(\exp_p^{-1} q)\big(\mathrm{P}^E_p(\exp_p^{-1} q)^{-1} \nabla^E_{e_i} \mathrm{P}^E_p(\omega^\varepsilon \circ \exp_p^{-1})(q)\big).
\end{aligned}
$$

Using the terminology introduced in [Br, Lemma 5.2] we obtain a unitary connection $(\mathrm{P}^E)^* \nabla^E =: \widetilde{\nabla}^E$ on $C^\infty(B_1(0), E_p)$ with the property that

$$
\nabla^E_{e_i} \widetilde{\omega^\varepsilon}(\exp_p s) = \mathrm{P}^E_p(s) \widetilde{\nabla}^E_{(T_s \exp_p)^{-1} e_i} \omega^\varepsilon(s). \tag{3.9}
$$

Writing $\tilde{e_i}(s) := (T_s \exp_p)^{-1}(e_i) \in T_s\mathbb{R}^m$, it follows that

$$
\mathrm{II} = \varepsilon^{\mu(I)} \widetilde{\alpha_p(e_I(p))} \otimes \nabla^E_{e_i} \widetilde{\omega^\varepsilon} = \Phi_{\varepsilon,p}\Big( e_I^\flat(p) \otimes (\widetilde{\nabla}^E_{\tilde{e}_i} \omega^\varepsilon)^{1/\varepsilon} \Big). \tag{3.10}
$$

Now we have in the normal coordinates defined by $(e_i)_{i=1}^m$

$$
\tilde{e_i}(s) =: \sum_{j=1}^{m} a_{ij}(s) \frac{\partial}{\partial s_j},
$$

with $a_{ij} \in C^\infty(B_1(0)), a_{ij}(0) = \delta_{ij}$. Thus we obtain, with certain $\hat{\Gamma}^i \in C^\infty(B_1(0), \mathrm{End} E_p)$,

$$
\Big( \widetilde{\nabla}^E_{\tilde{e}_i} \omega^\varepsilon \Big)^{1/\varepsilon}(s) = \varepsilon^{-1} \sum_{j=1}^{m} a_{ij}(\varepsilon s) \frac{\partial \omega}{\partial s_j}(s) + \hat{\Gamma}^i(\varepsilon s)\omega(s)
$$

or

$$\varepsilon\,\mathrm{II} = \Phi_{\varepsilon,p}\Big(e_I^\flat(p)\otimes\Big(\sum_{j=1}^m a_{ij}^{1/\varepsilon}\frac{\partial}{\partial s_j}+(\hat\Gamma^i)^{1/\varepsilon}\Big)\omega\Big). \tag{3.11}$$

Combining (3.8) and (3.11) we now find

$$\varepsilon\Phi_{\varepsilon,p}^{-1}\nabla_{e_i}^{S\otimes E(p)}\Phi_{\varepsilon,p} = \sum_{j=1}^m a_{ij}(\varepsilon s)\frac{\partial}{\partial s_j}+\varepsilon I_{\Lambda_p^*}\otimes\hat\Gamma^i(\varepsilon s) \tag{3.12}$$

$$-\tfrac{1}{8}\sum_{l=1}^m s_l\big(R^M_{iljk}(p)+\hat\Gamma^{il}_{jk}(p;\varepsilon s)\big)(w(e_j^\flat)-\varepsilon^2\lrcorner(e_j))(w(e_k^\flat)-\varepsilon^2\lrcorner(e_k))\otimes I_{E_p}.$$

The calculation of $\tau_{\varepsilon,p}$ is thus almost complete, in view of (3.1). Only the last term deserves some further attention:

$$\frac{1}{2}\sum_{i,j=1}^m c(e_i\cdot e_j)\otimes R^E_{e_i,e_j}\Phi_{\varepsilon,p}\Big(e_I^\flat(p)\otimes\omega\Big)(\exp_p s) \tag{3.13}$$

$$=\varepsilon^{\mu(I)}\frac{1}{2}\sum_{i,j=1}^m c(e_i\cdot e_j)\otimes R^E_{e_i,e_j}[\alpha_p(e_I(p))]^\sim\otimes[\omega^\varepsilon(\exp_p s)]^\sim$$

$$=\varepsilon^{\mu(I)}\frac{1}{2}\sum_{i,j=1}^m \mathrm{P}_p^{S\otimes E(p)}(s)\Big(\alpha_p\beta_p((w(e_i^\flat)-\lrcorner(e_i))(w(e_j^\flat)-\lrcorner(e_j))e_I^\flat(p)\otimes$$

$$\otimes\,\mathrm{P}_p^E(s)^{-1}R^E_{e_i,e_j}\mathrm{P}_p^E(s)(\omega(\varepsilon^{-1}s))\Big)$$

$$=\Phi_{\varepsilon,p}\Big(\frac{1}{2}\sum_{i,j=1}^m(\varepsilon^{-1}w(e_i^\flat(p))-\varepsilon\lrcorner(e_i(p)))(\varepsilon^{-1}w(e_j^\flat(p))-\varepsilon\lrcorner(e_j(p)))\otimes$$

$$\otimes\,((\mathrm{P}_p^E)^{-1}R^E_{e_i,e_j}\mathrm{P}_p^E)^{1/\varepsilon}\omega\Big)(\exp_p s).$$

From (3.1), (3.12), and (3.13) one now easily reads off the proof of Lemma 2.1. □

**4. Off-diagonal decay of heat traces.** We begin this section with some abstract results which are often used in the spectral analysis of elliptic operators, and which will lead to the proof of Theorem 2.2 and Lemma 2.6; many special versions are, of course, well known. But the point here is to use as little regularity as possible which forced us to present the treatment below. We work with the resolvent and transfer the results to the heat semigroup via Cauchy integrals.

Let $H$ be a Hilbert space, $\mathcal{D}$ a dense subspace, and $T : \mathcal{D} \to H$ a closed m-sectorial operator in $H$. By definition [Ka, p. 280] this means that the numerical range (and hence $\operatorname{spec} T$) is contained in a sector

$$\{z \in \mathbb{C} \mid |\arg(z-\gamma)| \leq \theta\}, \qquad \gamma \in \mathbb{R},\ 0 \leq \theta < \pi/2, \tag{4.1}$$

and that we have the resolvent estimate [Ka, p. 490]

$$\|(T-(\lambda+\gamma))^{-1}\| \leq C_{\theta'}|\lambda|^{-1}, \quad |\lambda| \geq 1,\ |\arg \lambda| \geq \theta' > \theta. \tag{4.2}$$

In what follows we assume for simplicity that

$$\gamma \geq 1. \tag{4.3}$$

We introduce the resolvent

$$G(z) := (T+z^2)^{-1},\ z \in Z_\delta := \{z \in \mathbb{C} \mid |z| \geq \delta,\ |\arg z| < (\pi-\theta')/2\},$$

where $\theta' > \theta$, and we use '$z^2$' in view of our applications to differential operators of second order. The class of closed m-sectorial operators is natural since it contains e.g. the Friedrichs extension of second order elliptic differential operators with scalar principal symbol.

To imitate the usual cut-off procedure we introduce a subspace, $\mathcal{S}$, of $\mathcal{L}(H)$ with the following properties (the "cut-off axioms"):

(CO1) For all $\phi \in \mathcal{S}$, $\phi(\mathcal{D}) \subset \mathcal{D}$.

(CO2) For all $\phi \in \mathcal{S}$, the operators

$$T^{-1/2}[T,\phi],\ [T,\phi]T^{-1/2} : \mathcal{D} \to H$$

extend to $H$ by continuity.

(CO3) There is a transitive relation '$<$' on $\mathcal{S}$ such that $\phi_1 < \phi_2$ implies

$$\phi_1\phi_2 = \phi_2\phi_1 = \phi_1, \tag{4.4a}$$

and that there is $\phi_3 \in \mathcal{S}$ with $\phi_1 < \phi_3 < \phi_2$ such that

$$\begin{aligned} T^{-1/2}[T,\phi_1](I-\phi_3) &= (I-\phi_3)[T,\phi_1]T^{-1/2} \\ &= T^{-1/2}[T,\phi_2]\phi_3 = \phi_3[T,\phi_2]T^{-1/2} = 0. \end{aligned} \tag{4.4b}$$

(CO4) If $I_H \notin \mathcal{S}$, then for all $\phi \in \mathcal{S}$ there is $\phi' \in \mathcal{S}$ with $\phi < \phi'$.

(CO5) $\mathcal{S}^* = \mathcal{S}$.

We remark that in view of (4.3) and [Ka, Ch V, §3.11] the operators $T^{-\alpha}, \alpha \in (0,1)$, are well defined bounded operators in $H$. We use again the von Neumann-Schatten classes of compact operators which we denote by $C_p(H), p > 0$, with norm $\|\cdot\|_p$.

**Lemma 4.1.** *Assume that a subspace $\mathcal{S}$ of $\mathcal{L}(H)$ satisfies the assumptions* (CO1) *through* (CO5), *and that in addition*

$$\phi T^{-1} \in C_{p_0}(H), \quad \text{for } \phi \in \mathcal{S} \text{ and some } p_0 > 0.$$

*Then, for all $\phi_1, \phi_2 \in \mathcal{S}$ with $\phi_1 < \phi_2$ and all $p > 0$ we have*

$$\phi_1 G(z)(I - \phi_2) \in C_p(H), \quad z \in Z_\delta.$$

*Moreover, we have the norm estimate*

$$\|\phi_1 G(z)(I - \phi_2)\|_p \leq C(\phi_1, \phi_2, p, N)|z|^{-N},$$

*valid for $z \in Z_\delta$ and all $N \in \mathbf{N}$.*

*Proof.* Using (4.4b) we determine $\phi_j \in \mathcal{S}, 1 \leq j \leq 4$, such that $\phi_1 < \phi_3 < \phi_4 < \phi_2$. Since $\phi G(z)u \in \mathcal{D}$ for $u \in H$, $\phi \in \mathcal{S}$, $z \in Z_\delta$, we obtain from

$$(T + z^2)\phi_3 G(z)(I - \phi_2) = [T, \phi_3]G(z)(I - \phi_2)$$

and (4.4b) the representation

(4.5)
$$\phi_1 G(z)(I - \phi_2) = (\phi_1 T^{-1/2})(T^{1/2}G(z)T^{1/2})(T^{-1/2}[T, \phi_3])\phi_4 G(z)(I - \phi_2).$$

Now we observe the easy estimates

$$\|G(z)\| \leq C|z|^{-2}, \tag{4.6a}$$

$$\|T^{1/2}G(z)T^{1/2}\| = \|I - z^2 G(z)\| \leq C, \tag{4.6b}$$

$$\|\phi G(z)\|_p \leq \|\phi T^{-1}\|_p + \|\phi T^{-1}\, z^2 G(z)\|_p \leq C\|\phi T^{-1}\|_p, \tag{4.6c}$$

where we have used the resolvent equation in (4.6c). Next we combine the integral representation [Ka, p. 281]

$$T^{-1/2} = \frac{1}{2\pi i} \int_\Gamma \xi^{-1/2} G((-\xi)^{1/2}) d\xi, \tag{4.7}$$

where $\Gamma$ is (e.g.) the wedge $|\arg z| = \theta'$, $\theta' > \theta$, traversed upward, with the interpolation inequality

$$\|A\|_{p'} \leq \|A\|^{1-p/p'} \|A\|_p^{p/p'}, \tag{4.8}$$

valid for $A \in C_p(H)$ and all $p' > p > 0$. Then we deduce that for $p_1 > 2p_0$

$$\|\phi G((-\xi)^{1/2})\|_{p_1} \leq C|\xi|^{p_0/p_1 - 1} \|\phi T^{-1}\|_{p_0}^{p_0/p_1} \leq C|\xi|^{p_0/p_1 - 1},$$

which gives with (4.7)

$$\phi T^{-1/2} \in C_{p_1}(H). \tag{4.9}$$

Using (4.9) (with $p_1 = 3p_0$), (4.6b), (CO2), and (4.6a) in (4.5) we derive

$$\|\phi_1 G(z)(I-\phi_2)\|_{3p_0} \leq C(\phi_1,\phi_2)|z|^{-2}, z \in Z_\delta. \tag{4.10a}$$

Next we use the Hölder inequality for $C_p$-norms in (4.5) to prove by induction on $L \in \mathbb{N}$

$$\|\phi_1 G(z)(I-\phi_2)\|_{3p_0/L} \leq C(\phi_1,\phi_2,L)|z|^{-2},$$

i.e. for $p > 0$

$$\|\phi_1 G(z)(I-\phi_2)\|_p \leq C(\phi_1,\phi_2,p)|z|^{-2}, \quad z \in Z_\delta. \tag{4.10b}$$

Finally, we rewrite (4.5) as

$$\phi_1 G(z)(I-\phi_2) = \phi_1 G(z) T^{1/2}(T^{-1/2}[T,\phi_3])\phi_4 G(z)(I-\phi_2)$$

and use the estimate

$$\|G(z)T^{1/2}\| \leq C|z|^{-1}, \quad z \in Z_\delta, \tag{4.10c}$$

which follows from (4.7) and the resolvent equation by a straightforward estimate. Then another induction will prove the assertion. □

By symmetry, the same reasoning also proves the following statement.

**Corollary 4.2.** *Assume that all the assumptions of Lemma 4.1 hold with the only change that now*

$$T^{-1}\phi \in C_{p_0}(H),$$

*for $\phi \in \mathcal{S}$ and some $p_0 > 0$. Then, for $\phi_1, \phi_2 \in \mathcal{S}$ with $\phi_1 < \phi_2$ and all $p > 0$ we have*

$$(I-\phi_2)G(z)\phi_1 \in C_p(H), \quad z \in Z_\delta,$$

*with norm estimate*

$$\|(I-\phi_2)G(z)\phi_1\|_p \leq C(\phi_1,\phi_2,p,N)|z|^{-N}, \quad z \in Z_\delta,\ N \in \mathbb{N}.$$

We will actually need the consequences of Lemma 4.1 for the heat operator,

$$e^{-tT} = \frac{1}{2\pi i}\int_\Gamma e^{-t\xi}(T-\xi)^{-1}d\xi. \tag{4.11}$$

Here we can choose $\Gamma$ as in (4.7) but, by holomorphy, we can replace it by $\Gamma_t$, the path obtained from $\Gamma$ by traversing the circle $|\xi| = t^{-1}$ outside $\Gamma$, and then $\Gamma$ for $|\xi| \geq t^{-1}$. This gives

**Lemma 4.3.** *Under the assumptions of Lemma 4.1 or Corollary 4.2 we have for $t \in (0,1]$ and $p > 0$*

$$\|\phi_1 e^{-tT}(I - \phi_2)\|_p \leq C(\phi_1, \phi_2, p, N)t^N \tag{4.12a}$$

*and*

$$\|(I - \phi_2)e^{-tT}\phi_1\|_p \leq C(\phi_1, \phi_2, p, N)t^N, \tag{4.12b}$$

*respectively.*

Now we turn to the class of operators which will be the main object of study in the remainder of this section, and to which we will apply the abstract results above. These are certain second order elliptic operators on Riemannian manifolds which are sectorial. Thus, consider a (smooth) connected Riemannian manifold $M$ of dimension $m$, a (smooth) hermitian bundle $F$ of rank $k$ over $M$, and a differential operator, $\tau$, on $C_0^\infty(F)$. We assume that, in a bundle chart $\varphi$ with local coordinates $s = (s_1, \ldots s_m) \in B_1(0)$, we have

$$\tau_\varphi := \varphi \circ \tau \circ \varphi^{-1} = -\sum_{|\gamma| \leq 2} \partial_s^\gamma B_\varphi^\gamma, \tag{4.13}$$

where the coefficients $B_\varphi^\gamma$ satisfy the assumptions (2.13a) and (2.13b″). These conditions are independent of the choice of $\varphi$. The following facts are proved by standard arguments.

**Lemma 4.4.** *Let $\tau$ satisfy the conditions* (4.13)
1) *$\tau$ has a formal adjoint on $C_0^\infty(F)$, $\tau^t$, which is also elliptic of second order on $C_0^\infty(F)$ and satisfies* (4.13).
2) *For $\phi \in C_0^\infty(M)$ and $u \in C_0^\infty(F)$ we have*

$$\|\phi u\|_{H^1_{\text{loc}}(F)} \leq C_3(\phi)(\operatorname{Re}(\tau u, u) + \|u\|^2), \tag{4.14}$$

*where $C_3$ also depends on a choice of norm in $H^1(F|\operatorname{supp}\phi)$.*
3) *For $\phi \in C^\infty(M)$ with $\operatorname{supp}(d\phi)$ compact and $u \in C_0^\infty(F)$ we have*

$$|(\tau\phi u, \phi u)| \leq C(\phi)(\operatorname{Re}(\tau u, u) + \|u\|^2). \tag{4.15}$$

With $\tau$ we associate the following sesquilinear forms on $C_0^\infty(F)$:

$$\mathfrak{t}[u,v] := (\tau u, v), \quad \mathfrak{t}[u] := \mathfrak{t}[u,u], \tag{4.16a}$$

$$\mathfrak{h}[u,v] := \left(\frac{1}{2}(\tau + \tau^t)u, v\right) =: (\mathcal{R}\tau u, v), \quad \mathfrak{h}[u] := \mathfrak{h}[u,u], \tag{4.16b}$$

$$\mathfrak{k}[u,v] := \left(\frac{1}{2i}(\tau - \tau^t)u, v\right) =: (\mathcal{I}\tau u, v), \quad \mathfrak{k}[u] := \mathfrak{k}[u,u], \tag{4.16c}$$

such that

$$\mathfrak{t}[u,v] = \mathfrak{h}[u,v] + i\mathfrak{k}[u,v].$$

Observe that $\mathcal{I}_T$ is a differential operator of first order.

To ensure sectoriality of $\mathfrak{t}$ (in the sense of [Ka, p. 310]) we require, in addition to the conditions (4.13), the estimates

$$\text{(4.17a)} \qquad \mathfrak{h}[u] \geq \gamma \|u\|^2,$$

$$\text{(4.17b)} \qquad |\mathfrak{k}[u]| \leq (\mathrm{tg}\theta)(\mathfrak{h}[u] - \gamma\|u\|^2), \ \theta \in [0, \pi/4),$$

for all $u \in C_0^\infty(F)$. Without loss of generality we may assume that $\gamma \geq 1$. Then $\mathfrak{h}$ is also sectorial and symmetric. So we can form the Friedrichs extension of both $\mathfrak{t}$ and $\mathfrak{h}$, to be denoted by $T$ and $H$, respectively (cf. [Ka, p. 325]). The forms $\mathfrak{t}$ and $\mathfrak{h}$ are closable, with closures $\tilde{\mathfrak{t}}$ and $\tilde{\mathfrak{h}}$, and it follows from (4.17b) that

$$\mathcal{D}(\tilde{\mathfrak{t}}) = \mathcal{D}(\tilde{\mathfrak{h}})\,.$$

Note that the estimates (4.14) and (4.15) extend to $\mathcal{D}(\tilde{\mathfrak{h}})$ by continuity.

In the setting just described we now have to verify the assumptions (CO1)–(CO5) for $T$, the Friedrichs extension of $\tau$ in $L^2(F)$. We put

$$\text{(4.18)} \qquad \mathcal{S} := C_0^\infty(M),$$

acting on $L^2(F)$ by multiplication.

For (CO1), we observe that by [Ka, p. 322] we have

$$\text{(4.19)} \qquad \mathcal{D}(T) = \{u \in \mathcal{D}(\tilde{\mathfrak{t}}) \mid |(u, \tau^t v)| \leq C_u\|v\| \text{ for all } v \in C_0^\infty(F)\}.$$

For $u \in \mathcal{D}(T)$ we can thus find a sequence $(u_n) \subset C_0^\infty(F)$ such that $u_n \to u$ in $L^2(F)$ and $\lim_{n,m\to\infty} \mathfrak{t}[u_n - u_m] = 0$. For $\phi \in \mathcal{S}$ we obtain $\phi u_n \to \phi u$ in $L^2(F)$ and, from (4.15),

$$\begin{aligned} |\mathfrak{t}[\phi(u_n - u_m)]| &= |(\tau\phi(u_n - u_m), \phi(u_n - u_m))| \\ &\leq C(\phi)(\mathfrak{h}[u_n - u_m] + \|u_n - u_m\|^2_{L^2(F)}) \\ &\leq C(\phi)(|\mathfrak{t}[u_n - u_m]| + \|u_n - u_m\|^2_{L^2(F)}) \to 0, \quad n, m \to \infty. \end{aligned}$$

Hence $\phi u \in \mathcal{D}(\tilde{\mathfrak{t}})$. Next, for $v \in C_0^\infty(F)$ we get

$$(\phi u, \tau^t v) = (u, [\phi, \tau^t]v) + (Tu, v);$$

since $[\phi, \tau^t]$ is a compactly supported differential operator of order at most one on $C^\infty(F)$, we obtain from (4.14) the estimate

$$|(\phi u, \tau^t v)| \leq C_u\|v\|,$$

as desired.

(CO2) we prove first for $H^{-1/2}$ in place of $T^{-1/2}$: for $u \in \mathcal{D}(\tilde{\mathfrak{h}}) = \mathcal{D}(H^{-1/2})$ we have in view of (4.14)

$$\|[T, \phi]u\|^2 \le C\|u\|^2_{H^1(F|\operatorname{supp}\phi)} \le C(\|H^{1/2}u\|^2 + \|u\|^2).$$

Setting $u := H^{-1/2}v$ gives the boundedness of $[T, \phi]H^{-1/2}$; applying the same argument to $T^*$ gives the boundedness of $H^{-1/2}[T, \phi]$. To relate $H^{-1/2}$ to $T^{-1/2}$ we choose $\psi \in C_0^\infty(M)$ with $\psi = 1$ in a neighborhood of $\operatorname{supp}\phi$ and note that $\psi\mathcal{D}(T) \subset \mathcal{D}(H)$, in view of (4.14) again and the identity

$$(\psi u, Hv) = (T\psi u + i\mathcal{I}_\tau\phi u, v),$$

valid for $u \in \mathcal{D}(T)$ and $v \in C_0^\infty(F)$. Then we derive the representation

$$\begin{aligned} \psi(T-\lambda)^{-1} &= (H-\lambda)^{-1}(\psi + ([\tau,\psi] - i\mathcal{I}_\tau\psi)(T-\lambda)^{-1}) \\ &=: (H-\lambda)^{-1}(\psi + A(\lambda)). \end{aligned} \tag{4.20}$$

The arguments given above show that

$$\|H^{-1/2}A(\lambda)\| \le C|\lambda|^{-1}, \quad \text{if } |\lambda| \ge 1 \text{ and } \operatorname{Re}\lambda \le 0. \tag{4.21}$$

Using the representation (4.7) we deduce from (4.20) and (4.21) that

$$\psi T^{-1/2} =: H^{-1/2}B,$$

for some bounded operator $B$. This gives the boundedness of $[T, \phi]T^{-1/2} = [T, \phi]\psi T^{-1/2}$; applying the argument to $T^*$ gives the boundedness of $T^{-1/2}[T, \phi]$.

The remaining properties, (CO3)–(CO5), are easy to see if we make the obvious choice for the relation '<':

$$\phi_1 < \phi_2 \iff \phi_2 = 1 \text{ in a neighborhood of } \operatorname{supp}\phi_1. \tag{4.22}$$

To apply Lemma 4.3 to $T$, it only remains to show that $\phi T^{-1} \in C_p(L^2(F))$ for $\phi \in C_0^\infty(M)$ and some $p > 0$. So choose a compact manifold with boundary, $M_1 \subset M$, such that $\operatorname{supp}\phi \subset \mathring{M}_1$. Denote by $H_{M_1}$ the Friedrichs extension of $\mathcal{R}_\tau$ in $L^2(F \mid \mathring{M}_1)$; then it is well known that

$$H_{M_1}^{-1} \in C_p(L^2(F \mid \mathring{M}_1)) \quad \text{for all } p > m/2.$$

But $\phi T^{-1}$ maps into $\mathcal{D}(H_{M_1})$ and

$$H_{M_1}\phi T^{-1} = [T, \phi]T^{-1} + \phi - i\mathcal{I}_\tau\phi T^{-1}$$

is bounded in view of the arguments proving (CO2).

We have proved:

**Lemma 4.5.** *If $T$ is the Friedrichs extension of a second order differential operator with the properties (4.13) and (4.17) then the estimates (4.12) hold, for all $\phi_1, \phi_2 \in C_0^\infty(M)$ with $\phi_2 = 1$ in a neighborhood of* $\operatorname{supp} \phi_1$.

With these preparations we are ready to prove Theorem 2.2 and Lemma 2.6.

*Proof of Theorem 2.2.* Choose $\chi_1, \chi_2 \in C_0^\infty(B_{1/2}(0))$ with $\chi_1 > \chi_2$. By elliptic regularity, we have for $u \in \mathcal{D}(T_{\varepsilon,p})$

$$\chi_1 P_{\varepsilon,p}(t)\chi_2 u \in \mathcal{D}(T_{\varepsilon,p}),$$

and an easy calculation gives

$$(4.23a)\quad (\partial_t + T_{\varepsilon,p})\chi_1 P_{\varepsilon,p}(t)\chi_2 u = [T_{\varepsilon,p}, \chi_1]P_{\varepsilon,p}(t)\chi_2 u =: v(t),\ t > 0,$$

$$(4.23b)\quad \lim_{t\to 0+} \chi_1 P_{\varepsilon,p}(t)\chi_2 u = \chi_2 u.$$

Now, for fixed $\varepsilon$ and $p$ we may write, in view of (CO2) above,

$$(4.24)\qquad v(t) = A[(D^{E(p)})^2, \tilde{\chi}_1]e^{-\varepsilon^2 t(D^{E(p)})^2}((D^{E(p)})^2 + 1)^{-1}\tilde{u}),$$

for some bounded operator $A : L^2(S \otimes E(p)|B_\varepsilon(p)) \to L^2(B_1(0), \Lambda_p^* \otimes E_p)$, some function $\tilde{\chi}_1 \in C_0^\infty(B_\varepsilon(p))$, and some $\tilde{u} \in L^2(S \otimes E(p))$. The spectral theorem then readily shows that $v$ is Hölder continuous in $[0,1]$ with exponent $1/2$; thus we can apply [Ka, Thm. IX, 1.27] to conclude the representation

$$\chi_1 P_{\varepsilon,p}(t)\chi_2 = e^{-tT_{\varepsilon,p}}\chi_2 + \int_0^t e^{-(t-u)T_{\varepsilon,p}}[T_{\varepsilon,p}, \chi_1]P_{\varepsilon,p}(u)\chi_2\, du$$

$$= e^{-tT_{\varepsilon,p}}\chi_2 + \int_0^t e^{-(t-u)T_{\varepsilon,p}}T_{\varepsilon,p}{}^{\frac{1}{2}}(T_{\varepsilon,p}{}^{-\frac{1}{2}}[T_{\varepsilon,p}, \chi_1])(I - \chi_3)P_{\varepsilon,p}(u)\chi_2\, du,$$

for any $\chi_2$ with $\chi_1 > \chi_3 > \chi_2$ (cf. (CO3)).

Now we deduce from the Cauchy-Dunford representation analogous to (4.11) the estimate

$$\|e^{-(t-u)T_{\varepsilon,p}}T_{\varepsilon,p}{}^{1/2}\| \le c(t-u)^{-1/2},$$

whereas the second factor in the integral is bounded by (CO2). The third factor can be written as

$$(I - \chi_3)P_{\varepsilon,p}(u)\chi_2 = \Phi_{\varepsilon,p}^{-1}(I - \tilde{\chi}_{3,\varepsilon})e^{-\varepsilon^2 u(D^{E(p)})^2}\tilde{\chi}_{2,\varepsilon}\Phi_{\varepsilon,p} =: \Phi_{\varepsilon,p}^{-1}Q_{\varepsilon,p}\Phi_{\varepsilon,p},$$

with $\tilde{\chi}_{2,\varepsilon} < \tilde{\chi}_{3,\varepsilon}$ in $C^\infty(M)$. Since the self-adjoint operator $(D^{E(p)})^2$ equals its Friedrichs extension from $C^\infty(S \otimes E(p))$ and satisfies (4.13) and also (4.17), by self-adjointness, we get from Lemma 4.5

$$\|(I-\chi_3)P_{\varepsilon,p}(u)\chi_2\|_{C_1} \leq \|\Phi_{\varepsilon,p}^{-1}\|\|\Phi_{\varepsilon,p}\|C(\varepsilon,p,N)u^N.$$

The proof is complete. □

*Proof of Lemma 2.6.* This proof is achieved by exactly the same arguments as above for Theorem 2.2, with only a few modifications. In $L^2(\mathbb{R}^m, \Lambda_p^*\otimes E_p)$ we introduce the integral operator, $P(t)$, with kernel

$$P(t;s_1,s_2) := (4\pi t)^{-m/2}P_1(t)\,P_2(t)\,P_3(t;s_1,s_2), \tag{4.25}$$

$$P_1(t) := \left(\det\frac{2\Omega(p)t}{\sinh 2\Omega(p)t}\right)^{1/2}, \tag{4.26}$$

(4.27)

$$P_2(t) := \exp\Big(-\frac{1}{4t}\big\langle\frac{2\Omega(p)t}{\tanh 2\Omega(p)t}(s_1-s_2),s_1-s_2\big\rangle + 4it\langle\Omega(p)s_1,s_2\rangle\Big),$$

$$P_3(t;s_1,s_2) := \exp(-\Omega^E(p)t), \tag{4.28}$$

where $\Omega = (\Omega_{ij})$ and $\Omega^E$ are defined in Lemma 2.1. Then, for $\chi_1 \in C_0^\infty(B_{1/2}(0))$, it is easy to see that $\chi_1 P(t)$ maps into $\mathcal{D}(T_{0,p})$. Moreover, it follows from [BeGeVe, Ho] that $P(t)$ solves the heat equation associated with $\tau_{0,p}$ so we derive the analogue of (4.23).

The explicit formula (4.25) ensures Hölder continuity and off-diagonal decay so we arrive at

$$\|\chi_1(e^{-tT_{0,p}} - P(t))\chi_2\|_{C_1} \leq C(N,\chi_1,\chi_2)t^N,$$

for all $N \in \mathbb{N}$.

The proof of the lemma is completed recalling (2.20) and observing that $[P(t;s,s)]_{\mathrm{top}}$ is actually independent of $t$, hence can be evaluated at $t = (2\pi\sqrt{-1})^{-1}$. □

**5. Estimates for $e^{-tT_{\varepsilon,p}}$.** It remains to prove Theorems 2.3 and 2.5 which is the goal of this final section. The main point is to find a suitable representation of $e^{-tT_{\varepsilon,p}}$ which allows to read off the uniform estimates (and continuity properties) required in the proof of Theorem 2.5. This is achieved by comparing $T_{\varepsilon,p}$ with the Friedrichs extension, $T^0_{\varepsilon,p}$, of the principal part, $\tau^0_{\varepsilon,p}$ (cf. (2.15)). The advantage of this operator over $T_{0,p}$ lies in the fact that it is diagonal with respect to $\Lambda_p^*$; the corresponding Neumann series has enough structure to carry through the estimates we need.

Thus we have to consider the operators $T_{\varepsilon,p}$ and $T^0_{\varepsilon,p}$ in the coordinate Hilbert space $\mathcal{H} = L^2(B_{2/3}(0), \Lambda^*_p \otimes E_p)$, the Friedrichs extensions of the differential operators $\tau_{\varepsilon,p}$ and $\tau^0_{\varepsilon,p}$ defined in (2.12) and (2.14), respectively. It follows from standard elliptic theory and the definition of the Friedrichs extension (4.19) that

$$\mathcal{D}(T_{\varepsilon,p}) = \mathcal{D}(T^0_{\varepsilon,p}) = H^2(B_{2/3}(0), \Lambda^*_p \otimes E_p) \cap H^1_0(B_{2/3}(0), \Lambda^*_p \otimes E_p). \tag{5.1}$$

Thus we obtain for $u \in \mathcal{H}$

$$(\partial_t + T_{\varepsilon,p})e^{-tT^0_{\varepsilon,p}}u = - \sum_{\beta,\gamma,|\gamma|\le 1} M_\beta \partial^\gamma_s A^{\beta\gamma}_{\varepsilon,p} e^{-tT^0_{\varepsilon,p}}u, \tag{5.2a}$$

$$\lim_{t\to 0} e^{-tT^0_{\varepsilon,p}}u = u. \tag{5.2b}$$

Since $T^0_{\varepsilon,p}$ generates a holomorphic semigroup, the standard a priori inequality for $u \in \mathcal{D}(T^0_{\varepsilon,p})$ (cf. Lemma 4.4) gives for $|\gamma| \le 1$

$$\|\partial^\gamma_s e^{-tT^0_{\varepsilon,p}}\|_{\mathcal{H}} \le Ct^{-|\gamma|/2}, \tag{5.3}$$

(with $C$ independent of $\varepsilon$ and $p$) such that we obtain, as in the proof of Theorem 2.2,

$$e^{-tT_{\varepsilon,p}} = e^{-tT^0_{\varepsilon,p}} + \sum_{\beta,|\gamma|\le 1} \int_0^t e^{-(t-u)T_{\varepsilon,p}} M_\beta \partial^\gamma_s A^{\beta\gamma}_{\varepsilon,p} e^{-uT^0_{\varepsilon,p}}\,du. \tag{5.4}$$

This leads us to define, as in Theorem 2.3,

$$R^0_{\varepsilon,p}(t) = e^{-tT^0_{\varepsilon,p}}, \tag{5.5a}$$

$$R^{\nu+1}_{\varepsilon,p}(t) = \sum_{\beta,\gamma,|\gamma|\le 1} \int_0^t R^\nu_{\varepsilon,p}(t-u) M_\beta \partial^\gamma_s A^{\beta\gamma}_{\varepsilon,p} e^{-tT^0_{\varepsilon,p}}\,du. \tag{5.5b}$$

One expects that the resulting Neumann series converges, as expressed in Theorem 2.3.

*Proof of Theorem 2.3.* The theorem follows if we prove that for positive constants $C_0, C_1$, independent of $\varepsilon$ and $p$, we have the estimate

$$\|R^\nu_{\varepsilon,p}(t)\| \le C_0 C_1^\nu (\nu!)^{-1/2} t^{\nu/2}, \quad \nu \ge 0,\ 0 < t \le 1. \tag{5.5c}$$

This is obvious for $\nu = 0$, from (5.5a) and (5.3).

If the estimate holds for some $\nu \geq 0$ then (5.5b) and (5.3) give

$$\begin{aligned}
\|R^{\nu+1}_{\varepsilon,p}(t)\|_{\mathcal{H}} &\leq C_2 \int_0^t \|R^{\nu}_{\varepsilon,p}(t-u)\|_{\mathcal{H}} u^{-1/2} du \\
&\leq C_2 C_0 C_1^{\nu} (\nu!)^{-1/2} \int_0^t (t-u)^{\nu/2} u^{-1/2}\, du \\
&\leq C_0 C_2 C_1^{\nu} (\nu!)^{-1/2} t^{(\nu+1)/2} \frac{\Gamma(\nu/2+1)\Gamma(1/2)}{\Gamma(\nu/2+3/2)} \\
&\leq C_0 C_1^{\nu} C_2 C_3 ((\nu+1)!)^{-1/2} t^{(\nu+1)/2},
\end{aligned}$$

where we have used the asymptotics of the $\Gamma$-function. The assertion follows with $C_1 := C_2 C_3$ and $\|e^{-tT^0_{\varepsilon,p}}\|_{\mathcal{H}} =: C_0$. □

We will need trace estimates, however, to achieve the proof of Theorem 2.5. In what follows we work with the von Neumann-Schatten classes of compact operators in $\mathcal{H}$; we denote them, as before, by $C_q(\mathcal{H})$ and write $\|\cdot\|_q := \|\cdot\|_{C_q(\mathcal{H})}$, for $q > 0$. In particular, we use $\|\cdot\|_\infty := \|\cdot\|_{\mathcal{H}}$.

**Lemma 5.1.** *There are positive constants $C_4, C_5$, independent of $\varepsilon$ and $p$, such that*

$$\|R^{\nu}_{\varepsilon,p}(t)\|_1 \leq C_4 C_5^{\nu} ((\nu+1)!)^{-1/2} t^{(\nu-m)/2}. \tag{5.6}$$

*Proof.* For $\nu = 0$, we recall that $T^0_{\varepsilon,p}$ is actually scalar and hence satisfies the kernel estimate, for $s_1, s_2 \in B_{2/3}(0)$,

$$|e^{-tT^0_{\varepsilon,p}}(s_1, s_2)| \leq C t^{-m/2} e^{-C'|s_1-s_2|^2/t}, \tag{5.7}$$

again with $C, C'$ independent of $\varepsilon$ and $p$ (cf. eg. [Da, p. 89]). This implies the estimates

$$\|e^{-tT^0_{\varepsilon,p}}\|_2 \leq C t^{-m/4}, \tag{5.8a}$$

$$\|e^{-tT^0_{\varepsilon,p}}\|_1 \leq \|e^{-\frac{1}{2}tT^0_{\varepsilon,p}}\|_2^2 \leq C t^{-m/2}, \tag{5.8b}$$

and with (5.3), for $|\gamma| \leq 1$,

$$\|\partial_s^{\gamma} e^{-tT^0_{\varepsilon,p}}\|_1 \leq \|\partial_s^{\gamma} e^{-\frac{1}{2}tT^0_{\varepsilon,p}}\|_\infty \|e^{-\frac{1}{2}tT^0_{\varepsilon,p}}\|_1 \leq C t^{-(m+|\gamma|)/2}. \tag{5.8c}$$

Hence the case $\nu = 0$ is settled by (5.8b).

If the assertion is true for some $\nu \geq 0$ we use (5.5b) again, now splitting the integral in order to avoid nonintegrable singularities:

$$\begin{aligned}
\|R^{\nu+1}_{\varepsilon,p}(t)\|_1 &\leq C_2\Big\{ \int_0^{t/2} \|R^{\nu}_{\varepsilon,p}(t-u)\|_1 u^{-|\gamma|/2}\,du \\
&\quad + \int_{t/2}^{t} C_0 C_1^{\nu}(\nu!)^{-1/2}(t-u)^{\nu/2}\|\partial_s^{\gamma} e^{-uT^0_{\varepsilon,p}}\|_1\,du\Big\} \\
&\leq C_2(\nu!)^{-1/2}\Big\{C_4C_5^{\nu}t^{(\nu-m+2-|\gamma|)/2}2^kB(\nu/2+1,1/2) \\
&\quad + C_0C_1^{\nu}t^{(\nu-m+2-|\gamma|)/2}2^kB(\nu/2+1,1/2)\Big\} \\
&\leq t^{(\nu+1-m)/2}((\nu+1)!)^{-1/2}C_2 2^k\big[C_3C_4C_5^{\nu} + C_0C_3C_1^{\nu}\big] \\
&\leq C_4C_5^{\nu+1}((\nu+1)!)^{-1/2}t^{(\nu+1-m)/2},
\end{aligned}$$

if $C_4, C_5$ are chosen appropriately. □

From Lemma 5.1, we derive the interesting consequence that

$$\begin{aligned}
\Big\|\eta_t \sum_{\nu=N}^{\infty} R^{\nu}_{\varepsilon,p}(t)\Big\|_1 &\leq C_4 t^{-m}\sum_{\nu\geq N}\frac{(C_5^2 t)^{\nu/2}}{(\nu!)^{1/2}}\|\eta\|_{L^\infty} \\
&\leq Ct^{1/2},
\end{aligned} \tag{5.9}$$

if $N > 2m$. Consequently, the proof of Theorem 2.5 depends on properties of only finitely many $R^{\nu}$. To examine those, we will now heavily invoke the weight structure of $\tau_{\varepsilon,p}$ as expressed in (2.13c). To use it, we will have to rewrite the basic recursion (5.5b) in a more complicated fashion. It has to incorporate the multiplication operators $M_\alpha$, a cut-off in the $t$-variable, and also 'retarded arguments'. More precisely, we propose to estimate inductively the $q$-norms of the operator family

$$\tilde{\eta}_t R^{\nu}_{\varepsilon,p}(u)M_\alpha, \quad 0 < u \leq t \leq 1,\ \varepsilon \in (0,1],\ p \in M,\ \alpha \in \mathbb{Z}^m_+,\ \nu \in \mathbb{Z}_+, \tag{5.10}$$

where we have written $\eta_t =: t^{-m/2}\tilde{\eta}_t$, i.e. $\tilde{\eta}_t(x) = \eta(x/\sqrt{t})$.

We will prove:

**Theorem 5.2.** *There are positive constants $C_\nu$, independent of $\varepsilon$ and $p$, such that*

$$\|\big[\tilde{\eta}_t R^{\nu}_{\varepsilon,p}(u)M_\alpha\partial_s^{\alpha'}\big]_{IJ}\|_q \leq C_{\nu\alpha q}t^{(|\mu(I)-\mu(J)|+|\alpha|)/2}(t/u)^{m/2q}u^{-|\alpha'|/2}, \tag{5.11}$$

*for $0 < u \leq t \leq 1$, $\varepsilon \in (0,1]$, $p \in M$, $\alpha,\alpha' \in Z^n_+$, $|\alpha'| \leq 1$, $I,J \subset \{1,\dots,m\}$, and $0 < q \leq \infty$.*

Let us remark that (5.11) does not follow from (5.5c) in the case $u = t$, $\alpha = 0$, $q = \infty$. Moreover, with Lemma 5.1 we have the following interesting estimate for the heat kernel.

**Corollary 5.3.** *For* $0 < u \le t \le 1$, $\alpha, \alpha' \in \mathbb{Z}_+^m$, $|\alpha'| \le 1$, *we have*

$$(5.12) \qquad \left\| \left[\tilde\eta_t e^{-uT^0_{\varepsilon,p}} M_\alpha \partial_s^{\alpha'}\right]_{IJ} \right\|_1 \le C_\alpha t^{(|\mu(I)-\mu(J)+|\alpha|)/2} (t/u)^{m/2} u^{-|\alpha'|/2},$$

*and, in particular,*

$$(5.13) \qquad \left\| \left[\eta_t e^{-tT_{\varepsilon,p}}\right]_{\mathrm{top}} \right\|_1 \le C.$$

For the proof of Theorem 5.2 we need several lemmas to which we turn now.

The presence of the factor $M_\alpha$ on the right of (5.5c) creates the operators $\partial_s^\gamma e^{-tT^0_{\varepsilon,p}} M_\alpha$ in the recursion, and we are forced to move $M_\alpha$ to the left. Thus we want to write (suppressing the dependence on $\varepsilon$, $p$ in the right hand side for simplicity)

$$(5.14) \qquad e^{-tT^0_{\varepsilon,p}} M_\alpha =: \sum_{\delta \le \alpha} M_\delta S_\delta^\alpha(t),$$

such that $S_0^0(t) = e^{-tT_{\varepsilon,p}} = S_\alpha^\alpha(t)$. Clearly, $S_\delta^\alpha(t)$ is an integral operator with kernel

$$(5.15) \qquad c_{\alpha\delta}(s_1 - s_2)^{\alpha-\delta} e^{-tT^0_{\varepsilon,p}}(s_1, s_2)$$

which allows to estimate the $C_q$-norms for $2 \le q \le \infty$, in view of (5.7). To incorporate derivatives, we argue as often before to derive the representation

$$M_\alpha e^{-tT^0_{\varepsilon,p}} = e^{-tT^0_{\varepsilon,p}} M_\alpha + \int_0^t e^{-(t-u)T^0_{\varepsilon,p}} [\tau^0_{\varepsilon,p}, M_\alpha] e^{-uT^0_{\varepsilon,p}}\, du = e^{-tT^0_{\varepsilon,p}} M_\alpha$$
$$- \sum_{\beta, |\gamma| = |\gamma'+\gamma''| = 2, |\gamma''| \ge 1} \int_0^t e^{-(t-u)T^0_{\varepsilon,p}} M_{\alpha-\gamma''} C_{\gamma\gamma'} \partial_s^{\gamma'} A^\gamma_{\varepsilon,p} e^{-uT^0_{\varepsilon,p}}\, du,$$

with certain constants $C_{\gamma\gamma'}$, or for $\delta < \alpha$

(5.16)

$$S_\delta^\alpha(t) = \sum_{\beta, |\gamma'+\gamma''| = 2, |\gamma''| \ge 1, \delta \le \alpha - \gamma''} \int_0^t S_\delta^{\alpha+\gamma''}(t-u) C_{\gamma\gamma'} \partial_s^{\gamma'} A^\gamma_{\varepsilon,p} e^{-uT^0_{\varepsilon,p}}\, du.$$

With this notation established, we now obtain from (5.5b) the following recursion:

$$(5.17) \qquad \left[\tilde\eta_t R^{\nu+1}_{\varepsilon,p}(u) M_\alpha \partial_s^{\alpha'}\right]_{IJ} =$$
$$= \sum_{\beta, |\gamma| \le 1, \delta, K} \int_0^u \left[\tilde\eta_t R^\nu_{\varepsilon,p}(u-v) M_\beta \partial_s^\gamma M_\delta\right]_{IK} \left[A^{\beta\gamma}_{\varepsilon,p} S^\alpha_\delta(v) \partial_s^{\alpha'}\right]_{KJ} dv$$
$$= \sum_{\beta, |\gamma| \le 1, \delta, K} \int_0^u \left[\tilde\eta_t R^\nu_{\varepsilon,p}(u-v)(M_{\beta+\delta} \partial_s^\gamma \right.$$
$$\left. + \langle\gamma, \delta\rangle M_{\beta+\delta-\gamma})\right]_{IK} \left[A^{\beta\gamma}_{\varepsilon,p} S^\alpha_\delta(v) \partial_s^{\alpha'}\right]_{KJ} dv.$$

Thus we have to establish estimates on $S^\alpha_\delta(v)\partial_s^{\alpha'}$ first.

**Lemma 5.4.** *We have the estimates, uniformly in $\varepsilon$ and $p$,*

$$\|S^\alpha_\delta(v)\partial_s^{\alpha'}\|_\infty \le C_\alpha v^{(|\alpha-\delta|-|\alpha'|)/2}, \tag{5.18}$$

$$\|\tilde\eta_t S^\alpha_\delta(v)\partial_s^{\alpha'}\|_q \le C_{\alpha q} t^{|\alpha-\delta|/2}(t/v)^{m/2q} v^{-|\alpha|/2}, \tag{5.19}$$

$$\|\tilde\eta_t S^\alpha_\delta(v)\partial_s^{\alpha'}(1-\tilde\eta^1_t)\|_q \le C_{\alpha q} t^{|\alpha-\delta|/2} v^{-|\alpha|/2}. \tag{5.20}$$

*Here $0 < q \le \infty$, $\delta, \alpha, \alpha' \in \mathbb{Z}^m_+$ with $\delta \le \alpha$ and $|\alpha'| \le 1$, $0 < v \le t \le 1$, and $\eta^1 > \eta$ in the sense of (4.22).*

*Proof.* (5.18) for $\alpha = 0$ follows from (5.3) by taking adjoints, and in general it follows easily, by induction on $|\alpha|$, from (5.16).

The estimates (5.19) and (5.20) are proved together by induction on $|\alpha|$. For (5.19) with $\alpha = 0$ we prepare the estimates

$$\|\tilde\eta_t e^{-uT^0_{\varepsilon,p}}\|_2 \le C(t/u)^{m/4}, \tag{5.21a}$$

$$\|\tilde\eta_t e^{-uT^0_{\varepsilon,p}}\tilde\eta^1_t\|_1 \le \|\tilde\eta_t e^{-u/2T^0_{\varepsilon,p}}\|_2 \|e^{-u/2T^0_{\varepsilon,p}}\tilde\eta^1_t\|_2 \le C(t/u)^{m/2}, \tag{5.21b}$$

$$\|\tilde\eta_t e^{-uT^0_{\varepsilon,p}}(1-\tilde\eta^1_t)\|_2 \le C_N(u/t)^N, \quad N \in \mathbb{N}, \tag{5.21c}$$

which are easy consequences of (5.7). To proceed we write the commutator $[\tilde\eta_t, e^{-uT^0_{\varepsilon,p}}]$ in the now familiar way to obtain the representation

$$\begin{aligned}\tilde\eta_t e^{-uT^0_{\varepsilon,p}}(1-\tilde\eta^1_t) &= -\sum_{\beta,|\gamma|=2}\int_0^u \tilde\eta_t e^{-(u-v)T^0_{\varepsilon,p}} M_\beta[\partial_s^\gamma, \tilde\eta^2_t] A^\gamma_{\varepsilon,p} e^{-vT^0_{\varepsilon,p}}(1-\tilde\eta^1_t)\,dv \\ &=: \sum_{\substack{\beta,|\gamma|=2\\ \delta\le\beta,|\gamma'|\le 1}}\int_0^u \tilde\eta_t e^{-(u-v)T^0_{\varepsilon,p}} M_\delta S^\beta_\delta((u-v)/2)\partial_s^{\gamma'}\tilde\eta^3_t C^{\gamma\gamma'}_{\varepsilon,p}(t) e^{-vT^0_{\varepsilon,p}}(1-\tilde\eta^1_t)\,dv\end{aligned} \tag{5.22}$$

where $\eta < \eta^2 < \eta^3 < \eta^1$ and

$$\|C^{\gamma\gamma'}_{\varepsilon,p}(t)\|_{L^\infty(B_{2/3},\operatorname{End}\Lambda^*_p\otimes E_p)} \le Ct^{(|\gamma|-2)/2}. \tag{5.23}$$

Next we get from (5.21a) by interpolation

$$\|\tilde\eta_t e^{-(u-v)/T^0_{\varepsilon,p}}\|_{2m} \le C(t/(u-v))^{1/4}. \tag{5.24}$$

Using (5.24), (5.21c), (5.18), and (5.23) together with the Hölder inequality for Schatten norms in (5.22) we arrive at

$$\|\tilde\eta_t e^{-uT^0_{\varepsilon,p}}(1-\tilde\eta^1_t)\|_{2m/(m+1)} \le C_N(u/t)^N. \tag{5.25}$$

Upon iteration we see that the same estimate holds for the norm of order $2m/(m+L)$, $L \in \mathbb{N}$, hence, for all $q > 0$,

$$\|\tilde{\eta}_t e^{-uT^0_{\varepsilon,p}}(1-\tilde{\eta}^1_t)\|_q \le C_{N,q}(u/t)^N. \tag{5.26}$$

Thus, with (5.21b),

$$\|\tilde{\eta}_t e^{-uT^0_{\varepsilon,p}}\|_1 \le C(t/u)^{m/2}. \tag{5.27}$$

Now we can iterate as in the derivation of (5.21b), using the Hölder inequality for Schatten norms, to derive (5.19) and (5.20) with $\alpha = \alpha' = 0$.

Next, from (5.19) (with $\alpha = \alpha' = 0$) and (5.3) we get

$$\begin{aligned}\|\tilde{\eta}_t e^{-uT^0_{\varepsilon,p}} \partial_s^{\alpha'}\|_q &\le \|\tilde{\eta}_t e^{-u/2T^0_{\varepsilon,p}}\|_q \|e^{-u/2T^0_{\varepsilon,p}}\partial_s^{\alpha'}\|_\infty \\ &\le C(t/u)^{m/2q}u^{-|\alpha'|/2},\end{aligned}$$

which is (5.19) for $\alpha = 0$.

For (5.20) with $\alpha = 0$, we use the representation following from (5.22) again, starting this time with the estimate

$$\|\tilde{\eta}_t e^{-uT^0_{\varepsilon,p}} \partial_s^{\alpha'}\|_{2m} \le C(t/u)^{1/4}u^{-|\alpha'|/2}.$$

This leads to

$$\|\tilde{\eta}_t e^{-uT^0_{\varepsilon,p}} \partial_s^{\alpha'}(1-\tilde{\eta}^1_t)\|_m \le Cu^{-|\alpha'|/2},$$

which gives (5.20) by iteration.

We turn to the inductive step and note that (5.19) with $q = \infty$ follows easily from (5.16) and the induction hypothesis. For general $q$, we use again (5.16) splitting it by writing $A^\gamma_{\varepsilon,p} = ((1-\tilde{\eta}^2_t)+\tilde{\eta}^2_t)A^\gamma_{\varepsilon,p}$. The $q$-norm of the first integral is estimated using (5.19) and the induction hypothesis for (5.20). The second integral is split at $u/2$, using (5.20) on the first and (5.18) on the second factor in the integral from 0 to $u/2$, and the other way around in the integral from $u/2$ to $u$. This, clearly, completes the induction for (5.19).

An entirely analogous estimate gives (5.20). □

With these preparations we can give the

*Proof of Theorem 5.2.* The proof is by induction on $\nu$, using the recursion (5.17), the induction hypothesis is formed by the estimate (5.11) and the parallel estimate

$$\left\|\left[\tilde{\eta}_t R^\nu_{\varepsilon,p}(u)M_\alpha \partial_s^{\alpha'}(1-\tilde{\eta}^1_t)\right]_{IJ}\right\|_q \le C_{\nu\alpha q}t^{(|\mu(I)-\mu(J)|+|\alpha|)/2}u^{-|\alpha'|/2}. \tag{5.11'}$$

For $\nu = 0$ we write

$$\tilde{\eta}_t R^0_{\varepsilon,p}(u) M_\alpha \partial_s^{\alpha'} = \sum_{\delta \leq \alpha} M_\delta \tilde{\eta}_t S^\alpha_\delta(u) \partial_s^{\alpha'}.$$

Since $T^0_{\varepsilon,p}$ is scalar, (5.11) and (5.11′) follow in this case from (5.19) and (5.20).

To establish the assertion (5.11) for $\nu + 1$, we insert $1 = 1 - \tilde{\eta}^1_t + \tilde{\eta}^1_t$ in (5.17) in front of $A^{\beta\gamma}_{\varepsilon,p}$, $\eta^1 > \eta$, and estimate the two resulting terms separately. For the $q$-norm of the first term we obtain the bound

$$\begin{aligned}
\sum_{\beta,\gamma,\delta \leq \alpha, K} \int_0^u & C_{\nu\alpha} t^{(|\mu(I)-\mu(K)|+|\beta+\delta|)/2} (u-v)^{-|\gamma|/2} v^{(|\alpha-\delta|-|\alpha'|)/2}\, dv \\
&\leq \sum_{\beta,\gamma,\delta \leq \alpha, K} C_{\nu\alpha} t^{(|\mu(I)-\mu(K)|+|\beta+\delta|)/2} u^{(|\alpha-\delta|-|\alpha'|-|\gamma|+2)/2} \\
&\leq \sum_{\beta,\gamma,\delta \leq \alpha, K} C_{\nu\alpha} t^{(|\mu(I)-\mu(K)|+|\alpha|+|\beta|-|\gamma|+2)/2} u^{-|\alpha'|/2} \\
&\qquad\qquad\leq C_{\nu\alpha} t^{(|\mu(I)-\mu(J)|+|\alpha|)/2} u^{-|\alpha'|/2},
\end{aligned}$$

since in a nonzero term in (5.17) we have, by (2.13c),

$$|\beta| - |\gamma| + 2 \geq |\mu(K) - \mu(J)|,$$

hence

$$|\mu(I) - \mu(K)| + |\beta| - |\gamma| + 2 \geq |\mu(I) - u(J)|.$$

In the second term we split again the integral at $u/2$, using the $q$-norm on the factor whose norm remains integrable. For the integral from 0 to $u/2$ the $q$-norm is bounded analogously by

$$\begin{aligned}
C_{\nu\alpha} \sum_{\beta,\gamma,\delta,K} \int_0^{u/2} & t^{(|\mu(I)-\mu(K)|+|\beta+\delta|)/2} (t/(u-v))^{m/2q} \cdot \\
&\cdot (u-v)^{-|\gamma|/2} t^{|\alpha-\delta|/2} v^{-|\alpha'|/2}\, dv \\
\leq C_{\nu\alpha} \sum_{\beta,\gamma,\delta,K} & t^{(|\mu(I)-\mu(K)|+|\alpha|+|\beta|)/2} (t/u)^{m/2q} u^{(2-|\gamma|-|\alpha'|)/2} \\
&\leq C_{\nu\alpha} t^{(|\mu(I)-\mu(J)|+|\alpha|)/2} (t/u)^{m/2q} u^{-|\alpha'|/2}.
\end{aligned}$$

The second integral is estimated analogously by the same quantity; this establishes (5.11) for $\nu + 1$.

The splitting with $\tilde{\eta}^1_t$ leads, by the same arguments, to (5.11′). This completes the proof. □

It remains to discuss the continuity properties of $R^\nu_{\varepsilon,p}$ as a function of $\varepsilon$. This we do first in the $C^2$ case; then an approximation argument will handle the $C^{1,1}$ case. We now use the fact that we have obtained estimates for all $q$-norms with $q > 0$. Thus we obtain from (4.8) and Theorem 5.2

$$\begin{aligned}(5.28)\quad &\left|\operatorname{tr}\left[\eta_t(R^\nu_{\varepsilon,p} - R^\nu_{0,p})(t)\right]_{\mathrm{top}}\right| \\ &\leq t^{-m/2}\|\tilde\eta_t(R^\nu_{\varepsilon,p} - R^\nu_{0,p})(t)\|^{1/2}_\infty \|\tilde\eta_t(R^\nu_{\varepsilon,p} - R^\nu_{0,p})(t)\|^{1/2}_{1/2} \\ &\leq C_\nu \|\tilde\eta_t(R^\nu_{\varepsilon,p} - R^\nu_{0,p})(t)\|^{1/2}_\infty t^{-m/4}.\end{aligned}$$

It is, therefore, enough to prove continuity in the operator norm of $R^\nu_{\varepsilon,p}$ at $\varepsilon = 0$; more precisely, we need

**Lemma 5.5.** *Under the regularity condition* (2.13b′), *there are positive functions* $C_{\nu\alpha}(\varepsilon)$, $\nu \in \mathbb{Z}^m_+$, $\varepsilon \in (0,1]$, *with* $\lim_{\varepsilon\to 0} C_{\nu\alpha}(\varepsilon) = 0$ *and*

$$(5.29)\quad \left\|\left[\tilde\eta_t(R^\nu_{\varepsilon,p} - R^\nu_{0,p})(u) M_\alpha \partial^{\alpha'}_s\right]_{IJ}\right\|_\infty \leq C_{\nu\alpha}(\varepsilon) t^{(|\mu(I)-\mu(J)|+|\alpha|)/2} u^{-|\alpha'|/2},$$

*for* $0 < u \leq t \leq 1$, $p \in M$, $I, J \subset \{1,\dots,m\}$, *and* $\alpha' \in \mathbb{Z}^m_+$ *with* $|\alpha'| \leq 1$.

*Proof.* A straightforward estimate of the right hand side in (5.17), using the smoothness assumption (2.13b), shows that the lemma follows inductively from the following facts:

(1) the estimate (5.19) with $\nu = 0$,
(2) the estimate

$$(5.30)\quad \|(S^\alpha_\delta(\varepsilon; v) - S^\alpha_\delta(0; v))\partial^{\alpha'}_s\|_\infty \leq C_\alpha(\varepsilon) v^{(|\alpha-\delta|-|\alpha'|)/2},$$

where the function $C_\alpha$ satisfies $\lim\limits_{\varepsilon\to 0} C_\alpha(\varepsilon) = 0$.

Using induction on $|\alpha|$ as before, we see that (5.29) follows readily from (5.16) once it is proved for $\alpha = 0$. To achieve this we invoke once more the solution of the inhomogeneous heat equation to obtain the identity

$$\begin{aligned}(5.31)\quad &(e^{-uT^0_{\varepsilon,p}} - e^{-uT^0_{0,p}})\partial^{\alpha'}_s \\ &= \sum_{|\gamma|=2} \int_0^u e^{-(u-v)T^0_{0,p}} \partial^\gamma_s (A^\gamma_{\varepsilon,p} - A^\gamma_{0,p}) e^{-vT^0_{\varepsilon,p}} \partial^{\alpha'}_s \, dv.\end{aligned}$$

Since $A^\gamma_{\varepsilon,p}$ is $C^1$ in $\varepsilon$ and since the operators

$$(T^0_{0,p})^{-1} \sum_{|\gamma|=2} \partial^\gamma_s (A^\gamma_{\varepsilon,p} - A^\gamma_{0,p}) \quad \text{and} \quad \sum_{|\gamma|=2} \partial^\gamma_s (A^\gamma_{\varepsilon,p} - A^\gamma_{0,p})(T^0_{\varepsilon,p})^{-1}$$

are bounded, we obtain the desired estimate by splitting again the integral at $u/2$.

It remains only to prove (5.29) for $\nu = 0$. We have

$$\tilde{\eta}_t(e^{-uT^0_{\varepsilon,p}} - e^{-uT^0_{0,p}})M_\alpha \partial_s^{\gamma'} = \sum_{\delta \le \alpha} M_\delta \tilde{\eta}_t(S^\alpha_\delta(\varepsilon; u) - S^\alpha_\delta(0; u))\partial_s^{\gamma'},$$

so that the estimate follows from (5.30). The lemma is proved. □

We conclude this section with filling the last gaps in the proof.

*Proof of Theorem 2.5 in the $C^2$ case.* From Corollary 5.3 we get

$$\|[\eta_t(\zeta_{\varepsilon,p} - 1)e^{-tT_{\varepsilon,p}}]_{\mathrm{top}}\|_1 \le C \sup_{p,s} |\zeta_{\varepsilon,p}(s) - 1| =: \phi_1(\varepsilon),$$

with $\lim_{\varepsilon \to 0} \phi_1(\varepsilon) = 0$. Furthermore, using (5.9) and (5.29) we find

$$\begin{aligned} &\left\| \left[\eta_t(e^{-tT_{\varepsilon,p}} - e^{-tT_{0,p}})\right]_{\mathrm{top}} \right\|_1 \\ &\qquad \le t^{-m/2} \sum_{\nu=0}^{2m} \left\| \left[\tilde{\eta}_t(R^\nu_{\varepsilon,p} - R^\nu_{0,p})(t)\right]_{\mathrm{top}} \right\|_1 + Ct^{1/2} \\ &\qquad\qquad \le \sup_{\nu \le 2m} C_{\nu_0}(\varepsilon) + Ct^{1/2} =: \phi_2(\varepsilon) + Ct^{1/2}, \end{aligned}$$

with $\lim_{\varepsilon \to 0} \phi_2(\varepsilon) = 0$. This completes the proof. □

We remark that in this case the convergence is even in $L^\infty(M)$.

Finally, we deal with the $C^{1,1}$ case using an approximation argument. For this, it is useful to rephrase the results of this section a little bit. So we look at all operators, $\tau$, satisfying the conditions (2.13) in $L^2(B_{2/3}(0), F)$, for some finite dimensional complex vector space $F$. We denote this set by $\mathcal{T}$ and introduce the norm

$$\|\tau\|_{1,1} := \sum_{\beta,\gamma,\delta \le \gamma} \|\frac{\partial^{|\delta|}}{\partial s^\delta} A^{\beta\gamma}(s)\|_{L^\infty(B_{2/3}(0))} + \lambda_1^{-1} + \lambda_2. \tag{5.32}$$

For $\tau \in \mathcal{T}$, we let $T$ again be the Friedrichs extension.

Then we have, actually, proved the following facts.

**Corollary 5.6.** (1) *For $\tau \in \mathcal{T}$ we have an estimate*

$$\|[\eta_t e^{-tT}]_{\mathrm{top}}\|_1 \le C = C(\|\tau\|_{1,1}), \quad \textit{uniformly in } t \le 1.$$

(2) *For any constants $\varepsilon > 0, C_1 > 0$ we can find $\delta > 0$ such that*

$$\|\tau\|_{1,1} + \|\tau'\|_{1,1} \le C_1$$

*and*

$$\|\tau - \tau'\|_{1,1} \le \delta$$

*imply, uniformly in* $t \le 1$,

$$\left\| \left[\eta_t(e^{-tT} - e^{-tT'})\right]_{\text{top}} \right\|_1 \le \varepsilon.$$

*Proof.* We simply have to go through the proofs of Theorem 5.2 and Lemma 5.5 to see that they work literally in this situation, too. □

This allows us to give the

*Proof of Theorem 2.5 in the* $C^{1,1}$ *case.* We choose a $C^2$ approximation of all data which locally approximates in the norm (5.32). This leads to functions $F_t^E$, $F_t^{n,E}$ according to (1.11) and also $F^E$, $F^{n,E}$ where

$$F^E(p) = F^E(p;t) = \text{tr}[\eta_t e^{-tT_{0,p}}]_{\text{top}},$$

and $F^{n,E}(p)$ is defined similarly. Then we derive from Corollary 5.6 that, for given $\varepsilon > 0$,

$$|F_t^E(p) - F_t^{n,E}(p)| \le \varepsilon, \tag{5.33a}$$

if $n \ge n(\varepsilon)$ and $t \le t(\varepsilon)$, uniformly in $p$. Next, it follows from the first case of Theorem 2.5 that

$$\lim_{t\to 0} |F_t^{n,E}(p) - F^{n,E}(p)| = 0, \tag{5.33b}$$

uniformly in $p$ but not necessarily in $n$. Finally, another application of Corollary 5.6 shows that

$$|F^E(p) - F^{n,E}(p)| \le C$$

for some constant $C$, uniformly in $n$ and a.e. in $p$, and

$$\lim_{n\to\infty} |F^E(p) - F^{n,E}(p)| = 0$$

for almost all $p$. Thus we derive from Lebesgue's theorem

$$\lim_{n\to\infty} \|F^E - F^{n,E}\|_{L^1(M)} = 0. \tag{5.33c}$$

The estimates (5.34) together yield

$$\lim_{t\to 0} F_t^E = F^E \quad \text{in} \quad L^1(M)$$

as desired. □

## References


[Al] L. Alvarez-Gaumé, *Supersymmetry and the Atiyah-Singer index theorem*, Comm. Math. Phys. **90** (1983), 161–173.

[AS] M. F. Atiyah and I. M. Singer, *The index of elliptic operators on compact manifolds*, Bull. Amer. Math. Soc. **69** (1963), 422–433.

[ABP] M. F. Atiyah, R. Bott and V. K. Patodi, *On the heat equation and the index theorem*, Inv. Math. **19** (1973), 279–330.

[Br] J. Brüning, *$L^2$-index theorems for complete manifolds of rank one type*, Duke Math. J. **66** (1992), 257–309.

[BeGeVe] N. Berline, E. Getzler and M. Vergne, *Heat kernels and Dirac operators*, Springer Verlag, Berlin, Heidelberg, New York, 1992.

[CSuT] A. Connes, D. Sullivan and N. Teleman, *Local formulas for topological Pontryagin classes*, C. R. Acad. Sci. Paris **317** (1993), 521–526.

[Da] E. B. Davies, *Heat kernels and spectral theory*, Cambridge Univ. Press, Cambridge, 1989.

[FW] D. Friedan and P. Windy, *Supersymmetric derivation of the Atiyah-Singer index theorem*, Nuclear Phys. B **235** (1984), 394–416.

[Ge] E. Getzler, *A short proof of the local Atiyah-Singer index theorem*, Topology **25** (1986), 111–117.

[Gi] P. B. Gilkey, *Invariance theory, the heat equation, and the Atiyah-Singer index theorem*, Publish or Perish, Wilmington, 1984.

[Gr] P. Greiner, *An asymptotic expansion for the heat equation*, Arch. Rat. Mech. Anal. **41** (1971), 163–218.

[H] J. Hadamard, *Lectures on Cauchy's problem in linear partial differential equations*, New York, 1952.

[Ka] T. Kato, *Perturbation theory for linear operators*, Springer Verlag, Berlin, Heidelberg, New York, 1976, Second edition.

[Ko] T. Kotake, *An analytic proof of the classical Riemann-Roch theorem*, Proc. Symp. Pure Math. **16** (1970), 137–146.

[LM] B. Lawson and M.-L. Michelsohn, *Spin geometry*, Princeton University Press, Princeton, 1989.

[McKS] H. P. McKean and I. M. Singer, *Curvature and the eigenvalues of the Laplacian*, J. Diff. Geom. **1** (1967), 43–69.

[MiPl] S. Minakshisundaram and Å. Pleijel, *Some properties of the eigenfunctions of the Laplace operator on Riemannian manifolds*, Canadian J. Math. **1** (1949), 242–256.

[MoWu] H. Moscovici and F.-B. Wu, *Localization of topological Pontrjagin classes via finite propagation speed*, Geom. Funct. Anal. **4** (1994), 51–92.

[P1] V. K. Patodi, *Curvature and the eigenforms of the Laplace operator*, J. Diff. Geom. **5** (1971), 233–249.

[P2] ———, *An analytic proof of the Riemann-Roch Hirzebruch theorem for Kähler manifolds*, J. Diff. Geom. **5** (1971), 251–283.

[Se] R. T. Seeley, *Complex powers of an elliptic operator* Proc. Symp. Pure Math. **10** (1967), 288–307.

[Si] B. Simon, *Patodi's proof of the Gauss-Bonnet-Chern theorem and superproofs of index theorems*, Chapter 12, pp. 238–299, of H. L. Cycon, R. G. Froese, W. Kirsch and B. Simon, *Schrödinger operators*, Springer Verlag, 1987.



Institut für Mathematik, Humboldt-Universität zu Berlin, Unter den Linden 6, D-10099 Berlin, Germany

*E-mail address*: bruening@mathematik.hu-berlin.de

# HOW IDEAS FROM MICROLOCAL ANALYSIS CAN BE APPLIED IN 2-D FLUID MECHANICS

JEAN-YVES CHEMIN

**Introduction.** In this text, we try to explain how some concepts developed in nonlinear microlocal analysis can be used for solving problems in 2-D fluid mechanics. First, we are going to present some classical and basic facts about 2-D fluid mechanics. After this, we are going to introduce the classical vortex patch problem. Then, before explaining why nonlinear microlocal analysis has something to do with fluid mechanics, we shall recall the way in which a part of microlocal analysis became nonlinear.

**1. A short introduction to Euler's system.** We want to describe the evolution of a perfect incompressible fluid between time $t_0$ and time $t_1$. A particle of the fluid located at a point $x$ at time $t_0$ is located at a point $\psi_1(x)$ at time $t_1$. The incompressibility of the fluid appears through the fact that the jacobian determinant of $\psi_1$ is one. In all this text, we shall neglect boundary effects, which means that we shall essentially consider fluids in the whole space with two dimensions.

To model a perfect incompressible fluid, let us consider the space $\mathcal{L}$ of all continuously differentiable functions from $[t_0, t_1] \times \mathbf{R}^d$ to $\mathbf{R}^d$ so that $\psi(t_0) = \mathrm{Id}$ and $\psi(t_1) = \psi_1$ and such that, at any time $t$, the function $\psi(t)$ is a diffeomorphism of $\mathbf{R}^d$ and the function $\partial_t \psi(t)$ is continuous from $[t_0, t_1]$ to $L^2$. Let us denote by $\mathcal{L}_0$ the space of all functions of $\mathcal{L}$ so that, at any time $t$, the diffeomorphism $\psi(t)$ preserves the Lebesgue measure on $\mathbf{R}^d$.

An a priori possible evolution of an incompressible fluid between time $t_0$ and time $t_1$ is described by a function $\psi$ belonging to the space $\mathcal{L}_0$. Now, let us define the functional of action.

**Definition 1.1.** The action is the functional $\mathcal{A}$ defined from $\mathcal{L}$ to $\mathbf{R}_+$ by

$$\mathcal{A}(\psi) = \frac{1}{2} \int_{t_0}^{t_1} \int_{\mathbf{R}^d} |\partial_t \psi(t,x)|^2 \, dx \, dt.$$

The action is a quadratic form. The space $\mathcal{L}$ is included in an affine space. We have

$$D\mathcal{A}(\psi)h = \int_{t_0}^{t_1} \int_{\mathbf{R}^d} \partial_t \psi(t,x) \partial_t h(t,x) \, dt \, dx. \tag{1}$$

The idea is to define a perfect incompressible fluid as an incompressible fluid whose evolution is described by an extremal point of the restriction of the action functional $\mathcal{A}$ to the space $\mathcal{L}_0$. This approach requires the definition of the set of infinitesimal variations (the "tangent" space) in a point of the space $\mathcal{L}_0$.

**Definition 1.2.** An infinitesimal variation at a point $\psi$ of $\mathcal{L}_0$ is the derivative at 0 of any function $\Theta$ continuously differentiable from $[0,1]$ to $\mathcal{L}_0$ so that $\Theta(0) = \psi$.

Now, we can state the mathematical definition of a perfect incompressible fluid.

**Definition 1.3.** An incompressible fluid is called perfect if and only if its evolution between time $t_0$ and time $t_1$ is described by a function $\psi$ of $\mathcal{L}_0$ so that, for all infinitesimal variations $\theta$ at the point $\psi$,

$$D\mathcal{A}(\psi)\,\theta = 0.$$

This definition makes sense only if we are able to describe the set of all infinitesimal variations. The proof of the following proposition is an easy exercise of calculus.

**Proposition 1.1.** *Let us denote by $\mathcal{T}$ the set of all vector fields $\tau$ whose coefficients are continuously differentiable on $[t_0,t_1]\times\mathbb{R}^d$, so that*

$$\tau(t_0) = \tau(t_1) = 0 \quad \text{and} \quad \forall\, t \in [t_0,t_1],\ \operatorname{div}\tau(t) = 0.$$

*Let $\theta$ be an infinitesimal variation at a point $\psi$ of $\mathcal{L}_0$; there exists a vector field $\tau$ belonging to $\mathcal{T}$ so that $\theta(t,x) = \tau(t,\psi(t,x))$. Conversely, let $\alpha$ be a smooth function compactly supported on the interval $]t_0,t_1[$ and $\tau$ a divergence free vector field whose coefficients belong to the space $\mathcal{S}$; if $\theta(t,x) = \alpha(t)\tau(t,\psi(t,x))$, then $\theta$ is an infinitesimal variation at $\psi$.*

We have described the evolution of an incompressible fluid by a curve in the space of diffeomorphisms that preserve the measure; this is the lagrangian point of view. An other way to describe a fluid is the knowledge of the speed of the particles of the fluid; this is the eulerian point of view. When the vector fields are smooth enough, the equivalence between the two points of view is nothing but the ordinary differential equations, which can be summarized by the following two relations:

$$v(t,x) = \partial_t\psi(t,\psi^{-1}(t,x)), \tag{2}$$

$$\begin{cases} \partial_t\psi(t,x) = v(t,\psi(t,x)) \\ \psi(0,x) = x. \end{cases} \tag{3}$$

Now, we can derive the familiar Euler system of an incompressible perfect fluid.

**Theorem 1.1.** *Let $\psi$ be an evolution of an incompressible perfect fluid and $v$ its associated vector field by relation (2). Then, there exists a tempered distribution $p$ such that, if we set $v \cdot \nabla = \sum_{i=1}^{d} v^i \partial_i$, we get*

$$\partial_t v + v \cdot \nabla v = -\nabla p \quad \textit{with} \quad \operatorname{div} v = 0. \tag{E}$$

The proof of this theorem is easy (see [34]). We shall denote by $v_0$ the initial value of the solution $v$.

By formal integrations by part, it is obvious that the kinetic energy is preserved, which means that

$$\frac{d}{dt}\|v(t)\|_{L^2}^2 = 0.$$

This is the analogue of a familiar fact in riemannian geometry; the geodesics are naturally given by a constant speed parametrization.

Here we have followed Arnold's point of view about fluid mechanics (see [5]). This approach has been generalized by E. Ebin and J. Marsden in [38]. More recently, works of Y. Brenier (see [23]), D. Serre (see [62]) and A. Shnirelman (see [63]) develop this point of view.

The pressure is an unknown of the Euler system (E). Assuming that the vector field $v$ is regular enough and using the fact that the divergence of $v$ is 0, we have,

$$-\Delta p = \sum_{j,k=1}^{d} \partial_j \partial_k (v^j v^k) = \sum_{j,k=1}^{d} \partial_j v^k \partial_k v^j. \tag{4}$$

As we have chosen to work on the space $\mathbf{R}^d$, we have to add a growth condition at infinity to determine the pressure. This will appear in the statement of Theorem 2.1.

The vorticity is the key quantity to understand Euler's system.

**Definition 1.4.** The vorticity of a vector field is the curl of this vector field.

**Convention.** *When the dimension is 2, we identify antisymmetric matrices with real numbers, so we denote by $\omega(v) = \partial_1 v^2 - \partial_2 v^1$ the vorticity of $v$. In higher dimension, we use the notation*

$$\Omega(v) = (\Omega_j^i(v))_{1 \leq i,j \leq d} \quad \textit{with} \quad \Omega_j^i(v) = \partial_j v^i - \partial_i v^j.$$

The importance of the vorticity in the study of Euler's system comes from the fact that vorticity determines the vector field, up to vector fields with harmonic coefficents. So, let us assume that the vector field $v$ vanishes at infinity. Then we get, for $d \geq 3$, the so-called Biot-Savart law

(5)
$$\widetilde{v}^i = \sum_k \partial_k E_d * \Omega_k^i(v) \quad \text{with} \quad E_d(x) = \frac{c_d}{|x|^{d-2}} \quad \text{and} \quad c_d = \frac{2\pi^{(d+1/2)}}{\Gamma((d+1)/2)}.$$

For $d = 2$ it takes the form

$$v(x) = \frac{1}{2\pi} \int_{\mathbf{R}^2} \frac{\overline{x-y}}{|x-y|^2} \omega(y) dy \quad \text{with} \quad \bar{z} = (-z_2, z_1). \tag{6}$$

Now, let us study the evolution of the vorticity. Let us differentiate the Euler system (E) and let us take the antisymmetric part of the result. It turns out that

$$\partial_t \Omega + v \cdot \nabla \Omega + \Omega \cdot \nabla v = 0 \quad \text{with} \quad (\Omega \cdot \nabla v)_j^i = \sum_{k=1}^d \Omega_j^k \partial_k v^i - \Omega_i^k \partial_k v^j. \tag{7}$$

But, when the dimension is 2, it follows from (7) that

$$\partial_t \omega + v \cdot \nabla \omega = 0. \tag{8}$$

This result is extremely important. It makes the huge difference between dimension 2 and dimension 3. This result, combined with the fact that the vector field $v$ is divergence free, says that all $L^p$ norms of the vorticity are preserved. This is the key to all the results about 2-D fluid mechanics that will be presented here.

**2. The case of regular initial data.** First of all, let us define the Hölder class.

**Definition 2.1.** Let $r$ be a positive noninteger real number. If $r \in ]0,1[$, we denote by $C^r$ the set of all functions such that

$$\|u\|_r \stackrel{\text{def}}{=} \|u\|_{L^\infty} + \sup_{\substack{(x,y)\in \mathbf{R}^d \\ x\neq y}} \frac{|u(x)-u(y)|}{|x-y|^r} < \infty.$$

If $r > 1$, then we denote by $C^r$ the set of all functions $u$ such that

$$\|u\|_r \stackrel{\text{def}}{=} \sum_{|\alpha|\leq [r]} \|\partial^\alpha u\|_{r-[r]} < \infty.$$

The basic statement is the following.

**Theorem 2.1.** *Let $r$ and $a$ be two real numbers strictly greater than 1 and $v_0$ a divergence free vector field belonging to the space $C^r$. Let us assume that $\nabla v_0$ belongs to $L^a$. Then there exists a unique maximal time $T^*$ and a unique solution $(v,p)$ of* (E) *such that $v \in L^\infty_{\rm loc}([0,T^*[;C^r)$ and $(\nabla v, \nabla p)$ belongs to the space $L^\infty_{\rm loc}([0,T^*[;L^a)$. Moreover, we have*

$$T^* < +\infty \Longrightarrow \int_0^{T^*} \|\Omega(t)\|_{L^\infty} dt = +\infty.$$

Using the conservation of vorticity stated in (8), we have

**Corollary 2.1.** *Let us suppose that $d = 2$ and consider real numbers $r$ and $a$ strictly greater than 1 and $v_0$ a divergence free vector field belonging to the space $C^r$. Let us assume that $\nabla v_0$ belongs to $L^a$. Then there exists a unique solution $(v,p)$ of* (E) *in the space $L^\infty_{\rm loc}(\mathbf{R}^+;C^r)$ such that $(\nabla v, \nabla p)$ belongs to the space $L^\infty_{\rm loc}(\mathbf{R}^+;L^a)$.*

The first part of Theorem 2.1 has been proved in 1933 by L. Lichtenstein (see [50]). See also the work of J. Leray [49]. Corollary 2.1 has been proved in 1933 by W. Wolibner (see [65]). A modern proof of the second part of Theorem 2.1 can be found in [8].

The proof of the first point of the above Theorem 2.1 is based on a standard iterative scheme. See for instance [34] for the details. The second point is based on two estimates, a dynamical one and a stationary one. The following dynamical estimate is proved in [34].

**Lemma 2.1.** *There exists a constant $C$ so that, if $r$ is a strictly positive number and $v$ a vector field belonging to the space $L^\infty([0,T];Lip \cap C^r)$ that is a solution of the Euler system, then, for all $t$ in the interval $[0,T]$, we have*

$$\|v(t)\|_r \leq \|v_0\|_r \exp\Big(\frac{C^{r+1}}{r}\int_0^t \|v(\tau)\|_{\rm Lip}\, d\tau\Big). \tag{9}$$

The stationary inequality is the following:

**Lemma 2.2.** *There exists a constant $C$ such that, for any $\epsilon \in ]0,1[$ and for any $a > 1$, we have*

$$\|\nabla v\|_{L^\infty} \leq Ca\|\Omega\|_{L^a} + \frac{C}{\epsilon}\|\Omega\|_{L^\infty} \log\Big(e + \frac{\|\Omega\|_\epsilon}{\|\Omega\|_{L^\infty}}\Big).$$

This lemma can proved in different ways. Here we are going to use the Littlewood-Paley theory. The basic idea of this theory is the localization in rings in the frequency space. To do so, we are going to use a dyadic partition of unity. Let us recall the following proposition.

**Proposition 2.1.** *Let us denote by $\mathcal{C}$ the ring of center 0, of small radius 3/4 and of big radius 8/3. There exist two positive radial functions $\chi$ and $\varphi$ belonging respectively to $C_0^\infty(B(0,4/3))$ and to $C_0^\infty(\mathcal{C})$ such that:*

$$\chi(\xi) + \sum_{q\geq 0} \varphi(2^{-q}\xi) = 1, \tag{10}$$

$$|p-q| \geq 2 \implies \operatorname{Supp}\varphi(2^{-q}\cdot) \cap \operatorname{Supp}\varphi(2^{-p}\cdot) = \emptyset, \tag{11}$$

$$q \geq 2 \implies \operatorname{Supp}\chi \cap \operatorname{Supp}\varphi(2^{-q}\cdot) = \emptyset, \tag{12}$$

*if $\widetilde{\mathcal{C}} = B(0,2/3) + \mathcal{C}$, then $\widetilde{\mathcal{C}}$ is a ring and we have*

$$|p-q| \geq 5 \implies 2^p\widetilde{\mathcal{C}} \cap 2^q\mathcal{C} = \emptyset. \tag{13}$$

For all the rest of this text, we choose two functions $\chi$ and $\varphi$ satisfying the properties (10)–(13). Now, we can define the general Hölder classes.

**Definition 2.2.** For any real number $r$ we denote by $C^r$ the set of all tempered distributions satisfying

$$\|u\|_r \stackrel{\text{def}}{=} \max\{\|\chi(D)u\|_{L^\infty}, \sup_{q\in\mathbf{N}} 2^{qr}\|\Delta_q u\|_{L^\infty}\} < \infty, \quad \Delta_q u = \varphi(2^{-q}D)u.$$

If $r \in \mathbf{R}^+ \setminus \mathbf{N}$, this definition defines the same space as Definition 2.1, and the two norms are equivalent.

*Proof of Lemma 2.2.* Let us go back to the proof of Lemma 2.2. The key point is the following inequality

$$\|f\|_{L^\infty} \leq \frac{C}{\epsilon}\|f\|_0 \log\left(e + \frac{\|f\|_\epsilon}{\|f\|_0}\right). \tag{14}$$

To prove it, we write $f$ as the sum of the $\varphi(2^{-q}D)f$; it turns out that

$$\|f\|_{L^\infty} \leq \|\chi(D)u\|_{L^\infty} + \sum_{q=0}^{N-1} \|\varphi(2^{-q}D)f\|_{L^\infty} + \sum_{q\geq N} \|\varphi(2^{-q}D)f\|_{L^\infty}.$$

Using the definition of norms in the Hölder spaces, we obtain that for any integer $N$,

$$\|f\|_{L^\infty} \leq (N+1)\|f\|_0 + \frac{2^{-(N-1)\epsilon}}{2^\epsilon - 1}\|f\|_\epsilon.$$

Choosing

$$N = 1 + \left[\frac{1}{\epsilon}\log_2 \frac{\|f\|_\epsilon}{\|f\|_0}\right],$$

we get

$$\|f\|_{L^\infty} \leq \frac{C}{\epsilon}\|f\|_0\Big(1+\log\frac{\|f\|_\epsilon}{\|f\|_0}\Big).$$

Now it is obvious that the Biot-Savart law implies that $\nabla v$ is computed from $\Omega$ with $C^\infty$ and degree 0 homogeneous Fourier multipliers. It is easy to prove that such operators map $C^r$ into itself. For the sake of simplicity, the low frequency terms, responsible for the term $Ca\|\Omega\|_{L^a}$ in the above inequality, have been omitted. The function $x \mapsto x\log(e+\alpha/x)$ is an increasing function on $\mathbf{R}_*^+$ for any $\alpha \geq 0$. This concludes the proof of the lemma.

*Proof of Theorem 2.1.* To prove the necessary condition for blow up, we first observe that, if

$$\int_0^{T^*} \|v(t)\|_{\mathrm{Lip}} dt < \infty,$$

then $T^*$ is not maximal. This follows from Lemma 2.1 which states that

$$\int_0^{T^*} \|v(t)\|_{\mathrm{Lip}} dt < \infty \implies v \in L^\infty([0,T^*];C^r).$$

Then very classical techniques say immediately that $T^*$ is not maximal. Ignoring low frequencies for sake of simplicity, we can write

$$\begin{aligned}\|v(t)\|_{Lip} &\leq \frac{C}{\epsilon}\|\Omega(t)\|_{L^\infty}\log\Big(e+\frac{\|v(t)\|_r}{\|\Omega(t)\|_{L^\infty}}\Big)\\ &\leq \frac{C}{\epsilon}(\|\Omega(t)\|_{L^\infty}+1)\log\Big(e+\|v_0\|_r\exp\int_0^t \|v(\tau)\|_{\mathrm{Lip}}\,d\tau\Big).\end{aligned}$$

Gronwall's lemma implies that $v \in L^1([0,T^*];\mathrm{Lip})$ if $\Omega \in L^1([0,T^*];L^\infty)$, so Theorem 2.1 is proved.

**3. Yudovich's theorem.** The goal of this section is the proof of a global wellposedness theorem for Euler's system for initial data with a bounded and compactly supported vorticity. Here it is not relevant to assume that the vector field itself belongs to $L^2$. So, let us state the following definition:

**Definition 3.1.** We call stationary and denote by $\sigma$ any smooth vector field of the form

$$\sigma = \Big(-\frac{x_2}{r^2}\int_0^r \rho g(\rho)d\rho, \frac{x_1}{r^2}\int_0^r \rho g(\rho)d\rho\Big) \quad \text{where} \quad g \in C_0^\infty(\mathbf{R}\setminus\{0\}).$$

It is very easily checked that $\sigma$ is a stationary smooth solution of Euler's system.

*Remark.* Let us assume that the above function $g$ is positive and not identically 0. It is then obvious that, if $|x|$ is big enough, we have $|\sigma(x)| \simeq C|x|^{-1}$. So, $\sigma$ does not belong to the space $L^2$. Perturbations in $L^2$ of such vector fields is a large enough class.

**Definition 3.2.** When $m$ is a real number we denote by $E_m$ the set of all divergence free vector fields $v$ in the plane such that there exists some $\sigma$ satisfying

$$\int_{\mathbf{R}^2} \omega(\sigma) = m \quad \text{and} \quad v - \sigma \in L^2.$$

We have the following lemma which is easily proved in [34].

**Lemma 3.1.** *Let us consider a bounded measure $\mu$ so that $(1 - |x|)\mu$ is also a bounded measure. If $\mu$ belongs to $H^{-1}(\mathbf{R}^2)$, there exists a unique vector field $v \in E_{\int d\mu}$ such that $\omega(v) = \mu$.*

This lemma implies in particular that $E_0 = L^2$. Let us state the theorem, originally proved by V. Yudovich in [66].

**Theorem 3.1.** *Let $m$ be a real number and $v_0$ be a vector field in the space $E_m$. Let us assume that $\omega_0$ belongs to $L^\infty \cap L^a$ with $1 < a < +\infty$. Then there exists a unique solution $(v, p)$ of Euler's system (E) in the space $C(\mathbf{R}; E_m) \times L^\infty_{\text{loc}}(\mathbf{R}; L^2)$ such that the vorticity $\omega$ of $v$ belongs to $L^\infty(\mathbf{R}^3) \cap L^\infty(\mathbf{R}; L^a(\mathbf{R}^2))$. Moreover, this vector field $v$ has a flow. This means that there exists a unique map $\psi$ continuous from $\mathbf{R} \times \mathbf{R}^2$ to $\mathbf{R}^2$ such that*

$$\psi(t, x) = x + \int_0^t v(s, \psi(s, x))\, ds.$$

*In addition, there exists a constant $C$ such that*

$$\psi(t) - \mathrm{Id} \in C^{\exp(-Ct\|\omega_0\|_{L^\infty \cap L^a})}.$$

*More precisely, if we set*

$$\|v(t)\|_{\mathrm{LL}} \stackrel{\text{def}}{=} \sup_{0<|x-x'|\leq 1} \frac{|v(t,x) - v(t,x')|}{|x - x'|(1 - \log|x - x'|)},$$

*then the flow $\psi$ has the property that, for any $t$, we have*

$$|x - y| \leq e^{1-\exp \int_0^t \|v(s)\|_{\mathrm{LL}}\, ds} \implies |\psi(t,x) - \psi(t,y)| \leq |x - y|^{\exp(-\int_0^t \|v(s)\|_{\mathrm{LL}}\, ds)} e^{1-\exp(-\int_0^t \|v(s)\|_{\mathrm{LL}}\, ds)}.$$

Existence and uniqueness of solutions can be deduced very easily from the following lemma.

**Lemma 3.2.** *For every real number $a > 1$ there exists a constant $C$ with the following properties:*

*Let $(v_1, p_1)$ and $(v_2, p_2)$ be two solutions of Euler's system (E). Let us assume that both belong to $L^\infty_{\rm loc}(\mathbf{R}; E_m) \times L^\infty_{\rm loc}(\mathbf{R}; L^2)$ and that $\omega_i$ belongs to $L^\infty \cap L^a$. Let us define*

$$\alpha(t) \stackrel{\rm def}{=} C\Big(\max_i \|v_i(0) - \sigma\|_{L^2} e^{t\|\nabla\sigma\|_{L^\infty}} + \max \|\omega_i\|_{L^\infty \cap L^a} + 1\Big)^{\frac{2}{a}} \quad \text{and}$$

$$\beta(t) = e\int_0^t \alpha(s)\,ds, \quad \text{where } \sigma \text{ is as in Definition 3.1.}$$

*Then we have the following inequality:*

$$\|v_1(0) - v_2(0)\|^2_{L^2} \leq e^{-a(\exp\beta(t)-1)}$$
$$\Longrightarrow \|v_1(t) - v_2(t)\|^2_{L^2} \leq \|v_1(0) - v_2(0)\|^{2\exp(-\beta(t))}_{L^2} e^{a(1-\exp(-\beta(t)))}.$$

We have to estimate the function

$$I(t) \stackrel{\rm def}{=} \|(v_1 - v_2)(t)\|^2_{L^2}.$$

Let us consider a $C^\infty_0$ function $\chi$ on $\mathbf{R}^2$ whose value is 1 near the origin, and a strictly positive number $\epsilon$, and define

$$I_\epsilon(t) \stackrel{\rm def}{=} \int_{\mathbf{R}^2} \chi(\epsilon x)|(v_1 - v_2)(t,x)|^2 dx.$$

First of all, let us notice that $v \cdot \nabla v$ belongs to $L^b$ for all $b \geq a$. But we know that $\partial_i p = -\partial_i \sum_{j,k} \partial_j \Delta^{-1}(v^k \partial_k v^j)$. It is well known that Fourier multipliers of degre 0 map $L^a$ into itself. We shall use the following precise result, proved for instance in [64].

**Theorem 3.2.** *There exists a constant $C$ satifying the following property: For all $a \in ]1, \infty[$ and for all divergence free vector fields $v$ whose gradient belongs to $L^a$, we have*

$$\|\nabla v\|_{L^a} \leq C\frac{a^2}{a-1}\|\Omega(v)\|_{L^a}.$$

*Proof of Lemma 3.2.* The fact that $(v_i, p_i)$ are solutions of Euler's system (E) implies that

$$\forall\, b \geq a\,,\ \partial_t v_i \in L^b.$$

Denoting $p \stackrel{\text{def}}{=} p_1 - p_2$, the following computations make sense:

$$\begin{aligned} I'_\epsilon(t) = &- \sum_j \int_{\mathbf{R}^2} \chi(\epsilon x) v_1^j(t,x) \partial_j |(v_1 - v_2)(t,x)|^2 \, dx \\ &+ \sum_i \int_{\mathbf{R}^2} \chi(\epsilon x)(v_1 - v_2)^i(t,x) \partial_i p(t,x) \, dx \\ &+ 2 \sum_{i,j} \int_{\mathbf{R}^2} \chi(\epsilon x)(v_1 - v_2)^i(t,x)(v_1 - v_2)^j(t,x) \partial_j v^i(t,x) \, dx. \end{aligned}$$

The vector fields $v_i$ are divergence free, so integrations by part imply that

$$\begin{aligned} I'_\epsilon(t) &= 2 \int_{\mathbf{R}^2} \chi(\epsilon x) |(v_1 - v_2)(t,x)|^2 |\nabla v_2(t,x)| \, dx + R_\epsilon(t) \quad \text{with} \\ R_\epsilon(t) &= \epsilon \sum_j \int_{\mathbf{R}^2} |(v_1 - v_2)(t,x)|^2 v_1^j(t,x)(\partial_j \chi)(\epsilon x) \, dx \\ &\quad + 2\epsilon \sum_i \int_{\mathbf{R}^2} (v_1 - v_2)^i(t,x)(\partial_i \chi)(\epsilon x) p(t,x) dx. \end{aligned}$$

The vector field $v_1 - v_2$ belongs to $L^\infty_{\text{loc}}(\mathbf{R}; L^2 \cap L^\infty)$ and the pressure to $L^\infty_{\text{loc}}(\mathbf{R}; L^2)$, so we get

$$R_\epsilon(t) \leq C(t)\epsilon \quad \text{with} \quad C \in L^\infty_{\text{loc}}(\mathbf{R}). \tag{15}$$

Hölder's inequality implies that we have

$$\begin{aligned} I'_\epsilon(t) \leq 2 \Big( \int_{\mathbf{R}^2} \chi(\epsilon x) |v_1(t,x) - v_2(t,x)|^{\frac{2a}{a-1}} dx \Big)^{1-\frac{1}{a}} \\ \times \Big( \int_{\mathbf{R}^2} |\nabla v_2(t,x)|^a dx \Big)^{\frac{1}{a}} + R_\epsilon(t). \end{aligned}$$

Then, it turns out that, for all $b \geq a$,

$$I'_\epsilon(t) \leq 2 \|v_1(t) - v_2(t)\|_{L^\infty}^{\frac{2}{b}} I_\epsilon(t)^{1-\frac{1}{b}} \|\nabla v_2(t)\|_{L^b} + R_\epsilon(t).$$

Using the Biot-Savart law, Theorem 3.2 and the conservation of vorticity (8), we obtain, for all $b \geq a$,

$$\|\nabla v_2(t)\|_{L^b} \leq Ca \|\omega_2(0)\|_{L^a}.$$

Moreover, we know that

$$\begin{aligned} \|v_i\|_{L^\infty} &\leq C \|\chi(D)(v_i - \sigma)\|_{L^\infty} + C\|\sigma\|_{L^\infty} + \|(\mathrm{Id} - \chi(D))v_i\|_{L^\infty} \\ &\leq C(\|v_i(t) - \sigma\|_{L^2} + \|\sigma\|_{L^\infty} + \|\omega_i(0)\|_{L^\infty}) \\ &\leq C(\|v_i(0) - \sigma\|_{L^2} e^{t\|\nabla\sigma\|_{L^\infty}} + \|\sigma\|_{L^\infty} + \|\omega_i(0)\|_{L^\infty}). \end{aligned}$$

So for all $b$ greater than $a$, we have

$$I'_\epsilon(t) \leq \alpha(t) b I_\epsilon(t)^{1-\frac{1}{b}} + R_\epsilon(t). \tag{16}$$

Now let us assume that $\|v_1(0) - v_2(0)\|^2_{L^2} < 1$. Let us consider a real number $\eta$ such that $0 < \eta < 1 - I(0)$. Let us define

$$J_{\epsilon,\eta}(t) \stackrel{\text{def}}{=} \eta + I_\epsilon(t).$$

All the following inequalities of this proof are valid if and only if $\eta + I(t) \leq 1$.

From inequality (16) we deduce that

$$J'_{\epsilon,\eta}(t) \leq \alpha(t) b J_{\epsilon,\eta}(t)^{1-\frac{1}{b}} + R_\epsilon(t).$$

Choosing $b = a - \log J_{\epsilon,\eta}(t)$, we obtain

$$\begin{aligned} J'_{\epsilon,\eta}(t) &\leq \alpha(t)(a - \log J_{\epsilon,\eta}(t)) J_{\epsilon,\eta}(t) \exp\Big(\frac{-\log J_{\epsilon,\eta}(t)}{a - \log J_{\epsilon,\eta}(t)}\Big) + R_\epsilon(t) \\ &\leq e\alpha(t)(a - \log J_{\epsilon,\eta}(t)) J_{\epsilon,\eta}(t) + R_\epsilon(t). \end{aligned}$$

It is easy to see, for any differentiable function $f$ from $\mathbf{R}$ to $]0,1[$, that, for all $\lambda \in ]0,1[$,

$$-\frac{1}{\lambda}\frac{d}{dt}\log(1 - \lambda \log f(t)) = \frac{f'(t)}{f(t)(1 - \lambda \log f(t))}. \tag{17}$$

From this, we deduce that

$$-\frac{d}{dt}\log(a - \log J_{\epsilon,\eta}(t)) \leq e\alpha(t) + \frac{R_\epsilon(t)}{a\eta}.$$

Let us define

$$\overline{\beta}_{\epsilon,\eta}(t) \stackrel{\text{def}}{=} \beta(t) + \int_0^t \frac{R_\epsilon(\tau)}{a\eta} d\tau.$$

After integration we obtain, applying the exponential twice,

$$J_{\epsilon,\eta}(t) \leq e^{a(1-e^{(-\overline{\beta}_{\epsilon,\eta}(t))})} J_{\epsilon,\eta}(0)^{\exp(-\overline{\beta}_{\epsilon,\eta}(t))}.$$

By definition of $R_\epsilon$, we have

$$R_\epsilon(t) \leq C(t)\epsilon \quad \text{with} \quad C \in L^\infty_{\text{loc}}.$$

Taking the limit when $\epsilon$ tends to 0, then taking the limit when $\eta$ tends to 0, we conclude the proof of Lemma 3.2.

The proof of the result on the flow $\psi$ uses very similar computations. For a more general method to study this type of ordinary differential equations, see [33].

Now we are going to exhibit a solution of Euler's system which will show that Yudovich's theorem is more or less optimal. We are going to construct a solution satisfying the following properties:

- the vorticity $\omega$ of the solution $v$ is at each time $t$ bounded and compactly supported,
- at any time $t$, the flow $\psi(t)$ of $v$ does not belong to the Hölder class $C^{\exp(-t)}$.

First of all, let us define the initial data. Let $\omega_0$ be the function in the plane $\mathbf{R}^2$ which vanishes outside $[-1,1]\times[-1,1]$ and is odd with respect to the two variables $x_1$ and $x_2$, and whose value is $2\pi$ on $[0,1]\times[0,1]$. Let us consider the vector field $v_0$ defined by

$$v_0(x_1,x_2)=\Big(-\frac{1}{2\pi}\int\frac{x_2-y_2}{|x-y|^2}\omega_0(y)dy,\ \frac{1}{2\pi}\int\frac{x_1-y_1}{|x-y|^2}\omega_0(y)dy\Big).$$

We have the following theorem, taken from [6].

**Theorem 3.3.** *Let $v$ be the solution of Euler's system associated to the above initial data $v_0$. At time $t$, the flow $\psi(t)$ of the solution $v$ does not belong to $C^\alpha$ for any $\alpha>\exp(-t)$.*

*Proof.* The proof of this result is based on symmetries. The vector field $v_0$ is symmetric with respect to the two axes of coordinates. Thus it is tangent to the two axes and therefore it vanishes at the origin. We are going to prove the existence of a constant $C$ such that, for all $x_1\in[0,C]$, we have

$$\exists C\ /\ \forall x\in[0,C],\ v_0^1(x_1,0)\geq -2x_1\log x_1. \tag{18}$$

Let us define $\widetilde{\omega}_0(x_1)\stackrel{\text{def}}{=}2H(x_1)-1$ ($H$ is the Heaviside function). Then we have

$$v_0^1(x_1,0)=\int_{-1}^1 dy_1\,\widetilde{\omega}_0(y_1)\int_0^1\frac{2y_2}{(x_1-y_1)^2+y_2^2}\,dy_2.$$

From a trivial computation, we deduce that, if $0\leq x_1<1$, then

$$v_0^1(x_1,0)=-4x_1\log x_1+f(x_1),$$

where $f$ is an odd and $C^\infty$ function on $]-1,1[$. This implies assertion (18).

By Theorem 3.1 the flow of $v$ is a continuous function of $(t,x)$. Moreover, we know that, at any time $t$, the vector field $v$ is symmetric with respect to the two axes of coordinates, so these two axes are globally invariant by the action of the flow. Hence the origin, which is their intersection point, is invariant by the flow $\psi$ of $v$. Then we have, for all $t$,

$$\psi(t,0)=0,\ \psi^1(t,0,x_2)=0 \quad \text{and} \quad \psi^2(t,x_1,0)=0. \tag{19}$$

Let $T$ be any strictly positive real number. The vorticity is preserved along the trajectories of $v$ (see equation (8)). The relation (19) above implies the existence of a neighbourhood $W$ of the origin such that, for all $t \in [0,T]$,

$$\omega(t)_{|W} = \omega_{0|W}.$$

The vector field $\widetilde{v}(t) \stackrel{\text{def}}{=} v(t) - v_0$ is symmetric with respect to the two axes of coordinates. Its vorticity is identically 0 on $W$. Hence there exists a constant $A$ such that, for all $t \in [0,T]$,

$$|v(t,x) - v_0(x)| \leq A|x|.$$

From inequality (18) we deduce the existence of a constant $C'$ such that, for $(t,x_1) \in [0,T] \times [0,C']$, we have

$$v(t,x_1,0) \geq -x_1 \log x_1.$$

Now let $x_1$ be a point in $[0,1[$ such that, for all $t \in [0,T]$, we have $\psi^1(t,x_1,0) \in [0,C']$. From the above inequality, we obtain

$$\psi^1(t,x_1,0) \geq x_1(t) \quad \text{with} \quad x_1'(t) = -x_1(t)\log x_1(t).$$

From this we deduce that $\psi^1(t,x_1,0) \geq x_1^{\exp(-t)}$. As we have $\psi(t,0)=0$, Theorem 3.3 is proved.

**4. The vortex patch problem.** In this section we are going to state the classical vortex patch problem which is the following. Let us assume that the vorticity of the initial data is the characteristic function of a bounded domain $D_0$ whose boundary is regular (this will mean in the Hölder class $C^{k+\epsilon}$ where $k$ is a strictly positive integer and $\epsilon$ a real number in the interval $]0,1[$). Thanks to Yudovich's Theorem 3.1, there exists a unique vector field satisfying the Euler system on $\mathbf{R}\times\mathbf{R}^2$, whose vorticity belongs to $L^\infty(\mathbf{R}^3)$. Then this solution is quasi-lipschitz. Again thanks to Yudovich's Theorem 3.1, this vector field has a flow $\psi$ such that $\psi(t)$ is a homeomorphism in the Hölder class $C^{\exp(-\alpha t)}$. By relation (8), the vorticity at time $t$ is the characteristic function of an open subset $D_t$ whose topology remains invariant. However, the boundary of this domain $D_t$ is a priori only in the Hölder class $C^{\exp(-\alpha t)}$. Now there are two very natural questions:

Does the boundary of the domain $D_t$ remain in the Hölder class $C^{k+\epsilon}$ for small time?

If yes, what happens for large time?

The following approach has been developed and studied by A. Majda in [51], P. Constantin and E. Titi in [35], S. Alinhac in [4] and P. Serfati in [59], especially for domains which are small perturbations of a circle.

Let $\gamma_0$ be an embedding of the circle $\mathbf{S}^1$ of class $C^{1+\epsilon}$ whose range is the boundary of the domain $D_0$. The vector field solution of Euler's system is completely determined by the boundary of $D_t$, so we are looking for a parametrization of this boundary. It is reasonable to choose the function $\gamma$ defined by

$$\partial_t \gamma(t,s) = v(t, \gamma(t,s)). \tag{20}$$

But, thanks to Biot-Savart's law, the vector field solution is defined by $v(t) = \nabla^\perp f(t)$ with

$$f(t,x) = \frac{1}{2\pi} \int_{D_t} \log |x-y|\, dy.$$

If we assume that $\gamma(t,\cdot)$ is an embedding of the circle of class $C^{1+\epsilon}$ whose range is $D_t$, it follows from Green's formula that

$$v(t,x) = \frac{1}{2\pi} \int_0^{2\pi} \log |x - \gamma(t,\sigma)|\, \partial_\sigma \gamma(t,\sigma)\, d\sigma.$$

By (20) we therefore have to solve, in the class of $C^{1+\epsilon}$-embeddings, the following equation:

$$\partial_t \gamma(t,s) = \frac{1}{2\pi} \int_0^{2\pi} \log |\gamma(t,s) - \gamma(t,\sigma)|\, \partial_\sigma \gamma(t,\sigma)\, d\sigma. \tag{21}$$

In [31] and in [32] we have proved the following theorem:

**Theorem 4.1.** *Let $\epsilon$ belong to the interval $]0,1[$ and $\gamma_0$ be a function in $C^{1+\epsilon}(\mathbf{S}^1;\mathbf{R}^2)$ which is one to one and whose derivative does not vanish. Then there exists a unique solution $\gamma(t,s)$ of the equation (21) belonging to $L^\infty_{\mathrm{loc}}(\mathbf{R}; C^{1+\epsilon}(\mathbf{S}^1;\mathbf{R}^2))$ and which is, for all time, an embedding of the circle.*

The approach we are going to explain here is different from the one that we have recalled before the statement. Our approach is based on the fact that this theorem is implied by a theorem on propagation of singularities for the Euler system studied as a quasilinear hyperbolic system. This propagation theorem (Theorem 6.2) is much more general than Theorem 4.1. It has been proved by a different method by P. Serfati in [60]. Let us mention that, after our proof, A. Bertozzi and P. Constantin proved Theorem 4.1 in [14] by our method, adapted to this very particular case.

Before explaining how microlocal analysis occurs in this problem, we are going to recall how microlocal analysis can be used in nonlinear partial differential equations.

**5. A brief survey of nonlinear microlocal analysis.** Microlocal analysis began at the very beginning of the seventies with the pioneer work of L. Hörmander (see [42]). This work was motivated by the understanding of linear partial differential equations, especially the description of the singularities. That was the beginning of an enormous development of the knowledge about linear partial differential equations. To get an idea of this, see the four books of L. Hörmander [43]. One of the key concepts of microlocal analysis is the wave front set of a distribution. Let us recall the definition.

**Definition 5.1.** Let $\Omega$ be an open subset of $\mathbf{R}^d$ and $u$ a distribution in $\Omega$. The wave front set of $u$, denoted by $WF(u)$, is the complement of the set of all $(x_0, \xi_0)$ in $\Omega \times \mathbf{R}^d \setminus \{0\}$ such that there exists a smooth function $f$, compactly supported in $\Omega$, with $f(x_0) \neq 0$, and an open cone $\Gamma$ containing $\xi_0$ such that

$$\forall N \in \mathbf{N}\,,\ \exists C > 0\,/\,\forall \xi \in \Gamma\,,\ |\widehat{fu}(\xi)| \leq C(1+|\xi|)^{-N}.$$

If $(x_0, \xi_0)$ does not belong to $WF(u)$, the distribution $u$ is said to be smooth "microlocally near $(x_0, \xi_0)$" and we write $u \in C^\infty_{(x_0,\xi_0)}$. We can define in the same way the concept of $H^s$-wave front set:

**Definition 5.2.** Let $\Omega$ be an open subset of $\mathbf{R}^d$ and $u$ a distribution in $\Omega$. The $H^s$-wave front set of $u$, denoted by $WF_{H^s}(u)$, is the complement of the set of all $(x_0, \xi_0)$ in $\Omega \times \mathbf{R}^d \setminus \{0\}$ such that there exist two distributions $u_1$ and $u_2$ in $\Omega$ such that

$$u = u_1 + u_2\,,\ u_1 \in H^s_{\mathrm{loc}}(\Omega) \quad \text{and} \quad u_2 \in C^\infty_{(x_0,\xi_0)}.$$

If $(x_0, \xi_0)$ does not belong to $WF_{H^s}(u)$, the distribution $u$ is said to be in $H^s$ "microlocally near $(x_0, \xi_0)$" and we write $u \in H^s_{(x_0,\xi_0)}$. It is very easily proved that

$$u \in H^s_{x_0} \iff \forall \xi \in R^d \setminus \{0\}\,,\ u \in H^s_{(x_0,\xi)}.$$

A fundamental theorem originally proved by L. Hörmander (see [42]) describes the nature of the wave front set of a solution $u$ to the equation

$$Pu \stackrel{\text{def}}{=} \sum_{|\alpha| \leq m} a_\alpha(x) \partial^\alpha u \in C^\infty,$$

where $P$ is a strictly hyperbolic operator, which means that

$$p_m(x,\xi) \stackrel{\text{def}}{=} \sum_{|\alpha|=m} a_\alpha(x)\xi^\alpha$$

is a strictly hyperbolic polynomial. Hörmander's propagation theorem is the following:

**Theorem 5.1 (on propagation of singularities).** *The $H^s$-wave front set of such solutions $u$ is exactly the union of a family of bicharacteristic curves of $P$ (i.e. integral curves of the hamiltonian field of $p_m$ included in $p_m^{-1}(0)$).*

The key result that opened the field of nonlinear partial differential equations to microlocal analysis is the following result, proved independently in 1978 by J.-M. Bony in [15], B. Lascar in [45] and J. Rauch in [54].

**Lemma 5.1.** *Let $s$ and $g$ be two real numbers such that*

$$s > \tfrac{1}{2}d \quad \text{and} \quad g \leq 2s - \tfrac{1}{2}d.$$

*For any $(x_0,\xi_0) \in \Omega \times \mathbf{R}^d \setminus \{0\}$, the set $H^s_{x_0} \cap H^g_{(x_0,\xi_0)}$ is an algebra.*

An important consequence of this lemma is the following theorem.

**Theorem 5.2.** *Let $P$ be a strictly hyperbolic equation of order $m$ and $u$ an $H^s$ solution of*

(SLE)

$$Pu = f(x, \partial^\alpha u)_{|\alpha|\leq m-2} \quad \textit{with} \quad s - \tfrac{1}{2}d - m + 2 \stackrel{\text{def}}{=} \rho_s > 0 \quad \textit{and} \quad f \in C^\infty.$$

*Then, for any $g \in [s, s+\rho_s]$, the $H^g$-wave front set is a union of bicharacteristic curves of $P$.*

An example constructed in 1982 by J. Rauch and M. Reed in [56] shows that nonlinear phenomena actually occur for the $H^g$-wave front set when $g$ is greater than $s+\rho_s$. Work of M. Beals (see [9] and [25] for some developments) showed in 1983 that it was possible to estimate (but not to describe as above) the $H^g$-wave front set up to $g \leq s + 2\rho_s$. An example built in [9] proved that there is no hope to go beyond this limit in dimension (in space-time) higher than 2. In the particular case of dimension 2, J. Rauch and M. Reed proved in [55] that it was possible to estimate the wave front set of any order for a solution of a strictly hyperbolic equation.

To go beyond this limit, J.-M. Bony made in 1979 a hypothesis about the nature of possible singularities of the solution. Let us take the following simple example. Consider an $H^s$ solution $u$ of

(SLWE) $$\partial_t^2 u - \Delta_{\mathbf{R}^{d-1}} u = f(u) \quad \text{with} \quad s > \tfrac{1}{2}d \quad \text{and} \quad f \in C^\infty.$$

Let us assume that the two initial data $u_{|t=0}$ and $\partial_t u_{|t=0}$ have their wave front sets included in the conormal bundle of some smooth hypersurface $\Gamma$. Moreover, assume that for any family $(Z_j^0)_{1\leq j\leq N}$ generating the tangent bundle of $\Gamma$, and for all integers $k$, we have

$$Z_{j_1}^0 \cdots Z_{j_k}^0 u \in H^s \quad \text{for any} \quad (j_l)_{1\leq l\leq k} \in \{1,\cdots,N\}^k.$$

**Theorem 5.3.** *Let $u$ be an $H^s$-solution of* (SLWE) *with $s > d/2$ associated to initial data as above. Let $\Sigma_1$ and $\Sigma_2$ be the two characteristic hypersurfaces (which means that the conormal bundle of $\Sigma_i$ is contained in $p_2^{-1}(0)$) containing $\Gamma$. Then, if $(Z_j)_{1\leq j\leq M}$ is a generating family of the space of all smooth vectors fields tangent to $\Sigma_1$ and $\Sigma_2$, we have, for all integers $k$,*

$$Z_{j_1} \cdots Z_{j_k} u \in H^s \quad \textit{for any} \quad (j_l)_{1\leq l\leq k} \in \{1,\cdots,M\}^k.$$

This theorem implies in particular that the wave front set of $u$ is included in the union of the conormal bundles of $\Sigma_1$, $\Sigma_2$ and $\Gamma$ and therefore the solution $u$ is smooth outside the union of $\Sigma_1$ and $\Sigma_2$. The key point of the proof is the study of the commutators between the vector fields $Z_j$ and the operator $P$.

This theorem was proved by J.-M. Bony in 1979 (see [16]). His method has been the origin of a lot of work. The geometry of the set where the singularitites are located may be much more complicated and singular than in the example above. Let us cite in particular the work of M. Beals (see [9]–[12]), J.-M. Bony (see [18]–[20]), G. Lebeau (see [46]–[48]), R. Melrose and N. Ritter (see [52]–[53]), A. Sá Baretto (see [57]–58]) and the author (see [24]).

All the papers cited above are devoted to semilinear partial differential equations. Semilinear partial differential equations are equations of the form

$$\text{(SLE)} \qquad \sum_{|\alpha|\leq m} a_\alpha(x)\partial^\alpha u = f(x,\partial^\beta u)_{|\beta|\leq m-1}.$$

But what about quasilinear equations, which are equations of the form

$$\text{(QLE)} \qquad \sum_{|\alpha|\leq m-1} a_\alpha(x,\partial^\beta u)_{|\beta|\leq m-1}\partial^\alpha u = f(x,\partial^\beta u)_{|\beta|\leq m-1}?$$

Theorem 5.2 has been generalized to this case by J.-M. Bony in [17]. The crucial tool that he introduced here is the paradifferential calculus. The basic idea of this calculus is the following. Using the dyadic partition of unity introduced in Section 2, we can write the product of two functions $u$

and $v$ in the following way, up to low frequencies which are neglected for sake of simplicity:

$$\begin{aligned} uv &= \sum_{q,q'} \varphi(2^{-q}D)u\,\varphi(2^{-q'}D)v \\ &= T_u v + T_v u + R(u,v) \quad \text{with} \\ T_u v &\stackrel{\text{def}}{=} \sum_{q'\leq q-2} \varphi(2^{-q'}D)u\,\varphi(2^{-q}D)v \quad \text{and} \\ R(u,v) &\stackrel{\text{def}}{=} \sum_{|q-q'|\leq 1} \varphi(2^{-q}D)u\,\varphi(2^{-q'}D)v. \end{aligned}$$

The operator $T$ is called the paraproduct and the operator $R$ the remainder. The paraproduct $T_u v$ is always defined for two compactly supported distributions $u$ and $v$, and its regularity is essentially determined by the one of $v$. The remainder operator is not always defined, but the regularity of $u$ and $v$ sum together. The idea has been used in 1981 by J.-M. Bony (see [17]) to construct an analogue of pseudodifferential calculus but with nonsmooth symbols. Bony's theorem in the case of quasilinear equations is the following:

**Theorem 5.4.** *Let $u$ be an $H^s$ solution of (QLE) with $\widetilde{\rho}_s \stackrel{\text{def}}{=} s - d/2 - m - 1 > 0$. Let us assume that*

$$p_m(x,\xi) \stackrel{\text{def}}{=} \sum_{|\alpha|=m} a_\alpha(x, \partial^\beta u(x))_{|\beta|\leq m-1}\xi^\alpha$$

*is a strictly hyperbolic polynomial. Then, for any $g \in [s, s+\widetilde{\rho}_s]$, the $H^g$-wave front set of $u$ is a union of bicharacteristic curves, i.e. integral curves of the hamiltonian of $p_m$.*

Again, the problem of the limitation of the order of the $H^g$-wave front set that we can describe is posed. The way to drop this limitation is the same as the one previously used in the semilinear case. The main difficulty of quasilinear equations comes from the fact that the geometrical objects that can be used for describing the singularities (in particular, here the characteristic hypersurfaces) depend on the solution itself and therefore they are not smooth. The techniques used to overcome this difficulty have been introduced by S. Alinhac in [1] and in [2] and developed by S. Alinhac in [3] and the author in [26]–[29]. For example, S. Alinhac proved in [3] the following theorem.

**Theorem 5.5.** *Let $u$ be an $H^s$ solution of*

$$\partial_t^2 u - \sum_{|\alpha|=2} a_\alpha(x,u,\nabla u)\partial^\alpha u = f(x,u,\nabla u) \quad \textit{with} \quad s > \tfrac{1}{2}d + 6.$$

*Assume that the initial data is as in Theorem 5.3 and that*

$$\sum_{|\alpha|=2} a_\alpha(x,u,\nabla u)\xi^\alpha \geq c|\xi|^2.$$

*Let $\Sigma_1$ and $\Sigma_2$ be the two characteristic hypersurfaces containing $\Gamma$. Then these two hypersurfaces are smooth outside $\Gamma$ and if $(Z_j)_{1\leq j\leq N}$ is a local basic of the tangent bundle to $\Sigma_j$, then, for any $k$ and any sequence $(j_l)_{1\leq l\leq k}$, we have $Z_{j_1}\cdots Z_{j_k}u \in H^s$.*

As in Theorem 5.3, we have that the wave front set of the solution is contained in the union of the conormal bundles of $\Sigma_1$, $\Sigma_2$ and $\Gamma$, hence the solution $u$ is smooth outside $\Sigma_1 \cup \Sigma_2$. In this particular case, the singularities are located just as if the equation was linear.

The reader interested in a more complete survey of microlocal analysis can read the two surveys of J.-M. Bony (see [21] and [22]).

**6. A theorem on propagation of singularities in Euler's system.** Now let us go back to the vortex patch problem. It is well known in the theory of quasilinear hyperbolic problems that the quantity which controls the equation is the lipschitz norm of the solution (see for instance the estimate (9)). The goal here is to find the condition on a bounded function $u$ which implies that the functions $\partial_i\partial_j\Delta^{-1}u$ are also bounded. In terms of fluid mechanics, this means that we search for a condition on the vorticity which implies that the associated vector field is lipschitz.

To do this, we have to solve the following problem: if we regularize the characteristic function of a bounded domain whose boundary is a $C^{1+\epsilon}$ curve, do we have some uniform estimates of the derivatives of the vector field associated to the regularized vorticity? Here uniform means independent of the parameter of regularization. The answer to this question is contained in the following statements.

In all of this section, $\epsilon$ is a real number in the interval $]0,1[$ and $\Sigma$ is a closed subset of the plane $\mathbf{R}^2$ (possibly empty).

**Definition 6.1.** Let $X = (X_\lambda)_{\lambda\in\Lambda}$ be a family of vector fields of class $C^\epsilon$ whose divergence is also of class $C^\epsilon$. Such a family is said to be complete outside $\Sigma$ if and only if we have

$$I(\Sigma, X) = \inf_{x\notin\Sigma}\sup_{\lambda\in\Lambda}|X_\lambda(x)| > 0.$$

Now let us define the concept of tangential regularity with respect to a complete family of vector fields. Let us take the following notation:

$$N_\epsilon(\Sigma,X) \stackrel{\text{def}}{=} \frac{1}{\epsilon}\sup_{\lambda\in\Lambda}\frac{\|X_\lambda\|_\epsilon + \|\operatorname{div} X_\lambda\|_\epsilon}{I(\Sigma,X)} \quad \text{and}$$

$$\|u\|^\epsilon_{\Sigma,X} \stackrel{\text{def}}{=} N_\epsilon(\Sigma,X)\|u\|_{L^\infty} + \sup_{\lambda\in\Lambda}\frac{\|X_\lambda(x,D)u\|_{\epsilon-1}}{I(\Sigma,X)}.$$

In the case of vortex patches with boundary without singularity, the set $\Sigma$ is empty and we can define

$$
\begin{aligned}
I(X) &\stackrel{\text{def}}{=} \inf_{x\in\mathbf{R}^2} \sup_{\lambda\in\Lambda} |X_\lambda(x)| \quad (>0),\\
N_\epsilon(X) &\stackrel{\text{def}}{=} \frac{1}{\epsilon} \sup_{\lambda\in\Lambda} \frac{\|X_\lambda\|_\epsilon + \|\operatorname{div} X_\lambda\|_\epsilon}{I(X)},\\
\|u\|_{\epsilon,X} &\stackrel{\text{def}}{=} N_\epsilon(X)\|u\|_{L^\infty} + \sup_{\lambda\in\Lambda} \frac{\| X_\lambda(x,D)u\|_{\epsilon-1}}{I(X)}.
\end{aligned}
$$

**Definition 6.2.** Let $X$ be a complete family of vector fields of class $C^\epsilon$ whose divergence is also $C^\epsilon$ (we assume that $\Sigma = \emptyset$) Then $C^\epsilon(\Sigma, X)$ denotes the set of all distributions $u$ belonging to $L^\infty$ such that, for all $\lambda$, we have

$$X_\lambda(x,D)u \stackrel{\text{def}}{=} \operatorname{div}(uX_\lambda) - u \operatorname{div} X_\lambda \in C^{\epsilon-1}.$$

This type of regularity has been used to prove theorems on propagation of singularities in quasilinear strictly hyperbolic equations like for instance Theorem 5.5 in [3] and [26]–[29]. Introduced first in [30] for studying Euler's system, this type of regularity allows to prove (see [31] and [32]) the following theorem.

**Theorem 6.1.** *Let $a$ be a real number greater than 1 and let $X_0 = (X_{0,\lambda})_{\lambda\in\Lambda}$ be a complete family of class $C^\epsilon$. We consider a vector field $v_0$ on $\mathbf{R}^2$ whose gradient is in $L^a$. If $\omega_0$ belongs to $C^\epsilon(X_0)$, then there exists a unique solution $v$ of* (E) *such that*

$$v \in L^\infty_{\text{loc}}(\mathbf{R}; \text{Lip}) \quad \textit{and} \quad \nabla v \in L^a.$$

*Moreover, if $\psi$ is the flow of $v$, then for all $\lambda$,*

$$X_{0,\lambda}(x,D)\psi \in L^\infty_{\text{loc}}(\mathbf{R}; C^\epsilon).$$

*Finally, if $X_{t,\lambda} = \psi(t)_* X_{0,\lambda}$, then the family $X_t = (X_{t,\lambda})_{\lambda\in\Lambda}$ is complete and we have*

$$N_\epsilon(X_t) \in L^\infty_{\text{loc}}(\mathbf{R}) \quad \textit{and} \quad \|\omega(t)\|_{\epsilon,X_t} \in L^\infty_{\text{loc}}(\mathbf{R}).$$

P. Serfati gave in [60] a new proof of this theorem. A 3-D and small time version of this theorem has been proved by P. Gamblin and X. Saint-Raymond in [40]. In certain particular cases, P. Serfati proved this theorem in [61] with additional results about analyticity in time.

It is easily checked that this theorem implies Theorem 4.1. To prove Theorem 6.1, let us regularize the initial data,

$$v_{0,n} = S_n v_0 \quad \text{and} \quad \omega_{0,n} = S_n \omega_0. \tag{22}$$

Theorem 2.1 tells us that there exists a global smooth solution $v_n$. The important point of the present proof consists in proving an a priori estimation of the lipschitz norm of a smooth solution. After this, passing to the limit is straightforward.

**Theorem 6.2.** *There exists a constant $C$ such that, for all real $\epsilon$ in the interval $]0,1[$, for all $a$ greater than 1, and for all closed subset $\Sigma$, we have the following properties:*

*Let $X$ be any family of vector fields of class $C^\epsilon$ whose divergence is $C^\epsilon$, and assume that $X$ is complete outside $\Sigma$. Let $\omega$ be a function belonging to $C^\epsilon(\Sigma, X)\cap L^a$. If $v$ is the vector field associated to $\omega$ by Biot-Savart's law (6), then the gradient of $v$ is bounded and we have*

$$\|\nabla v\|_{L^\infty(\Sigma^c)} \leq Ca\|\omega\|_{L^a} + \frac{C}{\epsilon}\|\omega\|_{L^\infty}\log\left(e + \frac{\|\omega\|^\epsilon_{\Sigma,X}}{\|\omega\|_{L^\infty}}\right). \tag{23}$$

There is a particular case where the proof of this theorem is particularly easy. Let us assume that the family $X$ is $\partial_1$ and (this second assumption is only a technical detail) that the support of the Fourier transform of $\omega$ does not contain the origin. It is obvious that, in this particular case, $\Sigma = \emptyset$ and that $\|X\|_\epsilon = I(\Sigma, X) = 1$. Applying inequality (14), it is clear that, for $j$ equal to 1 or 2,

$$\|\partial_1\partial_j\Delta^{-1}\omega\|_{L^\infty} \leq \frac{C}{\epsilon}\|\partial_1\partial_j\Delta^{-1}\omega\|_0 \log\left(e + \frac{\|\partial_1\partial_j\Delta^{-1}\omega\|_\epsilon}{\|\partial_1\partial_j\Delta^{-1}\omega\|_0}\right).$$

But homogeneous Fourier multipliers of degree $-1$ map $C^r$ into $C^{r+1}$, so we have

$$\|\partial_1\partial_j\Delta^{-1}\omega\|_{L^\infty} \leq \frac{C}{\epsilon}\|\omega\|_0 \log\left(e + \frac{\|\partial_1\omega\|_{\epsilon-1}}{\|\omega\|_0}\right). \tag{24}$$

For the estimate of $\|\partial_2^2\Delta^{-1}\omega\|_{L^\infty}$, we use the fact that $\partial_2^2 = \Delta - \partial_1^2$ to get

$$\|\partial_i\partial_j\Delta^{-1}\omega\|_{L^\infty} \leq \|\omega\|_{L^\infty} + \frac{C}{\epsilon}\|\omega\|_0 \log\left(e + \frac{\|\partial_1\omega\|_{\epsilon-1}}{\|\omega\|_0}\right).$$

Thus the theorem is proved in this very particular case.

For the general case, a serious difficulty appears. It comes from the lack of regularity of the vector fields $X_\lambda$ and of their possible tendancy to vanish. We overcome these difficulties using paradifferential calculus.

*Remark.* The example whose properties are described by Theorem 3.3 shows that the estimate (23) of Theorem 6.2 is optimal.

The dynamical part of the proof is based on estimates that look like estimates on propagation of singularities. The idea is very simple; we transport the geometrical data by the flow of the solution. More precisely, let us define the family $X_t = (X_{t,\lambda})_{\lambda\in\Lambda}$ by

$$X_{t,\lambda}(x) = \psi(t)_* X_{0,\lambda}(x) = (\psi(t)^{-1})^*\Big(X_{0,\lambda}(x,D)\psi(t)^*\Big). \tag{25}$$

The main point of the proof of Theorem 6.1 is the estimate of $\|\omega(t)\|_{\epsilon,X_t}$. Using paradifferential calculus and the following relations

$$\begin{cases} \partial_t X_{0,\lambda}(x,D)\psi(t,x) & = & \nabla v(t,\psi(t,x))X_{0,\lambda}(x,D)\psi(t,x), \\ X_{0,\lambda}(x,D)\psi(0,x) & = & X_{0,\lambda}(x), \end{cases}$$

$$\partial_t X_{t,\lambda} + v\cdot\nabla X_{t,\lambda} = X_{t,\lambda}(x,D)v,$$

$$\partial_t \operatorname{div} X_{t,\lambda} + v\cdot\nabla \operatorname{div} X_{t,\lambda} = X_{t,\lambda}(x,D)\operatorname{div} v,$$

we prove that

$$\frac{\|\omega(t)\|_{\epsilon,X_t}}{\|\omega(t)\|_{L^\infty}} \le C\frac{\|\omega_0\|_{\epsilon,X_0}}{\|\omega_0\|_{L^\infty}}\exp\Big(\frac{C}{\epsilon}\int_0^t \|\nabla v(\tau)\|_{L^\infty}\,d\tau\Big).$$

Let us apply Theorem 6.2. Thanks to the fact that the vector field $v$ is divergence free, we have

$$\begin{aligned}\|\nabla v(t)\|_{L^\infty} &\le Ca\|\omega_0\|_{L^a}\\ &\quad+\frac{C}{\epsilon}\|\omega_0\|_{L^\infty}\log\Big(e+\frac{\|\omega_0\|_{\epsilon,X_0}}{\|\omega_0\|_{L^\infty}}\exp\Big(\frac{C}{\epsilon}\int_0^t\|\nabla v(\tau)\|_{L^\infty}\,d\tau\Big)\Big).\end{aligned}$$

Then it is easily deduced that

$$\begin{aligned}\|\nabla v(t)\|_{L^\infty} &\le Ca\|\omega_0\|_{L^a}\\ &\quad+\frac{C}{\epsilon}\|\omega_0\|_{L^\infty}\log\frac{\|\omega_0\|_{\epsilon,X_0}}{\|\omega_0\|_{L^\infty}}+\frac{C}{\epsilon^2}\|\omega_0\|_{L^\infty}\int_0^t\|\nabla v(\tau)\|_{L^\infty}\,d\tau.\end{aligned}$$

Gronwall's lemma implies that

$$\begin{aligned}&\|\nabla v(t)\|_{L^\infty}\le \widetilde{N}(X_0,\epsilon,\omega_0)\exp\frac{Ct\|\omega_0\|_{L^\infty}}{\epsilon^2}\quad\text{with}\\ &\widetilde{N}(X_0,\epsilon,\omega_0)=Ca\|\omega_0\|_{L^a}+\frac{C}{\epsilon}\|\omega_0\|_{L^\infty}\log\frac{\|\omega_0\|_{\epsilon,X_0}}{\|\omega_0\|_{L^\infty}}.\end{aligned}\tag{26}$$

**7. Analytic type regularity.** In this section we look at the time regularity properties of the flow.

**Theorem 7.1.** *Let $r$ be a real number strictly greater than 1. Then there exists a constant $C$ such that, for all solution $v$ of the Euler's system* (E) *belonging to $L^\infty_{\rm loc}([0,T^*[;C^r)$, we have the following properties:*

*For any integer $k$, the function $(\partial_t + v\cdot\nabla)^k v$ belongs to $L^\infty_{\rm loc}([0,T^*[;C^r)$ and we have, for all $T < T^*$,*

$$(27)\qquad \|(\partial_t + v\cdot\nabla)^k v\|_{L^\infty([0,T];C^r)} \le C^k k!(\|v\|_{L^\infty([0,T];C^r)})^{k+1}.$$

*The function $t \mapsto \psi(t)$ is an analytic function from $[0,T^*[$ to $\mathrm{Id} + C^r$.*

This theorem has been proved by P. Serfati in [59] using a lagrangian approach. In Serfati's work, this theorem is proved with weaker hypotheses about the initial data (tangential regularity with respect to some particular families of vector fields). These results are announced in [61] and details are given in [59]. Let us mention that P. Serfati proved in [59] the analyticity for small time for the flow in the case of vortex patches which are small perturbations of the circle. For the problem of propagation of analyticity in the space variables see [7], [13] and [36]. In [39] P. Gamblin has proved Theorem 7.1 using a method that allows him to prove the following generalization.

**Theorem 7.2.** *There exists a constant $C$ with the following properties:*

*Let $v$ be a solution of Euler's system* (E) *in $\mathbf{R}^2$ belonging to the space of continuous functions from $\mathbf{R}$ to the space $E_m$ and whose vorticity belongs to $L^\infty(\mathbf{R};L^a(\mathbf{R}^2)) \cap L^\infty(\mathbf{R}^3)$. Then, for all strictly positive $\epsilon$ and for all integers $k$, the function $(\partial_t + v\cdot\nabla)^k v$ belongs to $L^\infty_{\rm loc}([0,T^*[;C^{1-\epsilon})$ and we have, for all $T < T^*$,*

$$(28)\quad \|(\partial_t + v\cdot\nabla)^k v\|_{L^\infty([0,T];C^{1-\epsilon})} \le \left(\frac{C}{\epsilon}\right)^{2k}(k!)^3(\|v\|_{L^\infty([0,T];C^1_*)})^{k+1}.$$

*For all strictly positive $\epsilon$, the function $t \mapsto \psi(t)$ is a function of Gevrey class 3 from $[0,T^*[$ to the space $\mathrm{Id} + C^r$.*

We refer to [39] for the details of the proofs of Theorems 7.1 and 7.2. Let us mention that the basic idea consists in a careful study of the repeated action of the operator $\mathcal{V} \stackrel{\rm def}{=} \partial_t + v\cdot\nabla$ on the gradient of the pressure, which can be written as

$$(29)\ \partial_k p(x) = \sum_{i,j}\int_{\mathbf{R}^d}(\partial_i\partial_j\partial_k E_d)(x-y)(v^i(x)-v^i(y))(v^j(x)-v^j(y))dy.$$

Let us now mention two corollaries of the two theorems above.

**Corollary 7.1.** *Let $r$ be a real number strictly greater than 1 and $v$ a solution of the Euler system (E) belonging to $L^\infty_{\text{loc}}([0,T^*[;C^r)$. Then we get*

$$WF_a(v) \subset \{(t,x;\tau,\xi)\,/\,\tau + (v(t,x)|\xi) = 0\}.$$

**Corollary 7.2.** *Let $v$ be a solution of Euler's system (E) in $\mathbf{R}^2$ belonging to the space of continuous functions from $\mathbf{R}$ to the space $E_m$ and whose vorticity belongs to $L^\infty(\mathbf{R}; L^a(\mathbf{R}^2)) \cap L^\infty(\mathbf{R}^3)$. In this case, we have*

$$WF_{G^3}(v) \subset \{(t,x;\tau,\xi)\,/\,\tau + (v(t,x)|\xi) = 0\}.$$

The proof of these two corollaries relies on the following theorem, proved by P. Gamblin in [39].

**Theorem 7.3.** *Let $\Omega$ be an open subset of $\mathbf{R}^d$, set $z(x,\xi) = -i\sum z^j(x)\xi_j$, and define $\mathcal{Z}$ by*

$$\mathcal{Z} \stackrel{\text{def}}{=} \sum z^j(x)\partial_j.$$

*Let us consider a function $u$, continuous on $\Omega$. Let us assume that there exists a smooth sequence $(z_n(x,\xi))_{n\in\mathbf{N}}$ with coefficients converging to those of $z$ in the space $L^\infty_{\text{loc}}(\Omega)$ and a sequence of smooth functions $(u_n)_{n\in\mathbf{N}}$ converging to $u$ in the space $L^\infty_{\text{loc}}(\Omega)$ and satisfying the following properties:*

*There exists a real number $s$ strictly greater than 1, such that, for all compact $K$ of $\omega$, there exists a constant $C$ such that, for all integers $k$ and $n$, we have*

$$\|\mathcal{Z}_n^k z_n^j\|_{L^\infty(K)} \leq C^{k+1}(k!)^s,$$
$$\|\mathcal{Z}_n^k \operatorname{div} z_n\|_{L^\infty(K)} \leq C^{k+1}(k!)^s,$$
$$\|\mathcal{Z}_n^k u_n\|_{L^\infty(K)} \leq C^{k+1}(k!)^s.$$

*Under these hypotheses we get*

$$WF_{G^s}(u) = \{(x,\xi) \in \Omega \times (\mathbf{R}^d \setminus \{0\})\,/\,z(x,\xi) = 0\}.$$

*Proof.* Let $(x_0,\xi_0)$ be in $\Omega \times \mathbf{R}^d \setminus \{0\}$ and assume that $z(x_0,\xi_0) \neq 0$. There exists a compact neighbourhood $K$ of $x_0$, a conic neighbourhood $\Gamma$ of $\xi_0$ and a constant $C$ such that, for all $(x,\xi)$ belonging to $K \times \Gamma$, we have

$$(30) \qquad |z_n(x,\xi)| \geq C|\xi|.$$

It is well known that there exists a sequence $(\chi_k)_{k\in\mathbf{N}}$ of smooth functions whose value is identically 1 near $K$ such that

$$\forall j \leq k \quad \sup_{|\alpha|=j} \|\partial^\alpha \chi_j\|_{L^\infty} \leq C(Ck)^j.$$

Define $u_{k,n} \stackrel{\text{def}}{=} \chi_k u_n$. The Fourier transform is defined by

$$\widehat{u}_{k,n}(\xi) = \int_{\mathbf{R}^d} e^{-i(x|\xi)} u_{k,n}(x)dx.$$

But we have $-\mathcal{Z}(e^{-i(\cdot|\xi)}) = z(\cdot,\xi)e^{-i(\cdot|\xi)}$; so if we define

$$R_{n,\xi}w \stackrel{\text{def}}{=} \mathcal{Z}\left(\frac{w}{z_n(\cdot,\xi)}\right) + i \operatorname{div} z_n \frac{w}{z_n(\cdot,\xi)}, \tag{31}$$

then integration by parts implies that

$$\widehat{u}_{k,n}(\xi) = \int_{\mathbf{R}^d} e^{-i(x|\xi)} R^k_{n,\xi}(u_{k,n})(x)dx. \tag{32}$$

Careful computations done in [39] show that

$$\|R^k_{n,\xi}(u_{k,n})\|_{L^\infty} \leq C^k(k!)^s|\xi|^{-k}. \tag{33}$$

As the function $R^k_{n,\xi}(\chi_k u_n)$ has its support included in the compact $K$, we have, for all integers $k$ and $n$ and for all $\xi$ belonging to $\Gamma$, $|\widehat{u}_{k,n}(\xi)| \leq C^k(k!)^s|\xi|^{-k}$. By hypothesis, we have $\lim_{n\to\infty} u_{k,n} = \chi_k u$ in $L^\infty$, therefore also in $L^1$. Passing to the limit, we get

$$|\widehat{\chi_k u}(\xi)| \leq C^k(k!)^s|\xi|^{-k},$$

which concludes the proof of the theorem.

**8. Vortex patches with singular boundary.** This section is devoted to the problem of vortex patches whose boundary has singular points like corners or cusps. Here we give only statements. The complete proofs are in [34]. We have the following theorem which is a generalization of Theorem 4.1.

**Theorem 8.1.** *Let $D_0$ be a bounded domain of the plane whose boundary is a curve of class $C^{1+\epsilon}$ outside a closed subset $\Sigma_0$. Let us consider the Yudovich solution $v$ associated to this initial data. At time $t$, the boundary $\gamma(t)$ of the domain $D(t) = \psi(t, D_0)$ is a curve of class $C^{1+\epsilon}$ outside the closed subset $\Sigma(t) = \psi(t, \Sigma_0)$.*

We follow the approach developed in Section 4. But here we use a technique coming from two related ideas introduced in nonlinear partial differential equations independently by R. Melrose and N. Ritter in [52] (the blow up) and by J.-M. Bony in [18] (the 2-microlocal analysis). Here it consists in working at distance $h$ from the singular set $\Sigma$, and in estimating carefully how all inequalities depend on $h$.

Let us recall that, if $A$ is a subset of the plane and $\alpha$ a strictly positive real number, $A_\alpha$ is the set of points whose distance to $A$ is smaller than $\alpha$. Moreover, we denote by $A^c_\alpha$ the set of the points whose distance to $A$ is greater than $\alpha$.

**Definition 8.1.** Let $\Sigma$ be a closed subset of the plane and $\Xi$ a triplet of real numbers $(\alpha, \beta, \gamma)$; we consider a family $\mathcal{X} = (X_{\lambda,h})_{(\lambda,h)\in\Lambda\times]0,e^{-1}]}$ of $C^\epsilon$ vector fields whose divergence is $C^\epsilon$. Let us define $\mathcal{X}_h \stackrel{\text{def}}{=} (X_{\lambda,h})_{\lambda\in\Lambda}$. The family $\mathcal{X}$ is said to be $\Sigma$-complete of index $\Xi$ if and only if the following three properties are satisfied:

$$\forall(\lambda,h) \in \Lambda\times]0,e^{-1}],\ \operatorname{Supp} X_{\lambda,h} \subset \Sigma^c_{h^\alpha}; \tag{34}$$

$$\mathcal{I}_\gamma(\Sigma,\mathcal{X}) \stackrel{\text{def}}{=} \inf_{h\in]0,e^{-1}]} h^\gamma I(\Sigma_h,\mathcal{X}_h) > 0; \tag{35}$$

$$\mathcal{N}_\epsilon(\mathcal{X}) \stackrel{\text{def}}{=} \sup_{h\in]0,e^{-1}]} h^{-\beta} N_\epsilon(\Sigma_h,\mathcal{X}_h) < \infty. \tag{36}$$

Now, let us introduce the concept of tangential regularity with respect to such a family.

**Definition 8.2.** Let $\Sigma$ be a closed subset of the plane and $\mathcal{X}$ a $\Sigma$-complete family of index $\Xi = (\alpha,\beta,\gamma)$. By definition, a function $u$ belongs to the space $C^{\epsilon,\beta}_{\Sigma,\mathcal{X}}$ if and only if $u$ is bounded and

$$\|u\|^{\epsilon,\beta}_{\Sigma,\mathcal{X}} \stackrel{\text{def}}{=} \sup_{h\in]0,e^{-1}]} h^{-\beta}\|u\|^\epsilon_{\Sigma_h,\mathcal{X}_h} < \infty.$$

**Theorem 8.2.** *Let $\Sigma_0$ be a closed subset and $m$ a real number. Let us consider a divergence free vector field $v_0$ belonging to $E_m$ whose vorticity $\omega_0$ is bounded and compactly supported. Let $v$ be a Yudovich solution, $\psi$ its flow and $\Sigma(t) \stackrel{\text{def}}{=} \psi(t,\Sigma_0)$. Let us consider a family $\mathcal{X}_0$ of class $C^\epsilon$ and $\Sigma_0$-complete of index $\Xi_0 = (\alpha_0,\beta_0,\gamma_0)$. We assume that $\omega_0 \in C^{\epsilon,\beta_0}_{\Sigma_0,\mathcal{X}_0}$.*

*For any time $t$, there exists a family $\mathcal{X}(t)$ which is of class $C^\epsilon$ and $\Sigma(t)$-complete of index $\Xi(t)$ such that $\omega(t) \in C^{\epsilon\ \beta(t)}_{\Sigma(t),\mathcal{X}(t)}$ with $\Xi(t)$ in $L^1_{\text{loc}}(\mathbf{R})$. Moreover, for any $(\lambda,h) \in \Lambda\times]0,e^{-1}]$, $X_{0,\lambda,\delta(t,h)}(x,D)\psi(t,x)$ is in $L^\infty_{\text{loc}}(\mathbf{R};C^\epsilon)$. Finally the vector field $v(t)$ is lipschitzian outside $\Sigma(t)$ and we have*

$$\sup_{h\in]0,e^{-1}]} \frac{\|\nabla v(t)\|_{L^\infty((\Sigma(t))^c_h)}}{-\log h} \in L^\infty_{\text{loc}}(\mathbf{R}).$$

The key point of the proof is that the Littlewood-Paley theory is almost local. The following two lemmas give a description of this.

**Lemma 8.1.** *There exists a constant $C$ with the following properties. Let $K$ be a closed subspace of $\mathbf{R}^d$, $g$ a function of $\mathcal{S}$ and $u$ a function in $L^b$ for every $b \in [a,+\infty[$ which vanishes outside $K$. Let us define*

$$g_\lambda(x) = \lambda^d g(\lambda x)\,,\ N(g) = \|(1+|\cdot|)g|_{L^1\cap L^\infty} \quad \text{and} \quad \|u\|_L \stackrel{\text{def}}{=} \sup_{b\geq a}\frac{\|u\|_{L^b}}{b}.$$

*For all* $(\lambda, h) \in [e, \infty[\times]0, e^{-1}]$, *we have*

$$\|g_\lambda * u\|_{L^\infty} \leq edN(g)\|u\|_L \log \lambda \quad \text{and}$$
$$\|g_\lambda * u\|_{L^\infty(K_h^c)} \leq -edN(g)\|u\|_L \log h.$$

**Lemma 8.2.** *Let $g$ be a function in $\mathcal{S}$ and $\lambda$ a strictly positive number. For any positive real number $N$ and any real number $r$, there exists a constant $C$ such that, for any closed subset $K$ of $\mathbf{R}^d$ and for any distribution $u$ in $C^r$ and supported in $K$, we have*

$$\forall (\lambda, h) \,/\, \lambda h \geq 1, \ \|g_\lambda * u\|_{L^\infty(K_h^c)} \leq C\lambda^{-r}(\lambda h)^{-N}\|u\|_r.$$

## References


1. S. Alinhac, *Paracomposition et opérateurs paradifférentiels*, Comm. in Partial Diff. Eq. **11** (1986), 87–121.
2. ———, *Évolution d'une onde simple pour des équations non linéaires générales*, Current topics in partial differential equations, Kinokuniya, Tokyo, 1986, pp. 63–90.
3. ———, *Interaction d'ondes simples pour des équations complètement non linéaires*, Ann. Sci. Éc. Norm. Sup. **21** (1988), 91–133.
4. ———, *Remarques sur l'instabilité du problème des poches de tourbillon*, J. Funct. Anal. **98** (1991), 361–379.
5. V. Arnold, *Sur la géométrie différentiable des groupes de Lie de dimension infinie et ses applications à l'hydrodynamique des fluides parfaits*, Ann. Inst. Fourier **16** (1966), 319–361.
6. H. Bahouri and J.-Y. Chemin, *Équations de transport relatives à des champs de vecteurs non-lipschitziens et mécanique des fluides*, Arch. Rat. Mech. Anal. **127** (1994), 159–182.
7. C. Bardos and S. Benachour, *Domaine d'analyticité des solutions de l'équation d'Euler dans un ouvert de $\mathbf{R}^n$*, Ann. Scuola Norm. Sup. Pisa **4** (1977), 647–687.
8. J. Beale, T. Kato and A. Majda, *Remarks on the breakdown of smoothness for the 3-D Euler equations*, Comm. Math. Phys. **94** (1984), 61–66.
9. M. Beals, *Self spreading and strength of singularities for solutions to semilinear wave equations*, Ann. of Math. **118** (1983), 187–214.
10. ———, *Vector fields associated with the nonlinear interaction of progressing waves*, Indiana Univ. Math. J. **3** (1988), 637–666.
11. ———, *Propagation and interaction of singularities in non linear hyperbolic problems*, Birkäuser Boston Inc., New York, 1989.
12. ———, *Regularity of nonlinear waves associated with a cusp*, Microlocal analysis and nonlinear waves (M. Beals, R. Melrose and J. Rauch, eds.), The I.M.A. volumes, **30** (1991), 9–28.
13. S. Benachour, *Analyticité des solutions des équations d'Euler*, Arch. Rat. Mech. Anal. **71** (1979), 271–299.
14. A. Bertozzi and P. Constantin, *Global regularity for vortex patches*, Comm. Math. Phys. **152** (1993), 19–26.
15. J.-M. Bony, *Localisation et propagation des singularités pour des équations non linéaires*, Actes des journées équations aux dérivées partielles de Saint Jean de Monts (1978).

16. ______, *Propagation des singularités pour des équations aux dérivées partielles non linéaires*, Séminaire équations aux dérivées partielles de l'École Polytechnique 1979–1980, exposé 22.
17. ______, *Calcul symbolique et propagation des singularités pour les équations aux dérivées partielles non linéaires*, Ann. Éc. Norm. Sup. **14** (1981), 209–246.
18. ______, *Interaction des singularités pour des équations de Klein-Gordon non linéaires*, Séminaire équations aux dérivées partielles de l'École Polytechnique 1983–1984, exposé 10.
19. ______, *Singularités des solutions de problèmes hyberboliques non linéaires*, Advances in microlocal analysis (Garnir, ed.), Reidel, NATO ASI Series, **128** (1985) 15–39.
20. ______, *Second microlocalization and interaction of singulartes for non linear partial differential equations*, Hyperbolic equations and related topics (S. Mizohata, ed.), Kinokuniya, 1986, pp. 11–49.
21. ______, *Analyse microlocale et singularités non linéaires*, Nonlinear Hyperbolic Problems, Springer Verlag, Proceedings Bordeaux, Springer Lecture Notes **1042** (1988), 114–128.
22. ______, *Analyse microlocale des équations aux dérivées partielles non linéaires*, Microlocal analysis and applications (Cattabriga and Rodino, eds.), Springer Verlag, Springer Lecture Notes **1495** (1991), 1–45.
23. Y. Brenier, *The least action principle and the related concept of generalized flow for incompressible perfect fluids*, J. Amer. Math. Soc. **2** (1989), 225–255.
24. J.-Y. Chemin, *Interaction de trois ondes dans les équations semi-linéaires strictement hyperboliques d'ordre 2*, Comm. Partial Diff. Eq. **12** (1987), 1203–1225.
25. ______, *Interaction contrôlée dans les équations aux dérivées partielles non linéaires*, Bull. Soc. Math. France **116** (1988), 341–383.
26. ______, *Calcul paradifférentiel précisé et application à des équations aux dérivées partielles non linéaires*, Duke Math. J. **56** (1988), 431–469.
27. ______, *Régularité de la solution d'un problème de Cauchy non linéaire à données singulières en un point*, Ann. Inst. Fourier **39** (1989), 101–122.
28. ______, *Évolution d'une singularité ponctuelle dans des équations strictement hyperboliques non linéaires*, Amer. J. Math. **112** (1990), 805–860.
29. ______, *Évolution d'une singularité ponctuelle dans un écoulement compressible*, Comm. Partial Diff. Eq. **15** (1990), 1237–1263.
30. ______, *Sur le mouvement des particules d'un fluide parfait, incompressible, bidimensionnel*, Inv. Math. **103** (1991), 599–629.
31. ______, *Persistance des structures géométriques liées aux poches de tourbillon*, Séminaire equations aux dérivées partielles de l'École Polytechnique 1990-1991, exposé 13.
32. ______, *Persistance de structures géométriques dans les fluides incompressibles bidimensionnels*, Ann. Éc. Norm. Sup. **26** (1993), 1–26.
33. J.-Y. Chemin and N. Lerner, *Flot de champs de vecteurs non-lipschitziens et équations de Navier-Stokes*, Prépublication de l'Ecole Polytechnique n° 1062, 1993, à paraître au J. Diff. Eq..
34. J.-Y. Chemin, *Fluides parfaits incompressibles*, Astéristique **230** (1995).
35. P. Constantin and E. Titi, *On the evolution of nearly circular vortex patches*, Comm. Math. Phys. **119** (1988), 177–198.
36. J.-M. Delort, *Estimations fines pour des opérateurs pseudo-differentiels analytiques sur un ouvert à bord de* $\mathbf{R}^n$*. Application aux équations d'Euler*, Comm. Partial Diff. Eq. **10** (1985), 1465–1525.

37. ______, *Conormalité des ondes semi-linéaires le long des caustiques*, Amer. J. Math. **113** (1991), 593–651.
38. D. Ebin and J. Marsden, *Group of diffeomorphism and the motion of an incompressible fluid*, Ann. of Math. **92** (1970), 102–163.
39. P. Gamblin, *Système d'Euler incompressible et régularité microlocale analytique*, Ann. Inst. Fourier **44** (1994), 1449–1475.
40. P. Gamblin and X. Saint-Raymond, *On three-dimensionnal vortex patches*, Prépublication de l'Université d'Orsay n° 93–02, 1993, à paraître au Bull. Soc. Math. France.
41. N. Hanges and F. Trêves, *On the analyticity of solutions of first order non linear partial differential equations*, Trans. Amer. Math. Soc. **331**, 627–638.
42. L. Hörmander, *Linear differential operators*, Actes du Congrès International des Mathématiciens de Nice **1** (1970), 121–133.
43. ______, *The analysis of linear partial differential equations*, Springer Verlag, 1983–1985.
44. T. Kato and G. Ponce, *Well-posedness of the Euler and Navier-Stokes equations in the Lebesgue spaces* $L^p_s(\mathbf{R}^2)$, Rev. Mat. Iberoamericana **2** (1986), 73–88.
45. B. Lascar, *Singularités des solutions d'équations aux dérivées partielles non linéaires*, C. R. Acad. Sci. Paris **287** (1978), 527–529.
46. G. Lebeau, *Problème de Cauchy semi-linéaire en trois dimensions d'espace*, J. Funct. Anal. **78** (1988), 185–196.
47. ______, *Équations des ondes semi-linéaires II. Contrôle des singularités et caustiques semi-linéaires*, Inv. Math. **95** (1989), 277–323.
48. ______, *Singularités de solutions d'équations d'ondes semi-linéaires*, Ann. Éc. Norm. Sup. **25** (1992), 201–231.
49. J. Leray, *Étude de diverses équations intégrales non linéaires and quelques problèmes que pose l'hydrodynamique*, J. Math. Pures Appl. **9** (1933), 1–82.
50. L. Lichtenstein, *Über einige Existenzprobleme der Hydrodynamik homogener unzusammendrückbarer, reibungsloser Flüßigkeiten und die Helmoltzschen Wirbelsätze*, Math. Z. **23** (1925), 89–154; **26** (1927), 196–323; **28** (1928), 387–415; **32** (1930), 608–725.
51. A. Majda, *Vorticity and the mathematical theory of an incompressible fluid flow*, Comm. Pure Appl. Math. **38** (1986), 187–220.
52. R. Melrose and N. Ritter, *Interaction of non linear waves for semilinear wave equations*, Ann. of Math. **121** (1985), 187–213.
53. ______, *Interaction of non linear waves for semilinear wave equations II*, Ark. Mat. **25** (1987), 91–114.
54. J. Rauch, *Singularities of solutions to semilinear wave equations*, J. Math. Pures Appl. **58** (1979), 299–308.
55. J. Rauch and M. Reed, *Propagation of singularities for semilinear hyperbolic wave equations in one space variable*, Ann. of Math. 111 (1980), 531–552.
56. ______, *Singularities produced by nonlinear hyperbolic Cauchy problems*, Comm. Partial Diff. Eq. **7** (1982), 1117–1133.
57. A. Sá Barreto, *Interaction of conormal waves for fully semilinear wave equations*, J. Funct. Anal. **89** (1990), 233–273.
58. ______, *Second microlocalellipticity and propagation of conormality for semilinear wave equations*, J. Funct. Anal. **102** (1991), 47–71.
59. P. Serfati, *Étude mathématique de flammes infiniment minces en combustion. Résultats de structure et de régularité pour l'équation d'Euler incompressible*, Thèse de l'Université de Paris VI, 1992.

60. ———, *Une preuve directe d'existence globale des vortex patches 2D*, Notes aux C. R. Acad. Sci. Paris **318** Série I (1993), 515–518.
61. ———, *Régularité stratifiée et équation d'Euler à temps grand*, Notes aux C. R. Acad. Sci. Paris **318** Série I (1993), 925–928.
62. D. Serre, *Sur le principe variationnel des équations de la mécanique des fluides parfaits*, Modélisation Mathématique et Analyse Numérique **27** (1993), 739–758.
63. A. I. Shnirelman, *On the geometry of the group of diffeomorphisms and the dynamics of an ideal incompressible fluid*, Math. USSR-Sb. **56**, 79–105.
64. A. Torchinsky, *Real variable methods in harmonic analysis*, Academic Press, New York, Pure and Applied Mathematics **123**.
65. W. Wolibner, *Un théorème d'existence du mouvement plan d'un fluide parfait, homogène, incompressible, pendant un temps infiniment long*, Math. Z. **37** (1933), 698–726.
66. V. Yudovich, *Non stationary flows of an ideal incompressible fluid*, Zh. Vychisl. Mat. i Mat. Fiz. **3** (1963), 1032–1066.
67. N. Zabusky, M. Hughues and K. Roberts, *Contour dynamics for the Euler equations in two dimensions*, J. Comput. Phys. **30** (1979), 96–106.

Laboratoire d'Analyse Numérique, URA 189 du CNRS, Université Paris 6, BP 187, F-75 232 Paris Cedex 05, France
*E-mail address*: chemin@ann.jussieu.fr

# $L^p$ SPECTRAL INDEPENDENCE FOR CERTAIN UNIFORMLY ELLIPTIC OPERATORS

E. B. DAVIES

**1. Heat Kernel Bounds.** We consider certain self-adjoint uniformly elliptic operators of order $2m$ acting on $L^p(\mathbf{R}^N)$ for values of $p$ in the interval $1 \leq p < \infty$. The kind of problem which we consider has been treated fairly thoroughly for operators with smooth coefficients, [K, R], and we discuss instead the case in which all of the coefficients, including those of highest order, are bounded and measurable [D3]. The theory described below has been extended to a Riemannian manifold context in [D3], but we only describe the Euclidean version here.

Let $\alpha$, $\beta$ denote multi-indices of the usual type, and $D^\alpha$ the corresponding partial differentiation operators. We consider the symmetric operator given formally by

$$Hf(x) := \sum_{\substack{|\alpha| \leq m \\ |\beta| \leq m}} (-1)^{|\alpha|} D^\alpha \{a_{\alpha,\beta}(x) D^\beta f(x)\}$$

where $a_{\alpha,\beta}(x) = \overline{a_{\beta,\alpha}(x)}$ are complex-valued bounded measurable functions on $\mathbf{R}^N$ for all $\alpha$, $\beta$. We define $H$ rigorously to be the self-adjoint operator associated with the quadratic form $Q$ defined on the Sobolev space $W^{m,2}$ by

$$Q(f,g) := \int_{\mathbf{R}^N} \sum_{\substack{|\alpha| \leq m \\ |\beta| \leq m}} a_{\alpha,\beta}(x) D^\beta f(x) \overline{D^\alpha g(x)}\, \mathrm{d}^N x.$$

The hypotheses on the coefficients are the minimum ones for the form to be closed and non-negative on $W^{m,2}$.

Even if one restricts attention to second order operators, the fact that we allow complex coefficients leads to a rather difficult theory, whose character depends upon the space dimension. One can prove Gaussian upper bounds on the absolute value of the heat kernel for $N \leq 2$, [AMT], but for $N \geq 5$ this is not possible, [Au]. For higher order operators our results are different depending upon whether $N < 2m$ or not, but the case $N = 2m$ may possibly be treated by a development of the methods of [AMT]. The main new difficulty with the higher order and complex coefficient theory is the fact that $t \to e^{-Ht}$ is not a contraction semigroup on $L^p$ for $p \neq 2$, and indeed it is not a priori clear that these operators are bounded on $L^p$ for $p \neq 2$ and $t > 0$. Our first main result concerns the integral kernels $K(t,x,y)$ of $e^{-Ht}$ for $t > 0$, and is not new if one assumes that the coefficients are uniformly smooth and uniformly elliptic [K,R].

**THEOREM.** *If $N < 2m$ then there exist positive constants $c_1$, $c_2$ and $c_3$ such that*

$$(1) \qquad |K(t,x,y)| \leq c_1 t^{-N/2m} \exp\left[-c_2|x-y|^{2m/(2m-1)} t^{-1/(2m-1)} + c_3 t\right]$$

*for all $t > 0$ and $x$, $y \in \mathbf{R}^N$.*

The proof in [D3] does not attempt to identify the coefficients $c_i$ in (1). In [BD], G. Barbatis and the author have obtained explicit expressions for $c_2$ and $c_3$ in this theorem in terms of the dimension, order and ellipticity constant of the operator $H$; no previous results of this type appear to exist even for operators with smooth coefficients. The expressions obtained are sharp (up to an arbitrarily small constant $\varepsilon > 0$) in the case in which $H := (-\Delta)^m$, for which independent calculations can be done by the use of Fourier transforms and the method of steepest descent.

The above bound can be extended to complex times $z$ such that $\mathrm{Re}(z) > 0$, and leads directly to the following semigroup theorem.

**THEOREM.** *If $N < 2m$ then the operators $e^{-Hz}$ on $L^2$ may be extended to bounded operators $T_p(z)$ on $L^p$ satisfying*

$$(2) \qquad \|T_p(z)\| \leq c(\cos\theta)^{-2N|\frac{1}{p}-\frac{1}{2}|} e^{kr\cos\theta}$$

*for all $1 \leq p \leq \infty$ and all $z = re^{i\theta}$ such that $r > 0$ and $|\theta| < \frac{\pi}{2}$.*

**2. $L^p$ Spectral Theory.** The first proof of $L^p$ spectral independence was given for Schrödinger operators with Kato potentials by Hempel and Voigt [HV]. Arendt [Ar] showed in a Euclidean space context that the proof depends only upon possessing suitable Gaussian upper bounds on the absolute value of the heat kernel. Sturm [St] has shown that if the volumes of balls in a complete Riemannian manifold are bounded uniformly subexponentially as the radius diverges to infinity, then the $L^p$ spectrum of the Laplace-Beltrami operator is independent of $p$. On the other hand it is known [DM, DST] that the $L^p$ spectrum of hyperbolic space and other non-compact hyperbolic manifolds does depend on $p$, and that it is not even real unless $p = 2$. Another approach to the proof of $L^p$ spectral independence based upon the Helffer-Sjöstrand formula, discussed below, has been given by the author in a context which allows application to higher order elliptic operators and to Schrödinger operators with magnetic fields [D2, D3].

**THEOREM.** *Let $N < 2m$ and let $-H_p$ be the generator of the holomorphic semigroup $T_p(z)$ on $L^p$. Then the spectrum of $H_p$ is independent of $p$ for $1 \leq p < \infty$. A similar result holds for $N \geq 2m$ in a certain open interval around $p = 2$.*

*Proof.* One first obtains $L^p$ operator bounds on the resolvent operators, by using (2) to estimate suitable integrals involving the complex time heat semigroups. From these calculations we learn that that the spectrum of $H_p$ is real for all $1 \leq p < \infty$ and that

$$\|(z - H_p)^{-1}\| \leq c|\operatorname{Im} z|^{-1}\left(\frac{\langle z\rangle}{|\operatorname{Im} z|}\right)^{\alpha} \tag{3}$$

for some $\alpha \geq 0$ and all $z \notin \mathbf{R}$, where $\langle z\rangle := (1 + |z|^2)^{1/2}$.

From this point on we use an abstract functional analytic argument, depending crucially upon the use of the Helffer-Sjöstrand formula [HS]:

$$f(H) := -\frac{1}{\pi}\int_{\mathbf{C}} \frac{\partial \tilde{f}}{\partial \bar{z}}(z - H)^{-1} \mathrm{d}x\mathrm{d}y \tag{4}$$

valid for all $f : \mathbf{R} \to \mathbf{C}$ in a certain algebra containing $C_c^\infty(\mathbf{R})$. In this formula $\tilde{f}$ is an almost analytic extension of $f$ to the complex plane, and lies in $C^\infty(\mathbf{C})$. The notion of almost analytic extension is due to Hörmander [Hö], and it is shown in [D1] that the operator $f(H)$ is independent of the choice of almost analytic extension within a certain specified class. The integral (4) is norm convergent even though the resolvent norm diverges as one approaches the real axis, and defines a functional calculus on the $L^p$ spaces; in other words the map $f \to f(H_p)$ is a bounded algebra homomorphism from a certain algebra of functions on $\mathbf{R}$ to the algebra of bounded linear operators on $L^p$, [D1]. The final step uses the consistency of the functional calculi in different $L^p$ spaces and the fact that if $x \in \mathbf{R}$ then $x \notin \operatorname{Spec}(H_p)$ if and only if there exists $f \in C_c^\infty$ such that $f(x) \neq 0$ but $f(H_p) = 0$, [D2].

There is no reason to confine attention to the action of the semigroup $\mathrm{e}^{-Ht}$ on the $L^p$ spaces. One can study the spectral behaviour of $H$ on a variety of other spaces, such as polynomially weighted $L^p$ spaces [D2]. The key inputs are the validity of a resolvent norm estimate of the type (3), and the consistency of the resolvents, and hence of the functional calculi defined by means of the H-S formula (4).

## References

[Ar] W. Arendt, *Gaussian estimates and interpolation of the spectrum in $L^p$*, Preprint 1993.

[AMT] P. Auscher, A. McIntosh and P. Tchamitchian, *Noyau de la chaleur d'opérateurs elliptiques complexes*, Math. Res. Lett. **1** (1994), 35–45.

[Au] P. Auscher, Private communication.

[BD] G. Barbatis and E. B. Davies, *Sharp bounds on heat kernels of higher order uniformly elliptic operators*, Preprint 1995.

[D1] E. B. Davies, *The functional calculus*, J. London Math. Soc., Preprint 1993 (to appear).

[D2] ———, *$L^p$ spectral independence and $L^1$ analyticity*, J. London Math. Soc. (to appear).

[D3] ———, *Uniformly elliptic operators with measurable coefficients*, J. Funct. Anal. (to appear).

[DM] E. B. Davies and N. Mandouvalos, *Heat kernel bounds on hyperbolic space and Kleinian groups*, Proc. London Math. Soc. (3) **57** (1988), 182–208.

[DST] E. B. Davies, B. Simon and M. Taylor, *$L^p$ spectral theory of Kleinian groups*, J. Funct. Anal. **78** (1988), 116–136.

[HS] B. Helffer and J. Sjöstrand, *Équation de Schrödinger avec champ magnetique et équation de Harper*, Schrödinger Operators (H. Holden and A. Jensen, eds.), Springer-Verlag, 1989, pp. 118–197, Lecture Notes in Physics, Vol. 345.

[HV] R. Hempel and J. Voigt, *The spectrum of a Schrödinger operator in $L^p(\mathbf{R}^N)$ is p-independent*, Comm. Math. Phys. **104** (1986), 243–250.

[Hö] L. Hörmander, *On the singularities of solutions of partial differential equations*, Proc. Intern. Conf. Tokyo 1969, Univ. of Tokyo Press, Tokyo, 1970, pp. 31–40.

[K] Yu. A. Kordyukov, *$L^p$-theory of elliptic differential operators on manifolds of bounded geometry*, Acta Appl. Math. **23** (1991), 223–260.

[MNP] V. G. Maz'ya, S. A. Nazarov and B. A. Plamenevskii, *Absence of the De-Giorgi-type theorems for strongly elliptic operators with complex coefficients*, J. Soviet Math. **28** (1985), 726–739.

[R] D. W. Robinson, *Elliptic Operators and Lie Groups*, Oxford University Press, 1991.

[St] K.-Th. Sturm, *On the $L^p$ spectrum of Laplace-Beltrami operators*, Preprint 1992.


Department of Mathematics, King's College, Strand, London WC2R 2LS, U.K.

*E-mail address*: udah210@kcl.ac.uk

# TRACE ASYMPTOTICS VIA ALMOST ANALYTIC EXTENSIONS

MOUEZ DIMASSI AND JOHANNES SJÖSTRAND

In this work we give a stationary approach to results on trace asymptotics with small remainder estimates, which in this degree of generality are due to Ivrii [I], who followed the more traditional method (for getting sharp estimates) of studying the trace of an associated unitary evolution group. The originality of Ivrii's results is that he does not require any parametrix constructions, and this permitted him to get very general results. For earlier results with parametrix constructions see Hörmander [H], Chazarain [C], Helffer-Robert [HeR]. That stationary methods can also produce sharp remainder estimates is known. See Métivier [M].

The following text is essentially a chapter in some lecture notes in preparation. In addition to the pleasure of giving some new approach, we also had in mind numerous situations that appear in mathematical physics and boundary value problems, where spectral problems can be reduced to similar problems in lower dimension, but where the dependence on the spectral parameter becomes non-linear. For the reduced problem, there does not seem to be natural well-posed evolution problems. It seems however that the methods of the present work easily apply to such more general problems and as a matter of fact, the first author has obtained such extensions in connection with the Schrödinger operator with slowly varying perturbations of periodic potentials, to be published elsewhere. Our approach is heavily based on the use of almost analytic extensions, introduced by Hörmander in his unpublished lecture notes from the Nordic Summer School on Tjörn in 1969, and since then used by a number of people, including the second author. Almost analytic extensions enter in a functional formula by Helffer-Sjöstrand [HeS]: If $f \in C_0^\infty(\mathbf{R})$, let $\widetilde{f} \in C_0^\infty(\mathbf{C})$ be an almost analytic extension, i.e. an extension of $f$, such that $\partial \widetilde{f}/\partial \overline{z}$ vanishes to infinite order on the real axis. Then, if $A$ is a self-adjoint operator on some Hilbert space, we have

$$f(A) = -\frac{1}{\pi} \int \frac{\partial \widetilde{f}}{\partial \overline{z}}(z - A)^{-1} L(dz), \tag{0}$$

where $L(dz)$ denotes the Lebesgue measure on $\mathbf{C}$. When $A$ is replaced by a scalar, this is just the fact that $(\pi z)^{-1}$ is a fundamental solution of the Cauchy-Riemann operator, and the formula is very easy to prove in the operator case. (See Dimassi [Di].) For extensions to the setting of Banach spaces, see Davies [D] and Jensen-Nakamura [JN]. This formula, also useful in the reductions mentioned above, will be our starting point. However, we will also use almost analytic extensions directly on our (microhyperbolic) $h$-pseudodifferential operators in a way reminiscent of the way Hörmander treated hyperbolic operators in his lecture notes mentioned above, although we do not try to obtain parametrices here. Neither do we appeal explicitly to propagation of singularities. Our proof keeps some fundamental ideas of Ivrii. For the associated evolution problem, he decomposes the time in various small intervals and estimates the corresponding traces, and shows that only a very small interval near zero counts in the end. In our approach this is done in the dual energy variable. We now explain the (known) result (due to Ivrii) and the (new) proof.

Let $\alpha < \beta$ be two real numbers, and let $P(h)$ be an $h$-pseudodifferential operator with symbol $p(x,\xi;h) \sim \sum_{j\geq 0} h^j p_j(x,\xi)$ in $S^0(\mathbf{R}^{2n};\mathcal{L}(\mathbf{C}^m,\mathbf{C}^m))$, $h \in ]0,h_0]$, $h_0 > 0$. Here $S^0(\mathbf{R}^{2n})$ is the space of $C^\infty$-functions which are bounded with all their derivatives, and we work with the Weyl quantization so that $P(h) = \mathrm{Op}(p(x,\xi;h)) = \mathrm{Op}_h(p) = p^w(x,hD;h)$. Sometimes we shall use the same letter for the symbol and for the corresponding $h$-pseudodifferential operator. We assume:

(H1) $P(h)$ is symmetric for every $h \in ]0,h_0]$. In particular $p_j(x,\xi)$ is a Hermitian matrix for every $j$ and every $(x,\xi) \in \mathbf{R}^{2n}$.

(H2) Let $\lambda_1(x,\xi) \leq \cdots \leq \lambda_m(x,\xi)$ be the eigenvalues of $p_0(x,\xi)$ arranged in increasing order. Then

$$\liminf_{|(x,\xi)|\to\infty} d(\lambda_j(x,\xi),[\alpha,\beta]) > 0$$

for all $j$, where $d$ denotes the standard distance on $\mathbf{R}$.

**Definition 1.** We say that $p_0(x,\xi)$ is microhyperbolic at $(\overline{x},\overline{\xi}) \in \mathbf{R}^{2n}$ in the direction $T \in \mathbf{R}^{2n}$, if there is a constant $C_0 > 0$ such that with standard norms and scalar products:

$$(\langle dp_0(x,\xi),T\rangle\omega|\omega) \geq \frac{1}{C_0}\|\omega\|^2 - C_0\|p_0(x,\xi)\omega\|^2,$$

for all $(x,\xi) \in \mathbf{R}^{2n}$ with $\|(x,\xi)-(\overline{x},\overline{\xi})\| \leq 1/C_0$ and for all $\omega \in \mathbf{C}^m$. (In this definition, we may replace $p_0$ by any other smooth function with values in the Hermitian matrices.)

For $\tau \in \mathbf{R}$, we put $\Sigma_\tau = \{(x,\xi) \in \mathbf{R}^{2n}; \det(p_0(x,\xi) - \tau) = 0\}$. We introduce the following assumption:

(H3) $p_0 - \tau$ is microhyperbolic (in some direction) at every point of $\Sigma_\tau$, when $\tau \in \{\alpha, \beta\}$.

The assumptions (H1), (H2) imply that if $\eta > 0$ is small enough, then (for $h > 0$ small enough) the spectrum of $P(h)$ in $[\alpha - \eta, \beta + \eta]$ is discrete. We choose $\eta$ small, so that (H3) still holds, when $\tau$ varies in $[\alpha - \eta, \alpha + \eta] \cup [\beta - \eta, \beta + \eta]$. Let $(\lambda_j(h))_{0 \leq j \leq N(h)}$ be the increasing sequence of eigenvalues of $P(h)$ in $[\alpha - \eta, \beta + \eta]$, repeated according to their multiplicities. Put

$$N_h(\alpha, \beta) = \#\{j; \lambda_j(h) \in [\alpha, \beta]\}.$$

Let $\mathcal{F}_h$ be the $h$-Fourier transform, defined by $\mathcal{F}_h u(t) = \int e^{-it\lambda/h} u(\lambda) d\lambda$, $u \in C_0^\infty(\mathbf{R})$, $t \in \mathbf{R}$. Let $\theta \in C_0^\infty(\mathbf{R})$ have its support in a small neighborhood of the origin and be equal to one near the origin. Let $f \in C_0^\infty(]\tau - \eta, \tau + \eta[)$, $\tau \in \{\alpha, \beta\}$. Under the assumptions (H1–3), we have:

**Theorem 2.** *For every integer $N \geq 1$, we have*

$$\operatorname{tr}(f(P)\mathcal{F}_h^{-1}\theta(\lambda - P)) = (2\pi h)^{-n}\Big(f(\lambda)\sum_{j=0}^{N-1} h^j \gamma_j(\lambda) + \mathcal{O}(h^N)\Big),$$

*uniformly for $\lambda \in [\tau - \eta, \tau + \eta]$. Moreover, $\gamma_j$ are smooth functions of $\lambda$, independent of $f$, $\theta$, and for $j = 0$ we have*

$$\gamma_0(\lambda) = -\frac{1}{2\pi i}\int \operatorname{tr}((\lambda + i0 - p_0(x,\xi))^{-1} - (\lambda - i0 - p_0(x,\xi))^{-1})\chi(x,\xi) dx d\xi.$$

*Here $\chi \in C_0^\infty(\mathbf{R}^{2n})$ is equal to 1 in a neighborhood of $\Sigma_{[\tau-\eta,\tau+\eta]} = \cup_{\sigma\in[\tau-\eta,\tau+\eta]}\Sigma_\sigma$.*

From Theorem 2 and standard arguments, we will get:

**Theorem 3.** *We have*

$$N_h(\alpha, \beta) = \frac{1}{(2\pi h)^n}\sum_{j=1}^{m}\iint_{\lambda_j^{-1}([\alpha,\beta])} dx d\xi + \mathcal{O}(h^{1-n}), \quad h \searrow 0.$$

The most essential step in the proof of Theorem 2 will be the following result:

**Proposition 4.** *There exists $C_0 > 0$, such that if $\theta \in C_0^\infty(]\frac{1}{2}, 1[)$, then for every $\delta \in ]0,1]$ we have*

$$\operatorname{tr}(f(P)\mathcal{F}_h^{-1}\theta_\epsilon(\lambda - P)) = \mathcal{O}(h^\infty), \tag{1}$$

*uniformly for $\lambda \in [\tau - \eta, \tau + \eta]$, $\epsilon \in [h^{1-\delta}, 1/C_0]$, where we have written $\theta_\epsilon(t) = \theta(t/\epsilon)$.*

*Proof.* Let $\widetilde{f}$ be an almost analytic extension of $f$, such that with $\overline{\partial} = \partial/\partial\overline{z}$:

$$\widetilde{f} \in C_0^\infty(\mathbf{C}), \tag{2}$$

$$\widetilde{f}(z) = f(z) \quad \text{for all } z \in \mathbf{R}, \tag{3}$$

$$\overline{\partial}\widetilde{f}(z) = \mathcal{O}(|\operatorname{Im} z|^N) \quad \text{for all } N \in \mathbf{N}. \tag{4}$$

Let $\psi(t) \in C^\infty(\mathbf{R}; [0,1])$ be equal to 1 for $t \le 1$ and equal to 0 for $t \ge 2$. Put

$$\psi_Y(z) = \psi(\operatorname{Im} z/Y), \quad z \in \mathbf{C},\ Y > 0.$$

We then get

$$\overline{\partial}(\widetilde{f}\psi_Y)(z) = \begin{cases} \mathcal{O}(|\operatorname{Im} z|^N), & \text{if } \operatorname{Im} z < 0 \\ \mathcal{O}\big(\psi_Y(z)|\operatorname{Im} z|^N + Y^{-1}1_{[1,2]}(\operatorname{Im} z/Y)\big), & \text{if } \operatorname{Im} z \ge 0. \end{cases} \tag{5}$$

Let $\chi \in C_0^\infty(\mathbf{R}^{2n}; [0,1])$ be equal to 1 in a small neighborhood of $\Sigma_{[\tau-\eta,\tau+\eta]}$ and put $\widetilde{P}(h) = P(h) + i\chi^w(x, hD_x)I$. By the definition of $\chi$, the operator $\widetilde{P} - z$ is elliptic for $z$ in a complex neighborhood $K$ of $[\tau - \eta, \tau + \eta]$. Let $\chi_1, \chi_2$ be two functions in $C_0^\infty(\mathbf{R}^{2n})$ which are equal to 1 in a neighborhood of $\operatorname{supp}\chi$. (From now on we shall sometimes use the same symbol for an $h$-pseudodifferential operator and for its Weyl-symbol.) Replacing $(z-P)^{-1}$ in the three last terms of the identity

$$\begin{aligned}(z-P)^{-1} &= \chi_1(z-P)^{-1}\chi_2 + (1-\chi_1)(z-P)^{-1}\chi_2 \\ &\quad + \chi_1(z-P)^{-1}(1-\chi_2) + (1-\chi_1)(z-P)^{-1}(1-\chi_2)\end{aligned}$$

by

$$(z-P)^{-1} = (z-\widetilde{P})^{-1} + (z-\widetilde{P})^{-1}(P-\widetilde{P})(z-P)^{-1},$$

and using the fact that $(z-\widetilde{P})^{-1}$ is holomorphic in $K$, as well as the fact that on the level of symbols $\operatorname{supp}(\widetilde{P} - P) \cap \operatorname{supp}(1-\chi_j) = \emptyset$ for $j = 1, 2$, we get, using (0) and the cyclicity of the trace:

$$\text{(6)}\quad \begin{aligned} &\operatorname{tr}(f(P)\mathcal{F}_h^{-1}\theta_\epsilon(\lambda - P)) \\ &\quad = -\operatorname{tr}\frac{1}{\pi}\int \overline{\partial}(\widetilde{f}\psi_Y)(z)\mathcal{F}_h^{-1}\theta_\epsilon(\lambda - z)(z-P)^{-1}L(dz) \\ &= -\operatorname{tr}\frac{1}{\pi}\int \overline{\partial}(\widetilde{f}\psi_Y)(z)\mathcal{F}_h^{-1}\theta_\epsilon(\lambda - z)\chi_1(z-P)^{-1}\chi_2 L(dz) + \mathcal{O}(h^\infty) \\ &\quad = -\operatorname{tr}\frac{1}{\pi}\int_{\operatorname{Im} z<0} \ldots L(dz) - \operatorname{tr}\frac{1}{\pi}\int_{\operatorname{Im} z>0} \ldots L(dz) + \mathcal{O}(h^\infty). \end{aligned}$$

Here $Y = M(h/\epsilon)\log(1/h)$ where $M > 0$ is a sufficiently large constant, to be fixed later. We have used that according to the Paley-Wiener theorem

$$\text{(7)}\qquad \mathcal{F}_h^{-1}\theta_\epsilon(\lambda - z) = \begin{cases} \mathcal{O}(\frac{\epsilon}{h}e^{\frac{\epsilon \operatorname{Im} z}{h}}) & \text{for } \operatorname{Im} z > 0, \\ \mathcal{O}(\frac{\epsilon}{h}e^{\frac{\epsilon \operatorname{Im} z}{2h}}) & \text{for } \operatorname{Im} z < 0, \end{cases}$$

and that the presence of $\chi_1$, $\chi_2$ assures that we have trace class operators. Indeed, $\chi_1$, $\chi_2$ are of trace class with trace norm $\mathcal{O}(h^{-n})$.

*Remark 5.* When $\operatorname{Im} z > 0$, we have

$$\text{(8)}\qquad \psi_Y(z)|\operatorname{Im} z|^N \le (2Y)^N$$

for all $N \ge 0$. Using also (6), (7), we will be able to neglect this contribution from (5) to (6).

The estimates (4) and (7) imply that the first term of the last member of (6) is $\mathcal{O}(h^\infty)$. In view of (5) and Remark 5, (1) will follow from

$$\text{(9)}\quad -\operatorname{tr}\frac{1}{\pi}\int_{\operatorname{Im} z>Y} \overline{\partial}(\widetilde{f}\psi_Y)(z)\mathcal{F}_h^{-1}\theta_\epsilon(\lambda - z)\chi_1(z-P)^{-1}\chi_2 L(dz) = \mathcal{O}(h^\infty).$$

If we choose $\chi_1$, $\chi_2$ of the form $\sum_i \chi_i(x,\xi)$ where each $\chi_i$ has its support in a small neighborhood of some point in $\Sigma_{[\tau-\eta,\tau+\eta]}$, we see that it suffices to show:

**Lemma 6.** *For every $(i,j)$ with $\operatorname{supp}\chi_i \cap \operatorname{supp}\chi_j \neq \emptyset$, and for every $N \in \mathbf{N}$, there exists $M(N) \ge 0$, such that for $Y = M(h/\epsilon)\log(1/h)$ and $M \ge M(N)$:*

(10)

$$-\frac{1}{\pi}\int_{\operatorname{Im} z>Y} \overline{\partial}(\widetilde{f}\psi_Y)(z)\mathcal{F}_h^{-1}\theta_\epsilon(\lambda - z)\operatorname{tr}(\chi_i(z-P)^{-1}\chi_j)L(dz) = \mathcal{O}(h^N).$$

In fact, by the cyclicity of the trace, it is easy to see that the corresponding integrals with $\operatorname{supp}\chi_j \cap \operatorname{supp}\chi_k = \emptyset$, are $\mathcal{O}(h^\infty)$. In the following, we fix $(i,j)$ as in the lemma. To prove Lemma 6, we need

**Lemma 7.** *Let $\widetilde{P} = \widetilde{p}^w(x, hD_x) + P(h) - p_0^w(x, hD_x)$ be self-adjoint, where $\widetilde{p} \in S^0(\mathbf{R}^{2n}; \mathcal{L}(\mathbf{C}^m, \mathbf{C}^m))$ and $\widetilde{p}(x,\xi) = p_0(x,\xi)$ in a small neighborhood of $\operatorname{supp}\chi_i \cup \operatorname{supp}\chi_j$. Then for every $N \in \mathbf{N}$, there exists $M(N) \geq 0$, such that for $Y = M(h/\epsilon)\log(1/h)$ and $M \geq M(N)$:*

$$(11)\quad \operatorname{tr}\frac{1}{\pi}\int_{\operatorname{Im} z > Y} \overline{\partial}(\widetilde{f}\psi_Y)(z)\mathcal{F}_h^{-1}\theta_\epsilon(\lambda - z)(\chi_i(z-P)^{-1}\chi_j)L(dz)$$
$$- \operatorname{tr}\frac{1}{\pi}\int_{\operatorname{Im} z > Y} \overline{\partial}(\widetilde{f}\psi_Y)(z)\mathcal{F}_h^{-1}\theta_\epsilon(\lambda - z)(\chi_i(z-\widetilde{P})^{-1}\chi_j)L(dz) = \mathcal{O}(h^N).$$

*Proof of Lemma 7.* Let $\widetilde{\chi} \in C_0^\infty(\mathbf{R}^{2n})$ be equal to one in a small neighborhood of $\operatorname{supp}\chi_i \cup \operatorname{supp}\chi_j$ and have its support contained in the interior of an $h$-independent set where the symbols of $P(h)$ and $\widetilde{P}(h)$ coincide. The identity

$$(12)\quad (z-P)\widetilde{\chi}(z-\widetilde{P})^{-1}\chi_j = \chi_j + \widetilde{\chi}(\widetilde{P}-P)(z-\widetilde{P})^{-1}\chi_j - [P,\widetilde{\chi}](z-\widetilde{P})^{-1}\chi_j + \mathcal{O}(h^\infty),$$

and the fact that $\chi_j\widetilde{\chi} = \chi_j + \mathcal{O}(h^\infty)$ in trace norm, imply

$$(13)\quad \chi_i(z-\widetilde{P})^{-1}\chi_j - \chi_i(z-P)^{-1}\chi_j$$
$$= \chi_i(z-P)^{-1}\widetilde{\chi}(\widetilde{P}-P)(z-\widetilde{P})^{-1}\chi_j$$
$$- \chi_i(z-P)^{-1}[P,\widetilde{\chi}](z-\widetilde{P})^{-1}\chi_j + \mathcal{O}\Big(\frac{h^\infty}{|\operatorname{Im} z|}\Big).$$

Here the first term of the second member is $\mathcal{O}(h^\infty|\operatorname{Im} z|^{-2})$ in trace norm. For the second term, we shall use the fact that modulo $\mathcal{O}(h^\infty)$ in trace norm $[P,\widetilde{\chi}]$ has a symbol with support in a compact set $\widetilde{K}$ such that $\operatorname{supp}\chi_i \cap \widetilde{K} = \emptyset$.

Let $G_0 \in C_0^\infty(\mathbf{R}^{2n})$ be real-valued and such that

$$G_0 = 1 \quad \text{near } \operatorname{supp}\chi_i, \qquad G_0 = 0 \quad \text{near } \widetilde{K}.$$

Put $G = \alpha G_0$, $\alpha > 0$. We notice that the symbol $a = e^{G\log\frac{1}{h}}$ is of class $S_\delta$ for every $\delta > 0$, uniformly for $\alpha$ in any fixed bounded interval. Here we define $S_\delta^m$ to be the space of functions $a(x,\xi;h)$ on $\mathbf{R}^{2n}\times]0,h_0]$, for some $h_0 > 0$, which are smooth in $(x,\xi)$ and satisfy $\partial^\alpha_{(x,\xi)}a = \mathcal{O}_\alpha(h^{-m-\delta|\alpha|})$ for every multiindex $\alpha$, and we let $S_\delta$ be the union over $m$ of the $S_\delta^m$. By $e^{G\log\frac{1}{h}}$ we also denote the corresponding $h$-pseudodifferential operator which is elliptic and has an inverse operator $(e^{G\log\frac{1}{h}})^{-1}$ with symbol in the same classes. It is then clear that

$$(14)\qquad (e^{G\log\frac{1}{h}})^{-1}(z-P)e^{G\log\frac{1}{h}} = z - P + \mathcal{O}(\alpha h\log\frac{1}{h}),$$

in operator norm, for $h \leq h(\alpha)$, where $h(\alpha) > 0$ is some continuous function. Using that $G = \alpha$ near $\operatorname{supp}\chi_i$ and that $G = 0$ near $\widetilde{K}$, we get in trace norm:

$$\begin{aligned}(15)\quad & e^{\alpha \log \frac{1}{h}} \chi_i (z-P)^{-1}[P, \widetilde{\chi}] \\ &= \chi_i e^{G \log \frac{1}{h}} (z-P)^{-1} (e^{G \log \frac{1}{h}})^{-1} [P, \widetilde{\chi}] + \mathcal{O}\Big(\frac{h^\infty}{|\operatorname{Im} z|}\Big) \\ &= \chi_i (z - e^{G \log \frac{1}{h}} P (e^{G \log \frac{1}{h}})^{-1})^{-1} [P, \widetilde{\chi}] + \mathcal{O}\Big(\frac{h^\infty}{|\operatorname{Im} z|}\Big).\end{aligned}$$

We choose $G$ with

$$\alpha = \min\Big(\frac{\operatorname{Im} z}{\widetilde{C} h \log \frac{1}{h}}, \mathcal{O}(1)\Big),$$

where $\mathcal{O}(1)$ is some arbitrarily large and fixed constant and $\widetilde{C}$ is sufficiently large. Then the expression (15) is $\mathcal{O}(1/(h^n |\operatorname{Im} z|))$ in trace norm, and we get with a new constant $C_2 > 0$:

$$\begin{aligned}\chi_i (z-P)^{-1}[P, \widetilde{\chi}] &= \mathcal{O}\big(h^{-n} |\operatorname{Im} z|^{-1} e^{-\alpha \log \frac{1}{h}}\big) \\ &= \mathcal{O}\big(h^{-n} |\operatorname{Im} z|^{-1} \max(h^{\mathcal{O}(1)}, e^{-\frac{\operatorname{Im} z}{C_2 h}})\big).\end{aligned}$$

Hence for $\frac{Mh}{\epsilon} \log \frac{1}{h} = Y \leq \operatorname{Im} z \leq 2Y$, we get, by using also (7):

$$\begin{aligned}(16)\quad & \mathcal{F}_h^{-1} \theta_\epsilon (\lambda - z) \chi_i (z-P)^{-1} [P, \widetilde{\chi}] \\ &= \mathcal{O}\big(\epsilon h^{-n-1} |\operatorname{Im} z|^{-1} \max(h^{\mathcal{O}(1)}, e^{-\frac{1}{h} \operatorname{Im} z (\frac{1}{C_2} - \frac{1}{C_0})})\big),\end{aligned}$$

where we recall that $\epsilon \leq 1/C_0$. We choose $C_0 > C_2$. With $Y = M(h/\epsilon)\log(1/h)$ the LHS of (11) is

$$\begin{aligned}& \mathcal{O}(h^\infty) + \int\limits_{\substack{Y \leq \operatorname{Im} z \leq 2Y \\ |\operatorname{Re} z| \leq \text{const.}}} \mathcal{O}(1) Y^{-2} \epsilon h^{-n-1} \max\big(h^{\mathcal{O}(1)}, e^{-\frac{1}{h} \operatorname{Im} z (\frac{1}{C_2} - \frac{1}{C_0})}\big) L(dz) \\ &= \mathcal{O}(h^\infty) + \mathcal{O}(1) \frac{\epsilon}{Mh \log \frac{1}{h}} \epsilon h^{-n-1} \max(h^{\mathcal{O}(1)}, e^{-C_0 M (\log \frac{1}{h})(\frac{1}{C_2} - \frac{1}{C_0})}).\end{aligned}$$

Choosing first $M$ sufficiently large and then the exponent in $h^{\mathcal{O}(1)}$ sufficiently large, we see that this is $\mathcal{O}(h^N)$ for any given $N$. □

*Proof of Lemma 6.* In the appendix, we show that we can replace $P$ by $\widetilde{P}$ as in Lemma 7, (provided that we take $\chi_i$, $\chi_j$ there with sufficiently small support,) with leading symbol $\widetilde{p}$, such that if $\tau \in \{\alpha, \beta\}$, $\tau - \widetilde{p}$ is globally and uniformly hyperbolic with respect to $H_\phi$, where $\phi$ is a linear function and $H_\phi$ the corresponding constant Hamilton field. After a linear canonical transformation, we may assume that $\phi = x_1$.

For $t \in \mathbf{R}$, we put $Q_t = e^{-it\phi/h} Q e^{it\phi/h}$, where $Q$ denotes one of $\chi_i$, $\chi_j$, $P$. On the symbol level: $Q_t(x,\xi) = Q((x,\xi) - tH_\phi)$. Let $\widetilde{Q}(x,\xi) \in S^0(\mathbf{C}^{2n})$ be an almost analytic extension (i.e. satisfying the Cauchy-Riemann equations to infinite order on $\mathbf{R}^n$,) and put $\widetilde{Q}_t(x,\xi) = \widetilde{Q}((x,\xi) - tH_\phi) \in S^0(\mathbf{R}^{2n})$ for $t \in \mathbf{C}$. Then on the operator level $\widetilde{Q}_t = e^{-i \operatorname{Re} t\phi/h} \widetilde{Q}_{i \operatorname{Im} t} e^{i \operatorname{Re} t\phi/h}$ and on the symbol level, we have the obvious result:

**Lemma 8.** *We have in $S^0$:*

$$\frac{\partial}{\partial \bar{t}} \widetilde{Q}_t = \mathcal{O}(|\operatorname{Im} t|^\infty), \tag{17}$$

$$\nabla_t^k \widetilde{Q}_t = \mathcal{O}(1), \quad k \in \mathbf{N}, \tag{18}$$

$$\widetilde{Q}_t = Q_t \text{ for } t \in \mathbf{R}. \tag{19}$$

*Proof of Lemma 6 continued.* We put

$$h(t) = \operatorname{tr}(\widetilde{\chi}_{i,t}(z - \widetilde{P}_t)^{-1} \widetilde{\chi}_{j,t}).$$

$h(t)$ is well defined in the zone $|\operatorname{Im} t| \leq \frac{1}{\mathcal{O}(1)} |\operatorname{Im} z|$ and has the following properties:

$$h(t) = \mathcal{O}(h^{-n} |\operatorname{Im} z|^{-1}),$$
$$\bar{\partial} h(t) = \mathcal{O}(h^{-n} |\operatorname{Im} t|^\infty |\operatorname{Im} z|^{-2}).$$

Since $h(t)$ only depends on $\operatorname{Im} t$ (by cyclicity of the trace and the fact that $\widetilde{\chi}_{i,t}(z - \widetilde{P}_t)^{-1} \widetilde{\chi}_{j,t}$ changes by conjugation when $\operatorname{Re} t$ is varied), it follows that

$$h(i \operatorname{Im} t) = h(0) + \mathcal{O}\Big(\frac{|\operatorname{Im} t|^\infty}{h^n |\operatorname{Im} z|^2}\Big).$$

We can then replace $\operatorname{tr}(\chi_i (z - P)^{-1} \chi_j)$ in the integral in the first member of (10) by $h(iY/C_1)$, where $C_1$ does not depend on the other constants.

By construction of $\widetilde{P}_t$ we see that the leading symbol of $z - \widetilde{P}_{i \operatorname{Im} t}$ is of the form

$$z - \widetilde{p}_{i \operatorname{Im} t}(x,\xi) = z - p_0(x,\xi) - i \operatorname{Im} t \frac{\partial p_0(x,\xi)}{\partial \xi_1} + \mathcal{O}(|\operatorname{Im} t|^2).$$

Put $\mu = \operatorname{Im} t = Y/C_1$ and let $-\frac{1}{2}\mu < \operatorname{Im} z < 0$. Using the microhyperbolicity (Definition 1, with $T = (0; (-1, 0, 0, \ldots, 0))$), we obtain:

$$\operatorname{Im}(z - \widetilde{p}_{i\mu}(x,\xi)) \geq (2\mu + \operatorname{Im} z) I - \mathcal{O}(1) \mu (\operatorname{Re} z - p(x,\xi))^* (\operatorname{Re} z - p(x,\xi))$$

in the sense of Hermitian matrices, (assuming for the simplicity of notation, that the factor $1/C_0$ there is equal to 3,) where the star indicates that we take the usual complex adjoint. Then, if $h$ is small enough, there exists a constant $C_2 > 0$ such that,

$$\text{(20)}\quad \operatorname{Im}(z - \widetilde{p}_{i\mu}(x,\xi)) + C_2\mu(z - \widetilde{p}_{i\mu}(x,\xi))^*(z - \widetilde{p}_{i\mu}(x,\xi)) \geq (\mu + \operatorname{Im} z)I.$$

Now the semi-classical version of the sharp Gårding inequality (see for instance [S2]) implies

$$\begin{aligned}\text{(21)}\quad & \operatorname{Im}(\mathrm{Op}_h(z - \widetilde{p}_{i\mu})u|u) + C_2\mu\|\mathrm{Op}_h^w(z - \widetilde{p}_{i\mu})u\|^2 \\ & \geq (\mathrm{Op}_h(\operatorname{Im}(z - \widetilde{p}_{i\mu}) + C_2\mu(z - \widetilde{p}_{i\mu})^*(z - \widetilde{p}_{i\mu}))u|u) - \mathcal{O}(1)h\mu\|u\|^2 \\ & \geq (\mu + \operatorname{Im} z)\|u\|^2 - \mathcal{O}(1)(h + \mu h)\|u\|^2 \geq \frac{\mu}{3}\|u\|^2,\end{aligned}$$

for all $u \in L^2(\mathbf{R}^n; \mathbf{C}^m)$, for $h$ small enough. (21) combined with the inequality $ab \leq \frac{\mu}{6}a^2 + \frac{3}{2\mu}b^2$ gives,

$$\begin{aligned}\frac{\mu}{3}\|u\|^2 &\leq \|\mathrm{Op}_h^w(z - \widetilde{p}_{i\mu})u\|\|u\| + C_1\mu\|\mathrm{Op}_h^w(z - \widetilde{p}_{i\mu})u\|^2 \\ &\leq \frac{\mu}{6}\|u\|^2 + (\frac{3}{2\mu} + C_1\mu)\|\mathrm{Op}_h(z - \widetilde{p}_{i\mu})u\|^2.\end{aligned}$$

Then there exists $C_2 > 0$ such that,

$$\text{(22)}\qquad \frac{\mu}{C_2}\|u\| \leq \|\mathrm{Op}_h^w(z - \widetilde{p}_{i\mu})u\|,$$

for all $u \in L^2(\mathbf{R}^n; \mathbf{C}^n)$. We conclude that $(z - \widetilde{P}_{iY/C_1})^{-1}$ extends as a holomorphic function of $z$ to the zone $\operatorname{Im} z \geq -Y/C_1^2$ and we deduce that the LHS of (10) is equal to

$$\frac{1}{\pi}\int_{\operatorname{Im} z<0} \overline{\partial}(\widetilde{f}\psi_{Y,Y/C_1^2})\mathcal{F}_h^{-1}\theta_\epsilon(\lambda - z)h(iY/C_1)L(dz) + \mathcal{O}(h^N),$$

where we define $\psi_{\alpha,\beta}(z) = \psi(-\operatorname{Im} z/\beta)\psi(\operatorname{Im} z/\alpha)$. Using Remark 5, which extends to the zone $\operatorname{Im} z < 0$, it is enough to estimate

$$\begin{aligned}&\frac{1}{\pi}\int_{\operatorname{Im} z<0} \widetilde{f}\overline{\partial}(\psi_{Y,Y/C_1^2})\mathcal{F}_h^{-1}\theta_\epsilon(\lambda - z)h(iY/C_1)L(dz) \\ &= \mathcal{O}(1)h^{-n}Y^{-2}\int_{\substack{\operatorname{Re} z\in]\tau-\eta,\tau+\eta[ \\ -2Y/C_1^2\leq\operatorname{Im} z\leq -Y/C_1^2}} \frac{\epsilon}{h}e^{\epsilon\operatorname{Im} z/2h}L(dz) \\ &= \mathcal{O}(1)\frac{\epsilon^2 h^{-n-2}}{M\log\frac{1}{h}}e^{\frac{-M}{2C_1^2}\log\frac{1}{h}} = \mathcal{O}(1)\frac{\epsilon^2 h^{-n-2+\frac{M}{2C_1^2}}}{M\log\frac{1}{h}}.\end{aligned}$$

To finish the proof, it suffices to choose $M$ sufficiently large. □

*Remark 9.* Proposition 4 remains valid if $\operatorname{supp}\theta \subset ]-1,-\frac{1}{2}[$.

*Proof of Theorem 2.* In the following we let $C$ be a large fixed constant and $\delta$ a constant in $]0,\frac{1}{2}[$. According to the preceding remark and Proposition 4, we have:

$$\begin{aligned}(23)\quad & \operatorname{tr}(\mathcal{F}_h^{-1}\theta(\lambda-P)f(P)) \\ &= -\operatorname{tr}\frac{1}{\pi}\int\overline{\partial}(\psi_{h^\delta}\widetilde{f})(z)\mathcal{F}_h^{-1}\theta_\epsilon(\lambda-z)(z-P)^{-1}\chi L(dz)+\mathcal{O}(h^\infty).\end{aligned}$$

Here $\chi$, $\theta$ are the functions occurring in Theorem 2, $\epsilon=h^{1-\delta}$ and $\psi_{h^\delta}=\psi(\frac{\operatorname{Im}z}{h^\delta})$, where now $\psi\in C_0^\infty(]-2,2[)$ is equal to 1 on $[-1,1]$. In fact, we can write $\theta-\theta_\epsilon$ as a sum of functions $\widetilde{\theta}_{\widetilde{\epsilon}}$, with $\widetilde{\theta}$, $\widetilde{\epsilon}$ as $\theta$, $\epsilon$ in Proposition 4 and Remark 9, in order to replace $\theta$ by $\theta_\epsilon$ in the formula

$$\operatorname{tr}(\mathcal{F}_h^{-1}\theta)(\lambda-P)f(P)=-\operatorname{tr}\frac{1}{\pi}\int\overline{\partial}\widetilde{f}(\mathcal{F}_h^{-1}\theta)(\lambda-P)(z-P)^{-1}L(dz),$$

and then insert $\psi_{h^\delta}$ and $\chi$ as before. Since $\overline{\partial}(\psi_{h^\delta}\widetilde{f})=\psi_{h^\delta}\overline{\partial}\widetilde{f}+\widetilde{f}\,\overline{\partial}\psi_{h^\delta}$, we can again use Remark 5 to reduce the domain of integration in (23) to $K_\delta=\{z\in\mathbf{C};|\operatorname{Im}z|\geq h^\delta\}$. Here we also use that $\mathcal{F}_h^{-1}\theta_\epsilon(\lambda-z)=\mathcal{O}(\frac{\epsilon}{h}e^{\frac{2\epsilon}{h}|\operatorname{Im}z|})$ by Payley-Wiener. It is easy to show, (see [S1],) that $(z-P)^{-1}$ is given in $K_\delta$ by:

$$(24)\qquad E_0(x,\xi,z)+hE_1(x,\xi,z)+\cdots+h^NE_N(x,\xi,z)+\mathcal{O}(h^{N(1-2\delta)}).$$

Here $E_j(x,\xi,z)$ is a finite sum of terms of the form

$$(z-p_0(x,\xi))^{-1}b_1(x,\xi,z)(z-p_0(x,\xi))^{-1}\cdots b_{k-1}(x,\xi,z)(z-p_0)^{-1}$$

with $k\leq 2j+1$ and $b_k\in S^0(\mathbf{R}^{2n};\mathcal{L}(\mathbf{C}^m,\mathbf{C}^m))$ uniformly in $z$. We then get:

$$(25)\qquad \operatorname{tr}((\mathcal{F}_h^{-1}\theta)(\lambda-P)f(P))=\frac{1}{(2\pi h)^n}\sum_{j=0}^N h^ja_j+\mathcal{O}(h^{N(1-2\delta)-n}),$$

where

$$(26)\qquad a_j=-\frac{1}{\pi}\int\operatorname{tr}\iint\overline{\partial}(\psi_{h^\delta}\widetilde{f})(z)\mathcal{F}_h^{-1}\theta_\epsilon(\lambda-z)E_j(x,\xi,z)\chi(x,\xi)dxd\xi L(dz).$$

**Lemma 10.** *The functions* $e_j(z) = \iint \operatorname{tr} E_j(x,\xi,z)\chi(x,\xi)dxd\xi$, *defined on* $]\tau-\eta,\tau+\eta[\,\pm i\,]0,\alpha[$ *for some constant* $\alpha > 0$, *extend to* $C^\infty$*-functions on* $]\tau-\eta,\tau+\eta[\,\pm i\,[0,\alpha[$.

*Proof.* For the simplicity of the notation, we only treat the case $j = 0$. Decomposing $\chi$ by a partition of unity, we may also assume that $\chi(x,\xi)$ has its support in some small neighborhood of some point in $\Sigma_{[\tau-\eta,\tau+\eta]}$. Let $U$ be a constant direction of microhyperbolicity for $p_0-\lambda$ in a neighborhood of $\operatorname{supp}\chi$.

Let $\widetilde{p}_0$ be an almost analytic extension of $p_0$. As with the formula (22), we show that $(z-\widetilde{p}_0((x,\xi)-itU))^{-1}$ exists and is of norm $\mathcal{O}(1/t)$ for $\operatorname{Re} z \in ]\tau-\eta,\tau+\eta[$, $\operatorname{Im} z \geq 0$, $0 \leq t \leq t_0$, $t_0 > 0$ sufficiently small, $\operatorname{Im} z + t > 0$. Stokes' formula gives for $\operatorname{Im} z > 0$:

$$\begin{aligned}
&\iint (z-p_0(x,\xi))^{-1}\chi(x,\xi)\,dx\,d\xi \\
&= \iint_{\mathbf{R}^{2n}+it_0U} (z-\widetilde{p}_0)^{-1}\widetilde{\chi}(x,\xi)dxd\xi \\
&+ \iiint_{\substack{0\leq t\leq t_0\\(x,\xi)\in\mathbf{R}^{2n}}} \Gamma^*\big(((z-\widetilde{p}_0)^{-1}\overline{\partial}\widetilde{p}_0(z-\widetilde{p}_0)^{-1}\widetilde{\chi}+(z-p_0)^{-1}\overline{\partial}\widetilde{\chi})\wedge dxd\xi\big),
\end{aligned}$$

where $\Gamma : (t,x,\xi) \mapsto (x,\xi)-itU$ and $\widetilde{\chi}$ is an almost analytic extension of $\chi$. Here the first term of the last member extends to a $C^\infty$ function on $]\tau-\eta,\tau+\eta[\,+i\,]0,\alpha[$. The same is true for the last term, since the integrand is $\mathcal{O}(t^\infty)$ as well as all its derivatives with respect to $z$. We have then proved that $e_0$ extends to a smooth function on $]\tau-\eta,\tau+\eta[\,+i\,[0,\alpha[$. Using the Hermitian property we get the extension to $]\tau-\eta,\tau+\eta[\,-i\,[0,\alpha[$ also. $\square$

*End of proof of Theorem 2.* From (26), Lemma 10 and Stokes' formula, we get

$$a_j = -\frac{1}{2\pi i}\int \widetilde{f}(\lambda')\mathcal{F}_h^{-1}\theta_\epsilon(\lambda-\lambda')(e_j(\lambda'+i0)-e_j(\lambda'-i0))d\lambda'. \tag{27}$$

Here

$$\mathcal{F}_h^{-1}\theta_\epsilon = \mathcal{F}_{h/\epsilon}^{-1}\theta = \mathcal{F}^{-1}\Big(\theta\Big(\frac{h}{\epsilon}\,\cdot\Big)\Big) = \delta_0 + \mathcal{O}(h^\infty)$$

in the sense of distributions, so (27) gives

$$a_j = -\frac{1}{2\pi i}f(\lambda)(e_j(\lambda+i0)-e_j(\lambda-i0)) + \mathcal{O}(h^\infty), \tag{28}$$

and we get Theorem 2 with

$$\gamma_j(\lambda) = -\frac{1}{2\pi i}(e_j(\lambda + i0) - e_j(\lambda - i0)) \tag{29}$$

□

Notice that for $f \in C_0^\infty(]\tau - \eta, \tau + \eta[)$, $(x, \xi) \in \operatorname{supp}\chi$:

$$-\frac{1}{2\pi i}\int f(\lambda)(\operatorname{tr}(\lambda + i0 - p_0(x,\xi))^{-1} - \operatorname{tr}(\lambda - i0 - p_0(x,\xi))^{-1})d\lambda$$
$$= \sum_{j=1}^{m} f(\lambda_j(x,\xi)).$$

Consequently,

$$\begin{aligned} \int f(\lambda)\gamma_0(\lambda)d\lambda &= \sum_{j=1}^{m} \iint f(\lambda_j(x,\xi))\chi(x,\xi)dxd\xi \\ &= \sum_{j=1}^{m} \iint f(\lambda_j(x,\xi))dxd\xi. \end{aligned} \tag{30}$$

*Proof of Theorem 3.* Let the functions $f_1 \in C_0^\infty(]\alpha - \eta, \alpha + \eta[; [0,1])$, $f_2 \in C_0^\infty(]\alpha + \frac{\eta}{2}, \beta - \frac{\eta}{2}[; [0,1])$, $f_3 \in C_0^\infty(]\beta - \eta, \beta + \eta[; [0,1])$ satisfy $f_1 + f_2 + f_3 = 1$ on $[\alpha - \frac{\eta}{2}, \beta + \frac{\eta}{2}]$. Let $\lambda_0(h) \leq \lambda_1(h) \leq .. \leq \lambda_{N(h)}$ be the eigenvalues of $P(h)$ (counted with their multiplicities) in $[\alpha - \eta, \beta + \eta]$. We have

$$\begin{aligned} N_h(\alpha,\beta) &= \sum_{\alpha \leq \lambda_j(h) \leq \beta} (f_1 + f_2 + f_3)(\lambda_j(h)) \\ &= \sum_{\alpha \leq \lambda_j(h)} f_1(\lambda_j(h)) + \sum f_2(\lambda_j(h)) + \sum_{\lambda_j(h) \leq \beta} f_3(\lambda_j(h)) \\ &= \sum_{\alpha \leq \lambda_j(h)} f_1(\lambda_j(h)) + \operatorname{tr} f_2(P(h)) + \sum_{\lambda_j(h) \leq \beta} f_3(\lambda_j(h)). \end{aligned} \tag{31}$$

Since the trace of an $h$-pseudodifferential operator with a sufficiently nice symbol $a(x, \xi; h)$ is equal to $(2\pi h)^{-n} \iint a(x, \xi)\, dx\, d\xi$, and since $f_2(P(h))$ has leading symbol $f_2(p_0(x, \xi))$, by a result of Helffer-Robert (see for instance [S1]), we have:

$$\operatorname{tr} f_2(P(h)) = (2\pi h)^{-n} \sum_{j=1}^{m} \iint f_2(\lambda_j(x,\xi))\, dx\, d\xi + \mathcal{O}(h^{1-n}). \tag{32}$$

Using a standard argument, we have:

$$\sum_{\alpha\leq\lambda_j(h)} f_1(\lambda_j(h)) = (2\pi h)^{-n}\int_\alpha^{+\infty} f_1(\lambda)\gamma_0(\lambda)\,d\lambda + \mathcal{O}(h^{1-n}), \tag{33}$$

$$\sum_{\lambda_j(h)\leq\beta} f_3(\lambda_j(h)) = (2\pi h)^{-n}\int_{-\infty}^{\beta} f_3(\lambda)\gamma_0(\lambda)\,d\lambda + \mathcal{O}(h^{1-n}). \tag{34}$$

Approaching $1_{[\alpha,\infty[}f_1$, $1_{]-\infty,\beta]}f_3$ from above and from below by $C_0^\infty$ functions and using that $\gamma_0(\lambda)$ is continuous, we get:

$$\int_{-\infty}^{\beta} f_3(\lambda)\gamma_0(\lambda)d\lambda = \sum_{j=1}^{m}\iint (f_3 1_{]-\infty,\beta]})(\lambda_j(x,\xi))dx d\xi \tag{35}$$

$$\int_{\alpha}^{\infty} f_1(\lambda)\gamma_0(\lambda)d\lambda = \sum_{j=1}^{m}\iint (f_1 1_{[\alpha,\infty[})(\lambda_j(x,\xi))dx d\xi, \tag{36}$$

and combining this with (31), (32), we get Theorem 3. □

## Appendix: Extension of a hyperbolic function.

Let $p(x)$ be smooth in a neighborhood of $0\in\mathbf{R}^n$ with values in the Hermitian $m\times m$-matrices, and assume that $p(x)$ is hyperbolic in the constant direction $t\in\mathbf{R}^n$, so that

$$t(p) \underset{\text{def}}{=} t(\partial_x)p \geq \frac{1}{C_0} - C_0(p(x))^2, \tag{A.1}$$

for some $C_0>0$, in the sense of Hermitian matrices. After a conjugation with a constant unitary transformation, we may assume that $p$ takes the block matrix form

$$p(x) = \begin{pmatrix} p_{11}(x) & p_{21}(x) \\ p_{12}(x) & p_{22}(x)\end{pmatrix}, \tag{A.2}$$

with $p_{11}(0)=0$, $p_{21}(0)=0$, $p_{12}(0)=0$ and with $p_{22}(0)$ bijective. Put

$$p_0(x) = \begin{pmatrix} \langle dp_{11}(0), x\rangle & 0 \\ 0 & p_{22}(0)\end{pmatrix}.$$

**Lemma A.1.** *$p_0(x)$ is uniformly hyperbolic on $\mathbf{R}^n$ in the direction $t$.*

*Proof.* In the corresponding block matrix form, let

$$\pi_0 = \begin{pmatrix} 1 & 0 \\ 0 & 0 \end{pmatrix}$$

Then for every $\delta \in ]0,1]$,

$$\begin{aligned} t(\partial_x)p_0 &= \langle t, dp(0)\rangle + \begin{pmatrix} 0 & \mathcal{O}(1) \\ \mathcal{O}(1) & \mathcal{O}(1) \end{pmatrix} \\ &\geq \frac{1}{C_0} - C_0(p(0))^2 - \mathcal{O}(1)(\delta\pi_0^2 + \frac{1}{\delta}(I-\pi_0)^2). \end{aligned}$$

Choosing $\delta > 0$ small enough, we get

$$t(\partial_x)p_0 \geq \frac{1}{2C_0} - \mathcal{O}(1)(p_0(x))^2, \tag{A.3}$$

uniformly with respect to $x$. □

Let $\chi \in C_0^\infty(\mathbf{R}^n;[0,1])$ be equal to one near 0 and put with a (new) sufficiently small $\delta > 0$:

$$p_\delta(x) = \chi(x/\delta)(p(x) - p_0(x)) + p_0(x). \tag{A.4}$$

If $|x| = \mathcal{O}(\delta)$, we have

$$p_\delta(x) = \begin{pmatrix} 0 & 0 \\ 0 & f(x) \end{pmatrix} + \mathcal{O}(\delta),$$

where $f(x)$, $f(x)^{-1} = \mathcal{O}(1)$, uniformly with respect to $x$. Hence

$$p_\delta(x)^2 = \begin{pmatrix} \mathcal{O}(\delta^2) & \mathcal{O}(\delta) \\ \mathcal{O}(\delta) & f(x)^2 + \mathcal{O}(\delta) \end{pmatrix} \geq \begin{pmatrix} -\mathcal{O}(\delta^2) & 0 \\ 0 & \frac{1}{|\mathcal{O}(1)|} \end{pmatrix}. \tag{A.5}$$

Using the fact that $p_\delta(x) = p_0(x)$ for $|x| \geq \mathcal{O}(\delta)$, we see that (A.5) extends uniformly to all $x \in \mathbf{R}^n$.

For $\delta$ small enough, we have,

$$t(\partial_x)p_\delta = \begin{pmatrix} \langle t, dp_{11}(0)\rangle + \mathcal{O}(\delta) & \mathcal{O}(1) \\ \mathcal{O}(1) & \mathcal{O}(1) \end{pmatrix},$$

and the argument that gave (A.3), now shows that

$$t(\partial_x)p_\delta \geq \frac{1}{3C_0}I - \begin{pmatrix} 0 & 0 \\ 0 & \mathcal{O}(1) \end{pmatrix}.$$

Thanks to (A.5), we get for $\delta$ small enough,

$$t(\partial_x)p_\delta \geq \frac{1}{4C_0}I - \mathcal{O}(1)p_\delta(x)^2. \tag{A.6}$$

We have then proved

**Lemma A.2.** *If $\delta > 0$ is small enough, then $p_\delta$ is uniformly hyperbolic in the direction $t$.*

In the following, we let $\delta$ be a small fixed constant. We need to further modify $p_\delta$ away from $x = 0$ in order to obtain a symbol of class $S^0$. Notice that all the derivatives of $p_\delta$ are uniformly bounded on $\mathbf{R}^n$. If $f \in C_0^\infty(\mathbf{R})$ it is easy to see that

$$f(p_\delta(x)) = -\frac{1}{\pi}\int \overline{\partial}\widetilde{f}(z)(z - p_\delta(x))^{-1}L(dz) \tag{A.7}$$

is in $S^0$, and if we let $f$ vary in a class of $C_0^\infty(\mathbf{R})$ functions with uniform bounds on the diameter of the supports and on the supremum of every derivative, then $f(p_\delta)$ varies in bounded set in $S^0$.

If $f \in S^0(\mathbf{R})$, then we can find such a class with the property that for every $x_0 \in \mathbf{R}^n$, we can find $f_1, .., f_m$ in the class such that

$$f(p_\delta(x)) = f_1(p_\delta(x)) + .. + f_m(p_\delta(x))$$

for $x$ in some neighborhood of $x_0$. It follows that $f(p_\delta) \in S^0(\mathbf{R}^n)$. Now choose $f \in S^0$ real-valued, such that $f(t) = t$ for $|t| \le 1$, $f(t) \ge 1$ for $t \ge 1$, $f(t) \le -1$ for $t \le -1$, and put $\widetilde{p}(x) = f(p_\delta(x))$. Without any loss of generality, we may assume that the norm of $p_\delta(x)$ is smaller than one for $x$ in some neighborhood of 0.

**Proposition A.3.**
(i) *$\widetilde{p}(x) = p(x)$ in a small neighborhood of 0,*
(ii) *$\widetilde{p} \in S^0(\mathbf{R}^n; \mathcal{L}(\mathbf{C}^m, \mathbf{C}^m))$,*
(iii) *$\widetilde{p}$ is globally and uniformly hyperbolic in the direction $t$.*

*Proof.* (i) follows from the construction and we have verified (ii).

For every $x_0 \in \mathbf{R}^n$, we can find $s \in [\frac{1}{|\mathcal{O}(1)|}, \frac{1}{2}]$, such that $\pm[s, s + \frac{1}{|\mathcal{O}(1)|}]$ are disjoint from the spectrum of $p_\delta(x)$ for $x$ in some neighborhood $W_{x_0}$ of $x_0$. (Here the $\mathcal{O}(1)$ is uniform with respect to $x_0$.) Then $\pi = \pi(x) = 1_{[-s,s]}(p_\delta(x))$ is $C^\infty$ in $W_{x_0}$ and

$$\widetilde{p} = \pi p_\delta \pi + (1 - \pi)a(1 - \pi), \quad [a, \pi] = 0, \tag{A.8}$$

where $0 < a$, $\partial_x^\alpha a$, $a^{-1}$, $\partial_x^\alpha \pi$ are $\mathcal{O}_\alpha(1)$ for $x \in W_{x_0}$, uniformly with respect to $x_0$.

Let $A$, $\Gamma$ be Hermitian matrices. Then for every $\alpha > 0$, we have:

$$A\Gamma + \Gamma A \ge -\alpha A^2 - \frac{1}{\alpha}\Gamma^2. \tag{A.9}$$

From (A.8), we get, writing $t(\partial_x) = t$

$$t(\widetilde{p}) = t(\pi)p_\delta\pi + \pi p_\delta t(\pi) + \pi t(p_\delta)\pi - t(\pi)a(1-\pi) - (1-\pi)at(\pi) + (1-\pi)t(a)(1-\pi). \tag{A.10}$$

Using (A.6), as well as the fact that $\pi^2 = \pi$, we get with a new constant $C_1 > 0$,

$$\pi t(p_\delta)\pi \geq \frac{\pi}{C_1} - C_1 \pi p_\delta^2 \pi. \tag{A.11}$$

From (A.8) and the properties of $a$, we obtain,

$$\pi t(p_\delta)\pi \geq \frac{1}{C_1} - \mathcal{O}(1)\widetilde{p}^2. \tag{A.12}$$

The argument used above, combined with (A.9), gives for arbitrary $\alpha > 0$,

$$\begin{aligned} \pi p_\delta t(\pi) + t(\pi)p_\delta\pi &\geq -\alpha - \frac{\mathcal{O}(1)}{\alpha}\widetilde{p}^2, \\ -t(\pi)a(1-\pi) - (1-\pi)at(\pi) &\geq -\alpha - \frac{\mathcal{O}(1)}{\alpha}\widetilde{p}^2, \\ (1-\pi)t(a)(1-\pi) &\geq -\mathcal{O}(1)\widetilde{p}^2. \end{aligned}$$

Choosing $\alpha$ sufficiently small, we obtain

$$t(\partial_x)\widetilde{p} \geq \frac{1}{2C_1}I - \mathcal{O}(1)\widetilde{p}^2. \tag{A.13}$$

□

## References


[C] J. Chazarain, *Spectre d'un hamiltonien quantique et mécanique classique*, Comm. Partial Diff. Eq. **6** (1980), 595–644.

[D] E. B. Davies, *Spectral theory and differential operators*, Cambridge University Press, 1995, Cambridge studies in advanced mathematics **42**.

[Di] M. Dimassi, *Développements asymptotiques des perturbations lentes de l'opérateur de Schrödinger périodique*, Comm. Partial Diff. Eq. **18** no 5–6 (1993), 771–803.

[HeR] B. Helffer and D. Robert, *Calcul fonctionnel par la transformation de Mellin et opérateurs admissibles*, J. Funct. Anal. **53** no 3 (1983), 246–268.

[HeS] B. Helffer and J. Sjöstrand, *Équation de Schrödinger avec champ magnétique et équation de Harper*, Springer Lecture Notes in Physics **345** (1989), 118–197.

[H] L. Hörmander, *The spectral function of an elliptic operator*, Acta Math. **121** (1968), 173–218.

[I] V. Ivrii, *Semiclassical microlocal analysis and precise spectral asymptotics*, Preprints from École Polytechnique, no 1 (1990), no 2-6 (1991), no 7-9 (1992), also avaiable through ftp://www.scar.toronto.edu/pub/math-preprints/ivrii/Book/.

[JN] A. Jensen and S. Nakamura, *Mapping properties of functions of Schrödinger operators between $L^p$-spaces and Besov spaces*, Advanced Studies in Pure Mathematics **23** (1994), 187–209.

[M] G. Métivier, *Estimation du reste en théorie spectrale*, Proceedings of Journées équations aux dérivées partielles, St Jean de Monts, June 1982, École Polytechnique, Université de Nantes, Université de Rennes 1.

[S1] J. Sjöstrand, *Microlocal analysis for the periodic magnetic Schrödinger equation and related questions*, Springer Lecture Notes in Math. **1495** (1991), 237–332, Lectures at CIME, Montecatini, July 1989.

[S2] ———, *Wiener type algebras of pseudodifferential operators*, Sém. d'équations aux dérivées partielles, École Polytechnique, 1994–95, exposé no 4.

Département de Mathématiques, Université de Paris Nord, Institut Galilée, Av. J. B. Clément, F-93430 Villetaneuse, France and URA 742, CNRS
*E-mail address*: dimassi@math.univ-paris13.fr

Centre de Mathématiques, École Polytechnique, F-91128 Palaiseau cedex, France, and URA 169, CNRS
*E-mail address*: johannes@math.polytechnique.fr

# A MICROLOCAL VERSION OF CONCENTRATION-COMPACTNESS

Patrick Gérard

**1. Introduction.** The concentration-compactness principle has been developed by P.-L. Lions [10] in order to study minimizing sequences of elliptic variational problems displaying some lack of compactness. It is closely related to the general problem of describing the sequences of functions on $\mathbb{R}^d$ which violate locally the compactness of the Sobolev imbedding

$$(1) \qquad H^s(\mathbb{R}^d) \hookrightarrow L^p(\mathbb{R}^d), \quad 0 < s < \frac{d}{2}, \quad \frac{1}{p} = \frac{1}{2} - \frac{s}{d}.$$

Let us briefly set the notation. Given $d, s, p$ as in (1), we consider a bounded sequence $(u_n)$ in $H^s(\mathbb{R}^d)$, weakly convergent to 0, satisfying the following assumption:

$$(2) \qquad \begin{array}{c} \text{there exists a compact subset } K \subset \mathbb{R}^d \text{ such that} \\ \|\langle D\rangle^s u_n\|_{L^2(\mathbb{R}^d \setminus K)} \to 0 \text{ as } n \to \infty. \end{array}$$

Here $\langle D\rangle^s$ denotes the Fourier multiplier associated to the symbol $\langle\xi\rangle^s = (1+|\xi|^2)^{s/2}$. Assumption (2) means that we are only interested in local obstructions to the compactness of imbedding (1).

The problem is to describe as precisely as possible those sequences $(u_n)$ which do not converge strongly to 0 in $L^p(\mathbb{R}^d)$. To this aim, the following objects are introduced. Since $|\langle D\rangle^s u_n|^2$ and $|u_n|^p$ are bounded in $L^1(\mathbb{R}^d)$, we can choose a subsequence $(u_{n(k)})$ such that

$$(3) \qquad |\langle D\rangle^s u_{n(k)}|^2 \rightharpoonup \alpha, \quad |u_{n(k)}|^p \rightharpoonup \beta,$$

for the weak $*$ topology of Radon measures on $\mathbb{R}^d$.

The concentration-compactness lemma is then the following

**Proposition 1 (P.-L. Lions, [10], Lemma I.1).** *The measures $\alpha$ and $\beta$ satisfy $\beta \ll \alpha$, and $\beta(x) = \sum_j c_j \delta(x - x_j)$ for some sequence $(x_j)$ of points in $\mathbb{R}^d$ and some sequence $(c_j)$ of nonnegative numbers.*

Observe that $u_{n(k)}$ converges strongly to 0 in $L^p$ if and only if $\beta = 0$. Hence an immediate corollary of Proposition 1 is

**Corollary 2.** *If $(u_n)$ is a bounded sequence in $H^s(\mathbb{R}^d)$, weakly convergent to 0, satisfying assumption (2) and such that $\|u_n\|_{L^p}$ does not converge to 0, then there exists a limit point $\alpha$ of $|\langle D\rangle^s u_n|^2$ in the space of Radon measures such that $\alpha(\{y\}) > 0$ for some $y \in \mathbb{R}^d$.*

In other words, sequences which violate locally the compactness of the Sobolev imbedding have to concentrate at some point. A typical example is of course

$$u_n(x) = \frac{1}{\varepsilon_n^{d/p}}\psi\Big(\frac{x-y}{\varepsilon_n}\Big), \tag{4}$$

where $\varepsilon_n \to 0$ and $\psi \in H^s(\mathbb{R}^d)$. In this case $\alpha(x) = \||D|^s\psi\|_{L^2}^2\delta(x-y)$.

Our purpose in this paper is to give a microlocal version of Corollary 2, having in mind applications to hyperbolic equations. In the next section we define microlocal analogues of the measures $\alpha$. The main result and its proof can be found in Section 3. Roughly speaking, it says that, "above" concentration points $y$, the sequence $(u_n)$ has to spread in the Fourier space, otherwise $u_n$ is strongly convergent in $L^p$. Applications to focusing sequences of solutions to linear and nonlinear wave equations can be found in Sections 4 and 5.

To conclude this introduction, let us indicate that most of the material in this paper can be found in details in the reference [3], with a somewhat different presentation.

**2. Microlocal measures.** Let $(f_n)$ be a bounded sequence in $L^2(\mathbb{R}^d)$, weakly convergent to 0, and satisfying assumption (2) with $s = 0$. For every pseudodifferential operator $A \in \operatorname{Op} S^0_{\mathrm{phg}}(\mathbb{R}^d)$ (see for instance [9]), denote by $\sigma_0(A)$ the principal symbol of $A$. Then, by the Rellich theorem and pseudodifferential $L^2$ estimates, we have

$$\overline{\lim_{n\to\infty}}\, \|Af_n\|_{L^2} \le C \sup_{|\alpha|+|\beta|\le m}\ \sup_{(x,\xi)\in K_1\times S^{d-1}} |\partial_x^\alpha\partial_\xi^\beta \sigma_0(A)(x,\xi)|, \tag{5}$$

where $K_1$ is a compact subset of $\mathbb{R}^d$, and $C, m$ are constants independent of $A$. Hence, by a diagonal extraction, we conclude that there exists a subsequence $(f_{n(k)})$ such that, for every $A$, the $L^2$ scalar product $(Af_{n(k)}, f_{n(k)})$ has a limit $\lambda(A)$, and that moreover $\lambda(A) = \langle\mu, \sigma_0(A)\rangle$, where $\mu$ is some distribution on $\mathbb{R}^d \times S^{d-1}$. We claim that $\mu$ is a positive measure. Indeed, if $\sigma_0(A) \ge 0$, then

$$|\operatorname{Im}(Af_n, f_n)| \le C(A)\|f_n\|^2_{H^{-\frac12}}, \quad \operatorname{Re}(Af_n, f_n) \ge -C(A)\|f_n\|^2_{H^{-\frac12}}, \tag{6}$$

where the second estimate is a consequence of the sharp Gårding inequality. Hence the claim follows from the Rellich theorem.

Such a measure $\mu$ is called a **microlocal defect measure** associated to $(f_n)$. It describes the microlocal obstruction to the strong convergence of $f_n$ in $L^2$. In particular, notice that

$$\int_{\xi\in S^{d-1}} \mu(x,d\xi) = \lim |f_{n(k)}(x)|^2 \tag{7}$$

for the weak topology of measures. These measures were first used by A. Shnirelman [14], where the $f_n$ are eigenfunctions of the Laplace operator on a Riemannian compact manifold. A more general setting can be found in [1] and [15].

It is also useful to introduce **semiclassical measures** associated to $(f_n)$ and to a "scale" $(\varepsilon_n)$, i.e. a sequence of positive numbers converging to 0 (see [16], [2], [8], [11]). Indeed, it is possible to replace, in the above analysis, the quantities $(A f_n, f_n)$, $A \in \mathrm{Op}\, S^0_{\mathrm{phg}}$ by $(a(x,\varepsilon_n D) f_n, f_n)$, where $a \in \mathcal{S}(\mathbb{R}^d\times\mathbb{R}^d)$. Similarly, there exists a subsequence $(f_{n(k)})$ such that, for every $a$, $(a(x,\varepsilon_{n(k)}D) f_{n(k)}, f_{n(k)})$ has a limit $l(a)$, and that moreover

$$l(a) = \int_{\mathbb{R}^d\times\mathbb{R}^d} a(x,\xi)\, dm(x,\xi), \tag{8}$$

where $m$ is a bounded positive measures on $\mathbb{R}^d\times\mathbb{R}^d$. Such a measure $m$ is called a semiclassical measure associated to $(f_n)$ and $(\varepsilon_n)$.

There is a simple relation between semiclassical and microlocal measures. If, say, $(f_n)$ admits a unique microlocal defect measure $\mu$, then every semiclassical measure $m$ associated to $(f_n)$ satisfies

$$\rho_*(\mathbf{1}_{\xi\neq 0} m) \le \mu, \tag{9}$$

where $\rho$ is the continuous map $(x,\xi)\in\mathbb{R}^d\times\mathbb{R}^d\backslash\{0\}\mapsto \left(x,\frac{\xi}{|\xi|}\right)\in\mathbb{R}^d\times S^{d-1}$.

Let us review briefly some examples. In the case of a WKB sequence,

$$f_n(x) = a(x) e^{iS(x)/\varepsilon_n}, \tag{10}$$

where $a\in L^2_{\mathrm{comp}}(\mathbb{R}^d)$, $S\in W^{1,1}_{\mathrm{loc}}(\mathbb{R}^d,\mathbb{R})$, $\nabla S(x)\neq 0$ a.e., one can show (see [2]) that

$$\begin{aligned} m(x,\xi) &= |a(x)|^2\, dx\ \ \delta(\xi - \nabla S(x)),\\ \mu(x,\xi) &= \rho_* m(x,\xi) = |a(x)|^2\, dx\ \ \delta\Big(\xi - \frac{\nabla S(x)}{|\nabla S(x)|}\Big). \end{aligned} \tag{11}$$

For a concentrating sequence,

$$f_n(x) = \frac{1}{\varepsilon_n^{d/2}}\psi\Big(\frac{x}{\varepsilon_n}\Big),\quad \psi\in L^2(\mathbb{R}^d), \tag{12}$$

$$m(x,\xi) = \delta(x)\,|\hat{\psi}(\xi)|^2 \,\frac{d\xi}{(2\pi)^d}, \quad \mu = \rho_* m. \tag{13}$$

Finally, one can mix concentration effects in $x$ and $\xi$ variables,

$$f_n(x) = \frac{1}{\delta_n^{d/2}}\,\psi\Big(\frac{x}{\delta_n}\Big)\,e^{ix\cdot\xi_0/\varepsilon_n}, \quad \varepsilon_n \ll \delta_n \ll 1, \tag{14}$$

$$m(x,\xi) = \|\psi\|_{L^2}^2\,\delta(x)\,\delta(\xi-\xi_0), \quad \mu = \rho_* m. \tag{15}$$

Of course, if, in examples (10), (12), (14), we compute semiclassical measures associated to another scale $(\varepsilon_n')$, we may find quite different results, in particular $m' = 0$ if $\varepsilon_n' \ll \varepsilon_n$, and $m'$ is supported by $\xi = 0$ if $\varepsilon_n' \gg \varepsilon_n$.

Let us close this section by two remarks. First, using example (14), it is not difficult to show that any bounded positive measure on $\mathbb{R}^d \times S^{d-1}$ (resp. $\mathbb{R}^d\times\mathbb{R}^d$), compactly supported in $x$, is the microlocal defect measure (resp. semiclassical measure, for some scale) of some sequence $(f_n)$.

Our second remark is that, as microlocal objects, these measures satisfy elliptic regularity and propagation theorems if the corresponding sequence satisfies a partial differential equation. Let us simply state an example of such a propagation result which will be useful in Sections 4 and 5. Let $(v_n)$ be a sequence of solutions to

$$\partial_t^2 v_n - \Delta v_n = 0, \quad t \in \mathbb{R}, \; x \in \mathbb{R}^d, \tag{16}$$

such that $(\partial_t v_n(0,\cdot))$ and $(\nabla_x v_n(0,\cdot))$ are bounded sequences of $L^2(\mathbb{R}^d)$, weakly convergent to 0, satisfying assumption (2) with $s = 0$. Set

$$f_n^{\pm} = \partial_t v_n \pm i\,|D_x|\, v_n. \tag{17}$$

Then, by using energy estimates and finite propagation speed, it is easy to show that, for every $t \in \mathbb{R}$, $(f_n^{\pm}(t,\cdot))$ is a bounded sequence in $L^2(\mathbb{R}^d)$, weakly convergent to 0, satisfying assumption (2) with $s = 0$.

**Proposition 3.** (i) *Assume $(f_n^{\pm}(0,\cdot))$ has a unique microlocal defect measure $\mu_0^{\pm}$. Then, for any $t$, $(f_n^{\pm}(t,\cdot))$ has a unique microlocal defect measure $\mu_t^{\pm}$, given by $\mu_t^{\pm}(x,\xi) = \mu_0^{\pm}(x \pm t\xi, \xi)$ and, for any $A$, the quantity $(A f_n^{\pm}(t,\cdot), f_n^{\pm}(t,\cdot))$ converges locally uniformly in $t$. Moreover,*

$$\begin{aligned}(A\,\partial_t v_n(t,\cdot), \partial_t v_n(t,\cdot)) + (A\,\nabla_x v_n(t,\cdot), \nabla_x v_n(t,\cdot))\\ \longrightarrow \frac{1}{2}\int_{\mathbb{R}^d\times S^{d-1}} \sigma_0(A)\, d(\mu_t^+ + \mu_t^-)\end{aligned} \tag{18}$$

*locally uniformly in t.*

(ii) *Assume $(f_n^{\pm}(0,\cdot))$ has a unique semiclassical measure $m_0^{\pm}$ for some scale $(\varepsilon_n)$. Then, locally uniformly in t, for any $a \in C_0^\infty(\mathbb{R}^d \times (\mathbb{R}^d \setminus \{0\}))$, we have*

$$\Big(a(x,\varepsilon_n D)\, f_n^{\pm}(t,\cdot), f_n^{\pm}(t,\cdot)\Big) \longrightarrow \int_{\mathbb{R}^d\times\mathbb{R}^d} a\, dm_t^{\pm}, \tag{19}$$

$$\begin{aligned}&\Big(a(x,\varepsilon_n D)\,\partial_t v_n(t,\cdot), \partial_t v_n(t,\cdot)\Big) + \Big(a(x,\varepsilon_n D)\,\nabla_x v_n(t,\cdot), \nabla_x v_n(t,\cdot)\Big)\\ &\qquad \longrightarrow \frac{1}{2}\int_{\mathbb{R}^d\times\mathbb{R}^d} a\, d(m_t^+ + m_t^-),\end{aligned} \tag{20}$$

*where $m_t^{\pm}(x,\xi) = m_0^{\pm}\big(x \pm t\,\frac{\xi}{|\xi|}, \xi\big)$.*

Observe the restriction on the test function $a$ in part (ii), due to the fact that $\xi = 0$ is a singular point.

**3. The main theorem.** We come back to the notation of the introduction, and shall deal with microlocal measures associated to $f_n = \langle D\rangle^s u_n$.

**Theorem 4.** *Let $(u_n)$ be a bounded sequence in $H^s(\mathbb{R}^d)$, weakly convergent to 0, satisfying assumption (2). Assume that $\|u_n\|_{L^p(\mathbb{R}^d)}$ does not converge to 0. Then*

(i) *there exists a microlocal defect measure $\mu$ associated to $(\langle D\rangle^s u_n)$, a point $y \in \mathbb{R}^d$, and a nonnegative Borel function $H$ on $S^{d-1}$, such that*

$$\mu(x,\xi) \geq \delta(x-y)\, H(\xi)\, d\sigma(\xi), \quad \int_{S^{d-1}} H\, d\sigma > 0, \tag{21}$$

*where $\sigma$ denotes the Lebesgue measure on $S^{d-1}$;*

(ii) *there exists a semiclassical measure $m$ associated to $(\langle D\rangle^s u_n)$ and some scale $(\varepsilon_n)$, a point $y \in \mathbb{R}^d$, and a nonnegative Borel function $h$ on $\mathbb{R}^d$, such that*

$$m(x,\xi) \geq \delta(x-y)\, h(\xi)\, d\xi, \quad \int_{\mathbb{R}^d} h\, d\xi > 0. \tag{22}$$

*Remarks.* **a)** Observe that, in view of formulae (12), (13), Theorem 4 is consistent with example (4).

**b)** In view of inequality (9), it is easy to show that part (i) in Theorem 4 follows from part (ii). Moreover, the following simple example shows that part (ii) is more precise than part (i). Let $\psi \in \mathcal{S}(\mathbb{R}^d)$ with $\int_{\mathbb{R}^d} \psi(x)\, dx = 1$. Then $n^d\psi(nx)$ converges weakly to the Dirac mass at the origin, hence is not bounded in $H^{-d/2}(\mathbb{R}^d)$. A simple calculation shows that the $H^{-d/2}$ norm of this function is equivalent to $c(\log n)^{1/2}$ with $c = |S^{d-1}|^{1/2}/(2\pi)^{d/2}$. Hence the sequence

$$u_n(x) = \langle D\rangle^{-s-\frac{d}{2}}\left((\log n)^{-1/2}\, n^d\, \psi(nx)\right) \tag{23}$$

converges weakly but not strongly to 0 in $H^s$. More precisely, the microlocal measure of $(\langle D\rangle^s u_n)$ is

$$\mu(x,\xi) = (2\pi)^{-d}\,\delta(x)\, d\sigma(\xi). \tag{24}$$

On the other hand,

$$\begin{aligned} \int_{\alpha_n\le|\xi|\le\beta_n} \langle\xi\rangle^{2s}|\hat{u}_n(\xi)|^2\,d\xi &= \int_{\frac{\alpha_n}{n}\le|\xi|\le\frac{\beta_n}{n}} \frac{|\hat{\psi}(\xi)|^2}{\left(\frac{1}{n^2}+|\xi|^2\right)^{d/2}}\,\frac{d\xi}{\log n} \\ &\le C\,\frac{\log\left(\frac{\beta_n}{\alpha_n}\right)}{\log n} \end{aligned} \tag{25}$$

which goes to 0 if $\frac{\beta_n}{\alpha_n}$ is bounded. Hence any semiclassical measure associated to $(\langle D\rangle^s u_n)$ and any scale is supported by $\xi = 0$. Hence the sequence $(u_n)$ does not satisfy property (ii), while it satisfies property (i) in view of (24).

**c)** The conclusion of Theorem 4 can be made more quantitative. In fact, our proof below shows that if $\overline{\lim}\, \|u_n\|_{H^s} \le M$, there exists a nonnegative continuous function $F$ on $[0,\infty)$, depending only on $M, s, d$, such that $F(0) = 0$ and

$$\overline{\lim}\, \|u_n\|_{L^p} \le F(\sigma),$$

where $\sigma$ denotes the supremum of all the quantities $\int_{\mathbb{R}^d} h\,d\xi$, where $h$ satisfies (22) and $y$ varies in $\mathbb{R}^d$, $m$ varies among the semiclassical measures associated to $(\langle D\rangle^s u_n)$ and any scale.

**d)** In Theorem 4, the assumption "$\|u_n\|_{L^p}$ does not converge to 0" can be replaced by "$|u_n|^p \rightharpoonup \beta$ as measures and $\beta(\{y\}) > 0$", with the same point $y$ in conclusions (i) and (ii).

Let us come to the proof of Theorem 4. First we observe that it is enough to prove it for some couple $(s,p)$ satisfying (1). Indeed, if $(t,q)$ is another couple satisfying (1) and $(u_n)$ is a bounded sequence in $H^t$, weakly convergent to 0, we can apply Theorem 4 with $(s,p)$ to $\tilde{u}_n = \langle D\rangle^{t-s}u_n$. Since $\langle D\rangle^s\tilde{u}_n = \langle D\rangle^t u_n$, the conclusion of Theorem 4 with $(t,q)$ and $(\tilde{u}_n)$

is equivalent to the conclusion of Theorem 4 with $(s,p)$ and $(u_n)$. On the other hand, if $t > s$, the Sobolev imbedding $W^{t-s,p} \hookrightarrow L^q$ implies

$$\|u_n\|_{L^q} \leq C\|\tilde{u}_n\|_{L^p}. \tag{26}$$

If $t < s$, the Gagliardo-Nirenberg interpolation inequality implies

$$\|u_n\|_{L^q} \leq C\|\tilde{u}_n\|^{\theta}_{L^p}\|\langle D\rangle^s\tilde{u}_n\|^{1-\theta}_{L^2} \leq C'\|\tilde{u}_n\|^{\theta}_{L^p} \tag{27}$$

for some $\theta \in ]0,1[$. In both cases, the strong convergence of $\tilde{u}_n$ in $L^p$ implies the strong convergence of $u_n$ in $L^q$. This proves the theorem for $(t,q)$.

We are led to prove Theorem 4 for one choice of $(s,p)$, say $p=4$, $s=\frac{d}{4}$. Our next step is to consider the special case where the Fourier transform of $u_n$ is supported in a ring.

**Lemma 5.** *Let $(u_n)$ be a bounded sequence in $H^{d/4}(\mathbb{R}^d)$. Assume there exists a scale $(\varepsilon_n)$ and positive constants $a,b$ such that*

$$\hat{u}_n(\xi) = 0 \ \text{ for } |\xi| \leq \frac{a}{\varepsilon_n} \ \text{ or } |\xi| \geq \frac{b}{\varepsilon_n}. \tag{28}$$

*Assume moreover that*

$$\frac{1}{\varepsilon_n^{3d/2}}\left|\hat{u}_n\left(\frac{\xi}{\varepsilon_n}\right)\right|^2 \rightharpoonup \lambda(\xi) \ \text{ weakly as measures.} \tag{29}$$

*Then*

$$\varlimsup_{n\to\infty}\|u_n\|_{L^4} \leq K(a,b,d)\left(\int_{\mathbb{R}^d} h(\xi)\,d\xi\right)^{1/2}, \tag{30}$$

*where $h$ is the Radon-Nikodym derivative of the measure $\lambda$.*

*Proof.* Set $U_n(x) = \varepsilon_n^{d/4}u_n(\varepsilon_n x)$. Then $\widehat{U}_n$ is supported in the fixed ring $R = \{a \leq |\xi| \leq b\}$, and $|\widehat{U}_n|^2 \rightharpoonup \lambda$ weakly. Moreover, the Hausdorff-Young inequality yields

$$\|u_n\|_{L^4} = \|U_n\|_{L^4} \leq C\|\widehat{U}_n\|_{L^{4/3}}, \tag{31}$$

hence it is enough to estimate the upper limit of $\|\widehat{U}_n\|_{L^{4/3}}$. By the Radon-Nikodym theorem, $\lambda = h\,d\xi + \lambda_s$ with $\lambda_s \perp d\xi$. Hence we can find two Borel subsets $A, B$ such that $A\cap B = \emptyset$ and $\mathbf{1}_A\lambda_s = \lambda_s$, $\mathbf{1}_B\,d\xi = d\xi$. Given $\delta > 0$, we choose two open subsets $U, V$ such that $A^c \subset U$, $B^c \subset V$, and

$$\lambda_s(U) \leq \delta, \quad \int_V d\xi \leq \delta. \tag{32}$$

Let $(\psi_1, \psi_2)$ be a continuous partition of unity related to the covering $(U, V)$ of $\mathbb{R}^d$. By Hölder's inequality, we have, for $j = 1, 2$,

(33)
$$\left(\int |\widehat{U}_n(\xi)|^{4/3}\psi_j(\xi)\,d\xi\right)^{3/4} \le \left(\int |\widehat{U}_n(\xi)|^2\psi_j(\xi)\,d\xi\right)^{1/2}\left(\int_R \psi_j(\xi)\,d\xi\right)^{1/4}.$$

By summing on $j = 1, 2$, we obtain

$$\overline{\lim_{n\to\infty}}\ \|\widehat{U}_n\|_{L^{4/3}} \le \left(\int_{\mathbb{R}^d} h\,d\xi + \delta\right)^{1/2} |R|^{1/4} + \lambda(\mathbb{R}^d)^{1/2}\delta^{1/4}, \tag{34}$$

where $|R|$ is the volume of $R$. The proof is completed by letting $\delta$ go to 0.

In order to come down to Lemma 5, we shall use dyadic decompositions. Let

$$1 = \psi(\xi) + \sum_{k=0}^{\infty}\varphi(2^{-k}\xi), \quad \psi \in C_0^\infty(\mathbb{R}^d),\ \varphi \in C_0^\infty(\mathbb{R}^d \setminus \{0\}), \tag{35}$$

be a dyadic partition of unity of $\mathbb{R}^d$. We denote by $S_0 = \psi(D)$, $\Delta_k = \varphi(2^{-k}D)$ the corresponding operators. The spaces $H^{d/4}(\mathbb{R}^d) \subset L^4(\mathbb{R}^d) \subset B^0_{44}(\mathbb{R}^d)$ are described as follows.

$$\|f\|_{H^{d/4}} \simeq \left(\|S_0 f\|^2_{L^2} + \sum_{k=0}^{\infty} 2^{kd/2}\|\Delta_k f\|^2_{L^2}\right)^{1/2}, \tag{36}$$

$$\|f\|_{L^4} \simeq \left\|\left(|S_0 f|^2 + \sum_{k=0}^{\infty}|\Delta_k f|^2\right)^{1/2}\right\|_{L^4}, \tag{37}$$

$$\|f\|_{B^0_{44}} = \left(\|S_0 f\|^4_{L^4} + \sum_{k=0}^{\infty}\|\Delta_k f\|^4_{L^4}\right)^{1/4}. \tag{38}$$

**Lemma 6.** *For some constant $C > 0$, we have, for every $f$,*

$$\|f\|_{L^4} \le C\|f\|_{B^0_{44}}\left[\log\left(2 + \frac{\|f\|_{H^{d/4}}}{\|f\|_{B^0_{44}}}\right)\right]^{1/4}. \tag{39}$$

*Proof.* Using (37), we have

$$\|f\|^4_{L^4} \le C\left(\|S_0 f\|^4_{L^4} + \sum_{k,l}\|\Delta_k f\,\Delta_l f\|^2_{L^2}\right). \tag{40}$$

We estimate $\|\Delta_k f\,\Delta_l f\|_{L^2}$ differently according to the size of $|k-l|$. Let $N$ be a positive integer. If $|k-l| \geq N$ and, say, $k \geq l$, we have

$$\|\Delta_k f\,\Delta_l f\|_{L^2} \leq \|\Delta_k f\|_{L^2}\,\|\Delta_l f\|_{L^\infty} \leq C\,2^{ld/2}\,\|\Delta_k f\|_{L^2}\,\|\Delta_l f\|_{L^2}, \tag{41}$$

since $\widehat{\Delta_l f}$ is supported in $|\xi| \leq C\,2^l$. This implies

$$\|\Delta_k f\,\Delta_l f\|_{L^2} \leq 2^{-Nd/4}\;2^{kd/4}\|\Delta_k f\|_{L^2}\;2^{ld/4}\|\Delta_l f\|_{L^2} \tag{42}$$

and, in view of (36),

$$\sum_{|k-l|\geq N} \|\Delta_k f\,\Delta_l f\|_{L^2}^2 \leq 2^{-Nd/2}\,C\,\|f\|_{H^{d/4}}^4. \tag{43}$$

If $|k-l| < N$, we just apply Hölder's inequality, so that

$$\begin{aligned}\sum_{|k-l|<N} \|\Delta_k f\,\Delta_l f\|_{L^2}^2 &\leq \sum_{|k-l|<N} \|\Delta_k f\|_{L^4}^2\,\|\Delta_l f\|_{L^4}^2 \\ &\leq 2N \sum_{k=0}^{\infty} \|\Delta_k f\|_{L^4}^4.\end{aligned} \tag{44}$$

Plugging (43) and (44) in (40), we obtain, in view of (38),

$$\|f\|_{L^4}^4 \leq C\Big(N\,\|f\|_{B^0_{44}}^4 + 2^{-Nd/4}\,\|f\|_{H^{d/4}}^4\Big). \tag{45}$$

By choosing $N$ properly, this proves Lemma 6.

Before completing the proof of Theorem 4, observe that, since $\widehat{\Delta_k f}$ is supported in $|\xi| \leq C\,2^k$, we have

$$\|\Delta_k f\|_{L^4} \leq C\,2^{kd/4}\,\|\Delta_k f\|_{L^2} \tag{46}$$

hence

$$\|f\|_{B^0_{44}} \leq C\bigg(\|S_0 f\|_{L^4} + \Big(\sup_k \|\Delta_k f\|_{L^4}^{1/2}\Big)\,\|f\|_{H^{d/4}}^{1/2}\bigg). \tag{47}$$

Given $(f_n)$ a bounded sequence in $H^{d/4}$, weakly convergent to 0 and satisfying assumption (2) with $s = \frac{d}{4}$, we conclude that

$$\overline{\lim}\,\|f_n\|_{B^0_{44}} \leq C\,\sigma_1[f_n]^{1/2}\;\overline{\lim}\,\|f_n\|_{H^{d/4}}^{1/2}, \tag{48}$$

where $\sigma_1[f_n]$ denotes the supremum of all $\overline{\lim}\ \|\Delta_{k_n} f_n\|_{L^4}$, as $(k_n)$ varies among sequences of integers going to infinity. For such a sequence, we can apply Lemma 5 to $\Delta_{k_n} f_n$ with $\varepsilon_n = 2^{-k_n}$. This leads to

$$\sigma_1[f_n] \leq K\,\sigma_2[f_n]^{1/2}, \tag{49}$$

where $\sigma_2[f_n]$ denotes the supremum of all $\int_{\mathbb{R}^d} h\,d\xi$, where $h$ varies among the Radon-Nikodym derivatives of all measures

$$\lambda(\xi) = |\varphi(\xi)|^2 \int_{\mathbb{R}^d} l(dx,\xi) \tag{50}$$

where $l$ is any semiclassical measure associated to $(\langle D\rangle^{d/4} f_n)$.

If $(u_n)$ satisfies the assumptions of the theorem with $p = 4$, we may suppose by Proposition 1 that there exists $y \in \mathbb{R}^d$ such that $|u_n|^4 \rightharpoonup \beta$ weakly and $\beta(\{y\}) > 0$. Then we apply (48), (49) to

$$f_n^\delta(x) = \theta\Big(\frac{x-y}{\delta}\Big)\, u_n(x), \tag{51}$$

with $y \in \mathbb{R}^d$, $\delta > 0$, $\theta \in C_0^\infty(\mathbb{R}^d)$, $\theta(0) = \sup|\theta| = 1$. We have

$$\overline{\lim}\ \|f_n^\delta\|_{H^{d/4}} \leq \overline{\lim}\ \|u_n\|_{H^{d/4}} \leq M, \tag{52}$$

and the semiclassical measures associated to $(\langle D\rangle^{d/4} f_n^\delta)$, are given by

$$l(x,\xi) = \left|\theta\Big(\frac{x-y}{\delta}\Big)\right|^2 m(x,\xi) \tag{53}$$

where $m$ is a semiclassical measure associated to $(\langle D\rangle^{d/4} u_n)$. The corresponding measure $\lambda$ given by (50) is then

$$\lambda(\xi) = |\varphi(\xi)|^2 \left[m(\{y\},\xi) + \int_{x\neq y} \left|\theta\Big(\frac{x-y}{\delta}\Big)\right|^2 m(dx,\xi)\right] \tag{54}$$

where $m(\{y\},\xi)$ is defined by $\mathbf{1}_{x=y}\, m(x,\xi) = \delta(x-y)\, m(\{y\},\xi)$. We may assume that $|\langle D\rangle^{d/4} u_n|^2 \rightharpoonup \alpha$. Then, since $|\varphi| \leq 1$, and by (7), (9),

$$\iint_{x\neq y} |\varphi(\xi)|^2 \left|\theta\Big(\frac{x-y}{\delta}\Big)\right|^2 m(dx,d\xi) \leq \int_{x\neq y} \left|\theta\Big(\frac{x-y}{\delta}\Big)\right|^2 \alpha(dx), \tag{55}$$

which goes to 0 as $\delta$ goes to 0. Hence, in view of (54), we have

$$\sigma_2[f_n^\delta] \leq \sigma + r^\delta, \tag{56}$$

where $r^\delta \to 0$ as $\delta \to 0$, and $\sigma$ denotes the supremum of $\int h(\xi)d\xi$ as $h$ varies among the Radon-Nikodym derivatives of $m(\{y\},\xi)$, where $m$ is any semiclassical measure associated to $(\langle D\rangle^{d/4} u_n)$. Our task is to prove that $\sigma > 0$. By (48), (49), (52), we have

$$\overline{\lim}\, \|f_n^\delta\|_{B^0_{44}} \leq C M^{1/2}(\sigma + r^\delta)^{1/4}, \tag{57}$$

and by inequality (39) in Lemma 6, we conclude

$$\int \left|\theta\Big(\frac{x-y}{\delta}\Big)\right|^4 \beta(dx) \leq C M^2 (\sigma + r^\delta) \log\Big(2 + \frac{M}{\sigma + r^\delta}\Big). \tag{58}$$

As $\delta$ goes to 0, we obtain finally

$$\beta(\{y\}) \leq C M^2 \sigma \log\Big(2 + \frac{M}{\sigma}\Big), \tag{59}$$

which completes the proof.

**4. Application to the linear wave equation.** Let $(v_n)$ be a sequence of solutions to the linear wave equation in $\mathbb{R} \times \mathbb{R}^d$,

$$\partial_t^2 v_n - \Delta v_n = 0, \quad v_n|_{t=0} = \varphi_n, \quad \partial_t v_n|_{t=0} = \psi_n. \tag{60}$$

We assume that $\varphi_n$ (resp. $\psi_n$) is bounded in $H^s(\mathbb{R}^d)$ (resp. $H^{s-1}(\mathbb{R}^d)$), weakly convergent to 0, satisfying assumption 2. Then energy estimates and finite propagation speed imply that the same properties hold for $v_n(t,\cdot)$ (resp. $\partial_t v_n(t,\cdot)$) for any $t$.

Assume $0 < s < \frac{d}{2}$, and consider the exponent $p$ given by (1). We are interested in the size of the following set of "focusing times",

$$F = \Big\{t : \underline{\lim} \int |v_n(t,x)|^p\, dx > 0\Big\}. \tag{61}$$

In the case $d = 1$, the example

$$v_n(t,x) = \frac{1}{\varepsilon_n^{1/p}} \Phi\Big(\frac{x-t}{\varepsilon_n}\Big), \quad \Phi \in C_0^\infty(\mathbb{R}), \tag{62}$$

shows that $F$ may be the whole line. Quite different is the situation $d \geq 2$, as suggested by the following generalized Strichartz inequalities (see e.g. Ginibre-Velo [6]),

$$\begin{gathered} \|v_n\|_{L^q(\mathbb{R},L^r(\mathbb{R}^d))} \leq C\big(\|\varphi_n\|_{H^s} + \|\psi_n\|_{H^{s-1}}\big), \quad \text{if} \\ \frac{1}{q} + \frac{d}{r} = \frac{d}{2} - s, \\ \frac{2}{q} \leq \min\Big((d-1)\Big(\frac{1}{2} - \frac{1}{r}\Big), 1\Big), \quad \Big(\frac{2}{q}, (d-1)\Big(\frac{1}{2} - \frac{1}{r}\Big)\Big) \neq (1,1). \end{gathered} \tag{63}$$

Observe that, if $d = 1$, these inequalities reduce to Sobolev inequalities in the $x$ variable, combined with energy estimates, while if $d \geq 2$, we can choose $r > p$ such that (63) holds. Let us fix a pair $(q, r)$ as in (63) with $r > p$. If we set $f_n^{(a)}(t) = \|v_n(t, \cdot)\|_{L^a}$, we have then, by Hölder's inequality,

$$f_n^{(p)}(t) \leq f_n^{(2)}(t)^\theta f_n^{(r)}(t)^{1-\theta} \tag{64}$$

for some $\theta \in ]0, 1[$. Hence there exist $\alpha > 0$, $\beta > 0$, $\gamma > 0$, $\delta > 0$ such that, for every $T > 0$,

$$\int_0^T f_n^{(p)}(t)^\alpha \, dt \leq \left( \int_0^T f_n^{(2)}(t)^\beta \, dt \right)^\gamma \left( \int_0^T f_n^{(r)}(t)^q \, dt \right)^\delta . \tag{65}$$

The second factor in the right hand side of (65) is bounded because of (63), while the first factor goes to 0 by the Rellich theorem and the dominated convergence theorem. Using Fatou's lemma, we conclude that $F$ has Lebesgue measure 0.

Using Theorem 4, we can actually obtain more.

**Theorem 7.** *Assume $d \geq 2$. Then the set $F$ is at most countable. More precisely, if $|\langle D\rangle^s \varphi_n|^2 + |\langle D\rangle^{s-1} \psi_n|^2$ has a weak limit in the space of measures, the set*

$$\tilde{F} = \left\{ t : \overline{\lim} \int |v_n(t, x)|^p \, dx > 0 \right\} \tag{66}$$

*is at most countable.*

*Remark.* By using a series of solutions which concentrate at given times, it is easy to check that Theorem 7 is optimal.

*Proof.* Since the set $F$ associated to $(v_n)$ is contained in the set $\tilde{F}$ associated to any subsequence of $(v_n)$, it is enough to prove the second statement. So let us assume that $|\langle D\rangle^s \varphi_n|^2 + |\langle D\rangle^{s-1} \psi_n|^2 \rightharpoonup e$ weakly. If $t \in \tilde{F}$, then Theorem 4 (i) implies that there exists a microlocal defect measure $\mu_t$ associated to $(\langle D\rangle^s v_n(t, \cdot))$ such that

$$\begin{aligned} &\mu_t(x, \xi) \geq \delta(x - y_t) \, H_t(\xi) \, d\sigma(\xi), \quad y_t \in \mathbb{R}^d, \\ &H_t \geq 0, \quad \int_{S^{d-1}} H_t \, d\sigma > 0. \end{aligned} \tag{67}$$

Applying Proposition 3 to $(\langle D\rangle^{s-1} v_{n(k)})$, where $(v_{n(k)}(t, \cdot))$ is the subsequence corresponding to $\mu_t$, we conclude that $e \geq \nu_t$, where $\nu_t$ is a Lebesgue density on the sphere of radius $t$ centered at $y_t$, of total mass $\int H_t \, d\sigma$. As

$t$ varies in $\tilde{F}$, the measures $\nu_t$ are pairwise mutually singular (since $d \geq 2$), hence, for any finite set $G$ of $\tilde{F}$, we have

$$e \geq \sum_{t \in G} \nu_t. \tag{68}$$

Since $e$ has a finite total mass, and every $\nu_t$ has a positive total mass, this completes the proof.

*Remark.* Using remark **c)** after the statement of Theorem 4, the proof above shows that, for any $\delta > 0$,

$$F_\delta = \left\{ t : \underline{\lim} \int |v_n(t,x)|^p \, dx \geq \delta \right\} \tag{69}$$

is finite.

**5. Application to semilinear wave equations with critical growth.** In this section we are interested in solutions to the following semilinear wave equations on $\mathbb{R} \times \mathbb{R}^d$,

$$\partial_t^2 u - \Delta u + |u|^{p-2} u = 0, \quad u|_{t=0} = \varphi \in H^1, \quad \partial_t u|_{t=0} = \psi \in L^2, \tag{70}$$

where $d \geq 3$, $p \leq \frac{2d}{d-2}$. In the subcritical case $p < \frac{2d}{d-2}$, Ginibre-Velo [4] proved existence and uniqueness of solutions in $C(\mathbb{R}, H^1) \cap C^1(\mathbb{R}, L^2)$. In addition these solutions satisfy conservation of energy,

$$\int \left( \frac{1}{2} \left( |\partial_t u(t,x)|^2 + |\nabla u(t,x)|^2 \right) + \frac{1}{p} |u(t,x)|^p \right) dx = \text{cst}, \tag{71}$$

and generalized Strichartz inequalities

$$\begin{gathered} \|u\|_{L^q(0,T;L^r(\mathbb{R}^d))} \leq C(T, \|\varphi\|_{H^1} + \|\psi\|_{L^2}), \\ \frac{1}{q} + \frac{d}{r} = \frac{d}{2} - 1, \quad q \geq \frac{d+1}{d-1}, \quad q > 2 \text{ if } d = 3, \end{gathered} \tag{72}$$

where $C(T,E)$ is a bounded function. Part of these results were generalized to the critical case by Shatah-Struwe [13], who proved existence and uniqueness of solutions in $C(\mathbb{R}, H^1) \cap C^1(\mathbb{R}, L^2) \cap L^q(\mathbb{R}, L^r(\mathbb{R}^d))$, with $q, r$ as in (72). It is clear that, for these solutions, conservation of energy (71) holds (for related results about the critical case, see also Ginibre-Soffer-Velo, [5]). Our purpose in this section is to study the influence of the exponent $p$ on the main features of the solution $u$. For this reason, we address the following problem. Let $(\varphi_n)$ (resp. $(\psi_n)$) be a bounded sequence in $H^1(\mathbb{R}^d)$ (resp. $L^2(\mathbb{R}^d)$), weakly convergent to 0, satisfying assumption (2) with $s = 1$ (resp. $s = 0$). Let $u_n$ be the corresponding solution to (70), and let $v_n$ be the solution to the linear wave equation with the same Cauchy data $(\varphi_n, \psi_n)$ at $t = 0$. Given a time interval $[0,T]$, does $u_n - v_n$ converge strongly to 0 in the energy space $C([0,T], H^1) \cap C^1([0,T], L^2)$? In this case, we shall say that sequence $(u_n)$ is **linearizable** on the time interval $[0,T]$.

**Proposition 8.** *If $p < \frac{2d}{d-2}$, the sequence $(u_n)$ is always linearizable on any compact time interval. If $p = \frac{2d}{d-2}$, the sequence $(u_n)$ is linearizable on $[0,T]$ if and only the corresponding solutions $v_n$ to the linear equation satisfy*

$$\|v_n\|_{L^\infty(0,T;L^p)} \longrightarrow 0. \tag{73}$$

This proposition shows that, for the critical exponent, there is a qualitative change in the dynamics of equation (70). The solutions which display such a nonlinear effect are characterized by the negation of condition (73) for the corresponding linear solution. It is then natural to describe as precisely as possible the Cauchy data $(\varphi_n, \psi_n)$ which give rise to such a phenomenon. This is the aim of the following result.

**Theorem 9.** *Assume $p = \frac{2d}{d-2}$. Then, if $(u_n)$ is not linearizable on $[0,T]$, there exists $\varepsilon \in \{\pm 1\}$, a semiclassical measure $m$ associated to $\psi_n + \varepsilon i|D|\varphi_n$ and some scale $(\varepsilon_n)$, a point $y \in \mathbb{R}^d$, a time $t \in [0,T]$ and a nonnegative Borel function $h$ on $\mathbb{R}^d$ such that*

$$m(x,\xi) \geq \delta\Big(x - y + \varepsilon t \frac{\xi}{|\xi|}\Big) h(\xi)\, d\xi, \quad \int_{\mathbb{R}^d} h(\xi)\, d\xi > 0. \tag{74}$$

As a consequence, if the weak limit $e(x)$ of $|\psi_n(x)|^2 + |\nabla\varphi_n(x)|^2$ is singular to Dirac masses and to superficial measures on spheres in $\mathbb{R}^d$, then $(u_n)$ is linearizable on any compact time interval.

Let us make some remarks about the proof of these results. Proposition 8 can be easily obtained by a combination of energy identities and generalized Strichartz inequalities. A detailed proof can be found in [3] in the case $d = 3$; the argument given there can be carried out to the general case. For Theorem 9, we apply Theorem 4 (ii) to a sequence $(u_n(t_n,\cdot))$ where $t_n \to t \in [0,T]$ and conclude by Proposition 3 (ii).

A natural open problem would be now to describe properly the solution $u_n$ after the first focusing time $t$ given by (74). It can be achieved in the following very particular example, where $(\psi_n, \varphi_n)$ are chosen so that, for some $t_0 > 0$, the linear solution is given by

$$v_n(t_0,x) = \frac{1}{\varepsilon_n^{d/2-1}} \Phi\Big(\frac{x}{\varepsilon_n}\Big), \quad \partial_t v_n(t_0,x) = \frac{1}{\varepsilon_n^{d/2}} \Psi\Big(\frac{x}{\varepsilon_n}\Big), \tag{75}$$

where $(\Phi,\Psi) \in H^1 \times L^2$. Then, at least if the energy of $(\Phi,\Psi)$ is small enough, the natural rescaling shows that $u_n - \tilde v_n$ goes strongly to 0 in $C^1([T_1,T_2],L^2) \cap C([T_1,T_2],H^1)$, $t_0 < T_1$, where $\tilde v_n$ is the solution to the linear equation given by

$$\tilde v_n(t_0,x) = \frac{1}{\varepsilon_n^{d/2-1}} \tilde\Phi\Big(\frac{x}{\varepsilon_n}\Big), \quad \partial_t \tilde v_n(t_0,x) = \frac{1}{\varepsilon_n^{d/2}} \tilde\Psi\Big(\frac{x}{\varepsilon_n}\Big), \tag{76}$$

and $(\tilde{\Phi}, \tilde{\Psi})$ are obtained from $(\Phi, \Psi)$ by the nonlinear scattering operator associated to equation (70) with $p = \frac{2d}{d-2}$ (see [7], Proposition 3.3). So far we do not know how such a result might be generalized to arbitrary nonlinearizable solutions.

## References


1. P. Gérard, *Microlocal Defect Measures*, Comm. Partial Diff. Eq. **16** (1991), 1761–1794.
2. P. Gérard and E. Leichtnam, *Ergodic Properties of Eigenfunctions for the Dirichlet Problem*, Duke Math. J. **71** (1993), 559–607.
3. P. Gérard, *Oscillations and Concentration Effects in Semilinear Dispersive Wave Equations*, Preprint Université de Paris-Sud, 1994.
4. J. Ginibre and G. Velo, *The Global Cauchy Problem for the Non Linear Klein-Gordon Equation*, Math. Z. **189** (1985), 487–505.
5. J. Ginibre, H. Soffer and G. Velo, *The Global Cauchy Problem for the Critical Non Linear Wave Equation*, J. Funct. Anal. **110** (1992), 96–130.
6. J. Ginibre and G. Velo, *Generalized Strichartz Inequalities for the Wave Equation*, Preprint LPTHE Orsay, 1994.
7. ———, *Scattering theory in the energy space for a class of non linear wave equations*, Comm. Math. Phys. **123** (1989), 535–573.
8. B. Helffer, A. Martinez and R. Robert, *Ergodicité et limite semi-classique*, Comm. Math. Phys. **109** (1987), 313–326.
9. L. Hörmander, *The Analysis of Linear Partial Differential Operators Vol. III*, Springer-Verlag, Berlin, Heidelberg, New York, Tokyo, 1985, Grundlehren Math. Wiss. **274**.
10. P.-L. Lions, *The Concentration-Compactness Principle in the Calculus of Variations. The Limit Case, Part 1*, Rev. Mat. Iberoamericana **1** (1985), 145–201.
11. P.-L. Lions and T. Paul, *Sur les Mesures de Wigner*, Rev. Mat. Iberoamericana **9** (1993), 553–618.
12. J. Shatah and M. Struwe, *Regularity Results for Nonlinear Wave Equations*, Ann. Math. **138** (1993), 503–518.
13. ———, *Well-Posedness in the Energy Space for Semilinear Wave Equations with Critical Growth*, IMRN **7** (1994), 303–309.
14. A. Shnirelman, *Ergodic properties of eigenfunctions*, Uspekhi Mat. Nauk **29** (1974), 181–182.
15. L. Tartar, *H-measures: a new approach for studying homogenization, oscillations and concentration effects in partial differential equations*, Proc. Roy. Soc. Edinburgh Sect. A, **115** (1990), 193–230.
16. E. Wigner, *On the quantum correction for thermodynamic equilibrium*, Phys. Rev. **40** (1932), 749–759.


Université de Paris-Sud, Département de Mathématiques, Bât. 425, F-91405 Orsay Cedex, France

*E-mail address*: Patrick.GERARD@math.u-psud.fr

# COMPLETE HEAT TRACE, RESOLVENT AND ZETA EXPANSIONS FOR GENERAL ATIYAH-PATODI-SINGER PROBLEMS

Gerd Grubb

The results presented in this talk were obtained jointly with R. T. Seeley in [GS1], [GS2].

**1. The general APS problem.** On a compact $n$-dimensional $C^\infty$ manifold $X$ with boundary $\partial X = X'$, consider a first-order elliptic differential operator

$$P : C^\infty(E_1) \to C^\infty(E_2)$$

between sections of vector bundles over $X$. $E_1$ and $E_2$ have Hermitian metrics, and $X$ has a smooth volume element, defining a Hilbert space structure on the sections. The restrictions of the $E_i$ to the boundary $X'$ are denoted $E'_i$. A neighborhood of $X'$ in $X$ has the form $X_c = X' \times [0, c]$, and there the $E_i$ are isomorphic to the pull-backs of the $E'_i$. Let $x_n$ denote the coordinate in $[0, c]$. Then our $P$ is represented in $X_c$ as

$$P = \sigma(\tfrac{\partial}{\partial x_n} + A + x_n P_1 + P_0), \tag{1.1}$$

where $\sigma$ is a unitary morphism from $E'_1$ to $E'_2$, independent of $x_n$, and $A$ is a fixed elliptic first order differential operator on $C^\infty(E'_1)$, selfadjoint with respect to the Hermitian metric in $E'_1$ and the volume element $v(x', 0)dx'$ on $X'$ induced by the element $v(x', x_n)dx'dx_n$ on $X$. The $P_j$ are smooth differential operators (in all variables) of order $\leq j$; they can be taken arbitrarily near $X'$, but for larger $x_n$, $P_1$ is subject to the requirement that $P$ be elliptic. (The generality of the setup is further discussed in [G2].) The case considered by Atiyah, Patodi and Singer in [APS] was the case where $P_1 = P_0 = 0$, so that $P$ is constant in $x_n$ on $X_c$; this is often called the *product case.*

Since $P$ is a first order system, it may not be possible to formulate a well-posed boundary value problem in terms of a *differential* boundary condition (a Dirichlet condition is too much, no boundary condition is too little, and the boundary bundle structure will not in general allow putting a Dirichlet condition on some "half" of the boundary data). But using pseudodifferential operators ($\psi$do's), one can get well-posedness:

**Definition 1.1. The APS boundary problem** consists of finding $u \in H^1(E_1)$ (Sobolev space), for given $f \in L_2(E_2)$, so that

$$Pu = f \text{ on } X, \quad B\gamma_0 u = 0 \text{ on } X'. \tag{1.2}$$

Here $\gamma_0 u = u|_{X'}$, and $B$ is an orthogonal projection in $L_2(E_1')$ commuting with $A$ and of the form $B = \Pi_> + B_0$, where $\Pi_>$ is the projection onto the sum of eigenspaces for $A$ with eigenvalues $\lambda > 0$, and $B_0$ ranges in the sum of eigenspaces for $A$ with $|\lambda| \le R$, for some $R \ge 0$.

The associated realization $P_B$ is defined as the operator from $L_2(E_1)$ to $L_2(E_2)$ acting like $P$ and with domain

$$D(P_B) = \{\, u \in H^1(E_1) \mid B\gamma_0 u = 0 \,\}; \tag{1.3}$$

it is a Fredholm operator called **the APS operator**, and **the APS index problem** consists of determining its index.

The Fredholm property of $P_B$ was shown by Seeley in [S], where it was moreover shown that the adjoint of $P_B$ is of a related type:

$$(P_B)^* = (P^*)_{B'}, \text{ with } B' = (I - B)\sigma^*. \tag{1.4}$$

The systems $\{P, B\gamma_0\}$ and $\{P^*, B'\gamma_0\}$ are injectively elliptic (overdetermined elliptic).

One of the ways to study the index of $P_B$ is to consider the "Laplacians"

$$\Delta_1 = P_B{}^* P_B, \quad \Delta_2 = P_B P_B{}^*, \tag{1.5}$$

and search for asymptotic expansions for $t \to 0$ (with $\varepsilon > 0$):

$$\operatorname{Tr}\exp(-t\Delta_i) = c_{-n,i}t^{-n/2} + \dots + c_{-1,i}t^{-1/2} + c_{0,i} + O(t^\varepsilon),\ i = 1,2. \tag{1.6}$$

When (1.6) holds, the index is determined by

$$\operatorname{index} P_B = \operatorname{Tr}\exp(-t\Delta_1) - \operatorname{Tr}\exp(-t\Delta_2) = c_{0,1} - c_{0,2}. \tag{1.7}$$

In [APS] it was shown in the product case, with $B = \Pi_\ge$, that

$$\operatorname{index} P_B = \int_X \alpha(x) - \tfrac{1}{2}\eta_A; \quad \eta_A = \eta(A,0) + \dim\ker A; \tag{1.8}$$

where $\alpha(x)$ is a certain form defined from the symbol of $P$, and $\eta(A,0)$ is the value at $s = 0$ of the eta function

$$\eta(A,s) = \operatorname{Tr}(|A|A^{-s-1}).$$

(Here $A^{-s-1}$ is defined as 0 on the nullspace of $A$, and meromorphic extension is used for $\operatorname{Re} s < n$.) Formula (1.8) was extended to the nonproduct case in [G2] as

$$\text{index}\, P_B = \int_X \alpha(x) + \int_{X'} \beta(x') - \tfrac{1}{2}\eta_A, \tag{1.9}$$

with a boundary form $\beta(x')$ defined from the symbols of $P$ and $B$. The forms defined from the symbols are regarded as *local contributions*, whereas the term $\eta_A$ depends on the setup in a global way.

Actually, [APS] did not calculate the two expressions $\operatorname{Tr}\exp(-t\Delta_1)$ and $\operatorname{Tr}\exp(-t\Delta_2)$ separately, but only their difference, which was set in relation to $\operatorname{Tr}[\exp(-t\widetilde{P}^*\widetilde{P})|_X - \exp(-t\widetilde{P}\,\widetilde{P}^*)|_X]$ (where $\widetilde{P}$ is an extension of $P$ to the double manifold $\widetilde{X}$), and to a similar operator on $X' \times \mathbb{R}_+$. Here the study was based on functional calculus for the selfadjoint operator $A$, which suffices for the treatment of the product case.

In [G2] the separate expansions (1.6) were proved with $\varepsilon = \frac{3}{8}$ in the nonproduct case with $B = \Pi_{\geq}$, by a combination of the general treatment of parameter-dependent $\psi$do boundary problems in [G1] with the special results from [APS]. It was shown that the global term $-\frac{1}{2}\eta_A$ enters in $c_{0,i}$ for both expansions, as $-\frac{1}{4}\eta_A$ for $i = 1$, resp. $\frac{1}{4}\eta_A$ for $i = 2$.

The subject of the present exposition is the study of asymptotic expansions beyond the zero'th term. In fact we can replace $O(t^\varepsilon)$ by full asymptotic expansions with arbitrarily high powers of $t$; then also logarithmic terms appear. The main result is the following [GS2]:

**Theorem 1.2.** *Let $\varphi$ be a $C^\infty$ morphism in $E_i$.*
(a) *There is a full expansion for $t \to 0$:*

$$\operatorname{Tr}[\varphi \exp(-t\Delta_i)] \sim \sum_{-n\leq j<0} c_{j,i}t^{j/2} + \sum_{j=0}^{\infty}(c_{j,i}t^{j/2}\log t + c'_{j,i}t^{j/2}). \tag{1.10}$$

*The coefficients $c'_{j,i}$ are in general globally determined, the $c_{j,i}$ local.*

We have two other formulations of the result that are essentially equivalent with (a):

(b) *The zeta function $\operatorname{Tr}(\varphi\Delta_i^{-s})$ is meromorphic on $\mathbb{C}$ and has the following pole structure:*

$$\Gamma(s)\operatorname{Tr}(\varphi\Delta_i^{-s}) \sim \sum_{-n\leq j<0} \frac{c_{j,i}}{s+\frac{j}{2}} + \frac{\nu_i}{s} + \sum_{j=0}^{\infty}\Big(\frac{-c_{j,i}}{(s+\frac{j}{2})^2} + \frac{c'_{j,i}}{s+\frac{j}{2}}\Big); \tag{1.11}$$

*here $\nu_i = \operatorname{Tr}(\varphi\Pi_0(\Delta_i))$.*

(c) *The trace of the resolvent (differentiated $m \geq n$ times) has an expansion for $\lambda \to \infty$ on rays in $\mathbb{C} \setminus \mathbb{R}_+$:*

$$\operatorname{Tr}(\varphi \partial_\lambda^m (\Delta_i - \lambda)^{-1}) \sim \sum_{-n \leq j < 0} a_{j,i}(-\lambda)^{\frac{-j}{2}-m-1} + \nu_i'(-\lambda)^{-m-1} + \sum_{j=0}^{\infty} (a_{j,i} \log \lambda + a'_{j,i})(-\lambda)^{\frac{-j}{2}-m-1}. \tag{1.12}$$

The coefficients in (c) are related in a systematic way to those in (b) (by multiplication by fixed gamma function factors). The rules for passage between the three versions are explained in detail in [GS1].

In [GS1] we show some more explicit results on these expansions in the product case. Here we can describe the coefficients directly in terms of $A$ and its zeta and eta functions. In particular, we find that for $n$ **even,** logarithms occur (besides possibly for $j = 0$) only at half-integers (values of $\frac{j}{2}$ where $j$ is odd), where the "global" coefficient $c'_{j,i}$ is proportional to the *second Laurent coefficient* of $\Gamma(s) \operatorname{Tr}(\varphi|_{X'} A^{-2s})$ (a zeta function of $A^2$) at $-\frac{j}{2}$. For $n$ **odd**, there are no logarithms at all; here the global coefficients $c'_{i,j}$ at the half-integers are proportional to the *values* of the zeta function of $A^2$ at $-\frac{j}{2}$.

The basic point of view in [GS1] is again to use functional calculus on $A$, which turns out to be particularly successful when one establishes (b) first, going directly to the zeta functions. (There are similar expansions of eta functions.)

In the nonproduct case, functional calculus is no longer sufficient, and new considerations on parameter-dependent pseudodifferential boundary problems are needed. The rest of the talk will describe this.

**2. Polyhomogeneous symbol classes.** Consider symbols $p(x, \xi, \mu)$, where $x$ and $\xi \in \mathbb{R}^n$, $\mu \in \Gamma$ (a sector of $\mathbb{C} \setminus \{0\}$). We shall say that:

$p$ is **strongly homogeneous** of degree $m$, when

$$p(x, t\xi, t\mu) = t^m p(x, \xi, \mu) \text{ for } |\xi|^2 + |\mu|^2 \geq 1,\ t \geq 1, \quad (\xi, \mu) \in \mathbb{R}^n \times (\Gamma \cup \{0\}); \tag{2.1}$$

$p$ is **weakly homogeneous** of degree $m$, when

$$p(x, t\xi, t\mu) = t^m p(x, \xi, \mu) \text{ for } |\xi| \geq 1,\ t \geq 1,\ (\xi, \mu) \in \mathbb{R}^n \times \Gamma. \tag{2.2}$$

**Example.** Let $a(x,\xi)$ be positive and $C^\infty$ on $\mathbb{R}^n \times \mathbb{R}^n$, and homogeneous in $\xi$ of degree $r$ for $|\xi| \ge 1$. Then $a(x,\xi)+\mu^r$ and $(a(x,\xi)+\mu^r)^{-1}$ extend to:

**strongly** homogeneous symbols of degree $r$, resp. $-r$, if $a$ is **polynomial** in $\xi$ (it is the symbol of a differential operator);

**weakly** homogeneous symbols of degree $r$, resp. $-r$, if $a$ is **not** polynomial in $\xi$ (it is the symbol of a genuine $\psi$do).

If for example $r = n = 2$, $a(x,\xi) = \xi_1^2 + \xi_2^2$ enters in the first case, and $a(x,\xi) = (\xi_1^4+\xi_2^4)/(\xi_1^2+\xi_2^2)$ (for $|\xi|\ge 1$) enters in the second case.

Both cases belong to the following symbol classes (where $(a(x,\xi) + \mu^r)^{-1} \in S^{-r,0} \cap S^{0,-r}$):

**Definition 2.1.** $S^{m,0}(\mathbb{R}^n,\mathbb{R}^n,\Gamma)$ consists of the functions $p(x,\xi,\mu)$ that are holomorphic in $\mu$ for $|(\xi,\mu)| \ge \varepsilon$ and satisfy, denoting $\frac{1}{\mu} = z$,

$$\begin{aligned}(2.3)\quad &\partial_z^j p(\cdot,\cdot,\tfrac{1}{z}) \text{ is in } S^{m+j}(\mathbb{R}^n,\mathbb{R}^n) \text{ for } \tfrac{1}{z}\in\Gamma, \text{ with}\\ &\qquad\text{uniform estimates for } |z|\le 1, \tfrac{1}{z}\in \text{closed subsectors of } \Gamma.\end{aligned}$$

Moreover, we set $S^{m,d}(\mathbb{R}^n,\mathbb{R}^n,\Gamma) = \mu^d S^{m,0}(\mathbb{R}^n,\mathbb{R}^n,\Gamma)$.

Here $S^m(\mathbb{R}^n,\mathbb{R}^n)$ denotes the standard $\psi$do symbol space consisting of the functions $p(x,\xi)\in C^\infty(\mathbb{R}^n\times\mathbb{R}^n)$ such that $\partial_x^\beta\partial_\xi^\alpha p$ is $O(\langle\xi\rangle^{m-|\alpha|})$ for all $\alpha\in\mathbb{N}^n$, $\beta\in\mathbb{N}^n$; here $\langle\xi\rangle = (1+|\xi|^2)^{\frac12}$ and $\mathbb{N} = \{\text{ integers } \ge 0\}$. The rules of calculus for such symbols are well known, see e.g. Hörmander [H1], Seeley [S], Shubin [Sh], Hörmander [H2] for various setups with local or global estimates in $x$. We call the symbols *classical*, when they moreover have expansions in series of homogeneous terms.

The symbols in $S^{m,d}(\mathbb{R}^n,\mathbb{R}^n,\Gamma)$ define $\psi$do's $P = \mathrm{OP}(p)$ (which depend on the parameter $\mu$) by the usual formula:

$$(2.4)\qquad \mathrm{OP}(p)f(x) = \int e^{ix\cdot\xi}p(x,\xi,\mu)\hat{f}(\xi)\,đ\xi, \quad f\in\mathcal{S}(\mathbb{R}^n),$$

with $đ\xi = (2\pi)^{-n}d\xi$. The definition extends to more general functions as in the nonparametrized case.

The operators have good composition rules, since $S^{m,d}\cdot S^{m',d'} \subset S^{m+m',d+d'}$, and since one can refer to the standard rules for $S^m$ symbol classes, which must here hold uniformly in $z$ as in (2.3). One finds for example that

$$(2.5)\qquad P\in\mathrm{OP}(S^{m,d}),\ P'\in\mathrm{OP}(S^{m',d'}) \implies PP'\in\mathrm{OP}(S^{m+m',d+d'})$$

(under the usual precautions on supports or global estimates), and the resulting symbol is described by the usual formula

(2.6)
$$(p \circ p')(x,\xi,\mu) \sim \sum_{\alpha \in \mathbb{N}^n} \tfrac{1}{\alpha!}\partial_\xi^\alpha p(x,\xi,\mu)(-i\partial_x)^\alpha p'(x,\xi,\mu) \text{ in } S^{m+m',d+d'}.$$

The expansion in (2.6) is an expansion in terms with decreasing $m$-exponents $m + m' - j$, $j \to \infty$ ($j = |\alpha|$). Such expansions enter in the theory as follows:

When $p_j \in S^{m_j,d}$ for a decreasing sequence $m_j \to -\infty$ (for $j \to \infty$, $j \in \mathbb{N}$), and $p \in S^{m_0,d}$, we say that $p \sim \sum_j p_j$ in $S^{m_0,d}$ if

$$p - \sum_{j<J} p_j \in S^{m_J,d} \text{ for any } J \in \mathbb{N}. \tag{2.7}$$

For any given sequence $p_j \in S^{m_j,d}$ with $m_j \searrow -\infty$, there exists a $p$ such that (2.7) holds.

For the present special symbols there is *another* type of expansion that is of great interest:

**Theorem 2.2.** *When $p \in S^{m,d}(\mathbb{R}^n,\mathbb{R}^n,\Gamma)$, then $p$ has an expansion in terms $\mu^{d-k}p_{(d,k)}(x,\xi)$ with $p_{(d,k)} \in S^{m+k}(\mathbb{R}^n,\mathbb{R}^n)$, such that for any $N$,*

$$p(x,\xi,\mu) - \sum_{0\le k<N} \mu^{d-k}p_{(d,k)}(x,\xi) \in S^{m+N,d-N}(\mathbb{R}^n,\mathbb{R}^n,\Gamma). \tag{2.8}$$

In the proof one reduces to the case $d = 0$ by multiplication by $\mu^{-d}$; then the expansion is essentially a Taylor expansion in $z = \frac{1}{\mu}$ at $z = 0$.

Note that in (2.8), the order of $p_{(d,k)}$ *increases* with increasing $k$, whereas the power of $\mu$ *decreases*.

**Corollary 2.3.** *When $p \in S^{-\infty,d}$, the kernel $K(x,y,\mu)$ of $\mathrm{OP}(p)$ has an expansion*

$$K(x,y,\mu) \sim \sum_{k\in\mathbb{N}} \mu^{d-k}K_k(x,y), \quad K_k \in C^\infty. \tag{2.9}$$

Definition 2.1 contains no homogeneity requirements, but we now define a polyhomogeneous subspace:

**Definition 2.4.** A symbol $p \in S^{m_0-d,d}$ is called **(weakly) polyhomogeneous**, when $p \sim \sum_{j\in\mathbb{N}} p_j$, with $p_j \in S^{m_j-d,d}$, $m_j \searrow -\infty$ for $j \to \infty$, $j \in \mathbb{N}$, such that the $p_j$ are weakly homogeneous of degrees $m_j$ (cf. (2.2)).

If the $p_j$ are in addition strongly homogeneous (cf. (2.1)), and

$$\partial_x^\beta \partial_\xi^\alpha \partial_\mu^k (p - \sum_{j<J} p_j) = O(\langle(\xi,\mu)\rangle^{m-J-|\alpha|-k}), \tag{2.10}$$

for all indices $\alpha, \beta, k, J$, then $p$ is called **strongly polyhomogeneous.**

*Remark 2.5.* One has in particular that *classical* $\psi$do symbols in $n+1$ cotangent variables $p(x,(\xi,\xi_{n+1}))$ give *strongly polyhomogeneous* symbols when $\xi_{n+1}$ is replaced by $\mu$.

The type of parameter-dependence mentioned in Remark 2.5 was used by Agmon and by Agranovič and Vishik in resolvent studies for differential operators; for $\psi$do's this is the kind of parameter-dependence studied e.g. in Shubin [Sh] and other works. It is a mild generalization that does not cover resolvents $(P-\lambda)^{-1}$ and parabolic operators such as $\partial/\partial t + P$ when $P$ is truly pseudodifferential (as treated in [G1]).

*Both* the expansion in Theorem 2.2 and the expansion in Definition 2.4 enter in the proof of:

**Theorem 2.6.** *Let $p$ be weakly polyhomogeneous as in Definition 2.4, with $m_0 - d < -n$. Then $\mathrm{OP}(p)$ has a continuous kernel $K_p(x,y,\mu)$ with an expansion on the diagonal*

$$K_p(x,x,\mu) \sim \sum_{j=0}^{\infty} a_j(x)\mu^{m_j+n} + \sum_{k=0}^{\infty}[c_k(x)\log\mu + c_k'(x)]\mu^{d-k}, \tag{2.11}$$

*for $|\mu| \to \infty$, uniformly for $\mu$ in closed subsectors of $\Gamma$. The coefficients $a_j(x)$ and $c_{d-m_j-n}(x)$ are determined by $p_j(x,\xi,\mu)$ for $|\xi| \ge 1$ (are "local"), while the $c_k'(x)$ are "global".*

In the proof one uses the general principle that "remainders contribute to $c_k'$ terms" (by Corollary 2.3). The $p_j$ contribute with

$$\begin{aligned} K_{p_j}(x,x,\mu) &= \int_{\mathbb{R}^n} p_j(x,\xi,\mu)\,đ\xi \\ &= \int_{|\xi|\ge|\mu|} p_j\,đ\xi + \int_{|\xi|\le 1} p_j\,đ\xi + \int_{1\le|\xi|\le|\mu|} p_j\,đ\xi = I_1 + I_2 + I_3, \end{aligned} \tag{2.12}$$

where $I_1$ gives part of the $a_j$ term, $I_2$ gives $c_k'$ terms, and $I_3$ gives the rest of $a_j$ and $c_{d-m_j-n}$ (if $d - m_j \in \mathbb{Z}$) and some $c_k'$ terms. One has of course to show that the contributions to the $c_k'$ pile up in a controlled way.

When the operator acts on a compact boundaryless manifold, integration of $K_p(x,x,\mu)$ in $x$ gives a similar expansion of the *trace* of $\mathrm{OP}(p)$.

The result applies in particular to expressions containing a differentiated resolvent:

$$P = Q\partial_\lambda^m (A-\lambda)^{-1}, \tag{2.13}$$

where $A$ is a classical elliptic $\psi$do of positive integer order $r$ with principal symbol $a_r(x,\xi)$ having no eigenvalues on $\mathbb{R}_-$, $Q$ is a classical $\psi$do of order $\omega$, and $m$ is chosen so that $\omega - r(1+m) < -n$. With $u = (-\lambda)^{1/r}$ for $\lambda$ in a narrow sector around $\mathbb{R}_-$, the symbol is in $S^{\omega - r(1+m),0} \cap S^{0,\omega - r(1+m)}$ and weakly polyhomogeneous. Then Theorem 2.6 gives an expansion of the diagonal kernel and the trace, generalizing the result of Agranovič [A] for $Q = I$.

**3. The APS resolvent.** We now return to the APS operator on a manifold with boundary, as described in Section 1.

To study the resolvents $R_i = (\Delta_i + \mu^2)^{-1}$, consider

$$\mathcal{P}_\mathcal{B} = \begin{pmatrix} 0 & -P_\mathcal{B}{}^* \\ P_\mathcal{B} & 0 \end{pmatrix}, \text{ the realization of } \mathcal{P} = \begin{pmatrix} 0 & -P^* \\ P & 0 \end{pmatrix} \tag{3.1}$$

under the following boundary condition on $u = \{u_1, u_2\}$ (cf. (1.4)):

$$\mathcal{B}\gamma_0 u = 0, \text{ where } \mathcal{B} = \begin{pmatrix} B & B' \end{pmatrix} : \begin{matrix} L_2(E_1') \\ \times \\ L_2(E_2') \end{matrix} \to L_2(E_1'). \tag{3.2}$$

The advantage of taking $P_\mathcal{B}$ and $P_\mathcal{B}{}^*$ together in this way is that $\mathcal{P}_\mathcal{B}$ is truly elliptic and skew-adjoint; and $\mathcal{R} = (\mathcal{P}_\mathcal{B} + \mu)^{-1}$ (defined for $\mu \in \mathbb{C} \setminus i\mathbb{R}$) satisfies

$$\mathcal{R} = (\mathcal{P}_\mathcal{B} + \mu)^{-1} = \begin{pmatrix} \mu R_1 & P_\mathcal{B}{}^* R_2 \\ -P_\mathcal{B} R_1 & \mu R_2 \end{pmatrix}, \tag{3.3}$$

allowing us to recover the resolvents we are looking for. (The off-diagonal terms are used to describe eta functions instead of zeta functions, cf. [GS1,2].)

A careful analysis of the structure leads to:

**Theorem 3.1.** *For large $\mu$, $\mathcal{R}$ can be written as*

$$\mathcal{R} = Q_+ + \mathcal{KST}. \tag{3.4}$$

*Here:*

- *$Q$ is a parametrix of $\mathcal{P} + \mu$ on an extension of the manifold $X$ to a compact boundaryless manifold $\widetilde{X}$, and $Q_+ = r^+ Q e^-$, where $e^+$ stands for extension by 0 on $\widetilde{X} \setminus X$, and $r^+$ denotes restriction to $X$.*

• $\mathcal{K}$ *is a Poisson operator and* $\mathcal{T}$ *is a trace operator in a parameter-dependent version of the Boutet de Monvel calculus, both* **strongly polyhomogeneous** *in a similar way as in Remark* 2.5.

• $\mathcal{S}$ *is a weakly polyhomogeneous* $\psi$*do on* $X'$, *with symbol in* $S^{-1,0} \cap S^{0,-1}$ *and with* $m$*'th* $\mu$*-derivatives having symbols in* $S^{-1-m,0} \cap S^{0,-1-m}$, *for all* $m$.

*Furthermore, with* $\varphi \in C^\infty(X)$ *one has for* $m \geq n$,

$$\operatorname{Tr}[\varphi \partial_\mu^m \mathcal{R}] = \operatorname{Tr}[\varphi \partial_\mu^m Q_+] + \operatorname{Tr}[\partial_\mu^m (\mathcal{T}\varphi\mathcal{K}\mathcal{S})], \tag{3.5}$$

*where the operator* $\partial_\mu^m(\mathcal{T}\varphi\mathcal{K}\mathcal{S})$ *is a* **weakly polyhomogeneous pseudo-differential operator** *on* $X'$ *with symbol in* $S^{-m-1,0} \cap S^{0,-m-1}$.

In the proof of (3.5) it is used that the trace is invariant under cyclic permutation, together with the rules of calculus for boundary operators. We find in particular that $\mathcal{T}\varphi\mathcal{K}$ is a *strongly polyhomogeneous* $\psi$do on $X'$, since its ingredients stem from the differential operator $\mathcal{P}$ and its strongly polyhomogeneous resolvent $(\mathcal{P}+\mu)^{-1}$.

A systematic calculus of *weakly polyhomogeneous boundary operators* can be developed, but in the present case where we are just studying the trace, the above suffices.

**Consequences.** Applying known results to $\operatorname{Tr}[\varphi\partial_\mu^m Q_+]$ and applying Theorem 2.6 to $\operatorname{Tr}[\partial_\mu^m(\mathcal{T}\varphi\mathcal{K}\mathcal{S})]$, we get a full asymptotic expansion for $\operatorname{Tr}[\varphi\partial_\mu^m\mathcal{R}]$:

$$\begin{aligned}\operatorname{Tr}[\varphi\partial_\mu^m(\mathcal{P}_B+\mu)^{-1}]\\ \sim \sum_{-n\leq j<0} a_j\mu^{-j-m-1} + \sum_{j=0}^{\infty}(a_j\log\mu + a_j')\mu^{-j-m-1},\end{aligned} \tag{3.6}$$

with local coefficients $a_j$ (determined by integrals over $X$ and $X'$ of densities defined from the homogeneous terms in the symbols), and global coefficients $a_j'$.

From this we deduce Theorem 1.2 and its variants:

• Full asymptotic expansions for $\operatorname{Tr}[\partial_\mu^m\varphi(\Delta_i+\mu^2)^{-1}]$, for $\mu\to\infty$,

• Full asymptotic expansions for $\operatorname{Tr}[\varphi\exp(-t\Delta_i)]$, for $t\to 0$,

• Meromorphic extension of $\Gamma(s)\operatorname{Tr}[\varphi\Delta_i^{-s}]$ on $\mathbb{C}$ and a full description of its poles.

Also differential operators can be allowed in the place of $\varphi$. There are similar results for eta functions such as $\Gamma(s)\operatorname{Tr}[\varphi P_B\Delta_1^{-s}]$, and for the related functions $\operatorname{Tr}[\varphi P_B\partial_\mu^m(\Delta_1+\mu^2)^{-1}]$ and $\operatorname{Tr}[\varphi P_B\exp(-t\Delta_1)]$.

## REFERENCES

[A] M. S. Agranovič, *Some asymptotic formulas for elliptic pseudodifferential operators*, Functional Anal. Appl. **21** (1987), 63–65.

[APS] M. F. Atiyah, V. K. Patodi and I. M. Singer, *Spectral asymmetry and Riemannian geometry, I*, Math. Proc. Cambridge Philos. Soc. **77** (1975), 43–69.

[G1] G. Grubb, *Functional Calculus of Pseudo-Differential Boundary Problems*, Progress in Math., Vol. 65, Birkhäuser, Boston, 1986.

[G2] ———, *Heat operator trace expansions and index for general Atiyah-Patodi-Singer boundary problems*, Comm. Partial Diff. Eq. **17** (1992), 2031–2077.

[GS1] G. Grubb and R. T. Seeley, *Zeta and eta functions for Atiyah-Patodi-Singer operators*, J. Geom. Anal. (to appear).

[GS2] ———, *Weakly parametric pseudodifferential operators and Atiyah-Patodi-Singer boundary problems*, Inventiones Math. **121** (1995) 481–529.

[H1] L. Hörmander, *Pseudo-differential operators and hypoelliptic equations*, Proc. Symp. Pure Math., vol. 10, 1967, pp. 138–183.

[H2] ———, *The analysis of linear Partial Differential Operators, III*, Springer Verlag, Heidelberg, 1985.

[S] R. T. Seeley, *Topics in pseudo-differential operators*, CIME Conference on Pseudo-Differential Operators 1968, Edizioni Cremonese, Roma, 1969, pp. 169–305.

[Sh] M. A. Shubin, *Pseudodifferential Operators and Spectral Theory*, Nauka, Moscow, 1978.

DEPARTMENT OF MATHEMATICS, UNIVERSITY OF COPENHAGEN, UNIVERSITETSPARKEN 5, DK-2100 COPENHAGEN, DENMARK.
*E-mail address*: grubb@math.ku.dk

# SEMI-CLASSICAL ANALYSIS FOR THE TRANSFER OPERATOR: WKB CONSTRUCTIONS IN DIMENSION 1

Bernard Helffer

**Abstract.** This lecture is devoted to problems coming from statistical mechanics. The transfer matrix approach consists in reducing the analysis of asymptotic properties of the energy of some statistical systems expressed in terms of Laplace integrals to the analysis of the spectral properties of transfer operators (matrices) appearing in the form of integral operators. Sometimes the new problem appears to have a semi-classical nature. Although the problem is similar to the semi-classical study of the Kac's operator presented in our paper with M.Brunaud [BruHe91] which was devoted to the study of $\exp(-\frac{v}{2})\cdot\exp(h^2\Delta)\cdot\exp(-\frac{v}{2})$ for $h$ small, new features appear for the model $\exp(-\frac{v}{2h})\cdot\exp(h\Delta)\cdot\exp(-\frac{v}{2h})$. The principal results we want to present concern the semi-classical analysis for this second operator. This lecture will only explain the constructions in the one-dimensional case. The proof of analogous results in dimension $> 1$ and further developments in large dimension will be given in [He95c]. (See also [He95a] and [He95b].)

**1. Introduction.** We consider a model generalizing the Kac model with nearest neighbors interaction:

$$\Phi(x) = \sum_{j\in\mathbb{Z}/m\mathbb{Z}} v(x_j) + \sum_{j\in\mathbb{Z}/m\mathbb{Z}} w(x_j, x_{j+1}), \tag{1.1}$$

for $x \in \mathbb{R}^{mp}$, $x = (x_1, \cdots, x_m)$, $x_j = (x_{j1}, \cdots, x_{jp})$ and we assume that

$$\Phi(-x) = \Phi(x). \tag{1.2}$$

The function $v$ is now a potential defined on $\mathbb{R}^p$ (which is usually invariant by circular permutation of the variables) and the ultimate goal would be to follow everything with respect to $p$ in order to take the limit $p \to \infty$ after having taken the limit $m \to \infty$. We are interested in the asymptotic properties of the measure

$$(\pi h)^{-\frac{mp}{2}} \exp\Big(-\frac{\Phi}{h}\Big)\, dx^{(mp)}$$

as $m$ and $1/h$ tend to $\infty$. More precisely we are interested in the behavior of the correlation, defined for $i$ and $j$ in $\mathbb{Z}/m\mathbb{Z}$,

$$\text{(1.3)}\quad \operatorname{Cor}^{(m)}(i,j) = \int x_{i1}x_{j1}\exp\Big(-\frac{\Phi}{h}\Big)\,dx^{(mp)} \Big/ \int \exp\Big(-\frac{\Phi}{h}\Big)\,dx^{(mp)},$$

as $\operatorname{dist}_{\mathbb{Z}/m\mathbb{Z}}(i,j)$ tends to $\infty$. As always in statistics the order in which we take the limit is quite important and it is our aim in this particular example to analyze the problem.

Following the general reduction called the "transfer matrix method", we have seen (for example in [He94b]) that we have to analyze the following transfer operator $K = K_v$ whose kernel is given by

$$K_v(x,y) = (\pi h)^{-\frac{p}{2}} \exp\Big(-\big(\frac{1}{2h}\cdot(v(x)+v(y))\big)\Big)\cdot\exp\Big(-h^{-1}w(x,y)\Big).$$

We recall indeed the following relations (see for example (12.19) in [Pa84] or [He94d], [He94e]) between the transfer operator and the pair correlations:

$$\text{(1.4)}\quad \Big(\frac{\mu_2}{\mu_1}\Big)^j\Big(\int_{\mathbb{R}^p} y_1^2 u_1(y)^2\,dy^{(p)}\Big) \geq \lim_{m\to\infty}\operatorname{Cor}^{(m)}(0,j) \geq \Big(\frac{\mu_2}{\mu_1}\Big)^j\Big[\int_{\mathbb{R}} y_1\,u_2(y)\,u_1(y)\,dy^{(p)}\Big]^2,$$

where $\mu_1 > \mu_2 \geq \mu_3 \cdots > 0$ is the decreasing sequence of eigenvalues of $K$ and $u_j$ is the corresponding basis of eigenvectors. We shall concentrate our study on the case when

$$w(x,y) = \frac{a}{4}(x-y)^2.$$

We shall sometimes analyze the dependence with respect to $a$ but are mainly interested in the case when $a > 0$ is small. As a typical $v$ we take

$$v^{(p)}(y) = \sum_{i=1}^{p} u(y_i) + b\sum_{i=1}^{p}|y_i - y_{i+1}|^2,$$

and we assume again that $u$ is a single or double well one dimensional potential with the symmetry

$$u(-t) = u(t).$$

We recognize indeed the "Kac" operator (up to a multiplicative constant) which was studied previously ([Ka66], [BruHe91], [He92b])

$$\exp\Big(-\frac{v}{2h}\Big)\cdot\exp\Big(\frac{h}{a}\Delta\Big)\cdot\exp\Big(-\frac{v}{2h}\Big),$$

but the semi-classical parameter is not inserted as in the standard semi-classical analysis. We note that the potential $V^{(m,p)}$ can in the case $a = b$ be written in the form

$$V^{(m,p)} = \sum_{(j,i)\in\mathbb{Z}/m\mathbb{Z}\times\mathbb{Z}/p\mathbb{Z}} u(x_{ji}) + a \sum_{(i,j)\sim(l,k)} |x_{ji} - x_{kl}|^2 ,$$

where $(j,i) \sim (k,l)$ means here that $(j,i)$ and $(l,k)$ are nearest neighbors in $\mathbb{Z}/m\mathbb{Z} \times \mathbb{Z}/p\mathbb{Z}$.

In the case when the potential $v$ is even and uniformly strictly convex, we can apply the universal estimate for this operator given in [He94b] and [He94c]

$$\mu_2/\mu_1 \leq \exp\Big( - \cosh^{-1}\Big(\frac{1}{a}\lambda_{\min}(\operatorname{Hess} v)(x) + 1\Big)\Big), \tag{1.5}$$

and this gives for the correlation the following majoration:

$$\begin{aligned} &\lim_{m\to\infty} \operatorname{Cor}^{(m)}(0,j) \\ &\quad \leq \exp\Big( - j\cosh^{-1}\Big(\frac{1}{a}\inf_x \lambda_{\min}(\operatorname{Hess} v)(x) + 1\Big)\Big)\Big(\int_{\mathbb{R}} u_1^2\, x^2\, dx\Big). \end{aligned} \tag{1.6}$$

When the minimum of $v$ is at 0, we then get that

$$\lim_{m\to\infty} \operatorname{Cor}^{(m)}(0,j) \leq Ch \exp\Big( - j\cdot\cosh^{-1}\Big(\frac{1}{a}\inf_x \lambda_{\min}(\operatorname{Hess} v)(x) + 1\Big)\Big). \tag{1.7}$$

This is interesting to compare with the results obtained by Helffer-Sjöstrand [HeSj93] and more recently by J. Sjöstrand [Sj94]. When the minimum of $v$ is at 0 we can naturally hope to replace $\inf_x \lambda_{\min}(\operatorname{Hess} v)(x)$ by $\lambda_{\min}(\operatorname{Hess} v)(0)$. This is simply the result obtained for the harmonic approximation.

We want here to analyze more in detail the one-well problem in the semi-classical context and look at the existence of WKB solutions. We shall actually limit ourselves to the case $p = 1$ but most of the results are not restricted to this case.

In addition to the semi-classical localization of the largest eigenvalue, the WKB solution is also one step together with the control of the decay of the first eigenfunctions toward the understanding of the tunneling effect between wells in the non-convex case. Some results in this direction were analyzed in [He95b] (see also [He95a]).

**2. Lower bound for the first eigenvalue.** We first prove a lower bound of the largest eigenvalue $\mu_1$ of $K$. We recall that the Perron-Frobenius theorem (more precisely, its extension to compact integral operators with strictly positive kernel) gives that the largest eigenvalue is simple and equal to the norm of the operator with kernel $K$. Moreover, the first eigenvector can be chosen strictly positive. Now we immediately obtain by using the minimax principle:

$$\mu_1 \geq \langle Ku \mid u \rangle / \|u\|^2, \quad \forall u \in L^2(\mathbb{R}),\ \|u\| \neq 0.$$

If we assume that

$$0 = v(0) \leq v(t) \leq C\, t^2, \quad \text{for some } C > 0,$$

we can for example take $u = h^{-\frac{1}{4}} \exp\big(-\frac{1}{h}t^2\big)$, and we get immediately a constant $\alpha > 0$ such that

$$\mu_1 \geq \alpha, \quad \forall h \in ]0, h_0].$$

In the double well case (with nondegenerate wells) we can take a gaussian localized at one of the minima of the potential $v$ (or better the eigenfunction corresponding to the quadratic approximation at one of the minima) and get the same result. One can also use Segal's Lemma giving a direct comparison between the largest eigenvalue of the Kac operator and the smallest eigenvalue $\lambda_1$ of the corresponding Schrödinger operator

$$-h\Delta + \frac{v}{h}.$$

This estimate gives simply

$$\mu_1 \geq \exp(-\lambda_1). \tag{2.1}$$

If $v \geq 0$, $\mu_1$ is bounded independently of $h$.

We have consequently the

**Lemma 2.1.** *Let $v$ satisfy the conditions (for some $C > 0$)*

$$v(t) \geq 0, \tag{2.2}$$

$$\frac{1}{C}t^2 - C \leq v(t) \leq Ct^2, \tag{2.3}$$

*and have (at least) one nondegenerate isolated minimum point $x_c$ with $v(x_c) = 0$. Then there exists a constant $\alpha > 0$ such that, for all $h \in ]0, h_0]$, the largest eigenvalue of $K_v$ satisfies*

$$\mu_1 \geq \alpha. \tag{2.4}$$

*Remark 2.2.* Although we do not need it here, it is possible to verify, by direct comparison with harmonic potentials, that

$$\lim_{h \to 0} \mu_1(h) = \mu^\infty.$$

Here $\mu^\infty$ is the largest eigenvalue of the operator whose kernel is

$$\pi^{-\frac{1}{2}} \exp\big(-\tfrac{1}{4}\big[v''(x_c)(x^2 + y^2) + a(x - y)^2\big]\big),$$

which is explicitly computable (see for example [He94c]).

**3. First properties of the first eigenfunction.** As recalled in Section 2, the first eigenfunction $u_1$ can be chosen strictly positive. We keep also the assumptions of the lemma and assume that $v$ has a finite number of nondegenerate minima where $v = 0$.

We now write that $u_1$ is the first normalized eigenfunction in the form

$$u_1(x;h) = \frac{1}{\mu_1(h)} \int K_v(x,y;h) u_1(y;h)\, dy. \tag{3.1}$$

Using Cauchy-Schwarz, we get immediately

$$0 < u_1(x;h) \leq \alpha^{-1} \Big( \int K_v(x,y)^2\, dy \Big)^{\frac{1}{2}}. \tag{3.2}$$

We get immediately the following lemma:

**Lemma 3.1.** *Under the assumption of Lemma 2.1 the first normalized positive eigenfunction satisfies, for some $h$-independent constant $C$,*

$$0 < u_1(x;h) \leq C \cdot h^{-\frac{1}{4}} \cdot \exp\Big( - \frac{v(x)}{2h} \Big). \tag{3.3}$$

This means that the eigenfunction is strongly localized near the minima of the potential $v$. This will permit to use cutoff functions in order to construct quasimodes attached to each well. We observe here that we are working with operators which are not local, but pseudolocal in a weak sense. The pseudolocality is obtained through the interaction term $\frac{a}{h} \cdot |x-y|^2$. This is a weaker pseudolocality than the pseudolocality observed in the semi-classical study of the Kac operator where the pseudolocality was obtained through an interaction in $\frac{1}{h^2} \cdot |x-y|^2$. This was analyzed in detail in [BruHe91] and in [He92b]. This will change the properties and the way of thinking for this problem. The case $a$ large will be more reminiscent of the semi-classical case as we shall see in Section 4.

**4. Improved decay estimates, comparison with the semi-classical results.** We just sketch some improvements and discuss for $a$ large the connexion with the Agmon distance estimating the decay in the semi-classical situation.

Our starting point is (3.3). We immediately observe that we can implement this upper bound in (3.1) and we get the estimate

$$0 < u_1(x;h) \leq C \cdot h^{-\frac{3}{4}} \exp(-\phi_1(x)/h),$$

with

$$\phi_1(x) = \frac{1}{2} v(x) + \inf_y \Big[ \frac{a}{4} (x-y)^2 + v(y) \Big].$$

We see that it is strictly better, and we observe the inequalities

$$\frac{1}{2}v(x) \leq \phi_1(x) \leq \frac{1}{2}v(x) + \frac{a}{4}\operatorname{dist}(x, \{\pm x_c\})^2.$$

If we try to iterate, we finally are interested in the largest function $\phi_\infty(x)$ which satisfies

$$|\phi_\infty(x) - \phi_\infty(y)| \leq \frac{1}{2}v(x) + \frac{a}{4}(x-y)^2 + \frac{1}{2}v(y). \tag{4.1}$$

This will be indeed the natural candidate for measuring the decay.

One has actually a natural construction in order to get this function. We define $\phi_0 = \frac{v}{2}$ and $\phi_1$ as before and introduce by recursion

$$\phi_n(x) = \frac{v(x)}{2} + \inf_y \Big[\frac{v(y)}{2} + \frac{a}{4}|x-y|^2 + \phi_{n-1}(y)\Big].$$

It is easy to see that this sequence is increasing with respect to $n$ and that

$$\frac{1}{2}v(x) \leq \phi_n(x) \leq \frac{1}{2}v(x) + \frac{a}{4}\operatorname{dist}(x, \{\pm x_c\})^2.$$

We can then define the limit $\phi_\infty(x)$ which satisfies

$$\phi_\infty(x) = \frac{v(x)}{2} + \inf_y \Big[\frac{v(y)}{2} + \frac{a}{4}|x-y|^2 + \phi_\infty(y)\Big].$$

This limit can be seen to be locally Lipschitzian. We observe indeed that $x \mapsto |x-y|^2$ has this property and that, for any compact $F$ in $\mathbb{R}$, there exists another compact $F'$ such that, for all $n$ and for all $x \in F$ the minimum $\inf_y(\frac{v(y)}{2} + \frac{a}{4}(x-y)^2 + \phi_n(y))$ is attained at a point $y_r$ in $F'$.

This function can play the role of the distance to the minima of $v$ and gives a good control of the decay of the first eigenfunction.

**The case $a$ large, or the reappearance of the Agmon distance.** The other case of interest is the case when $a$ is large. We now consider a Lipschitzian function $d_\infty(x)$ such that

$$(d'_\infty(x))^2 \leq v(x) \quad \text{a.e.}, \quad d_\infty(\pm x_c) = 0, \ d_\infty \geq 0.$$

We recognize here that the Agmon distance to the minima satisfies these conditions and we observe in this case that, for any $\epsilon > 0$, $(1-\epsilon)a^{\frac{1}{2}}d_\infty(x)$ satisfies (4.1) for $a$ large enough. We have indeed (assuming $d''$ bounded)

$$\begin{aligned} |d_\infty(x) - d_\infty(y)| &\leq \frac{1}{2}|x-y|(|d'(x)| + |d'(y)|) + \mathcal{O}(|x-y|^2) \\ &\leq \frac{v(x)}{2\sqrt{a}} + \frac{v(y)}{2\sqrt{a}} + \frac{\sqrt{a}}{4}|x-y|^2 + \mathcal{O}(|x-y|^2). \end{aligned}$$

Multiplying by $(1-\epsilon)a^{\frac{1}{2}}$, it is easy to get (4.1).

**The quadratic case.** The explicit computation when $v$ is quadratic ($v(x) = \frac{c}{2} \cdot x^2$) is also enlightening. We are looking for a solution of

$$\phi_\infty(x) = \inf_y \frac{1}{4}(cx^2 + a|x-y|^2 + cy^2) + \phi_\infty(y).$$

It is rather clear from our construction that $\phi_\infty(x)$ is quadratic and we are consequently looking for $b$ such that

$$\frac{1}{2} b\, x^2 = \inf_y \frac{1}{4}(c\, x^2 + a|x-y|^2 + c\, y^2) + \frac{1}{2} b\, y^2.$$

An immediate computation gives that the minimum is obtained for $y = ax/(a+c+2b)$, where $b$ is a solution of

$$2b(a+c+2b) = c(a+c+2b) + a(c+2b).$$

This is fortunately a second order equation whose unique positive solution is

$$b = \frac{1}{2}\sqrt{c(2a+c)}.$$

These computations are of course immediately related to the explicit computations concerning the harmonic Kac operator which can be seen as $\exp S$ where $S$ is an explicit harmonic oscillator (see [He94c]).

**5. Toward an adapted notion of Agmon distance.** We want to control the regularity in the preceding construction when we are near a nondegenerate minimum of the potential. We have the following identities assuming that everything is differentiable. We start from

$$\phi(x) = \inf_y \left[\Theta(x,y) + \phi(y)\right],$$

with

$$\Theta(x,y) = w(x,y) + \frac{v(x)}{2} + \frac{v(y)}{2}.$$

If we assume that the minimum is obtained near $x = 0$ for a unique regular $y(x)$ (and this will be a consequence of the implicit function theorem), we get

$$\phi(x) = \Theta(x, y(x)) + \phi(y(x)),$$

and

$$\partial_y \Theta(x, y(x)) + \phi'(y(x)) = 0.$$

Differentiating with respect to $x$, we first obtain

$$\phi'(x) = \partial_x \Theta(x, y(x)).$$

As it appeared clearly in a discussion with J. Sjöstrand, we are looking for a Lagrangian manifold $\Lambda_\phi = \{(x,\xi) \in \mathbb{R}^2 \mid \xi = \phi'(x)\}$ which is invariant under the canonical relation attached to $\Theta$:

$$C_\Theta = \{(x,\xi,y,\eta) \in \mathbb{R}^4 \mid \xi = \partial_x \Theta(x,y),\ \eta = -\partial_y \Theta(x,y)\}.$$

We recall that in the case of Schrödinger we met, in the construction of the WKB approximation of the first eigenfunction, a Lagrangian manifold contained in $\{(x,\xi) \mid p(x,\xi) =: \xi^2 - V(x) = 0\}$ constructed as an invariant outgoing Lagrangian manifold invariant under the flow $H_p$. Here we have the same phenomenon in a discrete version. Let us express the diffeomorphism $\kappa$ whose graph is $C_\Theta$ in the case when $w(x,y) = \frac{a}{4}|x-y|^2$. We find the equations

$$\begin{aligned} \xi &= v'(x) + \frac{a}{2}(x-y), \\ \eta &= -v'(y) + \frac{a}{2}(x-y), \end{aligned}$$

and this gives the expression for $\kappa$

$$\begin{aligned} y &= x + \frac{2}{a}v'(x) - \frac{2}{a}\xi \qquad &&:= y(x,\xi), \\ \eta &= -v'(y(x,\xi)) + \xi - v'(x) &&:= \eta(x,\xi). \end{aligned}$$

One gets also a similar expression for $\kappa^{-1}$.

Let us look at the fixed points of the diffeomorphism. We search for points such that $y(x,\xi) = x$, $\eta(x,\xi) = \xi$ and get $\xi = 0$, $v'(x) = 0$. We find as fixed points the points $(x,0)$ where $x$ is a critical point of $v$. In particular, if we assume that 0 is a nondegenerate minimum of $v$, $(0,0)$ is a fixed point and we have just to verify, in order to apply the general theory of Smale [Sm1], that we have an hyperbolic point for the diffeomorphism attached to $C_\Theta$:

$$(x,\xi) \mapsto (y,\eta)$$

with $(x,\xi,y,\eta) \in C_\Theta$. This property will be verified in the next section. The regularity of the stable manifold is an old result due to Smale [Sm1]. (See also [Sm2] (p.751) or [Irw].) The property that these stable manifolds are Lagrangian is obtained as in the continuous case. (See for example [HeSj84].) We have consequently obtained the

**Proposition 5.1.** *If $v$ is a $C^\infty$ potential defined in a neighborhood of 0 and admitting a nondegenerate minimum at 0 with $v(0) = 0$, then there exists a neighborhood of 0, a $C^\infty$ phase $\phi$ defined in this neighborhood and a $C^\infty$ function $x \mapsto y(x)$ which are solutions of*

$$\kappa(x, \phi'(x)) = (y(x), \phi'(y(x))), \tag{5.1}$$

*and satisfy*

$$0 < y'(0) < 1. \tag{5.2}$$

Once we have solved this problem, we arrive naturally to the study of the WKB construction: Find

$$a(x;h) \sim h^{-\frac{1}{4}}\Big(\sum_{j=0}^{\infty} a_j(x)h^j\Big) \tag{5.3}$$

and

$$E(h) \sim \sum_{j=0}^{\infty} E_j h^j \tag{5.4}$$

such that

$$\big(K_v a \exp\big(-\phi(\cdot)/h\big)\big)(x) \sim E(h)a(\cdot;h)\exp\big(-\phi(\cdot)/h\big). \tag{5.5}$$

We want the following formal identity

$$E(h)a(x;h) = h^{-\frac{1}{2}} \int a(y;h)\cdot \exp\Big(-\frac{1}{h}[\Theta(x,y) + \phi(y) - \phi(x)]\Big)\, dy.$$

Applying the Laplace integral method, we get for the main term

$$E_0 a_0(x) = c \cdot a_0(y(x))(\det[\partial^2_{yy}\Theta + \phi''])^{-\frac{1}{2}}(y(x)).$$

A necessary condition in order to solve is

$$E_0 = c\cdot(\det[\partial^2_{yy}\Theta + \phi''])^{-\frac{1}{2}}(0).$$

This actually determines $E_0$ and we find for $a_0$ the formula

$$a_0(x) = \prod_{n=1}^{\infty} \theta(y^{(n)}(x)),$$

where by definition

$$y^{(n)}(x) = y(y^{(n-1)}(x)),$$

and

$$\theta(y) = \left( \frac{\det[\partial^2_{yy}\Theta + \phi''(0)}{\det[\partial^2_{yy}\Theta + \phi''](y)} \right)^{\frac{1}{2}}.$$

An explicit computation shows that $\partial^2_{yy}\Theta(x,y) = \frac{1}{2}(v''(y) + a)$.

We observe here that for any $x$ sufficiently near the bottom 0 of $v$ we have

$$y^{(n)}(x) \to 0$$

and that the convergence is exponentially rapid.

In order to see what is going on, let us come back to the one dimensional quadratic case. We recall that we have

$$\phi(y) = \frac{b}{2}y^2$$

with

$$b = \frac{1}{2}\sqrt{c(2a+c)}$$

and

$$y(x) = \frac{a}{c+a+2b}x.$$

Of course we do not need the discussion in this very simple case. What will be relevant (for extension to larger dimension) is the property that

$$\|y(x)\| \leq k\|x\|,$$

with $0 < k < 1$ for a suitable norm (as can be seen in the proof of the stable manifold theorem). This is what will make the procedure convergent.

We can also solve in the same way successive transport equations.

**Theorem 5.2.** *If $v$ admits a nondegenerate minimum at 0, then there exists a formal WKB eigenfunction for $K_v$ satisfying (5.3), (5.4) and (5.5). As a consequence, the operator $K_v$ (with $v$ satisfying the condition of Lemma 2.1) admits one eigenvalue $\lambda(h)$ such that*

$$\lambda(h) = E(h) + \mathcal{O}(h^\infty).$$

**6. The analysis of the diffeomorphism in the quadratic case.** In order to verify this property of hyperbolicity for $\kappa$ at $(0,0)$, let us make explicit the computation of a diffeomorphism $\kappa_0$:

$$(y,\eta) = \kappa_0(x,\xi)$$

in the case when $v$ is quadratic (in our application $\kappa_0$ will be the tangent map to $\kappa$ at $(0,0)$). We assume that $v(x) = \frac{c}{2}x^2$. The map $\kappa_0$ is given by the $2 \times 2$ matrix

$$M_\kappa = \begin{pmatrix} \frac{(a+c)}{a} & -\frac{2}{a} \\ -\frac{1}{2a}c(c+2a) & \frac{(a+c)}{a} \end{pmatrix}.$$

The two eigenvalues are given by

$$\lambda_\pm = \frac{(a+c)}{a} \pm \frac{1}{a}\sqrt{c(c+2a)}$$

and one verifies that

$$-1 < \lambda_- < 1$$

and

$$\lambda_+ > 1.$$

We have of course the property: $\lambda_+ \cdot \lambda_- = 1$. We have consequently a hyperbolic point (saddle point situation) and a one dimensional subspace $\Lambda_-^0$ on which $\kappa_0$ is contracting, that is, the eigenspace attached to $\lambda_-$. One finds the equation

$$\xi = \frac{1}{2}\sqrt{c(c+2a)}x,$$

and, as in the case of the Schrödinger operator, one has

$$\xi = \phi_0'(x),$$

that is, $\phi_0(x) = \frac{1}{4}\sqrt{c(c+2a)}x^2$ is the generating function of the Lagrangian plane $\Lambda_-^0$.

**Acknowledgements.** I would like to thank the Johns Hopkins University (particularly S. Zelditch and M. Zworski) where this research was started (in the framework of a French-American program with the French Ministère des affaires étrangères).

I thank also the organizers of this Danish-Swedish meeting who gave me the opportunity to present this work.

I thank finally J. Sjöstrand for useful remarks around this work and particularly for his suggestion to consider this problem of finding the phase for the WKB solution as searching for an invariant Lagrangian manifold.

## References


[Ag] S. Agmon, *Lecture on exponential decay of solutions of second order elliptic equations*, Princeton University Press, 1982, Math. Notes 29.

[BruHe91] M. Brunaud and B. Helffer, *Un problème de double puits provenant de la théorie statistico-mécanique des changements de phase (ou relecture d'un cours de M. Kac)*, Preprint LMENS, March 1991.

[El85] R. S. Ellis, *Entropy, large deviations, and statistical mechanics*, Springer Verlag, New York, 1985, Grundlehren der mathematischen Wissenschaften 271.

[GlJa87] J. Glimm and A. Jaffe, *Quantum physics (a functional integral point of view)*, Springer Verlag, Berlin, 1987, Second edition.

[Ha80] E. M. Harrell, *On the rate of asymptotic eigenvalue degeneracy*, Comm. Math. Phys. **75** (1980), 239–261.

[He88] B. Helffer, *Introduction to the semiclassical analysis for the Schrödinger operator and applications*, Springer Verlag, 1988, Springer Lecture Notes in Math. 1336.

[He92a] ———, *Problèmes de double puits provenant de la théorie statistico-mécanique des changements de phase, II Modèles de Kac avec champ magnétique. Étude de modèles près de la température critique*, Preprint LMENS, March 1992.

[He92b] ———, *Décroissance exponentielle des fonctions propres pour l'opérateur de Kac: le cas de la dimension* > 1, Operator Theory: Advances and Applications, Birkhäuser Verlag, Basel, 1992, pp. 99–116, Proceedings of symposium on operator calculus and spectral theory, Lambrecht (Germany), December 1991, edited by M. Demuth, B. Gramsch and B.-W. Schulze.

[He93b] ———, *Spectral properties of the Kac operator in large dimension*, Proceedings on Mathematical Quantum Theory II: Schrödinger Operators, (August 1993), Edited by J. Feldman, R. Froese, and L. M. Rosen, CRM Proceedings and Lecture Notes.

[He94a] ———, *Around a stationary phase theorem in large dimension*, J. Funct. Anal. **119** (1994), 217–252.

[He94b] ———, *Universal estimate of the gap for the Kac operator in the convex case*, Comm. Math. Phys. **161** (1994), 631–642.

[He94c] ———, *Correlation decay and gap of the transfer operator*, Proceedings of the conference in the Euler Institute in St Petersburg, Submitted in April 1994.

[He94d] ———, *Around the transfer operator and the Trotter-Kato formula*, Proceedings of the conference in Holzhau, July 1994 (to appear).

[He94e] ———, *Recent results and open problems on Schrödinger operators, Laplace integrals and transfer operators in large dimension*, Advances in PDE, Akademie Verlag, Preprint November 1994 (to appear).

[He95a] ———, *Semiclassical analysis for Schrödinger operators, Laplace integrals and transfer operators in large dimension: an introduction*, Notes of a course in Orsay, January-April 1995.

[He95b] ———, *Laplace integrals and transfer operators in large dimension: examples in the non convex case*, Preprint 1995.

[He95c] ———, *Semi-classical analysis for the transfer operator: WKB constructions*, In preparation.

[HeSj84] B. Helffer and J. Sjöstrand, *Multiple wells in the semiclassical limit*, Comm. in PDE **9** (1984), 337–408.

[HeSj92a] ———, *Semiclassical expansions of the thermodynamic limit for a Schrödinger equation*, Astérisque **210** (1992), 135–181.

[HeSj92b] ———, *Semiclassical expansions of the thermodynamic limit for a Schrödinger equation II*, Helv. Phys. Acta **65** (1992), 748–765; *Erratum* **67** (1994), 1–3.

[HeSj93] ———, *On the correlation for Kac like models in the convex case*, J. Statist. Phys. **74** (1994), 147–187, Report No. 9, 1992-93, Institut Mittag-Leffler, March 1993.

[Irw] M. C. Irwin, *Smooth dynamical systems*, Academic Press, 1970.

[ItDr88] C. Itzykson and J. M. Drouffe, *Statistical field theory, Vol 1*, 1988, Cambridge Monographs on Mathematical Physics.

[Ka62] M. Kac, *Statistical mechanics of some one-dimensional systems*, Studies in mathematical analysis and related topics: Essays in honor of Georges Polya, Stanford University Press, Stanford, California, 1962, pp. 165–169.

[Ka66] ———, *Mathematical mechanisms of phase transitions*, Gordon and Breach, 1966, Brandeis lectures.

[HelKa63] M. Kac and E. Helfand, J. Math. Phys. **4** (1963), 1078–1088.

[KaTho66] M. Kac and C. Thompson, *On the mathematical mechanism of phase transition*, Proc. Nat. Acad. Sci. U. S. A. **55** (1966), 676—683; *Erratum* **56** (1966), 1625.

[KaTho69] ———, *Phase transition and eigenvalue degeneracy of a one dimensional anharmonic oscillator*, Stud. Appl. Math. **48** (1969), 257–264.

[Pa84] G. Parisi, *Statistical field theory*, Addison-Wesley Publishing Company, 1984, In Frontiers in Physics.

[Si83] B. Simon, *Instantons, double wells and large deviations*, Bull. Amer. Math. Soc. **8** (1983), 323–326.

[Si84] ———, *Semi-classical Analysis of low lying eigenvalues. I*, Ann. Inst. H. Poincaré **38** (1983), 295-307.; *II*, Ann. of Math. **120** (1984), 89–118.

[Si93] ———, *The statistical mechanics of lattice gases*, Princeton University Press, 1993, Princeton Series in Physics Volume 1.

[Sj93a] J. Sjöstrand, *Potential wells in high dimensions I*, Ann. Inst. H. Poincaré, Phys. Théor. **58** (1993), 1–41.

[Sj93b] ———, *Potential wells in high dimensions II, more about the one well case*, Ann. Inst. H. Poincaré, Sect. Phys. Théor. **58** (1993), 42–53.

[Sj93c] ———, *On ferromagnetic Laplace integrals and maximum principles*, Ann. Inst. Fourier **44** (1994), 601–628, Preprint de l'université d'Orsay, June 1993.

[Sj93d] ———, *Estimations sur les corrélations et principes du maximum*, Séminaire Equations aux dérivées partielles de l'Ecole Polytechnique, October 1993.

[Sj94] ———, *Correlation asymptotics and Witten Laplacians*, Preprint, June 1994.

[Sm1] S. Smale, *Stable manifolds for differential equations and diffeomorphisms*, Ann. Scuola Norm. Sup. Pisa **18** (1963), 97–116.

[Sm2] ———, *Differentiable dynamical systems*, Bull. Amer. Math. Soc. **73** (1967), 747–817.

[Tho72] C. Thompson, *Mathematical statistical mechanics*, The Macmillan company, New York, 1972.



UA762 du CNRS, DMI-ENS, 45 rue d'Ulm, F-75230 Paris Cédex, France
*E-mail address*: Bernard.Helffer@ens.fr

# SEVERAL RECENT RESULTS IN NONLINEAR GEOMETRIC OPTICS

Jean-Luc Joly, Guy Métivier and Jeffrey Rauch

**1. Introduction.** Many recent works are devoted to the study of high frequency oscillatory nonlinear waves, and to nonlinear geometric optics. Typical questions are the existence, the propagation, the interaction and the reflection of waves of the form

$$u^\varepsilon(t,x) \sim \underline{u}(t,x) + \varepsilon^\alpha \sum_{n\geq 0} \varepsilon^n U_n(t,x,\varphi(t,x)/\varepsilon). \tag{1.1}$$

The $U_n(t,x,\theta)$ are periodic (or quasi/almost periodic) in the $\theta$ variable and $\varphi$ is a real phase function. In the summation, $n$ runs in a countable set of nonnegative numbers. Typically, $n$ is an integer, a half integer..., $\underline{u}$ is a a background state, $\varepsilon$ is a wavelength and $\varepsilon^\alpha$ the magnitude of the oscillations (assuming that $\partial_\theta U_0 \neq 0$).

This is the natural generalization of phase-amplitude linear waves

$$u^\varepsilon(t,x) = a(t,x,\varepsilon)\, e^{i\varphi(t,x)/\varepsilon}, \quad a(t,x,\varepsilon) \sim \sum_{n\geq 0} \varepsilon^n a_n(t,x), \tag{1.2}$$

which includes the phenomenon of generation of harmonics by nonlinear interaction.

The waves $u^\varepsilon$ satisfy nonlinear equations of propagations, such as first order semilinear hyperbolic systems

$$\partial_t u + \sum_{j=1}^{d} A_j(t,x)\partial_{x_j} u + f(t,x,u) = 0, \tag{1.3}$$

or quasilinear first order hyperbolic systems,

$$\partial_t u + \sum_{j=1}^{d} A_j(t,x,u)\partial_{x_j} u + f(t,x,u) = 0, \tag{1.4}$$

or higher order equations such as nonlinear wave equations

$$\Box u + f(t, x, u, \nabla u) = 0. \tag{1.5}$$

The choice of scales, both in the expansion (1.1) and in the equations, is very important. Solutions of (1.3) of the form (1.1) model waves of wavelength $O(\varepsilon)$ and amplitude $O(\varepsilon^\alpha)$, which propagate at speed $O(1)$ during times $O(1)$. When $\alpha > 0$ the equations for all the $U_n$ are linear, and no nonlinear effect is revealed in the principal term $U_0$. For $\alpha < 0$, the problem is in general illposed: the domain of existence shrinks to $\{t = 0\}$ as $\varepsilon \to 0$. Thus the case $\alpha = 0$

$$u^\varepsilon(t, x) \sim \sum_{n \geq 0} \varepsilon^n U_n(t, x, \varphi(t, x)/\varepsilon), \tag{1.6}$$

is the most interesting for general semilinear first order systems. In (1.6) the background state $\underline{u}$ is incorporated in $U_0$. For general semilinear systems (1.3), this is the typical range of scales (up to rescaling) for the validity of the nonlinear geometric optics approximations.

For quasilinear first order systems, the similar critical exponent is $\alpha = 1$. The *weakly* nonlinear geometric optics, deals with solutions of (1.4) of the form

$$u^\varepsilon(t, x) \sim \underline{u}(t, x) + \varepsilon \sum_{n \geq 0} \varepsilon^n U_n(t, x, \varphi(t, x)/\varepsilon). \tag{1.7}$$

However, stronger waves with $\alpha = 0$ can be considered, when they are associated to a linearly degenerate mode, such as entropy waves in Euler's equations of fluid dynamics (see [Se 1], [Se 2], [He]).

For the semilinear wave equation (1.5), the similar critical exponent is $\alpha = 1$.

In the scalings above, the validity of the geometric optics approximations is in general limited to times $t = O(1)$. In general, solutions are shown to exist only for such times. For longer times, even if the solutions exist, other phenomena such as dissipation or dispersion can become important. The geometric optics approximations can be incorrect then, and other regimes such as paraxial approximations can occur (see [Do], [DR]).

In this paper, we will review several results about the justification of nonlinear geometric optics approximations. In §2, we discuss expansions (1.1) which model the propagation of a single wave. The nonlinear character of the equation models the interaction of the wave with the medium in which it propagates. Next, we consider the interaction of at least two oscillatory waves. The interesting phenomenon is the creation of a new outgoing oscillation of the same magnitude, from incoming oscillations. This happens when the phases $\varphi_j$ of the incoming waves and the phase $\varphi$ of the outgoing wave satisfy the *resonance relation*

$$\varphi = \sum_j \varphi_j, \tag{1.8}$$

or more generally

$$\alpha\varphi = \sum_j \alpha_j \varphi_j, \quad \alpha, \alpha_j \in \mathbb{Z}, \tag{1.9}$$

when harmonics are present. For semilinear problems, the analysis is made through multiphase expansions

$$u^\varepsilon(t,x) \sim \sum_{n\geq 0} \varepsilon^n U_n(t,x,\psi_1(t,x)/\varepsilon,\dots,\psi_m(t,x)/\varepsilon), \tag{1.10}$$

where the $U_j(t,x,\theta_1,\dots\theta_m)$ are periodic in the variables $\theta$. The leading term $U_0$ is often sought as a superposition of incoming and outgoing waves

$$U_0(t,x,\psi_1(t,x)/\varepsilon,\dots,\psi_m(t,x)/\varepsilon) = \sum_j S_j(t,x,\varphi_j(t,x)/\varepsilon). \tag{1.11}$$

Similar expansions, with oscillations of amplitude $O(\varepsilon)$, are used for quasilinear systems, in the regime of weakly nonlinear geometric optics. The nonlinear interaction is seen in the coupling of the equations for the $S_j$. *This coupling strongly depends on the existence or nonexistence of resonances* among the linear combinations of the $\varphi_j$.

A formal derivation of the equations of nonlinear geometric optics is given in [HK], [MR] and [HMR]. The justification of the asymptotic expansions for exact solutions happens to be delicate. Simple examples show that complete expansions (1.10) are not true in general. Instead, one has to study a larger class of solutions which satisfy the weaker estimate

$$u^\varepsilon(t,x) = U_0(t,x,\psi_1(t,x)/\varepsilon,\dots,\psi_m(t,x)/\varepsilon) + o(1), \tag{1.12}$$

or

$$u^\varepsilon(t,x) = \sum_j S_j(t,x,\varphi_j(t,x)/\varepsilon) + o(1), \tag{1.13}$$

with error terms which are not better than $o(1)$ and not $O(\varepsilon)$, in general.

In §3, we present the one space dimensional case where many questions are now understood. When the space dimension $d$ is larger, an important difficulty arises from the *focusing* of oscillations. This phenomenon is already present for linear equations, but in the nonlinear case, its consequences can be much more dramatic. For example, it may imply that the domain of existence of the (weak) solutions $u^\varepsilon$ shrinks to the empty set as $\varepsilon$ tends to zero (see [JMR 2], [JMR 7]). Examples of blow up, due to focusing of oscillations, are presented in §5.

A framework where all focusing is excluded, besides the one space dimensionnal case, occurs when one considers a *coherent* set of phases. A typical and important example is given by linear phases for a constant coefficient background system. Then, expansions like (1.10) or (1.12) can be justified, see [JMR 2], [JMR 3], [JMR 4], [Sch 2], [De]. Coherent multiphase expansions are presented in §4.

As described above, focusing of nonlinear oscillations can lead to blow up and nonexistence. In the extreme opposite direction, when it is combined with strongly dissipative mechanisms, focusing can lead to a complete absorbtion of oscillations, in finite time. The oscillations disappear when they reach the caustic set. An example of such a behaviour is given in [JMR 5] and presented in §6 below. To understand an intermediate case, we have considered in [JMR 7] semilinear systems with uniformly Lipschitzian nonlinearities (thus with at most linear growth at infinity). In that case, global weak solutions always exist, and the question of the behaviour of the oscillations after crossing the caustic makes sense. This problem is studied in [JMR 7] and the results are sketched in §7 below. The first order geometric optics approximation is justified in $L^2$ for all finite times, and in particular after the crossing of the caustic. Remark that this provides examples of *noncoherent* multiphase expansions. However, no precise behaviour near the caustic is given. In particular, a challenging open question, is the validity of the formal expansions near the caustic, computed in [HK 2].

*Further results.*

1. Another important problem is the reflection of oscillatory waves on boundaries. A formal approach is described in [MA]. The tranverse reflection at a boundary is studied in [Chi] for two speed equations, and in [Wi 1] under more general hypothesis. The possibility of glancing mode for flat boundaries and linear coherent phases, is studied in [Wi 2]. Oscillations near a diffractive point, for semilinear wave equations with globally Lipschitzian nonlinearity, are studied in [Che 4].

2. The equations of geometric optics expansions are simpler than the original full set of equations, and thus they give detailed descriptions of several mechanisms of nonlinear interaction. Applications to the study of qualitative behavior of solutions of systems of conservation laws are numer-

ous. See e.g. [Hu], [MR], [MRS], [JMR 9], [Wa].

3. The validity of geometric optics can be extended beyond the category of solutions of the form (1.1), to more general families of "oscillating" solutions $u^\varepsilon$. This is shown in [JMR 6] for semilinear systems either of size $3 \times 3$ or with quadratic interaction. This illustrates a link between geometric optics and compensated compactness, in the spirit of [DiP 1], [DiP 2], [T], [McLPT].

4. The equations (1.3) (1.4) and (1.5) are dispersionless. The corresponding eiconal equations are homogeneous. When $\varphi$ is a characteristic phase, all the harmonics $k\varphi$ are characteristic. Thus all the harmonics are expected in the leading term $U_0$ of the expansions, and they all propagate with the same group velocity. For applications to nonlinear optics, it is important to consider dispersive media. For example, expansions (1.6) for solutions of

$$\varepsilon\partial_t u^\varepsilon + \sum_{j=1}^{d} \varepsilon A_j(t,x)\partial_{x_j} u^\varepsilon + E(t,x)u^\varepsilon + \varepsilon f(t,x,u^\varepsilon) = 0, \tag{1.14}$$

yield a nonhomogeneous eiconal equation, since the zero'th order term $Eu^\varepsilon$ has no $\varepsilon$ in front of it. For such models, harmonics are rarely present in the principal term $U_0$, as expected in nonlinear optics. We refer to [Do] and [DR] for details, and for examples and applications to Maxwell's equations and nonlinear optics.

**2. One phase expansions.** We look for asymptotic solutions (1.6) of the first order system (1.3) which we assume here to be symmetric hyperbolic. Plugging the expansion into the equation leads to an infinite system of equations (see [CB])

$$L(t,x,d\varphi(t,x))\,\partial_\theta U_0 = 0, \tag{2.1}$$

and for $n \geq 0$

$$L(t,x,d\varphi(t,x))\,\partial_\theta U_{n+1} + L(t,x,\partial)U_n + F_n = 0. \tag{2.2}$$

Here $L(t,x,\partial) := \partial_t + \sum A_j(t,x)\,\partial_{x_j}$, and $F_n$ is a function of $(U_0, \dots U_n)$.

To solve (2.1), one requires that $\varphi$ satisfies the eiconal equation

$$\det L(t,x,d\varphi(t,x)) = 0, \tag{2.3}$$

and that the oscillation

$$U_0^*(t,x,\theta) := U_0(t,x,\theta) - \underline{U}_0, \quad \underline{U}_0(t,x) := \frac{1}{2\pi}\int_0^{2\pi} U_0(t,x,\theta)\,d\theta, \tag{2.4}$$

satisfies the polarization condition

$$\Pi U_0^* = U_0^*, \tag{2.5}$$

where $\Pi(t,x)$ is the orthogonal projector onto the kernel of $L(t,x,d\varphi(t,x))$.

For $n \geq 0$, take the average of equation (2.2) and apply $\Pi$ and $(\mathrm{Id}-\Pi)$ to its oscillation. This implies that (2.2) is equivalent to

$$L\underline{U}_n + \underline{F}_n = 0, \quad \Pi L \Pi U_n^* + \Pi F_n^* = -\Pi L(\mathrm{Id}-\Pi)U_n^* \tag{2.6}$$

and

$$(\mathrm{Id}-\Pi)U_{n+1}^* = -\partial_\theta^{-1} Q(LU_n^* + F_n^*). \tag{2.7}$$

The averages and oscillations of $U_n$ and $F_n$ are defined as in (2.4). $Q(t,x)$ is the partial inverse of $L(t,x,d\varphi(t,x))$ such that $QL = LQ = \mathrm{Id}-\Pi$, $Q\Pi = 0$, and $\partial_\theta^{-1}$ is the inverse of $\partial_\theta$ acting on periodic functions with vanishing mean value.

Of particular importance are the equations for $U_0$. They are given by (2.5) and (2.6). With $F_0 := f(t,x,U_0(t,x,\theta))$, they read

$$\left\{ \begin{array}{rcl} L\underline{U}_0 + \underline{F}_0 & = & 0, \\ (\mathrm{Id}-\Pi)U_0^* & = & 0, \\ \Pi L \Pi U_0^* + \Pi F_0^* & = & 0. \end{array} \right. \tag{2.8}$$

To simplify matters, suppose that $\lambda(t,x,\xi)$ is, for $\xi \neq 0$, a real eigenvalue of constant multiplicity of $\sum \xi_j A_j(t,x)$. Then a real solution of (2.3) is obtained by solving

$$\partial_t \varphi(t,x) + \lambda(t,x,\partial_x\varphi(t,x)) = 0. \tag{2.9}$$

Given a smooth solution with $d\varphi \neq 0$, the matrices $\Pi(t,x)$ and $Q(t,x)$ are $C^\infty$ functions. Moreover,

$$\Pi L \Pi = \partial_t + \sum_j \frac{\partial \lambda}{\partial \xi_j}(t,x,\partial_x\varphi)\,\partial_{x_j} + C(t,x). \tag{2.10}$$

Thus (2.8) is a nonlinear coupled system of a hyperbolic equation for $\underline{U}_0$ and a transport equation for $U_0^* = \Pi U_0^*$. Because of the integral term $\underline{F}_0$, the coupling is nonlocal in $\theta$. However, this system has the standard properties of symmetric hyperbolic systems, and the Cauchy problem has local smooth solutions.

The other terms of the expansion are found inductively. If $(U_0, \dots, U_n)$ are known, $(2.7)_n$ gives $(\mathrm{Id} - \Pi)\, U^*_{n+1}$. Then $(2.6)_{n+1}$ gives a coupled *linear* system for $(\underline{U}_{n+1}, \Pi U^*_{n+1})$ of the form

$$
(2.11) \qquad \begin{cases} L\,\underline{U}_{n+1} + \underline{F}_{n+1} &= 0 \\ \Pi\, L\, \Pi U^*_{n+1} + \Pi\, F^*_{n+1} &= \Pi G_n \\ F_{n+1} &= \partial_u f(t, x, U_0) \cdot U_{n+1} + H_n \end{cases}
$$

where $G_n$ and $H_n$ only depend on $(U_0, \dots, U_n)$. The Cauchy problem for (2.11) has smooth solutions. Moreover, since this problem is linear, the domain of existence of the solution only depends on the domain of existence for $U_0$.

Thus, given $C^\infty$ Cauchy data for $\varphi$ and $U_0$ with $\Pi U_0^* = U_0^*$, one can solve the eiconal equation (2.9) and the nonlinear profile equation (2.8) on a suitable domain of determinacy $\Omega$. Next, with arbitrary Cauchy data for $\underline{U}_n$ and $\Pi U^*_n$, one can solve (2.11) on $\Omega$. This provides formal solutions of the form (1.6).

**Theorem ([JR 1]).** *Given such a formal solution, if*

$$
(2.12) \qquad u_0^\varepsilon(x) \sim \sum_{n \geq 0} \varepsilon^n U_n(0, x, \varphi(0,x)/\varepsilon),
$$

*then the Cauchy problem for (1.3) with Cauchy data $u_0^\varepsilon$ has a unique solution $u^\varepsilon$ for $\varepsilon \leq 1$, defined and smooth on a domain $\Omega' \subset \Omega$ independent of $\varepsilon$, and satisfying*

$$
(2.13) \qquad u^\varepsilon(t, x) \sim \sum_{n \geq 0} \varepsilon^n\, U_n(t, x, \varphi(t,x)/\varepsilon)).
$$

*Moreover, $\Omega'$ can be taken arbitrarily close to $\Omega$ if $\varepsilon$ is restricted to be small enough.*

Remark that the existence of solutions on a domain independent of $\varepsilon$ is an important part of the statement, since the classical existence theorems of smooth solutions do not apply.

*Further results.*

1) Similar results are available for the quasilinear equations (1.4) and the expansions (1.7) ([G 2], [G 3]). The background state $\underline{u}$ is a solution of (1.4), and $\varphi$ satisfies the eiconal equation for the linearized operator on $\underline{u}$. The main novelty is that the transport operator (2.10) contains an additional nonlinear term in $U_0 \partial_\theta$. In particular, the transport equation for $U_0$ looks like a Burgers' equation and thus may develop shocks. The

other equations for $U_n$, $n \geq 1$, remain linear. The expansions are justified, as long as the phase $\varphi$ and the first profile $U_0$ exist and remain smooth.

2) Another interesting feature of [G 3] is the justification of expansions with *almost periodic profiles*. The construction sketched at the beginning of this paragraph breaks down immediately, since $\partial_\theta^{-1}$ does not act in the space of almost periodic functions with vanishing mean value. This indicates that the second term $U_1$ cannot be defined in general. However, the equations (2.8) still make sense and define the profile $U_0$.

[G 3] provides a construction of solutions which satisfy

$$u^\varepsilon(t,x) = U_0(t,x,\varphi(t,x)/\varepsilon) + o(1), \tag{2.14}$$

in place of (2.13). The error $o(1)$ tends to zero in an appropriate space, implying its convergence to 0 in $L^\infty$.

**3. One dimensional resonant expansions.** In this section we consider expansions with several phases, in one space dimension. For simplicity, we consider strictly hyperbolic semilinear $3 \times 3$ systems, and refer the reader to [JMR 1] for higher order semi- and quasi-linear systems. After a diagonalization and a change of dependent variables, the equations are

$$X_k u_k := \partial_t u_k + \lambda_k(t,x)\partial_x u_k = f_k(t,x,u_1,u_2,u_3), \quad k \in \{1,2,3\}. \tag{3.1}$$

The $\lambda_k$ are real and $\lambda_1 < \lambda_2 < \lambda_3$. The $f_k$ are smooth functions of the variables $(t,x,u) \in \mathbb{R} \times \mathbb{R} \times \mathbb{R}^3$.

The characteristic set is the union of the 3 lines $\tau + \lambda_k(t,x)\xi = 0$ and the characteristic phases $\varphi$ satisfy one of the equations $X_k\varphi = 0$. The corresponding eigenvector is the $k$-th vector of the canonical basis in $\mathbb{R}^3$. The oscillatory wave trains associated to the $j$-th eigenvalue $\lambda_j$ which are constructed in §2, are of the form

$$u_j^\varepsilon(t,x) = U_j(t,x,\varphi_j(t,x)/\varepsilon) + O(\varepsilon), \quad u_k^\varepsilon = O(\varepsilon) \quad \text{for } k \neq j, \tag{3.2}$$

where $\varphi_j$ satisfies

$$X_j\varphi_j = 0. \tag{3.3}$$

The interaction of such wave trains is described by a solution $u^\varepsilon$ of (3.1) such that

$$u_k^\varepsilon(t,x) = U_k(t,x,\varphi_k(t,x)/\varepsilon) + o(1), \quad k \in \{1,2,3\}. \tag{3.4}$$

The Cauchy data are assumed to be of the form

$$u_k^\varepsilon(0,x) = u_{0,k}^\varepsilon(x) := A_k(x,\varphi_k(0,x)/\varepsilon) + o(1), \quad k \in \{1,2,3\}. \tag{3.5}$$

For example, one can consider an initial phase $\psi(x)$ and oscillatory Cauchy data

$$u_0^\varepsilon := A(x, \psi(x)/\varepsilon). \tag{3.6}$$

Then $\varphi_j$ is the solution of (3.3) with initial data $\psi$ and the $A_k$ are the components of $A$.

The equations of nonlinear geometric optics are of the form

$$\begin{cases} X_k U_k(t,x,\theta) &= (\mathbf{E}_k F_k)(t,x,\theta) \\ U_k(0,x,\theta) &= A_k(x,\theta) \end{cases} \qquad k \in \{1,2,3\}, \tag{3.7}$$

where

$$F_k(t,x,\theta_1,\theta_2,\theta_3) := f_k\big(t,x,U_1(t,x,\theta_1),U_2(t,x,\theta_2),U_3(t,x,\theta_3)\big) \tag{3.8}$$

and $\mathbf{E}_k$ is an integral operator whose structure depends on the existence or nonexistence of resonances among the phases $\varphi_k$.

**Definition.** A resonance is a triple $(s_1, s_2, s_3)$ of functions such that

$$X_k s_k = 0 \quad \text{for all } k \in \{1,2,3\}, \quad \text{and} \quad \sum_{k=1}^{3} ds_k = 0. \tag{3.9}$$

The resonance is trivial when all the $ds_k$ are identically zero.

Remark that the definition does not depend on the vector fields themselves but on the foliations they define. The data of $N$ foliations by curves in $\mathbb{R}^2$ is called an $N$-web in differential geometry, and resonances correspond to abelian relations on webs (see [P], [BB]). The existence of resonances is a rare phenomenon: *for a 3-web, the dimension of the space of resonances (modulo constant functions) is equal to 0 or 1.* Moreover, for "generic" vector fields $X_k$, there are no nontrivial resonances. Therefore, in order to have resonances, first the vector fields $X_k$ must be suitably chosen and second, the phases $\varphi_k$ are also to be chosen carefully. However, resonance is a very important phenomenon. Three constant coefficient vector fields $X_k = \partial_t - \lambda_k \partial_x$, are resonant. The space of resonances is the space of $(s_1, s_2, s_3)$ such that $s_k = \alpha_k(x - \lambda_k t)$, with $\sum \alpha_k = 0$ and $\sum \lambda_k \alpha_k = 0$. We refer to [JMR 1] for a more general discussion of resonances and more examples.

We now describe the averaging operators $\mathbf{E}_k$ in a simple case. Consider an initial phase $\psi(x)$ on an interval $\omega$ and Cauchy data of the form (3.6). We assume that $d\psi \neq 0$ almost everywhere. Let $\Omega$ be a domain of determinacy of $\omega$. Let $\varphi_k$ be the solution of (3.3) on $\Omega$ with initial data equal to $\psi$ on

$\omega$. The idea is that the nonlinearities $f_k$ will reveal oscillations associated to all the phases $\sum \alpha_k\varphi_k$, but that only the resonant combinations will be propagated. Introduce

$$R := \{\alpha \in \mathbb{R}^3 \mid \sum_k \alpha_k\varphi_k \equiv 0 \text{ on } \Omega\}. \tag{3.10}$$

When $R = \{0\}$, the averaging operators are defined by

$$(\mathbf{E}_1 F)(t,x,\theta_1) := \lim_{T\to+\infty} \frac{1}{T^2}\int_0^T\!\!\int_0^T F(\theta_1,\theta_2,\theta_3)\,d\theta_2\,d\theta_3, \tag{3.11}$$

with similar definitions for $\mathbf{E}_2$ and $\mathbf{E}_3$. On the other hand, when there exists $\alpha \in R\setminus\{0\}$, then

$$(\mathbf{E}_1 F)(t,x,\theta_1) := \lim_{T\to+\infty} \frac{1}{T}\int_0^T F\Big(\theta_1, -\frac{\alpha_1\theta_1+\alpha_3\sigma}{\alpha_2}, \sigma\Big)\,d\sigma, \tag{3.12}$$

with similar formulas for $k = 2$ and $k = 3$. Since the vector fields are pairwise linearly independent, all three components $\alpha_k$ of $\alpha \in R\setminus\{0\}$ are not equal to 0. Thus the right hand side in (3.12) makes sense. Moreover, the dimension of $R$ is one. Therefore the definition does not depend on the choice of $\alpha \in R\setminus\{0\}$.

The definition of $\mathbf{E}_k$ depends on the local structure of resonances among the linear combinations of the phases. To avoid difficulties, we assume that this structure is the same near all the points $(t,x)\in\Omega$. More precisely, we assume that

$$\forall \alpha \notin R,\quad \sum_k \alpha_k\varphi_k \neq 0 \quad \text{almost everywhere on } \Omega. \tag{3.13}$$

**Theorem ([JMR 1]).** *Suppose that $A_k \in C^0(\omega\times\mathbb{R}/2\pi\mathbb{Z})$ and that $\{u^\varepsilon_{0,k}\}_{\varepsilon\in]0,1]}$ is a bounded family in $L^\infty(\omega)$ such that*

$$u^\varepsilon_{0,k}(\cdot) - A_k(\cdot,\psi(\cdot)/\varepsilon) \to 0 \quad \text{in } L^1(\omega), \quad \text{as } \varepsilon\to 0. \tag{3.14}$$

*Then there is $T>0$ such that:*

i) *for all $\varepsilon \in ]0,1]$, the Cauchy problem for (3.1) with Cauchy data $u^\varepsilon_{0,k}$ has a unique solution $u^\varepsilon$ in $L^\infty(\Omega_T)$, where $\Omega_T := \Omega\cap\{0<t<T\}$,*

ii) *the equations (3.7) have a unique solution $U \in C^0(\Omega_T\times\mathbb{R}/2\pi\mathbb{Z})$,*

iii)

$$u^\varepsilon_k(\cdot) - U_k(\cdot,\varphi_k(\cdot)/\varepsilon) \to 0 \quad \text{in } L^1(\Omega_T), \quad \text{as } \varepsilon\to 0. \tag{3.15}$$

*Remarks and further results.*

1. We refer the reader to [JMR 1] for the study of $N \times N$ systems, the justification of weakly nonlinear geometric optics for quasilinear systems, the introduction of almost periodic profiles and further results.

2. A formal derivation of the equations (3.7) is given in [HK], [MR] and [HMR]. Partial rigorous justifications were previously given by [T], [McLPT], [J], [Kal].

3. Even if the convergence in (3.14) holds in $L^\infty$, the approximation (3.15) is not always true in $L^\infty$. This is due to weak resonances, i.e. to linear combinations $\sum \alpha_k \varphi_k$ which are characteristic along isolated characteristic curves. They create new waves, of the form $V(t, \varphi/\varepsilon)$, with $V(t,\sigma)$ rapidly decreasing with respect to $\sigma \in \mathbb{R}$. These waves are small in $L^1$ but $O(1)$ in $L^\infty$. This phenomenon is studied in detail in [G 1].

4. The results quoted above concern sufficiently smooth solutions. For weak solutions of quasilinear systems of conservation laws, which may present shocks, the validity of weakly nonlinear geometric optics is now proved in a general setting in [Che 2] (see also [Che 3]), generalizing a result of [Sch 1]. Special cases were previously treated in [DiPM], [Liu], [Che 1]. The interaction of (strong) shock waves or contact discontinuities and small amplitude oscillations is described in [Co 1] and [Co 2].

**4. Coherent multiphase expansions.** For multidimensional problems, the description of resonances is much more complicated. To illustrate this difficulty, consider a semilinear system, with linear part $L$. Incoming oscillations with phases $\varphi_k$ produce, as source terms, oscillations with the phases $\varphi := \sum \alpha_k \varphi_k$. To understand how these new oscillations are propagated by $L$, one has to study the behaviour of the solution $u^\varepsilon$ of

$$Lu^\varepsilon = ae^{i\varphi/\varepsilon}. \tag{4.1}$$

When $\varphi$ is everywhere characteristic or everywhere noncharacteristic, the response $u^\varepsilon$ is given by a classical asymptotic expansion (see [La]). The difficulty comes from small sets where $\varphi$ satisfies the eiconal equation, and in particular from small sets where $d\varphi = 0$. In particular, $u^\varepsilon$ is not necessarily $O(1)$. Combined with nonlinearities, this mechanism can force the solutions to blow up, even at arbitrarily small times (see [JMR 2], [JMR 7] and §5 below). However, an interesting situation occurs when the dichotomy between characteristic and noncharacteristic phases is clear.

**Definition ([JMR 2]).** Let $L(t,x,\partial_t,\partial_x)$ be a linear first order $N \times N$ system on $\Omega$. A real vector space $\Phi \subset C^\infty(\Omega)$ is $L$-coherent when for all $\varphi \in \Phi \setminus \{0\}$, one of the following two conditions holds:

i) $\det L(t,x,d\varphi(t,x)) \equiv 0$ and $d\varphi(t,x) \neq 0$ at every point $(t,x) \in \Omega$,

ii) $\det L(t,x,d\varphi(t,x)) \neq 0$ at every point $(t,x) \in \Omega$.

An important example occurs when $L$ has constant coefficients and $\Phi$ is the space of linear functions. For simplicity, we now consider only this case. Suppose that $L = L(\partial_t, \partial_x)$ is a constant coefficient symmetric hyperbolic first order system and consider a semilinear problem

$$Lu^\varepsilon = f(u^\varepsilon). \tag{4.2}$$

We look for asymptotic solutions of the form

$$u^\varepsilon \sim \sum_{n\geq 0} \varepsilon^n U_n(t, x, t/\varepsilon, x/\varepsilon) \sim \sum_{n\geq 0} \varepsilon^n \sum_{\tau,\alpha} U_{n,\tau,\alpha}(t,x) e^{i(\tau t + \alpha x)/\varepsilon}. \tag{4.3}$$

The formal analysis is very similar to the analysis sketched in §2. One obtains a sequence of equations

$$L(\partial_T, \partial_X) U_0 = 0 \tag{4.4}$$

and for $n \geq 0$

$$L(\partial_T, \partial_X) U_{n+1} + L(\partial_t, \partial_x) U_n = F_n(U_0, \ldots, U_n). \tag{4.5}$$

Here $T$ and $X$ denote the fast variables, and the profiles $U_n(t, x, T, X)$ are periodic in these variables. Introduce $\Pi_{\tau,\alpha}$, the orthogonal projector in $\mathbb{C}^N$ on the kernel of $L(\tau, \alpha)$, and $\mathbf{E}$, the operator defined on (almost) periodic functions by

$$\mathbf{E}\Big(a e^{i(\tau T + \alpha X)}\Big) = (\Pi_{\tau,\alpha} a)\; e^{i(\tau T + \alpha X)}. \tag{4.6}$$

Applying $\mathbf{E}$ to (4.5) with $n = 0$ yields the following set of equations for $U_0$:

$$\mathbf{E} U_0 = U_0, \quad \text{and} \quad \mathbf{E}\big(L(\partial_t, \partial_x) U_0\big) = \mathbf{E}\big(f(U_0)\big). \tag{4.7}$$

Next, the Fourier coefficients of $(\mathrm{Id} - \mathbf{E}) U_1$ are determined by equations of the form

$$(\mathrm{Id} - \Pi_{\tau,\alpha}) L(\tau, \alpha)(\mathrm{Id} - \Pi_{\tau,\alpha}) U_{1,\tau,\alpha} = g_{1,\tau,\alpha}, \tag{4.8}$$

where the $g_{1,\tau,\alpha}$ only depend on $U_0$. The summation of the Fourier series can give rise to small divisor problems. However, *generically*, the summation of the Fourier series can be performed, defining $(\mathrm{Id} - \mathbf{E}) U_1$ as a smooth function. Next, $\mathbf{E} U_1$ is determined by equations analogous to (4.7). This procedure can be carried out to all orders $n$, and gives a complete formal solution ([JMR 3]) (see also [De] for the case of the wave equation). Using a general theorem from [G 2], this implies the existence of exact solutions satisfying (4.3).

**Theorem ([JMR 3]).** *Suppose that the* $U_n \in C^\infty(\Omega \times (\mathbb{R}/2\pi\mathbb{Z})^{1+d})$ *satisfy* (4.4) *and* (4.5), *where* $\Omega$ *is contained in the domain of determinacy of* $\omega \subset \mathbb{R}^d_x$. *Consider Cauchy data* $u_0^\varepsilon$ *such that*

$$u_0^\varepsilon(x) \sim \sum_{n\geq 0} \varepsilon^n U_n(0, x, 0, x/\varepsilon). \tag{4.9}$$

*Then there is* $T > 0$ *such that, for all* $\varepsilon \in ]0,1]$, *the Cauchy problem for* (4.2) *with Cauchy data* $u_0^\varepsilon$ *has a unique solution* $u^\varepsilon$ *on* $\Omega_T := \Omega \cap \{0 < t < T\}$. *Moreover,*

$$u_0^\varepsilon(t, x) \sim \sum_{n\geq 0} \varepsilon^n\, U_n(t, x, t/\varepsilon, x/\varepsilon). \tag{4.10}$$

The meaning of (4.9) is that for all $k$ and all $\nu \in \mathbb{Z}^d$

$$\partial_x^\nu \Big(u_0^\varepsilon(x) - \sum_{n=0}^{k} \varepsilon^n U_n(0, x, 0, x/\varepsilon)\Big) = O(\varepsilon^{k+1-\nu}) \quad \text{in} \quad L^\infty(\omega). \tag{4.11}$$

The meaning of (4.10) is similar.

*Remarks.*

1. No assumptions on the small divisors are necessary to define $U_0$. They are only needed to define $U_1$ and the higher order terms. When the small divisor assumptions are not satisfied, $U_0$ is still defined and one can look for solutions $u^\varepsilon$ such that

$$u^\varepsilon(t, x) = U_0(t, x, t/\varepsilon, x/\varepsilon) + o(1). \tag{4.12}$$

This program is worked out in different frameworks in [JMR 2], [Sch 2] and [JMR 4].

2. To recover the more usual form (1.13)

$$u^\varepsilon(t, x) = \sum_j S_j(t, x, (\tau_j t + \alpha_j x)/\varepsilon) + o(1) \tag{4.13}$$

which describes $u^\varepsilon$ as the superposition of wave trains associated with phases $\varphi_j(t,x) := \tau_j t + \alpha_j x$, one expands $U_0$ into its Fourier series

$$U_0(t, x, T, X) = \sum_{\tau,\alpha} U_{0,\tau,\alpha}(t, x) e^{i(\tau T + \alpha X)}, \tag{4.14}$$

and defines $S_j$ as the sum of terms when $(\tau, \alpha)$ runs over the set $\mathbb{R}(\tau_j, \alpha_j)$. Here we assume that the directions of $(\tau_j, \alpha_j)$ are pairwise different. In contrast with the one dimensional case, the number of characteristic directions

$(\tau_j, \alpha_j)$ is in general infinite and the form (4.12) contains an infinite number of oscillations propagating in an infinite number of directions. More surprisingly, three incoming phases can be sufficient to create, by nonlinear interaction, an infinite number of oscillations (see [JR 2], [JMR 2], [JMR 8]).

3. The analysis of coherent oscillations does not require $L$ to be hyperbolic with constant multiplicity. In particular, the results mentioned above, apply to conical refraction. This example is developed in [JMR 4].

4. Suppose that $L$ is a two speed system, which means that it has only two distinct eigenvalues. Suppose that $\Phi$ is generated by two phases which are everywhere linearly independent. Then $\Phi$ is $L$-coherent. Geometric optics expansions involving two phases, for two speed systems, were already considered in [JR 1].

**5. Focusing of oscillations and blow up.** Weakly nonlinear geometric optics for hyperbolic partial differential equations involve phases which are solutions of eiconal equations as in the linear theory. When the space has dimension at least equal to two, solutions of the eiconal equations exist only locally, the breakdown corresponding to focusing. The justifications of the weakly nonlinear geometric optics, quoted in the previous sections, are given on domains where focusing phenomena do not exist, either because the space dimension is equal to one, or thanks to the coherence assumption on the phases. However, it is an important issue to understand better the nonlinear effects of focusing.

When rays focus, amplitudes grow and, even in the linear case, one must change the asymptotic description. In the nonlinear case, the large amplitudes can be amplified by nonlinearities and therefore strongly nonlinear phenomena can occur. For example, the domain of existence of the (weak) solutions $u^\varepsilon$, may shrink to the empty set as $\varepsilon$ tends to zero (see [JMR 2] [JMR 7]). Focusing and blow up can be created by the principal oscillations themselves. This is called *direct focusing* in [JMR 2]. But nonlinear interactions make the problem much harder. Focusing and blow up can be created by phases not present in the principal term of the expansion, but which are generated after several interactions. This phenomenon is explored in detail in [JMR 2], where it is called *hidden focusing.*

In this section we give two examples of blow up caused by direct focusing of oscillations. The first one shows that blow up can occur before the first time of focusing. The second example shows solutions which cannot be extended, even in the weak sense, after focusing.

**Example 1.** Consider in space dimension $d = 3$ the equation

$$\Box u^\varepsilon - |\partial_t u^\varepsilon|^2 \, \partial_t u^\varepsilon = 0, \tag{5.1}$$

with oscillatory Cauchy data

$$u^\varepsilon(0,x) = \underline{u}_0(x) + \varepsilon U_0(x, |x|/\varepsilon), \qquad \partial_t u^\varepsilon(0,x) = U_1(x, |x|/\varepsilon), \tag{5.2}$$

where $\underline{u}_0$ is smooth and $U_0(x,\theta)$ and $U_1(x,\theta)$ are smooth, periodic in $\theta$ and vanish near $x=0$. For small times, the solutions satisfy ([JR 1]):

$$(5.3)\quad u^\varepsilon(t,x) = \underline{u}(t,x)+\varepsilon U_+(t,x,(|x|+t)/\varepsilon)+\varepsilon U_-(t,x,(|x|-t)/\varepsilon)+\varepsilon^2\ldots.$$

The ground state $\underline{u}$ and the oscillations $V_\pm := \partial_\theta U_\pm$ of the periodic profiles satisfy a coupled system of integro-differential equations. When $\underline{u}_0 = 0$ and $U_1 = \partial_\theta U_0$, with $U_0$ even in $\theta$, the equations simplify and

$$(5.4)\qquad u^\varepsilon(t,x) = \varepsilon U_+(t,x,(|x|+t)/\varepsilon) + \varepsilon^2\ldots.$$

Moreover, $V := \partial_\theta U_+$ satisfies

$$(5.5)\quad \partial_t V - \partial_{|x|}V - \frac{1}{|x|}V - \frac{1}{2}|V|^2 V = 0, \qquad V(0,x,\theta) = \partial_\theta U_0(x,\theta).$$

Then $W(t,x,\theta) := |x|^2|V(t,x,\theta)|^2$ satisfies

$$(5.6)\qquad W(t,x,\theta) = \frac{W(0,y,\theta)}{1-W(0,y,\theta)\dfrac{t}{|x|\,|y|}}, \qquad y := x + t\frac{x}{|x|}.$$

Therefore, if $V(0,y,\theta)\neq 0$, even if it is very small, $W$ and $V$ blow up on the ray $x = y - ty/|y|$ starting at $y$, *before the caustic set* $x=0$. Remark that the blow up occurs not only in $L^\infty$ but also in $L^1$.

The blow up is the consequence of two mechanisms. The first is due to the nonlinearity $-|V|^2V$, as in the ordinary differential equation $V' - |V|^2V = 0$. But for this equation, the blow up time depends on the size of the data. It is very large when the data is small. The second mechanism is the amplification caused by focusing, as in the linear transport equation $\partial_t V - \partial_{|x|}V - \frac{1}{|x|}V = 0$. Even if the data is small, it forces $V$ to be very large before one reaches the caustic. Then the first mechanism makes the solution blow up very quickly. This shows that focusing is an essential part of the mechanism which produces blow up in a finite time bounded independently of the smallness of the data.

When the profile $V$ blows up, the family of exact solutions $u^\varepsilon$ cannot remain bounded. Moreover, since the blow up of $V$ occurs in $L^1$, this implies that $u^\varepsilon$ is not bounded in any reasonable space in which weak solutions can be defined. For example, consider Cauchy data of the form (5.2). The first focusing time is the smallest value of $|y|$ such that $V(0,y,\cdot)$ is not identically zero.

**Theorem ([JMR 5]).** *There are Cauchy data of the form* (5.2) *and* $T_* > 0$, *strictly smaller than the first focusing time, such that*

i) *for all* $T < T_*$, *there is* $\varepsilon(T) > 0$ *such that for all* $\varepsilon < \varepsilon(T)$, *the solutions* $u^\varepsilon$ *exist and are smooth on* $[0,T] \times \mathbb{R}^3$,
ii) *if the family* $\partial_t u^\varepsilon$ *is bounded in* $L^3(]0,T[\times\mathbb{R}^3)$, *then* $T \leq T_*$.

**Example 2.** Consider in dimension $d = 5$, the equation

$$\Box u^\varepsilon = (\partial_t u^\varepsilon)^2 - |\nabla_x u^\varepsilon|^2 \tag{5.7}$$

with initial data

$$u^\varepsilon(0,x) = 0, \qquad \partial_t u^\varepsilon(0,x) = U_1(x, |x|/\varepsilon), \tag{5.8}$$

supported in the annulus $\{1 < |x| < 2\}$. Nirenberg's linearization, $v^\varepsilon := 1 - \exp(-u^\varepsilon)$, transforms (5.7) into the linear problem

$$\Box v^\varepsilon = 0, \qquad v^\varepsilon(0,x) = 0, \qquad \partial_t v^\varepsilon(0,x) = U_1(x, |x|/\varepsilon). \tag{5.9}$$

By finite speed of propagation, $u^\varepsilon = v^\varepsilon = 0$ in the cone $\mathcal{C} := \{|x| \leq 1 - t\}$. Let $\Omega_\varepsilon$ denote the connected open subset of $\{v^\varepsilon < 1\}$, which contains $\mathcal{C}$. Then the solutions $u^\varepsilon$ are defined and smooth on $\Omega_\varepsilon$. Moreover, $\partial u^\varepsilon = \partial v^\varepsilon/(1 - v^\varepsilon)$. If $dv^\varepsilon \neq 0$ near a point $(t,x) \in \partial\Omega_\varepsilon$, then $\partial\Omega_\varepsilon$ is a smooth manifold of codimension one near $(t,x)$, and $du^\varepsilon$ is not integrable up to the boundary in $\Omega_\varepsilon$, near $(t,x)$. Explicit computations, using the fundamental solution of $\Box$ in dimension $d = 5$, show that there are sequences $t_n \to 1$ and $\varepsilon_n \to 0$ such that $v^{\varepsilon_n}(t_n, 0) = 1$.

**Theorem ([JMR 5]).** *There are Cauchy data of the form* (5.8) *such that for all* $\delta > 0$, *there is* $\varepsilon_0$ *such that the ball* $B_\delta$, *of radius* $\delta$ *and centered at* $(t = 1, x = 0)$, *is not contained in* $\Omega_\varepsilon$ *for* $\varepsilon < \varepsilon_0$. *Moreover,* $(\partial_t u^\varepsilon)^2 - |\nabla_x u^\varepsilon|^2$ *is not integrable (up to the boundary) on* $\Omega_\varepsilon \cap B_\delta$.

In particular, $u^\varepsilon$ *cannot* be extended to $B_\delta$, even as a *weak* solution. Thus the first focusing time is the largest common lifetime of the (weak) solutions $u^\varepsilon$.

**6. Absorption of oscillations.** When combined with strongly dissipative mechanisms, focusing can lead to a complete absorption of oscillations, in finite time. The oscillations disappear when they reach the caustic set. The following example of such a behaviour is taken from [JMR 5].

Consider in dimension $d = 3$, the dissipative wave equation

$$\Box u + |\partial_t u|^2 \partial_t u = 0, \tag{6.1}$$

$$u^\varepsilon(0,x) = \underline{u}_0(x) + \varepsilon U_0(x, |x|/\varepsilon), \quad \partial_t u^\varepsilon(0,x) = U_1(x, |x|/\varepsilon). \tag{6.2}$$

There are global weak solutions $u^\varepsilon \in C^0([0,\infty[;H^1(\mathbb{R}^3)) \cap L^4([0,\infty[\times\mathbb{R}^4)$ (see [Li], [LS]). For small times, [JR 1] shows that $u^\varepsilon$ has an asymptotic expansion of the form (5.3). When $\underline{u}_0 = 0$, $U_0$ is even in $\theta$ and $U_1 = \partial_\theta U_0$, one has the simpler form

$$u^\varepsilon(t,x) = \varepsilon U_+(t,x,(|x|+t)/\varepsilon) + \varepsilon^2 \dots, \tag{6.3}$$

and $V := \partial_\theta U_+$ satisfies

$$\partial_t V - \partial_{|x|} V - \frac{1}{|x|} V + \frac{1}{2} |V|^2 V = 0. \tag{6.4}$$

The density of energy $W(t,x,\theta) := |x|^2 |V(t,x,\theta)|^2$ satisfies

$$W(t,x,\theta) = \frac{W(0,y,\theta)}{1 + W(0,y,\theta)\,\frac{t}{|x||y|}}, \qquad y := x + t\,\frac{x}{|x|}. \tag{6.5}$$

If $V(0,y,\theta) \neq 0$, then along the ray $x = y - t\,\frac{y}{|y|}$,

$$|V| \to +\infty, \quad \text{but} \quad W \to +0 \quad \text{as} \quad |x| \to 0. \tag{6.6}$$

This means that the amplitude $V$ is infinite at the caustic, but the energy transported by the oscillations tends to zero, that is, is entirely dissipated. This suggests that the oscillations are absorbed when they reach the caustic set. This can be rigorously proved.

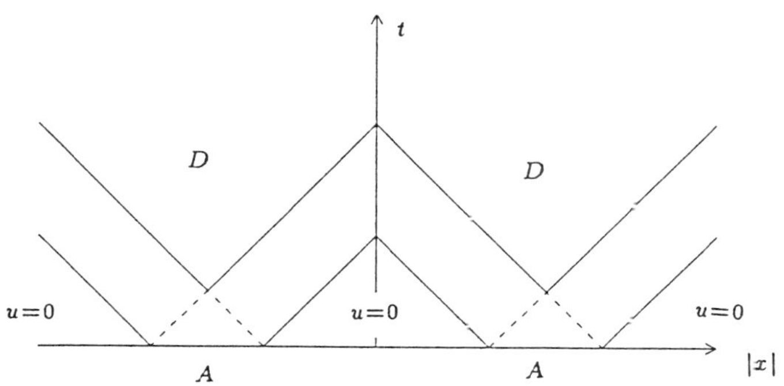


FIGURE 1

**Theorem ([JMR 5]).** *(See figure 1.) If the Cauchy data (6.2) are supported in the annulus $A$, then the family of solutions $u^\varepsilon$ is compact in $H^1$ on all compact subdomains of $D$.*

**7. Oscillations past caustics.** To understand an intermediate case between blow up and absorption, consider the semilinear wave equation

$$\Box u^\varepsilon + f(\partial_t u^\varepsilon) = 0, \tag{7.1}$$

with oscillatory Cauchy data

$$u^\varepsilon(0,x) = \underline{u}_0(x) + \varepsilon U_0(x, \psi(x)/\varepsilon), \qquad \partial_t u^\varepsilon(0,x) = U_1(x, \psi(x)/\varepsilon). \tag{7.2}$$

The main assumption is that $f$ is globally Lipschitzian and satisfies

$$f(0) = 0, \quad \big|f(v) - f(w)\big| \le K\big|v - w\big|, \tag{7.3}$$

with $K$ independent of $v$ and $w$. This immediately implies that there are global weak solutions $u^\varepsilon \in C^0([0,\infty[\,; H^1(\mathbb{R}^3))$. For small times, the solutions satisfy ([JR 1]):

$$u^\varepsilon(t,x) = \underline{u}(t,x) + \varepsilon\, U_+(t,x,\varphi_+(t,x)/\varepsilon) + \varepsilon U_-(t,x,\varphi_-(t,x)/\varepsilon) + \varepsilon^2 \ldots, \tag{7.4}$$

where $\varphi_\pm$ are the two solutions of the eiconal equation for $\Box$ with Cauchy data $\psi$:

$$\partial_t \varphi_\pm = \pm|\partial_x \varphi_\pm|, \quad \varphi_\pm|_{t=0} = \psi. \tag{7.5}$$

The ground state $\underline{u}$ and the oscillations $V_\pm := \partial_\theta U_\pm$ of the periodic profiles satisfy a coupled system of integro-differential equations. Moreover, [JR 1] shows that asymptotically, the solutions satisfy (7.4) *as long as the eiconal equations* (7.5) *have smooth solutions.*

The question is now to describe the solutions $u^\varepsilon$ for arbitrarily large times. In the linear theory, one idea to study the behaviour of oscillations near caustics, is to replace phase-amplitude expansions like (7.4) by *Lagrangian integrals* (see [Du], [Hö 1], [Hö 2]). In the context of wave equations, they are of the form

$$(2\pi\varepsilon)^{-d} \iint e^{i\Phi_\pm(t,x,y,\xi)/\varepsilon} a(t,y)\, dy\, d\xi, \tag{7.6}$$

with

$$\Phi_\pm(t,x,y,\xi) := \pm t|\xi| + (x-y)\cdot\xi + \psi(y). \tag{7.7}$$

These integrals are associated to Lagrangian manifolds, $\Lambda_\pm \subset T^*\mathbb{R}^{1+d}$, where $\Lambda_\pm$ is the set of points $(t,x,\tau,\xi)$ such that there exists $y$ such that

$$x = y \mp t\frac{\xi}{|\xi|}, \quad \xi = d\psi(y), \quad \tau = \pm|\xi|. \tag{7.8}$$

Remark that (7.8) provides a global parametrization of the Lagrangians with $(t, y)$ as parameters. In particular, we can consider $a(t, y)$ as the *symbol* of the integral (7.6).

When harmonics are present, the definition (7.6) has a clear extension. For

$$A(t, y, \theta) = \sum_{n \neq 0} a_n(t, y) e^{in\theta} \tag{7.9}$$

introduce

$$I_{\pm}^{\varepsilon}(A) = \sum_{n \neq 0} \frac{|n|^d}{(2\pi\varepsilon)^d} \iint e^{in\Phi_{\pm}(t,x,y,\xi)/\varepsilon} a_n(t, y)\, dy\, d\xi. \tag{7.10}$$

The local behaviour of such integrals is given by stationary phase expansions. Introduce the hessian matrix of $\Phi_{\pm}$ with respect to the variables $(y, \xi)$, evaluated on $\Lambda_{\pm}$, that is, for $x$ and $\xi$ as in (7.8). Denote the square root of the absolute value of the determinant by $\Delta_{\pm}(t, y)$, and the signature by $2m_{\pm}(t, y)$. The caustic set is $\mathcal{C} := \mathcal{C}_+ \cup \mathcal{C}_-$, where

$$\mathcal{C}_{\pm} := \{(t, x) \mid \exists y : x = y \mp tg(y) \text{ and } \Delta_{\pm}(t, y) = 0\}, \tag{7.11}$$

with $g(y) := d\psi(y)/|d\psi(y)|$. *The classical stationary phase theorem* implies that for $(t, x) \notin \mathcal{C}$

$$I_{\pm}^{\varepsilon}(A)(t, x) \sim J_{\pm}^{\varepsilon}(A)(t, x) := \sum_{y | y \mp tg(y) = x} \frac{1}{\Delta_{\pm}(t, y)} (\mathcal{H}^{m(t,y)} A)(t, y, \psi(y)/\varepsilon), \tag{7.12}$$

where $\mathcal{H}$ is the Hilbert transform defined on Fourier series by

$$\mathcal{H}\Big(\sum_{n \neq 0} a_n e^{in\theta}\Big) := \sum_{n \neq 0} i^{\operatorname{sign} n} a_n e^{in\theta}. \tag{7.13}$$

Note that the phase shift $m(t, y)\pi/2$ changes to $-m(t, y)\pi/2$ when passing from positive to negative harmonics. This explains the presence of the Hilbert transform.

Locally there is a regular labelling $y_{\pm,k}$ of points $y$ such that $y \mp tg(y) = x$. Then

$$\varphi_{\pm,k}(t, x) := \psi(y_{\pm,k}) \tag{7.14}$$

are solutions of the eiconal equation, and $J_{\pm}^{\varepsilon}(A)$ is of the form

$$J_{\pm}^{\varepsilon}(A) = \sum_k U_{\pm,k}(t, x, \varphi_{\pm,k}(t, x)/\varepsilon), \tag{7.15}$$

with

$$U_{\pm,k}(t,x,\theta) := \frac{1}{\Delta_\pm(t,y_{\pm,k})}\big(\mathcal{H}^{m(t,y_{\pm,k})}A\big)(t,y_{\pm,k},\theta). \tag{7.16}$$

The linear theory (see e.g. [Du], [Hö 2]), shows that

$$u^\varepsilon \sim \underline{u} + \varepsilon I_+^\varepsilon(A_+) + \varepsilon I_-^\varepsilon(A_-) \sim \underline{u} + \varepsilon J_+^\varepsilon(A_+) + \varepsilon J_-^\varepsilon(A_-). \tag{7.17}$$

The phase-amplitude representation of $u^\varepsilon$ outside the caustic then follows from (7.15). In fact, for linear equations, the results are much more precise and refined stationary phase expansions yield precise descriptions of $u^\varepsilon$ near $\mathcal{C}$. ([Lu], [Du], [H2]).

For the nonlinear wave equation (7.1) with globally Lipschtzean nonlinearity $f$, one can show, with suitable assumptions, that the solution $u^\varepsilon$ still admits an asymptotic description of the form (7.17) (see [JMR 7]).

*Assumptions on the initial phase $\psi$.*

1. $\psi \in C^\infty(\overline{\omega})$, $\omega$ open subset of $\mathbb{R}^d$, and the measure of $\partial\omega$ in $\mathbb{R}^d$ is equal to 0.
2. $d\psi(y) \neq 0$ for all $y \in \overline{\omega}$.
3. The number of solutions $y$ of $y \pm tg(y) = x$ is bounded independently of $(t,x)$.
4. **Nonresonance.** For all domains $\Omega$ which do not intersect $\mathcal{C}$, the local phases $\varphi_{\pm,k}$ (7.14) do not resonate, which means that for all families $\alpha_{\pm,k} \in \mathbb{Z}$ with at least two nonvanishing components, $\sum \alpha_{\pm,k}\varphi_{\pm,k}$ is noncharacteristic a.e. on $\Omega$.

**Examples and Remarks.**

1. Examples of phases which satisfy these assumptions are

   a) $\psi(y) = |y|$, away from the origin.
   b) $\psi(y) = a_1 y_1^2 + a_2 y_2^2$, on $\mathbb{R}^2 \setminus \{0\}$.
   c) $\psi(y) = y_d + y_1^2 + \cdots + y_{d-1}^2$ in $\mathbb{R}^d$.

2. The second assumption can be relaxed, to cover the case where $\psi$ has isolated critical points as in example b), or isolated singular points as in the first example. The third condition requires that the mapping $y \to y \pm tg(y)$ is not too complicated.

3. Much more important is the nonresonance condition 4. This assumption is *much weaker than coherence.* Coherence would require that the phases $\sum \alpha_{\pm,k}\varphi_{\pm,k}$ are *nowhere* characteristic. Moreover, the coherence assumption *is not satisfied* in the examples b) or c) above, while the nonresonance condition 4 is satisfied in these examples.

We refer the reader to [JMR 7] for further examples and discussions.

*Assumptions on the Cauchy data.*

Suppose that $\underline{u}_0 \in H^1(\mathbb{R}^d)$, that $U_0(x,\theta)$ and $U_1(x,\theta)$ are smooth enough, periodic in $\theta$ and vanish for $x \notin \omega$. Error terms $r_0^\varepsilon$ and $r_1^\varepsilon$ can be added to the Cauchy data, provided that they converge to zero in $H^1(\mathbb{R}^d)$ and $L^2(\mathbb{R}^d)$ respectively.

The idea is that Lagrangian integrals can be used to study the asymptotic behaviour of the solutions $u^\varepsilon$ of the Cauchy problem (7.1), (7.2). We sketch the main steps, referring for details to [JMR 7].

A) *$L^2$ estimates of oscillatory integrals.* For $A$ smooth enough, $I_\pm^\varepsilon(A)$ and $J_\pm^\varepsilon(A)$ are bounded families in $L^\infty([0,T], L^2(\mathbb{R}^d))$ and

$$\limsup_{\varepsilon\to 0} \| I_\pm^\varepsilon(A)(t)\|_{L^2(\mathbb{R}^d)} \le \|A(t)\|_{L^2(\omega\times\mathbb{R}/\mathbb{Z})},$$
$$\limsup_{\varepsilon\to 0} \| J_\pm^\varepsilon(A)(t)\|_{L^2(\mathbb{R}^d)} \le C\|A(t)\|_{L^2(\omega\times\mathbb{R}/\mathbb{Z})}.$$

The limsup's are uniform on bounded intervals of time. Moreover, for all $T$,

$$\|I_\pm^\varepsilon(A) - J_\pm^\varepsilon(A)\|_{L^2([0,T]\times\mathbb{R}^d)} \to 0 \quad \text{as } \varepsilon \to 0.$$

A stronger convergence in $L^\infty([0,T], L^2(\mathbb{R}^d))$ holds, when additional transversality conditions are satisfied (see [JMR 7]). This property is satisfied in the three examples of phases given above.

B) *Nonlinear functions of Lagrangian integrals.* Suppose that $\Omega$ does not intersect $\mathcal{C}$. Let $\varphi_{\pm,k}$ denote the local phases on $\Omega$ defined in (7.14). Suppose that

$$v^\varepsilon \sim \underline{v} + J_+^\varepsilon(B_+) + J_-^\varepsilon(B_-). \tag{7.18}$$

Then, by (7.15), $v^\varepsilon$ is of the form $v^\varepsilon \sim \underline{v} + \sum V_{\pm,k}(\cdot, \varphi_{\pm,k}/\varepsilon)$. Thus,

$$f(v^\varepsilon(t,x)) \sim \mathcal{F}(t,x,\varphi_*(t,x)/\varepsilon), \tag{7.19}$$

where

$$\mathcal{F}(t,x,\theta_*) := f\Big(\underline{v}(t,x) + \sum V_{\pm,k}(t,x,\theta_{\pm,k})\Big). \tag{7.20}$$

We have used the notation $\theta_* := \{\theta_{\pm,k}\}$, $\varphi_* := \{\varphi_{\pm,k}\}$.

The idea is now to extract from (7.19) the characteristic oscillations. Here we use the nonresonance assumption. Introduce first the total average

$$\mathcal{E}_0(\underline{v}, B_+, B_-)(t,x) := \int \mathcal{F}(t,x,\theta_*)\, d\theta_*. \tag{7.21}$$

Next, for $(t, y)$ such that $x := y - tg(y) \notin \mathcal{C}$, label the $y_{\pm,k}$ so that $y = y_{+,1}$. Introduce the average of $\mathcal{F}$ with respect to the variables $\theta_*$ except the first one $\theta_{+,1}$:

$$\mathcal{F}_{+,1}(\underline{v}, B_+, B_-)(t, y, \theta_{\pm,1}) := \int \mathcal{F}(t, x, \theta_*)\, d\theta_{+,2} \dots d\theta_{-,1} \dots . \tag{7.22}$$

Its oscillatory part is

$$\mathcal{F}^{\mathrm{osc}}_{+,1}(t, y, \theta) := \mathcal{F}_{+,1}(\underline{v}, B_+, B_-)(t, y, \theta) - \mathcal{E}_0(\underline{v}, B_+, B_-)(t, x). \tag{7.23}$$

To lift it to the Lagrangian, introduce

$$\mathcal{E}_+(\underline{v}, B_+, B_-)(t, y, \theta) := \Delta_+(t, y)\big(\mathcal{H}^{-m(t,y)} \mathcal{F}^{\mathrm{osc}}_{+,1}\big)(t, y, \theta). \tag{7.24}$$

The definition of $\mathcal{E}_-(\underline{v}, B_+, B_-)$ is similar.

The Lipschitz property (7.3) for $f$ and the $L^2$ estimates for $J^\varepsilon$, imply that the nonlinear operators $\mathcal{E}_0$ and $\mathcal{E}_\pm$ act nicely in $L^2$. Moreover, expanding $\mathcal{F}(t, x, \eta_*)$ into its Fourier series and *using the non resonance assumption*, one proves that

$$f(v^\varepsilon) \sim \mathcal{E}_0 + J^\varepsilon_+(\mathcal{E}_+) + J^\varepsilon_-(\mathcal{E}_-) + g^\varepsilon, \tag{7.25}$$

where $g^\varepsilon$ has *no characteristic oscillations.*

C) *The profile equations.*

Now take $v^\varepsilon = \partial_t u^\varepsilon$ and consider $f(v^\varepsilon)$ as a source term for $\Box$. Only the characteristic oscillations are propagated. Thus the terms $g^\varepsilon$ in (7.25) are expected to contribute small errors in $H^1$. On the other hand, standard computations for oscillatory integrals show that

$$\Box\big(\varepsilon I^\varepsilon_\pm(A_\pm)\big) \sim I^\varepsilon_\pm(2\partial_t B_\pm) \quad \text{and} \quad \partial_t\big(\varepsilon I^\varepsilon_\pm(A_\pm)\big) \sim I^\varepsilon_\pm(B_\pm),$$

with $B_\pm := \pm\, |d\psi| \partial_\theta A_\pm$. Assuming that $u^\varepsilon$ is asymptotically given by (7.17), these considerations lead, for $\underline{u}$ and $B_\pm := \pm\, |d\psi|\partial_\theta A_\pm$, to the following set of equations:

$$\begin{cases} \Box \underline{u} & + \quad \mathcal{E}_0(\partial_t \underline{u}, B_+, B_-) = 0, \\ 2\partial_t B_+ & + \quad \mathcal{E}_+(\partial_t \underline{u}, B_+, B_-) = 0, \\ 2\partial_t B_- & + \quad \mathcal{E}_-(\partial_t \underline{u}, B_+, B_-) = 0. \end{cases} \tag{7.26}$$

The natural initial conditions are

$$\begin{cases} \underline{u}_{|t=0}(x) = \underline{u}_0(x), \quad \partial_t \underline{u}_{|t=0}(x) = \underline{u}_1(x) := \dfrac{1}{2\pi} \displaystyle\int_0^{2\pi} U_1(x, \theta)\, d\theta, \\ B_{\pm|t=0}(x, \theta) = \tfrac{1}{2}\Big(U_1(x, \theta) - \underline{u}_1(x) \pm |d\psi(x)| \partial_\theta U_0(x, \theta)\Big). \end{cases} \tag{7.27}$$

Remark that the data of $B_\pm$ uniquely determines $A_\pm$, since both are periodic in $\theta$ and their mean values are zero. $B_\pm$ corresponds to the time derivatives of $\varepsilon I^\varepsilon_\pm(A_\pm)$. For the spatial derivatives, one has

(7.28)
$$\partial_{x_j}\big(\varepsilon I^\varepsilon_\pm(A_\pm)\big) \sim I^\varepsilon_\pm(B_{\pm,j}) \quad \text{with} \quad B_{\pm,j} := \partial_{y_j}\psi\partial_\theta A_\pm = \pm\,\frac{\partial_{y_j}\psi}{|d\psi|}B_\pm .$$

D) *The main result.*

**Theorem ([JMR 7]).**

i) *The profile equations* (7.26) *with Cauchy data* (7.27) *have a unique solution.*

ii) *For all* $T > 0$, *the solutions* $u^\varepsilon$ *of* (7.1) *with Cauchy data* (7.2) *satisfy*

$$(7.28)\qquad \begin{cases} u^\varepsilon - \underline{u} \to 0 \\ \partial_t u^\varepsilon - \big(\partial_t \underline{u} + I^\varepsilon_+(B_+) + I^\varepsilon_-(B_-)\big) \to 0 \\ \partial_{x_j} u^\varepsilon - \big(\partial_{x_j}\underline{u} + I^\varepsilon_+(B_{+,j}) + I^\varepsilon_-(B_{-,j})\big) \to 0 \end{cases}$$

*strongly in* $C^0([0,T], L^2(\mathbb{R}^d))$.

iii) *In addition, for all* $T > 0$,

$$(7.29)\qquad \begin{cases} \partial_t u^\varepsilon - \big(\partial_t \underline{u} + J^\varepsilon_+(B_+) + J^\varepsilon_-(B_-)\big) \to 0 \\ \partial_{x_j} u^\varepsilon - \big(\partial_{x_j}\underline{u} + J^\varepsilon_+(B_{+,j}) + J^\varepsilon_-(B_{-,j})\big) \to 0 \end{cases}$$

*strongly in* $L^2([0,T]\times\mathbb{R}^d)$.

The convergences in (7.28) and (7.29) are the precise statement corresponding to the equivalence

$$(7.30)\qquad u^\varepsilon \sim \underline{u} + \varepsilon I^\varepsilon_+(A_+) + \varepsilon I^\varepsilon_-(A_-) \sim \underline{u} + \varepsilon J^\varepsilon_+(A_+) + \varepsilon J^\varepsilon_-(A_-).$$

The form (7.15) of $J^\varepsilon_\pm(A_\pm)$ provides a phase amplitude expansion for $u^\varepsilon$ outside the caustic set.

## References


[BB] W. Blaschke and G. Bol, *Geometrie der Gewebe*, Springer-Verlag, 1938.

[CB] Y. Choquet-Bruhat, *Ondes asymptotiques et approchées pour les systèmes d'équations aux dérivées partielles non linéaires*, J. Math. Pure Appl. **48** (1969), 117–158.

[Che 1] C. Cheverry, *Oscillations de faible amplitude pour les systèmes* 2 × 2 *de lois de conservation*, Asymptotic Analysis (to appear).

[Che 2] ———, *Optique géométrique faiblement non linéaire, le cas général*, preprint.

[Che 3] ———, *Justification de l'optique géométrique non linéaire pour un système de lois de conservation*, Séminaire de l'École Polytechnique, année 94-95, exposé no 15.

[Che 4] ———, *Propagation d'oscillations près d'un point diffractif*, J. Math. Pures et Appl. (to appear).

[Chi] J. Chikhi, *Sur la réflexion des oscillations pour un système à deux vitesses*, C. R. Acad. Sc. Paris **313** (1991), 675–678.

[Co 1] A. Corli, *Weakly non-linear geometric optics for hyperbolic systems of conservation laws with shock waves*, Asymptotic Analysis **10** (1995), 117–172.

[Co 2] ———, *Asymptotic analysis of contact discontinuities*, preprint.

[De] J. M. Delort, *Oscillations semi-linéaires multiphasées compatibles en dimension 2 et 3 d'espace*, Comm. in Partial Diff. Eq. **16** (1991), 845–872.

[DiP 1] R. Di Perna, *Convergence of approximate solutions to conservation laws*, Arch. Rat. Mech. Anal. **82** (1983), 27–70.

[DiP 2] ———, *Compensated compactness and general systems of conservation laws*, Trans. Amer. Math. Soc. **292** (1985), 383–420.

[DiPM] R. Di Perna and A. Majda, *The validity of nonlinear geometric optics for weak solutions of conservation laws*, Comm. Math. Phys. **98** (1985), 313–347.

[Do] P. Donat, *Quelques contributions mathématiques en optique non linéaire*, thèse, École Polytechnique, 1994.

[DR] P. Donat and J. Rauch, *Dispersive nonlinear geometric optics*, preprint.

[Du] J. J. Duistermaat, *Oscillatory integrals, Lagrangian immersions and unfolding of singularities*, Comm. Pure Appl. Math. **27** (1974), 207–281.

[DH] J. J. Duistermaat and L. Hörmander, *Fourier Integral Operators II*, Acta Math. **128** (1972), 183-269.

[G 1] O. Gues, *Ondes solitaires engendrées par l'interaction d'ondes oscillantes non linéaires*, J. Math. Pures et Appl. **74** (1995), 199–252.

[G 2] ———, *Développements asymptotiques de solutions exactes de systèmes hyperboliques quasilinéaires*, Asymptotic Anal. **6** (1993), 241-269.

[G 3] ———, *Ondes multidimensionnelles $\varepsilon$-stratifiées et oscillations*, Duke Math. J. **68** (1992), 401–446.

[He] A. Heibig, *Error estimates for oscillating solutions to hyperbolic systems of conservation laws*, Comm. in Partial Diff. Eq. **18** (1993), 281–304.

[Hö 1] L. Hörmander, *Fourier Integral Operators I*, Acta Math. **127** (1971), 79–183.

[Hö 2] ———, *The Analysis of Linear Partial Differential Operators*, Springer-Verlag, 1991.

[Hu] J. Hunter, *Strongly nonlinear hyperbolic waves*, Nonlinear Hyperbolic Equations (Ballman J. and Jeltsch R., eds.), Vieweg Verlag, Braunschweig, 1989, Notes on numerical fluid dynamics, Vol 24.

[HK 1] J. Hunter and J. Keller, *Weakly nonlinear high frequency waves*, Comm. Pure Appl. Math. **36** (1983), 547–569.

[HK 2] ———, *Caustics of nonlinear waves*, Wave Motion **9** (1987), 429-443.

[HMR] J. Hunter, A. Majda, and R. Rosales, *Resonantly interacting weakly nonlinear hyperbolic waves II: several space variables*, Stud. Appl. Math. **75** (1986), 187–226.

[J] J.-L. Joly, *Sur la propagations des oscillations semi-linéaires en dimension 1 d'espace*, C. R. Acad. Sc. Paris **296** (1983).

[JMR 1] J.-L. Joly, G. Metivier and J. Rauch, *Resonant one dimensional nonlinear geometric optics*, J. Funct. Anal. **114** (1993), 106–231.

[JMR 2] ———, *Coherent and focusing multidimensional nonlinear geometric optics*, Ann. Éc. Norm. Sup. **28** (1995), 51–113.

[JMR 3] ———, *Generic rigorous asymptotic expansions for weakly nonlinear multidimensional oscillatory waves*, Duke Math. J. **70** (1993), 373–404.

[JMR 4] ———, *Coherent nonlinear waves and the Wiener algebra*, Ann. Inst. Fourier **44** (1994), 167–196.

[JMR 5] ———, *Focusing at a point and absorption of nonlinear oscillations*, Trans. Amer. Math. Soc. (to appear).

[JMR 6] ———, *Trilinear compensated compactness*, Ann. of Math. **142** (1995), 121–169.

[JMR 7] ———, *Nonlinear oscillations beyond caustics*, Comm. Pure Appl. Math. (to appear).

[JMR 8] ———, *Dense oscillations for the compressible 2-D Euler equations*, Non Linear Partial Differential Equations and their Applications (H. Brézis and J. L. Lions, eds.), Pitman Publ. Corp (to appear).

[JMR 9] ———, *A nonlinear instability for* $3\times 3$ *systems of conservation laws*, Comm. Math. Phys. **162** (1994), 47–59.

[JR 1] J. L. Joly and J. Rauch, *Justification of multidimensional single phase semilinear geometric optics*, Trans. Amer. Math. Soc. **330** (1992), 599–625.

[JR 2] ———, *Nonlinear resonance can create dense oscillations*, Microlocal Analysis and Nonlinear Waves (M. Beals, R. Melrose and J. Rauch, eds.), Springer Verlag, IMA volumes in Math. and Appl. **30** (1991), 113–124.

[Kal] L. A. Kalyakin, *Long wave asymptotics, integrable equations as asymptotic limit of nonlinear systems*, Russian Math. Surveys **44** no 1 (1989), 3–42.

[K] J. Keller, *Corrected Bohr-Sommerfield quantum conditions for nonseparable systems*, Ann. Physics **4** (1958), 180–188.

[La] P. Lax, *Asymptotic solutions of oscillatory initial value problems*, Duke Math. J. **24** (1957), 627–645.

[Li] J. L. Lions, *Quelques méthodes de résolution de problémes aux limites non linéaires*, Dunod, Gauthier-Villars, Paris, 1969.

[LS] J. L. Lions and W. Strauss, *Some nonlinear evolution equations*, Bull. Soc. Math. France **93** (1965), 43–96.

[Liu] T. P. Liu, *Decay to N-waves of solutions of general systems of nonlinear hyperbolic conservation laws*, Comm. Pure Appl. Math. **30** (1977), 585–610.

[Lu] D. Ludwig, *Uniform asymptotic expansions at a caustic, Comm. Pure Appl. Math* **13** (1966), 85–114.

[MA] A. Majda and M. Artola, *Nonlinear geometric optics for hyperbolic mixed problems*, Analyse Mathématique et Applications, Gauthier-Villars, Paris, 1988.

[McLPT] D. W. McLaughlin, G. Papanicolaou and L. Tartar, *Weak limits of semi-linear hyperbolic systems with oscillating data*, Macroscopic Modelling of Turbulent Flow, Lecture Notes in Physics **230** (1985), 277–298.

[MR] A. Majda and R. Rosales, *Resonantly interacting weakly nonlinear hyperbolic waves I: a single space variable*, Stud. Appl. Math. **71** (1984), 149–179.

[MRS] A. Majda, R. Rosales and M. Schonbeck, *A canonical system of integrodifferential equations arising in resonant nonlinear acoustics*, Stud. Appl. Math. **79** (1988), 205–262.

[P] Henri Poincaré, *Sur les surfaces de translation et les fonctions Abeliennes*, Bull. Soc. Math. France **29** (1901), 61–86.

[Se 1] D. Serre, *Oscillations non linéaires des systèmes hyperboliques: méthodes et résultats qualitatifs*, Ann. Inst. Henri Poincaré **8** (1991), 351–417.

[Se 2] ———, *Oscillations non linéaires de haute fréquence*, Nonlinear variational problems and partial differential equations (A. Marino and M. K. V. Murthy, eds.), Pitman Publ. Corp., London, Pitman Research Notes **320** (1995), 245–294.

[Sch 1] S. Schochet, *Resonant nonlinear geometric optics for weak solutions of conservation laws*, J. Diff. Eq. **113** (1994), 473–504.

[Sch 2] ———, *Fast singular limits of hyperbolic partial differential equations*, J. Diff. Eq. **114** (1994), 474–512.

[T] L. Tartar, *Solutions oscillantes des équations de Carleman*, Sém. Goulaouic-Meyer-Schwartz, École Polytechnique, Paris, année 1980-1981, exposé no 12.

[Wa] T. Wagenmaker, *Analytic solutions and resonant solutions of hyperbolic partial differential equations*, Ph. D. Thesis, University of Michigan, 1993.

[Wi 1] M. Williams, *Resonant reflection of multidimensional semilinear oscillations*, Comm. in Partial Diff. Eq. **18** (1993), 1901–1959.

[Wi 2] ———, *Nonlinear geometric optics for reflecting and glancing oscillations*, preprint.

CEREMAB, Université de Bordeaux I, F-33405 Talence, France
*E-mail address*: joly@math.u-bordeaux.fr

IRMAR, Université de Rennes I, F-35042 Rennes, France
*E-mail address*: Guy.Metivier@univ-rennes1.fr

Department of Mathematics, The University of Michigan, Ann Arbor MI 48109, USA
*E-mail address*: rauch@math.lsa.umich.edu

# STABILIZATION OF THE WAVE EQUATION BY THE BOUNDARY

GILLES LEBEAU AND LUC ROBBIANO

**1. Introduction.** We consider here problems of stabilization for the wave equation on a connected manifold with a compact boundary. The stabilization, i.e. the decrease in energy, will be obtained by a dissipative boundary condition.

We know (HARAUX [Ha]) that the energy of any solution tends to zero as soon as there is dissipation on a nonempty open subset of the boundary. Since the work of BARDOS, LEBEAU & RAUCH [B.L.R] we have a condition (called geometric control) which is essentially necessary and sufficient for the decrease in energy to be exponential for any finite energy data.

In the case of stabilization in the interior LEBEAU [L] has shown that when the geometric control hypothesis is not satisfied, the energy decreases at least logarithmically if the data are in the domain of $A^k$, where $A$ is the infinitesimal generator of the evolution equation.

Here we obtain related results for stabilization by the boundary. As in [L] we apply an interpolation inequality which is obtained from Carleman's inequalities. We will need a Carleman inequality which takes into account the boundary conditions. This is done by a slight extension of the results from [L.R]. As in [L] we obtain a lower bound for the spectrum of $A$. In [L], the rate of decrease is obtained from this lower bound. Here we cannot proceed in this manner and we majorize the energy after a transformation of F.B.I. type which allows us to have an almost optimal result.

**2. Results.** Let $(M, g)$ be a compact connected Riemannian manifold with a nonempty $C^\infty$ boundary. We denote by $\Delta$ the laplacian on $M$ for the metric $g$. We are interested in the decrease of energy for the following problem

$$(1)\qquad \begin{aligned} (\partial_t^2 - \Delta)u &= 0 \\ u|_{t=0} &= u_0 \in H^1(M) \\ \partial_t u|_{t=0} &= u_1 \in L^2(M) \\ \partial_n u + a(x)\partial_t u &= 0 \text{ on } \partial M \times \mathbb{R}, \end{aligned}$$

where $\partial_n$ is the exterior normal derivative to $\partial M$, $a \in C^\infty(\partial M, \mathbb{R}_+)$, and $\{x \in \partial M; a > 0\} \neq \emptyset$.

Let $H = H^1(M) \oplus L^2(M)$,

$$A = \begin{pmatrix} 0 & \mathrm{Id} \\ \Delta & 0 \end{pmatrix},$$

and

$$D(A) = \Big\{u = (u_0, u_1) \in H; Au \in H, \partial_n u_0 + a u_1 = 0 \text{ on } \partial M\Big\}.$$

The injection $D(A) \hookrightarrow H$ is compact, the half plane $\operatorname{Re}\lambda > 0$ is contained in the resolvent set of $A$, since we have $a(x) \geq 0$, and the spectrum of $A$ is discrete. The spectrum always contains 0, $u =$ constant being a solution of (1).

For a solution $u(x,t)$ of (1), we denote

$$E(u,t) = \frac{1}{2}\int_M (|\partial_t u(\cdot,t)|^2 + |\nabla_x u(\cdot,t)|^2)\,dx.$$

We have, after a simple integration by parts,

$$E(u,T_1) - E(u,T_2) = \int_{T_1}^{T_2}\int_{\partial M} a(x)\Big|\frac{\partial u}{\partial s}(x,s)\Big|^2\,dx\,ds. \tag{2}$$

**Theorem 1.** *There exists a constant $C > 0$ such that if $\lambda \in \operatorname{Sp}(A) \setminus \{0\}$ then $\operatorname{Re}\lambda < -e^{-C|\operatorname{Im}\lambda|}/C$, and if*

$$\operatorname{Re}\lambda \in \left[\frac{-e^{-C|\operatorname{Im}\lambda|}}{C}, 0\right], \quad |\lambda| > 1$$

*then one has*

$$\|(A-\lambda)^{-1}\|_{\mathcal{L}(H,H)} \leq C e^{C|\operatorname{Im}\lambda|}.$$

**Theorem 2.** *For any $k > 0$ and any $\delta > 0$, there exists $C > 0$ such that for every $\underline{u} = (u_0, u_1) \in D(A^k)$ the solution $u$ of (1) (with $(u_0,u_1)$ as initial data) satisfies*

$$[E(u,t)]^{1/2} \leq \frac{C\|\underline{u}\|_{D(A^k)}}{[\log(2+t)]^{k-\delta}} \qquad \forall t \geq 0.$$

*Remarks.* Theorems 1 and 2 are the analogues of Theorem 1 of Lebeau [L]. Theorem 1 is obtained using the method of [L]. In [L], Theorem 2 for the evolution problem is a consequence of the estimate of $(A-\lambda)^{-1}$. In the case of interior stabilization, $A$ is a compact perturbation of a self-adjoint operator and the spectrum is contained in $\{z \in \mathbb{C}; \operatorname{Re} z \in [-C,0]\}$ for a $C>0$. In the case of the boundary stabilization, the perturbation is not compact and we have very little information about the spectrum. To prove Theorem 2 we use a transformation of F.B.I. type that is better adapted than the one used in [R].

We remark that in the case where $M$ is the annulus of the plane $M=\{x\in\mathbb{R}^2,\ 1<|x|<2\}$, with $a\equiv 0$ when $|x|=2$, $a\equiv 1$ when $|x|=1$, an explicit calculation in polar coordinates using Bessel and Hankel functions shows that there exists a constant $C>0$ such that the set $\{\operatorname{Re}\lambda \in \big[-e^{-C|\operatorname{Im}\lambda|},0\big]\}$ contains an infinity of spectral values. Theorem 1 is then optimal, and in this case the rate of uniform decrease of the energy with data in $D(A^k)$ is at most $O\big((\log t)^{-k}\big)$ and hence Theorem 2 is in general almost optimal. We note however that a deeper study of the spectral localization and the decrease rates of the energy for regular data should include the global dynamics of the generalized geodesic flow on $M$. Theorems 1 and 2, thus give only the "a priori" bounds that can be obtained without any hypothesis on the dynamics since they use exclusively the Carleman inequalities which express the "tunnel effect".

The interpolation inequality that we use is given by Theorem 3.

Let $X=]-1,+1[\times M$, $\partial X=]-1,+1[\times\partial M$ and let $v$ be a solution of

$$(3)\qquad \begin{aligned} (\partial_{x_0}^2+\Delta)v &= v_0 \text{ on } X\\ (\partial_n - ia\partial_{x_0})v &= v_1 \text{ on } \partial X. \end{aligned}$$

We denote $Y=\big]-\frac{1}{2},\frac{1}{2}\big[\times M$ and for $\delta>0$, $\Gamma_\delta=\{x\in\partial X, a(x)>\delta\}$, $\tilde\Gamma_\delta=]-1,1[\times\Gamma_\delta$, and we suppose that $\Gamma_\delta\neq\emptyset$ .

**Theorem 3.** *For $\delta>0$ small, there exists $C>0$, $\mu\in]0,1[$, such that for any solution $v$ of (3), we have*

$$\|v\|_{H^1(Y)} \le C\Big(\|v_0\|_{L^2(X)}+\|v_1\|_{L^2(\partial X)} + \|v\|_{L^2(\tilde\Gamma_\delta)}+\|\partial_{x_0}v\|_{L^2(\tilde\Gamma_\delta)}\Big)^{\mu}\Big(\|v\|_{H^1(X)}\Big)^{1-\mu}.$$

## REFERENCES

[B.L.R] C. Bardos, G. Lebeau and J. Rauch, *Sharp sufficient conditions for the observation, control and stabilization of waves from the boundary*, SIAM J. Control Optim. **30** (1992), 1024–1065.

[Ha] A. Haraux, *Stabilization of trajectories for some weakly damped hyperbolic equations*, J. Diff. Equations **59** (1985), 145–154.

[Hö] L. Hörmander, *The Analysis of Linear Partial Differential Operators*, Springer Verlag, 1985.

[L] G. Lebeau, *Equations des ondes amorties*, Algebraic and Geometric Methods in Mathematical Physics, Kluwer's Maths-Physics Book Series (to appear).

[L.R] G. Lebeau and L. Robbiano, *Contrôle exact de l'équation de la chaleur*, Comm. Partial Diff. Equations **20** (1995), 335–356.

[R] L. Robbiano, *Fonction de coût et contrôle des solutions des équations hyperboliques*, Asymptotic Analysis **10** (1995), 95–115.

[T] D. Tataru, *A priori estimates of Carleman's type in domains with boundary*, J. Math. Pures et Appl. **73** (1994), 355–387.

Université de Paris-Sud, Département de Mathématiques, Bât. 425, F-91405 Orsay Cedex, France
*E-mail address*: lebeau@matups.matups.fr

Université de Paris-Val de Marne, UFR de Sciences, Av. du Gal De Gaulle, 94010 Créteil Cedex et Université de Paris-Sud, Département de Mathématiques, Bât. 425, F-91405 Orsay Cedex, France
*E-mail address*: robbiano@matups.matups.fr

# ABOUT SMALL-POWER SEMILINEAR WAVE EQUATIONS

HANS LINDBLAD AND CHRISTOPHER D. SOGGE

Let us consider semilinear Cauchy problems of the form

$$(1)\qquad \begin{cases} \Box u = c|u|^p\,, & (t,x) \in \mathbb{R}^{1+n}_+ \\ u(0,x) = f(x)\,, & \partial_t u(0,x) = g(x)\,. \end{cases}$$

Here $c \in \mathbb{R}$ is a constant, and $\Box = \partial_t^2 - \Delta$ denotes the d'Alembertian on $\mathbb{R}^{1+n}_+ = \mathbb{R}_+ \times \mathbb{R}^n$. In any of the results to follow, one can replace the nonlinearity $c|u|^p$ by $F_p(u)$, if $F_p \in C^{[p]}(\mathbb{R})$ and $(\partial/\partial u)^j F_p(u) = O(|u|^{p-j})$, $j = 0, 1$, where $[p]$ denotes the integer part of $p$.

In 1979 John [4] showed that, if $(f_0, g_0) \in C^3(\mathbb{R}^3) \times C^2(\mathbb{R}^3)$ with compact support are fixed, and if the Cauchy data is $(f, g) = \varepsilon(f_0, g_0)$, then, when $n = 3$, (1) always has a global $C^2$ solution for small $\varepsilon > 0$ if $p > 1 + \sqrt{2}$, but not when $p < 1 + \sqrt{2}$. Some time later it was shown by Schaeffer [13] that there can also be blowup for small data when $p = p_c = 1 + \sqrt{2}$.

The strange number $1 + \sqrt{2}$ seems to have first arisen in the work of Strauss [18] on scattering for semilinear Schrödinger equations with small data. Based on this, he conjectured in [19] that, when $n \geq 2$, the corresponding problem in $(1+n)$-dimensions should always have a global solution with appropriate regularity if $\varepsilon > 0$ and $p > p_c = p_c(n)$, where $p_c$ is that positive root of the quadratic equation

$$(2)\qquad (n-1)p^2 - (n+1)p - 2 = 0\,.$$

This conjecture was soon verified when $n = 2$ by Glassey [1], and recently by Zhou [22] when $n = 4$, in which case $p_c = 2$. John's blowup result was extended to other dimensions by Sideris [14], showing that, even under the assumption of spherical symmetry there can be blowup for arbitrarily small data if $p < p_c$.

In this paper we wish to investigate what happens in other dimensions. We are also interested in improving the smallness assumption in John's theorem and allowing data without compact support. In the case where

The authors were supported in part by the National Science Foundation.

the data is assumed to be spherically symmetric we shall also see that (1) has a global (weak) solution under minimal regularity assumptions on the data.

Our main results are that, for small data, if $p > p_c$, then (1) always has a global (weak) solution for any $n$, under the assumption of spherical symmetry, and, in the non-radial case, a global solution always exists if $n \leq 8$.

To formulate the smallness hypotheses, we need to introduce mixed-norms. If, say, $h(x,y)$ is a function of $(x,y) \in \mathbb{R}^d \times \mathbb{R}^m$, then we set

$$\|h\|_{L^p_x L^q_y} = \Big( \int \|h(x,\,\cdot\,)\|^p_{L^q_y}\, dx \Big)^{1/p} = \Big( \int \Big( \int |h(x,y)|^q \, dy \Big)^{p/q} dx \Big)^{1/p}.$$

Similar notation will be used if one of the variables is $\omega \in S^{n-1}$.

Our result for spherically symmetric solutions of (1) then is the following:

**Theorem 1.** *Let $n \geq 3$. Assume that $p_c < p < (n+3)/(n-1)$ and that $f$ and $g$ are spherically symmetric. Then there is an $\varepsilon_0 = \varepsilon_0(p) > 0$ so that if*

$$\|r^{(n-1)/2} f\|_{L^q_r(\mathbb{R}_+)} + \|\partial_r^{-1}(r^{(n-1)/2} g)\|_{L^q_r(\mathbb{R}_+)} < \varepsilon_0(p)\,, \qquad q = q(p) = 2(p-1)/((n+3)-(n-1)p)\,, \tag{3}$$

*then* (1) *has a unique (weak) global solution verifying*

$$\sup_{t>0} \|r^{(n-1)/2} u(t,\,\cdot\,)\|_{L^q_r} + \|r^{(n+1)/2p} u\|_{L^{pq}_t L^p_r} < \infty\,. \tag{4}$$

Here we have of course identified, for instance, $u(t,x)$ with $u(t,r)$, $r = |x|$. Also, in (3), $\partial_r^{-1}$ denotes the anti-derivative vanishing at $\infty$:

$$\partial_r^{-1} h(r) = \int_r^\infty h(s)\, ds\,.$$

Since $q > 1$, by Hardy's inequality, the condition on $g$ in (3) can be replaced by the stronger one saying that $\|r^{(n+1)/2} g\|_{L^q(\mathbb{R}_+)}$ is small.

For odd dimensions, a version of Theorem 1 involving somewhat stronger assumptions on the data has recently been obtained independently by Kubo [6] using different techniques.

In Theorem 1 we have only considered powers smaller than the conformally invariant power $p_{\text{conf}} = (n+3)/(n-1)$, since it is well known that, even in the non-radial case, there is global existence for (1) for appropriate small data if $p \geq p_{\text{conf}}$. See, e.g., Lindblad and Sogge [10]. In this paper it was also shown that if

$$p_h = 1 + 4n/(n+1)(n-1),$$

then global solutions of (1) exist in the non-radial case if $p_h < p < p_{\text{conf}}$. The power $p_h$ arises in a natural way if one considers estimates for the d'Alembertian which do not involve mixed-norms. Specifically, Harmse [2] showed that for the inhomogeneous wave equation $\Box w = F$ with zero Cauchy data one has

$$\|w\|_{L^q(\mathbb{R}^{1+n}_+)} \leq C_q \, \|F\|_{L^p(\mathbb{R}^{1+n}_+)},$$

if and only if

$$(n+1)\left(\frac{1}{p} - \frac{1}{q}\right) = 2\,, \text{and} \quad \frac{2n}{n-1} < q < \frac{2n(n+1)}{(n-1)^2}\,.$$

The power $p_h$ arises since the exponents in Harmse's inequality satisfy

$$\frac{q}{p} < \frac{2n}{n-1} \Big/ \frac{2n(n+1)}{n^2+4n-1} = p_h\,.$$

Because of this, without using mixed-norm estimates, one seems to only be able to handle in a straightforward manner nonlinearities involving powers larger than $p_h$.

Nonetheless, by using mixed-norm estimates to follow, we can handle all other possible smaller powers when $n \leq 8$:

**Theorem 2.** *Let $3 \leq n \leq 8$ and assume that $p_c < p \leq p_h$, where $p_h$ is as above. Then, there is an $\varepsilon_0(p) > 0$ so that, if we write $x = r\omega$, $r > 0$, $\omega \in S^{n-1}$, then* (1) *always has a global (weak) solution verifying*

$$\sup_{t>0} \|r^{(n-1)/2} u(t,\,\cdot\,)\|_{L^q_r L^{2n/(n-1)}_\omega} + \sum_{|\alpha|\leq 1} \|r^{(n+1)/2p} \partial^\alpha_\omega u\|_{L^{pq}_t L^p_r L^{2n/(n-1)}_\omega} < \infty\,,$$

*if $q$ is as in* (3) *and*

$$(3') \qquad \sum_{|\alpha|\leq 3} \|r^{(n-1)/2} \partial^\alpha_\omega f\|_{L^q_r L^2_\omega} + \sum_{|\alpha|\leq 2} \|r^{(n+1)/2} \partial^\alpha_\omega g\|_{L^q_r L^2_\omega} < \varepsilon_0(p)\,.$$

One can actually get by with less regularity than that in (3′). If $p > 2$ one could use one less derivative, while if $p < 2$ one can use a condition involving $\Lambda^{2n/(n-1),\infty}_p(S^{n-1})$ spaces. (See [17] for definitions.)

If one assumes more regularity on the data, then $u$ may be smoother. For instance, we shall see later that, if $n = 3$, then a global $C^2$ solution of (1) exists if $(f, g)$ are in $C^3(\mathbb{R}^3) \times C^2(\mathbb{R}^3)$ and satisfy conditions somewhat weaker than those given by (3′). This will quantify the smallness assumption needed in John's theorem.

Before we indicate how these results are proved, let us try to explain the role of $p_c$, and, at the same time, motivate our assumptions on the data. To do this, let us assume that, instead of assuming (3), we wished to show that global solutions always exist if the data have small Sobolev norms of the appropriate order. To take into account scaling considerations, it is natural to use homogeneous Sobolev spaces and measure the data using norms of the form

$$\begin{aligned} \|(f,g)\|_{\dot\alpha} &= \|f\|_{\dot H^\alpha(\mathbb{R}^n)} + \|g\|_{\dot H^{\alpha-1}(\mathbb{R}^n)} \\ &= \big\| \, |D_x|^\alpha f \, \big\|_{L^2(\mathbb{R}^n)} + \big\| \, |D_x|^{\alpha-1} g \, \big\|_{L^2(\mathbb{R}^n)} . \end{aligned}$$

To proceed, notice that, if $u(t,x)$ solves (1) with data $(f,g)$, then, for a given $\varepsilon > 0$, $u_\varepsilon(t,x) = \varepsilon^{-2/(p-1)} u(t/\varepsilon, x/\varepsilon)$ solves the same equation, but with data $(f_\varepsilon, g_\varepsilon) = \varepsilon^{-2/(p-1)}(f(x/\varepsilon), \varepsilon^{-1} g(x/\varepsilon))$. Note that

$$\|(f_\varepsilon, g_\varepsilon)\|_{\dot\alpha} \, / \, \|(f,g)\|_{\dot\alpha} = \varepsilon^{n/2-2/(p-1)-\alpha} .$$

If $c = 1$ it is easy to construct spherically symmetric $C_0^\infty$ data for which $u$ blows up in finite time. Hence, letting $\varepsilon \to 0$ or $\varepsilon^{-1} \to 0$, one sees that a necessary condition on $\alpha$ which might possibly allow that (1) admits a global solution if $\|(f,g)\|_{\dot\alpha}$ is small is that $\alpha$ must equal

$$\alpha(p) = n/2 - 2/(p-1) .$$

On the other hand, even in the radial case, no such result can hold if $p < p_c$. This is easy to see. For if $0 \le \chi \in C_0^\infty(\mathbb{R})$ equals one near 0, then

$$g(x) = \chi(|x|)\, |x|^{-n/2+\delta-1} \in \dot H^{\alpha-1} , \forall\, \alpha < \delta . \tag{5}$$

On the other hand, it is possible to see that the solution to the free wave equation $\Box u_0 = 0$ with data $(0, \varepsilon g)$ is dominated by $\varepsilon \, |\, t - |x|\, |^{\delta - 1/2}$ if $t \approx 1$ and $(t,x)$ is near the forward light cone $(|x|, x)$, $x \in \mathbb{R}^n$, and hence

$$|u_0|^p \notin L^1_{\mathrm{loc}}(\mathbb{R}^{1+n}_+) \text{ if } \ (\delta - 1/2)p \le -1 . \tag{6}$$

It turns out that for data of this type $u \ge u_0$ near the light cone if $u$ is a weak solution of (1) with $c > 0$. Consequently we cannot even find a local distributional solution with this data if $\delta \le 1/2 - 1/p$, because, in order for the equation to hold in the sense of distributions, one needs the nonlinear term to belong to $L^1_{\text{loc}}$. Recalling (5), we conclude that, even for local existence, in the radial case one must assume that $(f,g) \in \dot{H}^{\alpha}(\mathbb{R}^n) \times \dot{H}^{\alpha-1}(\mathbb{R}^n)$ with

$$\alpha > 1/2 - 1/p\,.$$

If we combine this with our observations regarding the scaled equations, we see that, even in the spherically symmetric case, if a theorem saying that $\|(f,g)\|_{\dot{\alpha}(p)} < \varepsilon_0(p)$ implies global existence for (1), then one must have that

$$n/2 - 2/(p-1) = \alpha(p) > 1/2 - 1/p\,. \tag{2'}$$

This turns out to be equivalent to the condition that $p > p_c$ since it can be rewritten as the condition that

$$(n-1)p^2 - (n+1)p - 2 > 0\,.$$

When $n \le 4$ one can use Theorem 1 and one-dimensional Sobolev estimates to see that, under the assumption of spherical symmetry, if $p > p_c$ then there must be an $\varepsilon_0(p) > 0$ so that if

$$\|(f,g)\|_{\dot{\alpha}(p)} < \varepsilon_0(p)\,, \tag{3''}$$

then (1) must have a global solution. Without the assumption of spherical symmetry more regularity is needed even for local existence as was shown in [10]. In higher dimensions $n > 4$, it turns out that $p_c < 2$ and so for $p$ close to $p_c$, $\alpha(p) < 0$. For this reason a theorem involving assumptions regarding the Sobolev norms on the data seems less natural here. On the other hand, our assumption (3), which is weaker than (3″) when $p > 2$, involves norms which scale exactly like those in (3″).

A related remark is that, in both Theorem 1 and Theorem 2, our decay assumptions on the data are of the best possible nature. Notice that (3) is always satisfied, if, for a given $\delta > 0$,

$$(1+r)^{2/(p-1)+\delta}\left(|f(r)| + r|g(r)|\right) < \varepsilon_0(p,\delta)\,,$$

and hence for data of this type global solutions to (1) always must exist if $p > p_c$. On the other hand, the above arguments can be used to show that if $\delta < 0$ no such result can hold, since there cannot be a global solution to (1) if $f = 0$ and $g = \varepsilon \log(1+|x|)(1+|x|)^{-2/(p-1)}$. For the same reasons, the assumptions in Theorem 2 are natural and both results allow the right type of local singularities and of course are weaker than pointwise estimates of the above form since they involve $L^q$ rather than $L^\infty$ norms.

The proof of Theorems 1 and 2 relies on certain inequalities, which, at least in the spherically symmetric case, can be thought of as generalizations of the energy inequality for the d'Alembertian.

To be more specific, let us recall that if $\Box u_0 = 0$ and if $u_0$ has Cauchy data $u_0|_{t=0} = f$ and $\partial_t u_0|_{t=0} = g$, then

$$\|u_0'(t,\cdot)\|_{L^2(\mathbb{R}^n)} \le \|\partial_x f\|_{L^2(\mathbb{R}^n)} + \|g\|_{L^2(\mathbb{R}^n)}\,.$$

Since $|D_x|^{-1} = (-\Delta_x)^{-1/2}$ commutes with $\Box$, this leads to the following equivalent version:

$$\|u_0(t,\cdot)\|_{L^2(\mathbb{R}^n)} \le \|f\|_{L^2(\mathbb{R}^n)} + \|\,|D_x|^{-1} g\,\|_{L^2(\mathbb{R}^n)}\,. \tag{7}$$

Since the volume element $dx$ is $r^{n-1}dr d\omega$ in polar coordinates $x = r\omega$, it is not surprising that, under the assumption of spherical symmetry,

(7′)

$$\|\,r^{(n-1)/2}u_0(t,\cdot)\|_{L^2(dr)} \le C\|r^{(n-1)/2}f\|_{L^2(dr)} + C\|\,\partial_r^{-1}(\,r^{(n-1)/2}g\,)\,\|_{L^2(dr)}\,.$$

Our main inequality says that this extends to other Lebesgue spaces:

$$\begin{aligned} &\|\,r^{(n-1)/2}u_0(t,\cdot)\|_{L^q(\mathbb{R}_+)} \\ &\quad\le C_q\|r^{(n-1)/2}f\|_{L^q(\mathbb{R}_+)} + C_q\|\,\partial_r^{-1}(\,r^{(n-1)/2}g\,)\,\|_{L^q(\mathbb{R}_+)}\,, \quad 1<q<\infty\,. \end{aligned} \tag{7''}$$

The proof of Theorem 1 of course also requires estimates for the inhomogeneous wave equation. Specifically, if $w$ has zero Cauchy data at $t=0$ and if $\Box w = F$, then, under the assumption of spherical symmetry,

$$\begin{aligned} &\|r^{(n+1)/2p}w\|_{L_t^{pq}L_r^p(\mathbb{R}_+\times\mathbb{R}_+)} + \sup_{t>0}\|r^{(n-1)/2}w(t,\cdot)\|_{L_r^q(\mathbb{R}_+)} \\ &\qquad\le C_p\|r^{(n+1)/2}F\|_{L_t^q L_r^1(\mathbb{R}_+\times\mathbb{R}_+)}\,, \end{aligned} \tag{8}$$

provided that $p_c < p < p_{\text{conf}}$ and $q = q(p)$ is as in (3).

It turns out that we can also dominate the $L_t^{pq}L_r^p$ norm of $r^{(n+1)/2p}u_0$ by the right side of (7″). If we combine this along with (7″) and (8), we get our main estimate:

**Theorem 3.** *Let $n \ge 3$ and assume that $f(x)$, $g(x)$, and $F(t,x)$ are spherically symmetric functions of $x \in \mathbb{R}^n$, and let $v(t,x)$ be the solution of the inhomogeneous Cauchy problem*

$$\begin{cases} \Box v = F\,, & (t,x)\in\mathbb{R}_+^{1+n} \\ v|_{t=0} = f\,, & \partial_t v|_{t=0} = g\,. \end{cases} \tag{9}$$

*Then, if* $p_c < p < p_{\text{conf}}$ *and if* $q = 2(p-1)/((n+3)-(n-1)p)$,

$$(10)\quad \|r^{(n+1)/2p}v\|_{L_t^{pq}L_r^p} + \sup_{t>0}\|r^{(n-1)/2}v(t,\cdot)\|_{L_r^q} \le C_p\|r^{(n-1)/2}f\|_{L_r^q} + C_p\|\partial_r^{-1}(r^{(n-1)/2}g)\|_{L_r^q} + C_p\|r^{(n+1)/2}F\|_{L_t^qL_r^1}.$$

A related inequality is due to Klainerman and Machedon [5]. The dual version of their inequality says that when $n=3$ the second term in the left side of (8) is controlled by the right side if $q=2$. A similar inequality was also obtained by the first author in [9].

Let us explain the main ideas that are required for the proof of (10). For the sake of simplicity, let us assume that $F=0$, since the proof of the estimates (8) for the solution of the inhomogeneous wave equation with zero Cauchy data is similar to the corresponding ones for $u_0$. We shall also assume, for the sake of notation, that $g=0$, since the argument for $g\neq 0$ is similar. The argument for odd $n$ is more straightforward, so let us discuss this case first.

If $n$ is odd and $g$ vanishes then

$$(11)\quad 2r^{(n-1)/2}u_0(t,r) = (r+t)^{(n-1)/2}f(r+t) + (r-t)^{(n-1)/2}f(|r-t|) - \int_{|t-r|}^{t+r} P'_{(n-3)/2}(\mu)\,\frac{t}{r\rho}\,\rho^{(n-1)/2}f(\rho)\,d\rho\,,$$

where $P'_k$ denotes the derivative of the $k$th Legendre polynomial,

$$P_k(\mu) = \frac{1}{k!2^k}\frac{d^k}{d\mu^k}(1-\mu^2)^k\,,$$

and

$$\mu = \mu(t,r) = (r^2+\rho^2-t^2)/2r\rho\,.$$

Note that $-1\le\mu\le 1$ on the support of the integral in (11). Therefore, if we wish to prove (7″), it suffices to see that if $h(\rho)=\rho^{(n-1)/2}|f(\rho)|$, then

$$(12)\quad \Big\|\int_{|t-r|}^{t+r}\frac{t}{r\rho}h(\rho)\,d\rho\,\Big\|_{L_r^q} \le C_q\|h\|_{L_r^q}\,,\ 1<q<\infty\,,$$

since clearly the $L_r^q$ norm of each of the first two terms in the right side of (11) is smaller than $2\|r^{(n-1)/2}f\|_{L^q}$.

To prove (12), we need to consider two cases: $2r>t$ and $2r<t$. Since $t/r\rho < 2/\rho$ in the first case, (12) follows here from the Hardy inequality:

$$(13)\quad \Big\|\int_r^\infty \rho^{-1}h(\rho)\,d\rho\,\Big\|_{L^q(\mathbb{R}_+)} \le C_q\|h\|_{L^q(\mathbb{R}_+)}\,,\ 1\le q<\infty\,.$$

For the other case, we note that if $r < t/2$, then, in the support of the integral in (12), $\rho > t - r > t/2$, and so the estimate here follows from the Hardy-Littlewood maximal theorem, since the integral in (12) is $\le 2\mathcal{M}h(t-r)$, where

$$\mathcal{M}h(s) = \sup_{r>0} \frac{1}{r} \int_s^{s+r} h(\rho)\, d\rho\,,$$

denotes the Hardy-Littlewood maximal function of $h$.

The fact that we can also control the $L_t^{pq} L_r^q$ norm of $r^{(n+1)/2p} u_0(t,r)$ follows from the above arguments and the Hardy-Littlewood inequality for fractional integrals. To be more specific, let

$$\sigma = \sigma(p) = (n-1)/2 - (n+1)/2p\,,$$

so that $r^{(n+1)/2p} u_0 = r^{-\sigma} r^{(n-1)/2} u_0$. Then a calculation gives that

$$p\sigma = 1 - (p/q - 1/q) > 0\,. \tag{14}$$

If we call the last term in (11) $Tf(t,r)$, then the above arguments yield

$$\begin{aligned} \|r^{-\sigma} Tf\|^p_{L_t^{pq} L_r^p} &= \|r^{-\sigma p}(Tf)^p\|_{L_t^q L_r^1} \\ &\le C \Big\| \int_0^\infty r^{-\sigma p} \Big( \int_{|t-r|}^\infty \rho^{-1} h(\rho)\, d\rho \Big)^p dr \Big\|_{L_t^q} \\ &\quad + C \Big\| \int_0^\infty r^{-\sigma p} (\mathcal{M}h(t-r))^p\, dr \Big\|_{L_t^q}\,, \end{aligned}$$

if as before $h = \rho^{(n-1)/2}|f(\rho)|$. Since $1 < q/p < q < \infty$, the Hardy-Littlewood inequality for fractional integrals says that convolution with $r^{-\sigma p}$ maps $L^{q/p}$ into $L^q$, on account of (14). Therefore, the right side of the last inequality is dominated by

$$\begin{aligned} &\Big\| \Big( \int_r^\infty \rho^{-1} h(\rho)\, d\rho \Big)^p \Big\|_{L_r^{q/p}} + \big\| (\mathcal{M}h(r))^p \big\|_{L_r^{q/p}} \\ &\qquad = \Big\| \int_r^\infty \rho^{-1} \rho^{(n-1)/2} |f(\rho)|\, d\rho \Big\|^p_{L_r^q} + \big\| \mathcal{M}(\rho^{(n-1)/2} f)(r) \big\|^p_{L_r^q}\,. \end{aligned}$$

Consequently, if we apply Hardy's inequality and the Hardy-Littlewood maximal inequality, we conclude that $r^{-\sigma}$ times the last term in (11) satisfies the desired mixed norm estimates. Since the Hardy-Littlewood fractional integral estimate implies that the other two terms in the right side of (11) satisfy the same bounds, this completes our proof of (10) when $n$ is odd and both $F$ and $g$ vanish.

The proof of the corresponding inequality for even spatial dimensions is a bit more involved due to the lack of strong Huygen's principle for this case. It turns out, though, that the main term that one has to estimate involves the terms arising in the odd-dimensional case, plus another which looks like the last term in (11), except that $\rho^{(n-1)/2} f$ must be replaced there by the Hilbert transform of $\rho^{(n-1)/2} f$. Hence, the bound for the "main term," which involves an integral over $|t-r| < \rho < t+r$, follows from the above considerations and the boundedness of the Hilbert transform. There is also another term that involves an integral over $0 < \rho < |t-r|$, which is not present for odd $n$ because of the strong Huygen's principle there. This term can be estimated by arguments similar to those given for odd $n$, except here one needs to use more singular maximal inequalities of the type

$$\begin{aligned}(15)\quad \|\sup_{r>0} \frac{1}{r}\Big|\int_s^{s+r} \left(\frac{r}{s+r-\rho}\right)^{\delta} h(\rho)\, d\rho\,\Big|\,\|_{L^q(\mathbb{R})} \\ \le C_{\delta,q}\, \|h\|_{L^q(\mathbb{R})}\,,\ 1<q<\infty\,,\ \delta<1/q\,.\end{aligned}$$

This inequality follows easily from Hölder's inequality and the case $\delta = 0$, which is the Hardy-Littlewood maximal inequality.

The above arguments can easily be adapted to prove related estimates for non-radial functions. Specifically, we have the following

**Theorem 4.** *Let $n \ge 3$ and suppose that $v$ solves the inhomogeneous Cauchy problem* (9). *Then, if we write $x = r\omega$, $r > 0$, $\omega \in S^{n-1}$, and if $p_c < p < p_h$ and $q = 2(p-1)/((n+3)-(n-1)p)$, then*

(10′)

$$\begin{aligned}&\|r^{(n+1)/2p} v\|_{L_t^{pq} L_r^p L_\omega^{2n/(n-1)}} + \sup_{t>0} \|r^{(n-1)/2} v(t,\,\cdot\,)\|_{L_r^q L_\omega^{2n/(n-1)}} \\ &\le C_p \sum_{|\alpha|\le 1} \|r^{(n-1)/2} \partial_\omega^\alpha f\|_{L_r^q L_\omega^{2n/(n+1)}} + C_p \|r^{(n+1)/2} g\|_{L_r^q L_\omega^{2n/(n+1)}} \\ &+ C_p \|r^{(n+1)/2} F\|_{L_t^q L_r^1 L_\omega^{2n/(n+1)}}\,.\end{aligned}$$

The restriction that $p < p_h = 1 + 4n/(n+1)(n-1)$ seems to be natural in view of the bounds for $\mathcal{A}_\mu$ below. When $p = p_h$ a variant of (10′) holds where in the last term the exponent $2n/(n+1)$ is replaced by any larger one. This turns out to be adequate for proving the part of Theorem 2 where $p = p_h$.

The proof of (10′) is similar to that of (10) if one uses known bounds for the angular part of the solution operator. For instance, if, say $n$ is odd and both $g$ and $F$ vanish, then $v = u_0$, where

(11′)

$$2r^{(n-1)/2}u_0(t,r\omega) = (r+t)^{(n-1)/2}f((r+t)\omega) + (r-t)^{(n-1)/2}f((r-t)\omega) - \int_{|t-r|}^{t+r} \frac{t}{r\rho}\,(\partial_\mu \mathcal{A}_\mu f)(t,r,\rho;\omega)\,\rho^{(n-1)/2}\,d\rho\,,$$

with the angular operator involved given by

$$(\mathcal{A}_\mu f)(t,r,\rho;\omega) = \frac{1}{2(2\pi)^{(n-1)/2}} \int_{S^{n-1}} \delta^{((n-3)/2)}(\omega\cdot\sigma-\mu) f(\rho\sigma)\,d\sigma\,,$$

if $\mu$ is as above. Here $\delta^{(j)}$, $j = 0,1,2,\ldots$, denotes the $j$th derivative of the Dirac delta distribution. Note that, using this formula, one can derive the special case (11).

For $-1 < \mu < 1$, it turns out that the integral operator on $S^{n-1}$ with kernel $\delta^{((n-3)/2)}(\omega\cdot\sigma-\mu)$ is a Fourier integral operator of order $-1/2$, and hence its $\mu$-derivative, $\delta^{((n-1)/2)}(\omega\cdot\sigma-\mu)$, has order $1/2$. They share a canonical relation which is a canonical graph, and their singular support is a smooth hypersurface in $S^{n-1}\times S^{n-1}$ of codimension 1. Because of this, one can use arguments found for instance in [15], to obtain that, for fixed $t$, $r$, $\rho$,

$$\begin{aligned} &\|\mathcal{A}_\mu f\|_{L^{2n/(n-1)}_\omega} \le C\,(1-\mu^2)^{-(n-1)/2n}\,\|f\|_{L^{2n/(n+1)}_\omega}\,, \\ &\|\partial_\mu \mathcal{A}_\mu f\|_{L^{2n/(n-1)}_\omega} \le C\,(1-\mu^2)^{-1+2/n} \sum_{|\alpha|\le 1} \|\partial^\alpha_\omega f\|_{L^{2n/(n+1)}_\omega}\,. \end{aligned} \tag{16}$$

The first estimate is used in proving the estimate (8) for $w$ and the corresponding estimates for the part of $u_0$ coming from $g$, while the second is needed to handle the part coming from $f$.

To apply these bounds, one uses singular variants of the Hardy-Littlewood maximal theorem like (15), along with our earlier arguments, to show that

$$(T_\sigma h)(t,r) = \int_{|t-r|}^{t+r} \frac{t}{r\rho}\,(1-\mu^2)^{-\sigma}\,h(\rho)\,d\rho$$

satisfies

$$\|T_\sigma h\|_{L^{pq}_t L^p_r} + \sup_{t>0} \|T_\sigma h(t,\cdot)\|_{L^q_r} \le C_\sigma\,\|h\|_{L^p_r}\,, \quad \sigma < 1\,.$$

Combining this with the estimates for the angular operator yields the bounds for $u_0$. As before, the bounds for $w$ follow from similar arguments.

Also, the bounds for even spatial dimensions follow from similar arguments, but the proof is a bit more technical due to the lack of strong Huygen's principle.

Since the estimates in Theorem 3 respect the nonlinearity in our equation $\Box u = c|u|^p$, the global existence results under the assumption of spherical symmetry follow immediately, using standard iteration arguments. Our estimates for the non-radial case are less favorable, due to the fact that we have $L^{2n/(n-1)}_\omega$ in the left side of (10′) and $L^{2n/(n+1)}_\omega$ in the right, and, for large $n$,

$$\frac{2n}{n-1}\Big/\frac{2n}{n+1} = \frac{n+1}{n-1} \approx 1+\frac{2}{n} < p_c \approx 1+\frac{4}{n}\,.$$

Nonetheless, if we use Sobolev's lemma for $S^{n-1}$, we can obtain existence results for small $n$, using the fact that

$$\sum_{|\alpha|\le 1} \| \, r^{(n+1)/2}\partial^\alpha_\omega(|u|^p)\, \|_{L^q_t L^1_r L^{2n/(n+1)}_\omega} \le C \sum_{|\alpha|\le 1} \|r^{(n+1)/2p}\partial^\alpha_\omega u\|^p_{L^{pq}_t L^p_r L^{2n/(n-1)}_\omega}$$

if the norms are taken over any strip $[0,T]\times\mathbb{R}^n$ with $n\le 6$ and $p\le p_h$. Related arguments handle the case where $n=7$ or $n=8$.

As we have indicated, the above arguments are quite standard. Rather than presenting them here, let us instead show how, when $n=3$, a variant of (10′) can be used to obtain a generalization of John's existence theorem for $\mathbb{R}^3$ which better quantifies his smallness assumption.

The inequality we shall require says that, if $v$ solves the equation $\Box v = F$ in $\mathbb{R}_+\times\mathbb{R}^3$ with Cauchy data $(f,g)$, then

$$(17)\quad \|r^{2/p}v\|_{L^{pq}_t L^p_r L^\infty_\omega} + \sup_{t>0}\|rv(t,\,\cdot\,)\|_{L^q_r L^\infty_\omega}$$
$$\le C_p\|r^2F\|_{L^q_t L^1_r L^\infty_\omega} + C_p\|r^2 g\|_{L^q_r L^\infty_\omega} + C_p\sum_{|\alpha|\le 1}\|r\partial^\alpha_\omega f\|_{L^q_r L^\infty_\omega}\,,$$

assuming, as before, that, when $n=3$, $q=q(p)=(p-1)/(3-p)$. The proof of this result follows from that of (10′) if one merely replaces (16) in the argument with

$$(16')\quad \begin{aligned} &\|\mathcal{A}_\mu h\|_{L^\infty(S^2)} \le C\|h\|_{L^\infty(S^2)} \\ &\|\mathcal{A}'_\mu h\|_{L^\infty(S^2)} \le C(1-\mu^2)^{-1/2}\|\partial_\omega h\|_{L^\infty(S^2)}\,, \end{aligned}$$

with $\mathcal{A}^{(j)}_\mu$ denoting the integral operator on $S^2$ whose kernel is $\delta^{(j)}(\omega\cdot\sigma-\mu)$. As a side remark, we should point out that the corresponding $L^\infty_\omega$ estimates in higher dimensions are much less favorable, since for $S^{n-1}$, with $n$ odd, say, one would have to sum over derivatives of order $0 \le |\alpha| \le (n-3)/2$ in the right side of the analogue of the first estimate, and sum over derivatives of order $1 \le |\alpha| \le (n-1)/2$ in the the right side of the analogue of the second. For this reason, it seems that, in higher dimensions, one cannot expect to take supremum norms for the angular variables in the proof of the existence theorems considered here.

Using (16) we can obtain the following extension of John's theorem:

**Theorem 5.** *Suppose that $1+\sqrt{2} < p < 3$. Then there is an $\varepsilon_0 = \varepsilon_0(p) > 0$ so that (1) has a global $C^2$ solution if the Cauchy data $(f,g)$ is in $C^3(\mathbb{R}^3) \times C^2(\mathbb{R}^3)$ and satisfies*

$$\|r^2 g\|_{L_r^{(p-1)/(3-p)} L_\omega^\infty} + \sum_{|\alpha|\le 1} \|r\partial_\omega^\alpha f\|_{L_r^{(p-1)/(3-p)} L_\omega^\infty} < \varepsilon_0 . \tag{18}$$

As before, the decay assumptions are of the best possible nature since they allow data satisfying bounds of the form

$$(1+|x|)^{2/(p-1)+\delta}\big(\,|x|\,|g(x)| + \sum_{|\alpha|\le 1} |\partial_\omega^\alpha f(x)|\,\big) < \varepsilon_0(p,\delta)\,, \tag{18$'$}$$

if $\delta > 0$. Also, notice that the only quantities which enter this condition (or the weaker version (18)) are the ones which seem to allow blow-up. This is because John's blow-up construction in [4] shows that if the free solution $\Box u_0 = 0$ with the same data are too large and positive then there must be blow-up, and the quantities in (18) and (18′) are exactly those which determine the size of $u_0$. In other words, as suggested by John's blow-up result, for existence, the size of first and second order derivatives of $g$ and second and third order derivatives of $f$ is irrelevant.

To prove Theorem 5, we may, for the sake of notation, of course assume that $c = \pm 1$ in (1). More important, we may also assume that the data have compact support. If not, we can obtain a global solution by a limiting argument since the value of $u$ at $(t,x)$ depends only on the part of the data inside the ball of radius $t$ centered at $x$. Under the assumption of compactly supported data, standard local existence theorems for semilinear equations in $\mathbb{R}_+ \times \mathbb{R}^3$ (see, e.g., [8], [16]) say that there is always a $C^2$ solution $u$ in some half-open strip $[0,T_*) \times \mathbb{R}^3$. Moreover, if $u$ cannot be extended to a $C^2$ solution in a larger strip, one must have that $u \notin L^\infty([0,T_*) \times \mathbb{R}^3)$, that is, $u$ must blow-up pointwise. Thus, if we wish to prove Theorem 5, it suffices to show that if $u$ solves (1) in the above strip, then

$$u \in L^\infty([0,T_*) \times \mathbb{R}^3)\,, \tag{19}$$

provided that (18) holds for our compactly supported data, with $\varepsilon_0$ sufficiently small, but independent of the support.

To show that (19) must hold if the data are small enough, we shall use a standard continuity argument. If $C = C_p$ denotes the constant in (17), then, assuming, as we may, that $C > 1$, we claim that for $0 < T < T_*$

$$A(T) = \|r^{2/p}u\|_{L_t^{pq}L_r^pL_\omega^\infty(S_T)} + \sup_{0<t<T} \|ru(t,\,\cdot\,)\|_{L_r^qL_\omega^\infty} \le 2C\varepsilon_0\,, \tag{20}$$

provided that the data satisfy (18) with $\varepsilon_0$ sufficiently small. Since $u$ is $C^2$ and has relatively compact support in $[0, T_*) \times \mathbb{R}^3$, clearly (20) must hold for small $T > 0$. Thus, if we could show that bounds of the form $A(T) < 4C\varepsilon_0$ imply (20) for small $\varepsilon_0$ we would get our claim, since the set of $T \in [0, T_*)$ for which (20) holds would have to be both open and relatively closed, and hence would have to be the entire interval.

Using (17) with $F = \pm|u|^p$ we see that if $A(T) < 4C\varepsilon_0$, then

$$\begin{aligned} A(T) &\le C\varepsilon_0 + C\|r^2|u|^p\|_{L_t^qL_r^1L_\omega^\infty(S_T)} \\ &\le C\varepsilon_0 + C\|r^{2/p}u\|^p_{L_t^{pq}L_r^pL_\omega^\infty(S_T)} \\ &= C\varepsilon_0 + C(A(T))^p \le C\varepsilon_0 + C(4C\varepsilon_0)^{p-1}A(T)\,. \end{aligned}$$

Hence, if $C(4C\varepsilon_0)^{p-1} < 1/2$, we obtain (20), as claimed.

If $\varepsilon_0$ is small enough we can also bound the norms of first and second order spatial derivatives. In fact, if we use Hölder's inequality and (17), we see that whenever $0 < T < T_*$,

$$\begin{aligned} &\|r^{2/p}\nabla_x u\|_{L_t^{pq}L_r^pL_\omega^\infty(S_T)} + \sup_{0<t<T}\|r\nabla_x u(t,\,\cdot\,)\|_{L_r^qL_\omega^\infty} \\ &\le C\|r^2\nabla_x g\|_{L_r^qL_\omega^\infty} + C\sum_{|\alpha|\le 1}\|r\partial_\omega^\alpha\nabla_x f\|_{L_r^qL_\omega^\infty} + C\|r^2|u|^{p-1}\nabla_x u\|_{L_t^qL_r^1L_\omega^\infty(S_T)} \\ &\qquad \le C\|r^2\nabla_x g\|_{L_r^qL_\omega^\infty} + C\sum_{|\alpha|\le 1}\|r\partial_\omega^\alpha\nabla_x f\|_{L_r^qL_\omega^\infty} \\ &\qquad\qquad + C(A(T))^{p-1}\|r^{2/p}\nabla_x u\|_{L_t^{pq}L_r^pL_\omega^\infty(S_T)}\,. \end{aligned}$$

The terms involving $f$ and $g$ need not be small; however, they are finite since we are assuming that $(f, g) \in C^3 \times C^2$ have compact support. Hence, if $C(2C\varepsilon_0)^{p-1} < 1/2$, we conclude from (20) and the last inequality that

$$\begin{aligned} &\|r^{2/p}\nabla_x u\|_{L_t^{pq}L_r^pL_\omega^\infty(S_T)} + \sup_{0<t<T}\|r\nabla_x u(t,\,\cdot\,)\|_{L_r^qL_\omega^\infty} \\ &\qquad \le 2C\|r^2\nabla_x g\|_{L_r^qL_\omega^\infty} + 2C\sum_{|\alpha|\le 1}\|r\partial_\omega^\alpha\nabla_x f\|_{L_r^qL_\omega^\infty}\,. \end{aligned} \tag{21}$$

To bound second derivatives, let us note that

$$\sum_{|\alpha|\le 2} \big|\,\partial_x^\alpha(\,|u|^p\,)\,\big| \le C_0|\nabla_x u|^2\,|u|^{p-2} + C_0 \sum_{|\alpha|\le 2} |u|^{p-1}|\partial_x^\alpha u|\,.$$

Therefore, if we argue as above, we see that

$$\begin{aligned}\sum_{|\alpha|\le 2} \|r^{2/p}\partial_x^\alpha u\|_{L_t^{pq}L_r^pL_\omega^\infty(S_T)} &+ \sup_{0<t<T} \|r\partial_x^\alpha u(t,\,\cdot\,)\|_{L_r^qL_\omega^\infty} \\ &\le 2C\sum_{|\alpha|\le 2}\|r^2\partial_x^\alpha g\|_{L_r^qL_\omega^\infty} + 2C\sum_{|\gamma|\le 1}\sum_{|\alpha|\le 2}\|r\partial_\omega^\gamma\partial_x^\alpha f\|_{L_r^qL_\omega^\infty} \\ &\quad + 2CC_0\|r^{2/p}u\|_{L_t^{pq}L_r^pL_\omega^\infty(S_T)}\,\|r^{2/p}\nabla_x u\|_{L_t^{pq}L_t^pL_\omega^\infty(S_T)}\,.\end{aligned}$$

The right side is bounded by a uniform constant because of (20), (21) and our assumptions on the data. Hence, in particular, we have

$$\sup_{0<t<T_*}\sum_{|\alpha|\le 2}\|r\partial_x^\alpha u(t,\,\cdot\,)\|_{L_r^qL_\omega^\infty} < \infty\,. \tag{22}$$

Since the data have compact support, $u(t,x)$ must vanish if $|x|$ is larger than a fixed constant depending on $T_*$ if $0<t<T_*$. Consequently, since $q>p>1+\sqrt{2}>2$, we can use Hölder's inequality in the $r$-variable to see that (22) gives

$$\sup_{0<t<T_*}\sum_{|\alpha|\le 2}\|\partial_x^\alpha u(t,\,\cdot\,)\|_{L^2(\mathbb{R}^3;\,dx)} < \infty\,.$$

By Sobolev's lemma, this in turn implies (19), which completes the proof of Theorem 5. □

We are grateful to Y. Zhou for helpful conversations on his work and related matters. We would also like to thank E. M. Stein for helping us see how the estimates in Theorem 4 can be used to prove the existence results for non-radial functions when $n=7$ and 8.

Further results and more detailed proofs will be given in a later paper.

## References

1. R. Glassey, *Existence in the large for* $\Box u = F(u)$ *in two dimensions*, Math. Z. **178** (1981), 233-261.
2. J. Harmse, *On Lebesgue space estimates for the wave equation*, Indiana Univ. Math. J. **39** (1990), 229–248.
3. L. Hörmander, *Non-linear hyperbolic differential equations*, Lund Univ. Lecture Notes, 1988.

4. F. John, *Blow-up of solutions of nonlinear wave equations in three space dimensions*, Manuscripta Math. **28** (1979), 235–265.
5. S. Klainerman and M. Machedon, *Space-time estimates for null forms and the local existence theorem*, Comm. Pure and Appl. Math **46** (1993), 1221–1268.
6. H. Kubo, *On the critical decay and power for semilinear wave equations in odd space dimensions*, Hokkaido Univ. preprint.
7. T.T. Li and X. Yu, *Life-span of classical solutions to fully nonlinear wave equations*, Comm. Partial Diff. Equations **16** (1991), 909-940.
8. H. Lindblad, *Blow-up for solutions of* $\Box u = |u|^p$ *with small initial data*, Comm. Partial Diff. Equations **15** (1990), 757–821.
9. ———, *A sharp counterexample to local existence of low regularity solutions to nonlinear wave equations*, Duke Math. J. **72** (1993), 503–539.
10. H. Lindblad and C. D. Sogge, *On existence and scattering with minimal regularity for semilinear wave equations*, J. Funct. Anal. **130** (1995), 357–426.
11. H. Pecher, *Nonlinear small data scattering for the wave and Klein-Gordan equations*, Math. Z. **185** (1984), 261–270.
12. M. A. Rammaha, *Finite-time blow-up for nonlinear wave equations in high dimensions*, Comm. Partial Diff. Equations **12** (1987), 677-700.
13. J. Schaeffer, *The equation* $\Box u = |u|^p$ *for the critical value of* $p$, Proc. Roy. Soc. Edinburgh **101** (1985), 31-44.
14. T. Sideris, *Nonexistence of global solutions to semilinear wave equations in high dimensions*, J. Diff. Equations **52** (1984), 378–406.
15. C. D. Sogge, *Fourier integrals in classical analysis*, Cambridge Univ. Press, Cambridge, 1993.
16. ———, *Lectures on nonlinear wave equations*, International Press (to appear).
17. E. M. Stein, *Singular integrals and differentiability properties of functions*, Princeton Univ. Press, Princeton, NJ, 1970.
18. W. Strauss, *Nonlinear scattering theory*, Scattering theory in mathematical physics, Reidel, Dordrect, 1979, pp. 53–78.
19. ———, *Nonlinear scattering at low energy*, J. Funct. Anal. **41** (1981), 110-133.
20. H. Takamura, *Blow-up for semilinear wave equations with slowly decaying data*, Math. Z. **217** (1994), 567-576.
21. Y. Zhou, *Blow up of classical solutions to* $\Box u = |u|^{1+\alpha}$ *in three space dimensions*, J. Diff. Equations **5** (1992), 21-32.
22. ———, *Cauchy problem for semilinear wave equations with small data in four space dimensions*, preprint.



PRINCETON UNIVERSITY
*E-mail address*: hans@math.princeton.edu

UNIVERSITY OF CALIFORNIA, LOS ANGELES
*E-mail address*: sogge@math.ucla.edu

# THE FADDEEV APPROACH TO INVERSE SCATTERING FROM A MICRO-LOCAL PERSPECTIVE

ANDERS MELIN

**Abstract.** One of the basic ideas in Faddeev's approach to inverse scattering for the Schrödinger operator in three dimensions was to consider direction dependent solutions to the perturbed Helmholtz equation. This led to the concept of generalized scattering matrix, obtained essentially by complexifying the momentum variables. It turned out that this object solved a non-linear system of $\overline{\partial}$-equations and that such systems could be used to characterize the scattering data. Also, the $\overline{\partial}$-methods gave a procedure to reconstruct the potential, and it made it possible also to establish some new uniqueness results in inverse scattering.

Although Faddeev's ideas gave rise to much progress in inverse scattering it is not easy to explain, on for example more physical grounds, the relevance of some objects considered in his theory. In my lecture I will try to give a justification of these methods, based on a microlocal decomposition of the wave operators. At the same time I will illustrate how the equations developed in the $\overline{\partial}$-approach may be formulated without passing to complex momentum. Instead I will use some simple arguments from PDE theory.

**Introduction.** We shall consider Schrödinger operators $H_V = H_0 + V$ on $\mathcal{H} = L^2(\mathbf{R}^n)$ when $n \geq 3$. Here $H_0 = -\Delta = \sum_1^n D_j^2$, where $D_j = -i\partial/\partial x_j$, and the potential $V$ is a bounded self-adjoint operator which may be non-local. It will satisfy some decay conditions that we are going to state later on. These conditions will guarantee the existence and completeness of the wave operators

$$W_\pm = \lim_{t\to\pm\infty} \exp(itH_V)\exp(-itH_0).$$

By the completeness of the wave operators we mean that the ranges of $W_+$ and $W_-$ are equal to the subspace of absolute continuity of $H_V$. Hence the scattering operator

$$S = S(V) = W_+^* W_-$$

is unitary. It commutes with $H_0$ and after conjugation by the Fourier transformation it may be represented as a direct integral $S(V) = \int_{k>0} S(V,k)\,dk$ of unitary operators (scattering matrices) on $L^2(S^{n-1})$.

Let $\mathcal{V}$ be some space of potentials satisfying the conditions above. Then we have a non-linear mapping, the scattering mapping

$$\mathcal{V} \ni V \mapsto S(V) \in \mathcal{U}(\mathcal{H}),$$

where $\mathcal{U}(\mathcal{H})$ is the set of unitary operators on $\mathcal{H}$. The inverse scattering problem is to compute the inverse of this mapping. This requires, of course, that the scattering mapping itself is well understood. First of all one has to prove that it is injective. In the case of local potentials this can often be done by a high frequency analysis of the scattering operator. In fact, when $k \to \infty$ then some matrix elements of $S(V,k)$ turn out to converge to the Fourier transform of $v$ when $V$ is multiplication by $v$. (See Faddeev [F1].) In the language of physics this means that two different potentials may be distinguished from each other in a scattering experiment if the energy is sufficiently large. However, the proof of injectivity does not give much insight into how the potential can be computed effectively from the scattering data, and it is of great interest to develop mathematical methods which allow one to obtain as much information as possible from an analysis of the scattering matrix in a bounded interval of the frequency variable $k$.

The task to characterize the image of $\mathcal{V}$ under the scattering mapping offers a much more complicated problem than the injectivity questions. The existing results about characterizations of scattering matrices (see [HN, Section 1.3] and [We]) are of little use in practical computations. Also, lack of stability results brings considerable difficulties to numerical investigations of the inverse problem.

When discussing the scattering mapping further it is convenient to introduce a metric in the set of Schrödinger operators. We say that $H_{V_1}$ and $H_{V_2}$ are within distance $\varepsilon$ from each other if $\varepsilon \geq 2$, or if $\varepsilon < 2$ and there is a unitary operator $U$ on $L^2(\mathbf{R}^n)$ such that

$$UH_{V_1} = H_{V_2}U, \quad \text{and} \quad \|U - I\| \leq \varepsilon.$$

The $\delta$-distance, $\delta(H_{V_1}, H_{V_2})$, between $H_{V_1}$ and $H_{V_2}$ is then the infimum of all $\varepsilon$ such that $H_{V_1}$ and $H_{V_2}$ are within distance $\varepsilon$ from each other. We let $(\mathcal{V}, \delta)$ denote the metric space where the distance between potentials is defined as the $\delta$-distance between the corresponding Schrödinger operators.

It is more or less obvious from the definition of the $\delta$-distance and the wave operators that the mapping $V \mapsto S(V)$ is Lipschitz continuous from $(\mathcal{V}, \delta)$ to $\mathcal{U}(\mathcal{H})$ with the uniform operator topology. It turns out that under suitable restrictions on the set $\mathcal{V}$ it is true that potentials of finite rank form a dense set in $(\mathcal{V}, \delta)$. Some precise results will be stated in Section 3. In the class of finite rank potentials it is relatively easy to solve the inverse scattering problem, and various attempts have been made to solve the inverse scattering problem also for local potentials by arguing as if the scattering matrix belonged to a finite rank perturbation of $H_0$ (see for example Holst [Ho]). However, the absence of results on stability properties for the inverse problem makes it difficult to develop these methods further.

The inverse scattering problem for a local potential is overdetermined in the sense that the integral kernel of the scattering operator depends on $2n - 1$ real parameters while the potential itself is a function of $n$ such parameters. Thus in order to recover the potential one should only need some restricted scattering data. In fact, it has been proved by Eskin and Ralston (see [ER1], [ER2] and the references given there) that for a generic set of potentials in dimensions $n = 2$ or $n = 3$ it is possible to determine the potential from the so-called backscattering data. These depend on $n$ parameters. We also mention that the inverse backscattering problem may be viewed as a non-linear perturbation of the inverse Radon transform. (See [M2, Section 7].)

However, most attempts to solve the inverse scattering problem have been carried out in strictly overdetermined situations. This was needed in order to compensate for lack of stability. One strategy for solving the inverse problem is that one in a first step replaces the scattering operator by some generalized scattering data which may depend on additional parameters. In a second step one tries to compute the potential from the new data. As an example of such methods we mention the $\overline{\partial}$-approach in which a generalized scattering matrix was computed by complexification of the momentum variables (see Section 5). From these data the potential could be computed by residue calculus, or simply by letting the momentum tend to infinity. Such methods were first introduced for systems of ordinary differential equations by Beals and Coifman [BC] and then for multi-dimensional inverse scattering by Ablowitz and Nachman [AN]. (See also Henkin-Novikov [HN], Newton [Ne], and Weder [We].) These methods also lead to some characterization of the scattering matrix. Although the inversion problem becomes easy in this approach, the computation of the generalized scattering matrix seems to be rather complicated, at least when it comes to numerical computations. In Section 5 we will discuss an alternative method which leads to the characterization results mentioned above.

One of the most important ingredients in Faddeev's theory of inverse scattering [F2] was the introduction of direction dependent fundamental solutions to Helmholtz' equation. This was also the starting point for the $\overline{\partial}$-approach. (We want to cite here also a recent paper of Isozaki [Is], where Faddeev's ideas have been worked out in a micro-local framework.) This allowed Faddeev to construct a family $\{A_\theta\}$ of intertwining operators between $H_V$ and $H_0$ parametrized by the unit sphere. We shall discuss such operators after analyzing the dynamics of free wave packets with concentrated momentum, and we shall see in Section 2 that the family $\{A_\theta\}$ may be viewed as some kind of micro-local decomposition of the wave operators $W_\pm$. This gives us a new motivation for considering the operators $A_\theta$ although we have not been able to motivate them from considerations in physics alone.

The intertwining operators $A_\theta$, which will play a central rôle in our presentation, may also be viewed as the natural generalizations to higher dimensions of the intertwining operators considered in one-dimensional scattering. Their importance there is obvious from the Gelfand-Levitan theory (see [Ma], [M1]). However, the Volterra technique used to construct such operators in the one-dimensional case has no simple counterpart in higher dimensions, where the constructions are technically much more complicated. In Section 3 we present some existence results for intertwining operators. (Complete proofs may be found in [M4].) The method we shall use is a plane wave decomposition of a family $E_\theta(x, y)$, $\theta \in S^{n-1}$, of fundamental solutions to the ultra-hyperbolic operator. This will be discussed to some extent in Section 4. The Fourier transforms of the $A_\theta$ with respect to the variable $y$ are the generalized solutions to Helmholtz' equation mentioned above. In Section 5 we present a system of integro-differential equations for the $A_\theta$ when these objects are treated as operator valued functions of $\theta \in S^{n-1}$. These equations are closely linked to the $\overline{\partial}$-equations in scattering theory and give an alternative approach to the characterization of the generalized scattering matrices in the papers [AN], [HN], [Ne], [We] and [Is]. Complete proofs of the results stated in this paper can be found in [M4].

Since scattering theory is a mathematical model for an experiment in physics, it is desirable that the mathematical analysis of the inverse problem also produces methods that can be applied in for example numerical computations. In the mathematical modelling one should therefore try to work as much as possible with quantities that have some meaning in physics. One of the flaws of Faddeev's approach as well as of the presentation given here is the lack of motivation for the intertwining operators considered here.

In Section 1 below we introduce some classes of operators that are natural to consider in scattering theory and we also present the decay and regularity assumptions on the potentials that we going to work with.

**1. Notation and definitions.** The operators we are going to consider will always be continuous from $C_0^\infty$ to $\mathcal{D}'$ at least. In view of the kernel theorem of Schwartz we may identify the space of such operators with the corresponding space of distribution kernels, and we shall often use the same notation for an operator and its distribution kernel. In these notations the intertwining equation

$$H_V A = A H_0$$

takes the form

$$(\Delta_x - \Delta_y) A(x,y) = (V \circ A)(x,y). \tag{1.1}$$

and the distribution kernel of the Fourier transform

$$u \mapsto \widehat{u}(\xi) = (\mathcal{F}u)(\xi) = \int e^{-i\langle x,\xi\rangle} u(x)\, dx,$$

is the function $(\xi, x) \mapsto e^{-i\langle x,\xi\rangle}$. The transpose and adjoint of an operator $U$ are defined by

$$U^{\tau}(x,y) = U(y,x), \quad U^*(x,y) = \overline{U(y,x)}.$$

We shall use the notation $\Sigma$ for the Schur class of operators that are continuous on $L^p$ for every $p \in [1,\infty]$ . (The corresponding distribution kernels are measures on $\mathbf{R}^n \times \mathbf{R}^n$.) The norm $\|U\|_\Sigma$ of $U \in \Sigma$ is the maximum over $p$ of the norm of $U$ as an operator from $L^p$ to $L^p$. Let $\mathcal{M}$ be the largest subspace of $\Sigma$ which is invariant under commutation with constant coefficient vector fields. Since we have identified operators with their distribution kernels we may also describe $\mathcal{M}$ as the set of $U$ such that $(\partial_x + \partial_y)^\alpha U \in \Sigma$ for every multi-index $\alpha$. We observe that $\Sigma$ is a Banach algebra while $\mathcal{M}$ is a Fréchet algebra.

We are going to consider localizations at infinity of operators in $\Sigma$ and $\mathcal{M}$. Set $\tau_z f(x) = f(x+z)$ when $f$ is a function and set

$$T_z A = \tau_{-z} \circ A \circ \tau_{-z}$$

when $A$ is an operator. This means that the distribution kernel of $T_z A$ equals $A(x-z, y+z)$. Let $\Sigma_{\text{loc}}$ be the space of operators that are continuous from $L^\infty$ to $L^\infty_{\text{loc}}$ and from $L^1_{\text{cpt}}$ to $L^1$. This is a Fréchet space under the semi-norms $\|\chi_K A\|_{L^\infty \mapsto L^\infty} + \|A\chi_K\|_{L^1 \mapsto L^1}$, where $\chi_K$ is (multiplication by) the characteristic function of a compact set $K$. The space $\mathcal{M}_{\text{loc}}$ is defined in a similar way.

**Definition 1.1.** Let $\theta \in S^{n-1}$. Then $\mathcal{M}_\theta$ is the set of all $A \in \mathcal{M}$ such that $T_z A \to 0$ in $\mathcal{M}_{\text{loc}}$ as $|z| \to \infty$ and $z/|z| \to \theta$.

We shall also need the definition of a *dominated family* parametrized by $\theta$.

**Definition 1.2.** Let $\{A_\theta\}_{\theta\in S^{n-1}}$ be a family of operators in $\mathcal{M}$. We say that it is dominated if for each multi-index $\alpha$ there is a $B_\alpha \in \Sigma$ such that $|(\partial_x + \partial_y)^\alpha A_\theta| \le B_\alpha$ for every $\theta$. (Here the left-hand side is interpreted as the absolute value of a measure.)

In addition to assuming that our potential operators are in $\mathcal{M}$ we shall also have to impose some decay conditions at infinity. If $V \in \mathcal{M}$, then $V(x, x+y)$ and $V(x-y, x)$ are smooth functions of $x$ with values in the Banach space $M = M(\mathbf{R}^n)$ of bounded measures on $\mathbf{R}^n$. We use the notations $V^\rho(x)$ and $V^\ell(x)$ for these measures and introduce the following decay conditions:

**Definition 1.3.** Let $0 \le \sigma \le 1$. Then $\mathcal{V}^\sigma$ is the set of all $V \in \mathcal{M}$ such that

$$\int_{\mathbf{R}^n} (1+|x|)^{\sigma(2-n+|\alpha|)} \left(\|\partial_x^\alpha V^\rho(x)\|_M + \|\partial_x^\alpha V^\ell(x)\|_M\right) dx < \infty$$

for every $\alpha$.

The spaces $\mathcal{V}^\sigma$ are Fréchet spaces. We use the notation $V^\sigma_{\mathrm{loc}}$ for the subspaces of local potentials. These are multiplications by functions $v$ such that

$$\int (1+|x|)^{\sigma(2-n+|\alpha|)} |v^{(\alpha)}(x)|\, dx < \infty.$$

If for example $\sigma = 1$ then it is easily seen that $|x|^2 v(x)$ is a bounded function. Hence our decay conditions are rather restrictive.

Finite rank perturbations of $H_0$ are important to consider in numerical investigations of the Schrödinger operator. The finite rank operators considered here will always be finite linear combinations of tensor products $f(x)g(y)$ where $f^{(\alpha)}, g^{(\alpha)} \in L^1(\mathbf{R}^n)$ for every $\alpha$. The space of such operators is denoted $\mathcal{V}^{\mathrm{fr}}$. This is obviously contained in $\mathcal{V}^0$.

**2. Expressing the wave operators in terms of the intertwining operators.** Let $H_V$ be a Schrödinger operator, where $V$ is bounded and self-adjoint on $L^2(\mathbf{R}^n)$. We say that a family $\{A_\theta\}_{\theta\in S^{n-1}}$ in $\mathcal{M}$ is *adapted* to $H_V$ if the following conditions are fulfilled:

(2.i) The $A_\theta$ are intertwining operators so that $H_V A_\theta = A_\theta H_0$ for every $\theta$.

(2.ii) The family $\{A_\theta\}$ is dominated in the sense of Definition 1.2, and $A_\theta$ is a continuous function of $\theta$ with values in $\mathcal{M}$.

(2.iii) $A_\theta - I \in \mathcal{M}_\theta$ for every $\theta$.

**Theorem 2.1.** *Assume that* $\{A_\theta\}$ *and* $V$ *are as above. Then the wave operators*

$$W_\pm = \lim_{t\to\pm\infty} \exp(itH_V)\exp(-itH_0)$$

*exist, at least in the weak sense, and the distribution kernels of* $W_\pm \circ \mathcal{F}^*$ *are given by*

$$(W_\pm \circ \mathcal{F}^*)(x, s\theta) = \int A_{\pm\theta}(x,y)e^{is\langle y,\theta\rangle}\,dy, \quad s > 0, \tag{2.1}$$

*where the integral is to be interpreted as the integral of a measure.*

*Proof.* It suffices for us to prove that $W_\pm f$ exist when $\widehat{f} \in C_0^\infty(\mathbf{R}^n \setminus 0)$, for these functions form a dense set in $L^2$. Choose $\chi \in \mathcal{S}(\mathbf{R}^n)$ so that $\chi(0) = 1$. Then $\chi(t^{-1/2}\cdot)f \to f$ in $L^2$. Set $\chi_{t,\eta}(x) = e^{i\langle x,\eta\rangle}\chi(t^{-1/2}x)$ when $t > 0$. It follows from Fourier's inversion formula that

$$e^{-itH_0}f \sim (2\pi)^{-n}\int \widehat{f}(\eta)e^{-itH_0}\chi_{t,\eta}\,d\eta, \tag{2.2}$$

in the sense that the difference between the two sides tends to 0 in $L^2$ as $t \to +\infty$. If $h = e^{-H_0}\chi$, then $h \in \mathcal{S}$, and a simple computation shows that

$$|e^{-itH_0}\chi_{t,\eta}(x)| = |h(t^{-1/2}x - 2t^{1/2}\eta)|. \tag{2.3}$$

We write the right-hand side of (2.2) as a sum $p_t + q_t$, where

$$p_t(x) = (2\pi)^{-n}\int \widehat{f}(\eta)A_\eta e^{-itH_0}\chi_{t,\eta}\,d\eta,$$

and

$$q_t(x) = (2\pi)^{-n}\int \widehat{f}(\eta)R_\eta e^{-itH_0}\chi_{t,\eta}\,d\eta.$$

Here $I = A_\eta + R_\eta$, and $A_\eta$ is extended from the unit sphere to a homogeneous function of degree 0 in $\eta$.

We have

$$e^{itH_V}p_t = (2\pi)^{-n}\int \widehat{f}(\eta)A_\eta\chi_{t,\eta}\,d\eta. \tag{2.4}$$

for the intertwining property (2.i) implies that $e^{itH_V}A_\eta = A_\eta e^{itH_0}$. When $t \to +\infty$ the right-hand side of (2.4) converges in $L^\infty_{\mathrm{loc}}$ to the corresponding expression with $\chi_{t,\eta}(x)$ replaced by $e^{i\langle x,\eta\rangle}$.

Set $h_{t,\eta}(x) = |h(t^{-1/2}x - 2t^{1/2}\eta)|$ and choose $0 \le R \in \Sigma$ so that $|R_\eta| \le R$ for every $\eta$. Then

$$|q_t| \le (2\pi)^{-n}\int |\widehat{f}(\eta)|Rh_{t,\eta}\,d\eta.$$

Since

$$t^{-n/2}\sup_{\eta}\|h_{t,\eta}\|_{L^1}+t^{n/2}\sup_{x}\int h_{t,\eta}(x)\,d\eta\le 2\|h\|_{L^1},$$

we have the estimates

$$t^{-n/2}\|q_t\|_{L^1}\le C\|h\|_{L^1},\quad t^{n/2}\|q_t\|_{L^\infty}\le C\|h\|_{L^1},$$

where $C = C_f$ is independent of $t$ and $h$. The first inequality can be improved. In fact,

$$t^{-n/2}\|q_t\|_{L^1}\le(2\pi)^{-n}\int|\widehat{f}(\eta)|\,\|t^{-n/2}R_\eta h_{t,\eta}\|_{L^1}\,d\eta,$$

where

$$\sup_{\eta\in\mathbf{R}^n,\,t\ge1}\|t^{-n/2}R_\eta h_{t,\eta}\|_{L^1}<\infty.$$

After subtracting from $h$ a function with an arbitrarily small norm in $L^1$ one achieves that $h_{t,\eta}$ is for large $t$ supported in a ball $B_{t,\eta}$ which is contained in an arbitrarily small conic neighbourhood of $\mathbf{R}_+\eta$ and escapes to infinity as $t\to+\infty$. Since $R_\eta\in\mathcal{M}_{\eta/|\eta|}$ it is easily seen that $\|t^{-n/2}R_\eta h_{t,\eta}\|_{L^1}\to 0$ for every $\eta$ as $t\to\infty$. It follows therefore from Lebesgue's theorem that $\|t^{-n/2}q_t\|_{L^1}\to 0$ as $t\to+\infty$. Combining this with the estimate for $q_t$ in $L^\infty$ we have proved therefore that $q_t\to 0$ in $L^2$ as $t\to+\infty$. Hence $W_+$ exists as a weak limit, and the distribution kernel $W_+\mathcal{F}^*$ satisfies the equation in (2.1). This completes the proof, for the assertion about $W_-$ is proved in exactly the same way.

**3. On the existence of intertwining operators.** The following result combined with the remarks made in the introduction about distances between operators is of interest in numerical computations of scattering data, since it implies that it is often possible to reduce such computations to the case of finite rank potentials.

**Theorem 3.1.** *Assume that $V\in\mathcal{V}^\sigma$ is self-adjoint and let $\varepsilon$ be an arbitrary positive number. Then one may find a unitary operator $U\in\mathcal{M}$ such that $\|U-I\|<\varepsilon$ and*

$$U^*H_VU=H_0+K,$$

*with $K\in\mathcal{V}^{\mathrm{fr}}$.*

One may choose $U$ so that $K$ depends continuously on $V$ when $V$ belongs to a bounded set in $\mathcal{V}^\sigma$. It is also possible to choose $U$ at the same time so that $K=0$ when $V$ is small. (For this to be true it is important that $n\ge3$.)

For small potentials, i.e. for potentials that are sufficiently close to the origin with respect to the Fréchet space topology in $\mathcal{V}^\sigma$, it is true that one may find a family $A_\theta$ satisfying the conditions (2.i)–(2.iii) in the previous section.

**Theorem 3.2.** *Assume that $V \in \mathcal{V}^\sigma$ and that $V$ is sufficiently small. Then there exists a family $\{A_\theta\}_{\theta \in S^{n-1}}$ which is adapted to $H_V$ in the sense that the conditions (2.i)–(2.iii) are fulfilled. Moreover, one may choose the $A_\theta$ so that these operators are invertible in $\mathcal{M}$.*

*Remark.* We do not have to assume that $V$ is self-adjoint. The operators $A_\theta$ are not uniquely determined by the conditions of Theorem 3.2. To see this we take $a \in \mathcal{S}(\mathbf{R}^n)$ and form

$$u(x,y) = \int_{\mathcal{O}(n)} a(x - Ty)\, dT,$$

where $\mathcal{O}(n)$ is the orthogonal group and the Haar measure $dT$ is normalized so that its total measure equals 1. It can be proved that

$$\int |u(x,y)|\, dy \to \int \Big| \int_{z \perp \phi} a(z + \tau\phi)\, dz \Big|\, d\tau$$

when $|x| \to \infty$, $x/|x| \to \phi$. If we choose $a$ so that $\hat{a}(s\theta) = 0$ when $s \in \mathbf{R}$ it follows that $U = u \circ u^* \in \mathcal{M}_\theta$. Since $(\Delta_x - \Delta_y)u = 0$ it is true that $(\Delta_x - \Delta_y)U = 0$. By letting $a$ depend smoothly on $\theta$ we may then construct a non-trivial dominated family $\{U_\theta\}$ in $\mathcal{M}$ so that $U_\theta \in \mathcal{M}_\theta$, $\theta \mapsto U_\theta$ is continuous and $[H_0, U_\theta] = 0$. But then $A'_\theta = A_\theta \circ (I + U_\theta)$ will satisfy conditions (2.i)–(2.iii) if this is true for the family $\{A_\theta\}$. By Theorem 3.3 it is possible to choose the $A_\theta$ invertible when $V \in \mathcal{V}^\sigma_{\text{loc}}$ is small and then $A'_\theta \neq A_\theta$ for some $\theta$.

In order to define a unique intertwining family of operators $A_\theta$ one has to impose extra conditions that are not easily motivated from physical considerations alone. The additional conditions we are going to introduce are support conditions on the distribution kernels of the $A_\theta$, and when doing this it will be necessary to assume that the potential is local.

**Theorem 3.3.** *Assume that $V \in \mathcal{V}^\sigma_{\text{loc}}$ and that $V$ is small. Then for every $\theta \in S^{n-1}$ there is a unique $A_\theta \in \mathcal{M}$ satisfying the following conditions:*

(3.i) *The intertwining condition $H_V A_\theta = A_\theta H_0$ is fulfilled.*
(3.ii) *If $(x,y) \in \operatorname{supp}(A_\theta)$ then $\langle x - y, \theta \rangle \geq 0$.*
(3.iii) $A_\theta - I \in \mathcal{M}_\theta$.

*The operator $A_\theta$ is invertible. Moreover, the family $\{A_\theta\}_{\theta \in S^{n-1}}$ is dominated in $\mathcal{M}$, and $A_\theta$ depends continuously on $\theta$. Hence $\{A_\theta\}$ is adapted to $H_V$.*

We shall discuss the existence proof in the next section. The uniqueness assertion is a consequence of the following lemma about solutions to the ultra-hyperbolic equation.

**Lemma 3.4.** *Assume that $U \in \mathcal{D}'(\mathbf{R}^n \times \mathbf{R}^n)$ solves the ultra-hyperbolic equation $(\Delta_x - \Delta_y)U = 0$ and that $\chi(x-y)U(x,y)$ is in $\mathcal{S}'(\mathbf{R}^n \times \mathbf{R}^n)$ when $\chi \in C_0^\infty(\mathbf{R}^n)$. If $\langle x-y, \theta\rangle \geq 0$ in the support of $U$ and $U(x+t_j\theta, y+t_j\theta) \to 0$ in $\mathcal{D}'$ for some sequence $(t_j)$, then $U = 0$.*

In order to see that the uniqueness assertion of Theorem 3.3 follows from the lemma we use the existence part of that theorem to construct some invertible $B$ in $\mathcal{M}$ which satisfies (3.i)–(3.iii) after $\theta$ has been replaced by $-\theta$. The uniqueness follows then if we prove that $U = 0$ when $U = B^\tau A_\theta - I$ and $A_\theta$ satisfies (3.i)–(3.iii). We observe that $\langle x - y, \theta\rangle \geq 0$ in the support of $U$, and it is easily verified that $U \in \mathcal{M}_\theta$. Since $(\Delta_x - \Delta_y)U(x,y) = 0$, by the intertwining equations for $A_\theta$ and $B$, it follows now from the lemma that $U = 0$.

We shall use the notation $A_{\theta,V}$ for the operators constructed in Theorem 3.3. It follows from the proof of the construction of the $A_{\theta,V}$ that $A_{\theta,zV}$ is an analytic function of $z \in \mathbf{C}$ with values in $\mathcal{M}$ when $z$ is small. It is clear that this function cannot in general be extended to an entire analytic function with values in $\mathcal{M}$, since the existence of an invertible intertwining operator between $H_{zV}$ and $H_0$ in that space excludes bound states of $H_{zV}$. However, if instead we consider $A_{\theta,zV}$ as a distribution valued function of $z$, then we have an object which is entire analytic in $z$:

**Theorem 3.5.** *Assume that $V \in \mathcal{V}^\sigma_{\text{loc}}$. Then $A_{\theta,zV}$, which a priori is defined for $z$ small, extends to an entire analytic function of $z$ with values in $\mathcal{D}'(\mathbf{R}^n \times \mathbf{R}^n)$.*

It follows from this theorem that $A_{\theta,V}$ is defined for arbitrary $\theta \in S^{n-1}$ and $V \in \mathcal{V}^\sigma_{\text{loc}}$. In order to describe the properties of the corresponding operators we let $\mathcal{M}_{\theta,\lambda}$ be the set of all $U \in \mathcal{D}'(\mathbf{R}^n \times \mathbf{R}^n)$ such that $e^{-\lambda\langle x-y,\theta\rangle}U \in \mathcal{M}_\theta$. Here $\lambda$ may be an arbitrary number $\geq 0$. We observe that $\mathcal{M}_{\theta,\lambda}$ is a Fréchet algebra, and its elements are continuous linear operators on the spaces $L^p_{\theta,\lambda}$ consisting of all $f$ such that $e^{-\lambda\langle x,\theta\rangle}f \in L^p$. Hence the $\mathcal{M}_{\theta,\lambda}$ are the analogues of $\mathcal{M}_\theta$ obtained after introducing exponential weight functions in the $L^p$ spaces.

**Theorem 3.6.** *Assume that $V \in \mathcal{V}^\sigma_{\text{loc}}$ and that $\lambda$ is large. Then $A_{\theta,V} - I \in \mathcal{M}_{\theta,\lambda}$ when $\theta \in S^{n-1}$.*

*Remark.* It follows by analytic continuation that $A_{\theta,V}$ is an isomorphism on $L^p_{\theta,\lambda}$ with inverse $A^\tau_{-\theta,V}$ when $\lambda$ is sufficiently large.

**4. Construction of the intertwining operators.** We shall construct $A_{\theta,V}$ first when $\theta$ is kept fixed. It is not necessary to require that $V \in \mathcal{V}^\sigma$ is local, but we assume that $\langle x - y, \theta\rangle \geq 0$ in the support of its distribution kernel. It is natural to make the 'Ansatz'

$$A_{\theta,V} = U_0 + U_1 + \dots,$$

where the $U_j$ are $j$-linear forms in $V$ with values in $\mathcal{S}'(\mathbf{R}^n \times \mathbf{R}^n)$ and $U_0(x,y) = \delta(x-y)$. If $A_{\theta,zV}$ is analytic in $z$ with values in $\mathcal{M}_\theta$ when $|z|$ is small and if (3.i)–(3.iii) holds, then it is necessary that

$$(\Delta_x - \Delta_y)U_j = V \circ U_{j-1}, \quad j = 1, 2, \dots \tag{4.1}$$

and that $U_j \in \mathcal{M}_\theta$ when $j > 0$. Also, it is necessary that $\langle x-y, \theta\rangle \geq 0$ in the support of $U_j$. It follows from Lemma 3.4 that there is at most one solution to (4.1) satisfying these conditions. We shall solve (4.1) by convolving the right-hand side by a fundamental solution $E_\theta$ of the ultra-hyperbolic operator which satisfies appropriate growth and support conditions. One has the following result:

**Theorem 4.1.** *Let $\theta \in S^{n-1}$. Then there is a unique $E_\theta \in \mathcal{S}'(\mathbf{R}^n \times \mathbf{R}^n)$ which satisfies the following conditions:*

(4.i) *$E_\theta$ is a fundamental solution of $\Delta_x - \Delta_y$.*
(4.ii) *The inequality $\langle x-y, \theta\rangle \geq 0$ holds in the support of $E_\theta$.*
(4.iii) *$E_\theta(x+s\theta, y+s\theta) \to 0$ in $\mathcal{D}'(\mathbf{R}^n \times \mathbf{R}^n)$ as $s \to \infty$.*

Our search for an $E_\theta$ satisfying these conditions was partly motivated by the Fourier decomposition of the free wave packets discussed in Section 2. The uniqueness assertion follows from Lemma 3.4. In order to construct $E_\theta$ in arbitrary dimension $n \geq 3$ we shall make use of the following decomposition of the Dirac measure into plane waves (see Section I.3.2 of [GS]):

$$\delta(x) = (n-1)!(2\pi i)^{-n} \int_{S^{n-1}} \frac{d\omega}{(\langle x,\omega\rangle - i0)^n}. \tag{4.2}$$

It is convenient here, as often in high frequency analysis of operators in scattering theory, to blow up the diagonal of $\mathbf{R}^n \times \mathbf{R}^n$ by considering the mapping

$$\gamma : \mathbf{R}^n \times \mathbf{R} \times S^{n-1} \ni (x, r, \omega) \mapsto (x - r\omega/2, x + r\omega/2).$$

Let $\dot{\mathcal{D}}'(\mathbf{R}^n \times \mathbf{R} \times S^{n-1})$ be the set of distributions supported in the set where $r \geq 0$. We have a surjective push-forward mapping on distributions

$$\gamma_* : \dot{\mathcal{D}}'(\mathbf{R}^n \times \mathbf{R} \times S^{n-1}) \mapsto \mathcal{D}'(\mathbf{R}^n \times \mathbf{R}^n)$$

defined by $\langle f, \gamma_* b\rangle = \langle f \circ \gamma, b\rangle$ when $f \in C_0^\infty(\mathbf{R}^n \times \mathbf{R}^n)$ and $b \in \dot{\mathcal{D}}'(\mathbf{R}^n \times \mathbf{R} \times S^{n-1})$. One may now prove the following lemma, which will allow us to separate variables in our construction of fundamental solutions.

**Lemma 4.2.** *Let $F \in \mathcal{D}'(\mathbf{R})$ and $G \in \mathcal{D}'(S^{n-1})$. Set*

$$b(x,r,\omega) = Y_+(r)F(\langle x,\omega\rangle)G(\omega)$$
$$c(x,r,\omega) = \delta(r)F'(\langle x,\omega\rangle)G(\omega)$$

*where $Y_+$ is the Heaviside function. Then*

$$(\Delta_x - \Delta_y)\gamma_* b = -2\gamma_* c,$$

*and $\gamma_* b$ is a fundamental solution for $\Delta_x - \Delta_y$ if*

$$\int G(\omega)F'(\langle x,\omega\rangle)\,d\omega = -2^{-1}\delta(x). \tag{4.3}$$

Now we construct a fundamental solution by letting $G = G_\theta$ be the characteristic function for the set where $\langle\omega,\theta\rangle \le 0$ and choosing $F$ so that $F'$ is even and

$$\int_{S^{n-1}} F'(\langle x,\omega\rangle)\,d\omega = -\delta(x).$$

Then (4.3) is fulfilled. It follows from (4.2) that we can take $F(s)$ equal to a positive constant times $(s+i0)^{-(n-1)} + (-1)^n(s-i0)^{-(n-1)}$. Let us set

$$\lambda_\nu(s) = 2^{-1}\left(\frac{1}{s+i0} + (-1)^\nu \frac{1}{s-i0}\right)$$

when $\nu$ is an integer and consider

$$e_\theta(x,r,\omega) = (2\pi i)^{-n}Y_+(r)\lambda_n^{(n-2)}(\langle x,\omega\rangle)Y_+(-\langle\theta,\omega\rangle).$$

Then one verifies that

$$E_\theta = \gamma_* e_\theta \tag{4.4}$$

satisfies the conditions of Theorem 4.1.

There exist many other ways to construct $E_\theta$. One may for example examine $b(x^2-y^2)Y_+(\langle x-y,\theta\rangle)$ when $b$ is a homogeneous distribution of degree $2-2n$, using some arguments from Sections 3.2 and 6.2 of [Hö]. The advantage of the method described here is that it gives us at the same time the decomposition

$$E_\theta = \int_{\langle\omega,\theta\rangle\ge 0} L_\omega\,d\omega, \tag{4.5}$$

where the distribution $L_\omega$ is defined by

$$(4.6)\quad \langle u, L_\omega\rangle$$
$$= C_n \int_{r\geq 0} \Big( \int_{\mathbf{R}^n} ((\omega, \partial_x + \partial_y)^{n-2} u)(x + r\omega, x - r\omega)\lambda_n(\langle x, \omega\rangle)\, dx\Big)\, dr.$$

Here $C_n$ is a constant and the integral with respect to $x$ is integration against a distribution. The explicit formula for $L_\omega$ will be important when deriving the necessary estimates for the $U_j$ in (4.1).

We shall now briefly sketch how (4.1) is solved and the intertwining operators are constructed. It will be assumed that $V \in \mathcal{V}^0$, which will simplify the discussions. An inspection of (4.6) allows us to define the convolution between $L_\omega$ and an arbitrary element in $\mathcal{V}^0$. This gives us a continuous mapping

$$\mathcal{V}^0 \ni V \mapsto L_\omega * V \in \mathcal{M}.$$

We have also a continuous composition mapping

$$\mathcal{V}^0 \times \mathcal{M} \ni (V, U) \mapsto V \circ U \in \mathcal{V}^0.$$

Combining these mappings we obtain for each $\omega \in S^{n-1}$ a continuous mapping $T_\omega : \mathcal{M} \mapsto \mathcal{M}$, defined as

$$T_\omega U = L_\omega * (V \circ U).$$

If $\Omega = (\omega^1, \ldots, \omega^j) \in (S^{n-1})^j$ we set $T_\Omega = T_{\omega^j} \circ \ldots \circ T_{\omega^1}$. It can be proved that $T_\Omega U$ depends continuously on $\Omega$, at least in the weak sense when $T_\omega$ is considered as a distribution only. Hence the integral

$$U_j = \int_{(S_\theta^{n-1})^j} T_\Omega U_0 d\omega^1 \ldots d\omega^j$$

is defined, where $S_\theta^{n-1} = \{\omega \in S^{n-1}; \langle \omega, \theta\rangle \geq 0\}$, and it solves (4.1).

We observe that the distributions $L_\omega$ have positive order. When $j$ becomes large it might happen that some of the derivatives that enter in the definition of $U_j$ through the expression $T_\Omega U_0$ hit a factor $V$ there a large number of times. This would make it impossible to estimate $U_j$ when $j \to \infty$. The problem is taken care of by partial integrations so that each factor $V$ is differentiated exactly $n - 2$ times. In order to give the reader some insight into how this can be carried out we consider the case when $n = 3$ and $V$ is multiplication by some $v \in C_0^\infty$.

Let $L_\omega^0$ be the distribution obtained from $L_\omega$ when the factor $\langle \omega, \partial_x + \partial_y\rangle$ is removed. Then

$$L_\omega = \langle \omega, \partial_x + \partial_y\rangle L_\omega^0$$

and

$$L^0_\omega = C \int_0^\infty \delta(\langle x+y,\omega\rangle)\delta(x-y-r\omega)\,dr.$$

We may write

$$L^0_\omega(x,y) = K^+_\omega(x,y-x) = K^-_\omega(y,x-y),$$

where $K^\pm_\omega(x,y)$ are measures in $y$ which are weakly continuous in $x$, and

$$\sup_x \int |K^\pm_\omega(x,y)|\,dy < \infty. \tag{4.7}$$

From these estimates it is easily seen that $T_\omega$ is continuous from $\mathcal{M}$ to $\mathcal{M}$. The same is true for $T^0_\omega$ obtained when $L_\omega$ is replaced by $L^0_\omega$, and there is a positive constant $C$, which is independent of $v$ and $\omega$, such that

$$\|T^0_\omega U\|_\Sigma \le C\|\nabla v\|_{L^1}\|U\|_\Sigma \tag{4.8}$$

when $U \in \mathcal{M}$.

Next we observe that

$$\langle\omega, \partial_x + \partial_y\rangle T^0_\phi U = 0 \quad \text{if } \omega \perp \phi \text{ and } U \in \mathcal{M},$$

since convolution by $L^0_\phi$ produces distributions that are annihilated by $\langle\omega, \partial_x + \partial_y\rangle$. Let us assume that $j \ge 2$ and write

$$\omega^j = \alpha^j + \beta^j,$$

where $\alpha^j \perp \omega^j$ and $\beta^j \perp \omega^{j-1}$. This is possible for $\omega^{j-1}$ and $\omega^j$ in general position, and one may choose $\beta^j$ so that

$$|\beta^j| = \mu(\omega^{j-1},\omega^j) \quad \text{where } \mu(\omega,\phi) = (1-\langle\omega,\phi\rangle^2)^{-1/2}.$$

If $v(x)$ is considered as a function of $(x,y)$ on $\mathbf{R}^n \times \mathbf{R}^n$ and $U \in \mathcal{M}$ it is true that

$$\begin{aligned} T_{\omega^j}T_{\omega^{j-1}}U &= \langle\beta^j, \partial_x+\partial_y\rangle\left(L^0_{\omega^j} * (vT_{\omega^{j-1}}U)\right) \\ &= L^0_{\omega^j} * \left((\langle\beta^j,\partial_x\rangle v)T_{\omega^{j-1}}U\right). \end{aligned}$$

Hence the derivatives sitting in $L_{\omega^j}$ will hit only the first vector $v$ in $T_\Omega U_0$ and not the others where $v$ has already been differentiated. Since $\mu(\omega,\phi)$

is symmetric in $(\omega, \phi)$ and $\sup_\omega \int \mu(\omega, \phi)\, d\phi < \infty$ one obtains from (4.8) and a simple argument of induction the estimate

$$\text{(4.9)} \quad \|U_j\|_\Sigma \leq C^j \Big( \int \cdots \int \mu(\omega^j, \omega^{j-1}) \ldots \mu(\omega^2, \omega^1) d\omega_1 \ldots d\omega_j \Big) \|\nabla v\|_{L^1}^j \leq (C')^j \|\nabla v\|_{L^1}^j .$$

There are similar estimates for the other semi-norms in $\mathcal{M}$ with higher derivatives of $v$ in the right-hand side, and it follows that $\sum U_j$ converges in $\mathcal{M}$ when the norm of $\nabla v$ in $L^1$ is small. Also, when $n > 3$ one can essentially argue in the same way, although the case when $n$ is even is a little more complicated. There are also some additional problems when $\theta$ is allowed to vary or when $\mathcal{V}^0$ is replaced by an arbitrary $\mathcal{V}^\sigma$.

The case when $V$ is large can be handled by first reducing $H_V$ to a finite rank perturbation of $H_0$. When doing this reduction one has to carry out further integrations by parts, and if the potential is large one must compensate by cutting off low frequencies from the intertwining operator. The natural 'Ansatz' in that case is to consider the series $\sum U_j \circ (I - \chi(\varepsilon D))^j$, where $\chi$ is a function in $C_0^\infty$ which is equal to 1 in a neighbourhood of the origin and the $U_j$ are as above. The size of $\varepsilon$ depends on the potential.

**5. A functional equation for the $A_\theta$ and the generalized scattering amplitude.** In this section we consider a potential $V \in \mathcal{V}^0_{\text{loc}}$, and we let $A_\theta = A_{\theta,V}$ be the corresponding intertwining operators discussed in Section 3. The $A_\theta$ are uniquely determined from $V$, and conversely, every $A_\theta$ determines $V$. Hence $A_\theta$ determines $A_{\theta'}$ for arbitrary $\theta$ and $\theta'$. It is therefore natural to ask if there is some system of equations which is solved by $A_\theta$ when this is considered as an operator valued function of $\theta$. We shall see that this is in fact true, and that such systems of operators defined by a local potential can be characterized by a system of operator valued integro-differential equations on the sphere.

In order to give a motivation for the equations we are going to study we consider again the fundamental solutions $E_\theta$ constructed in Section 4. One can prove that the $E_\theta$ form a smooth family of distributions and also that

$$\partial_{\theta_j} E_\theta * \mathcal{V}^0 \subset \mathcal{M}.$$

(This inclusion is not true with $\mathcal{M}$ replaced by $\mathcal{M}_\theta$.) Let us also assume for a while that $V$ is small. Then it follows from the preceding constructions that

$$\text{(5.1)} \qquad A_\theta = U_0 + E_\theta * (V \circ A_\theta),$$

where $U_0(x,y) = \delta(x-y)$. It turns out that $A_\theta$ is a $C^1$ function of $\theta$ in the weak sense and that (the components of) its gradient is (are) in $\mathcal{M}$. The last term in (5.1) obeys Leibniz' rule for differentiation with respect to $\theta$ so that

$$\partial_\theta A_\theta = (\partial_\theta E_\theta) * (V \circ A_\theta) + E_\theta * (V \circ \partial_\theta A_\theta). \tag{5.2}$$

Set

$$Q_\theta = A_\theta^{-1} \partial_\theta A_\theta - (\partial_\theta E_\theta) * (V \circ A_\theta). \tag{5.3}$$

We claim that $Q_\theta = 0$. First one observes that the intertwining property of $A_\theta$ together with the equation $(\Delta_x - \Delta_y)\partial_\theta E_\theta = 0$ implies that we have $(\Delta_x - \Delta_y)Q_\theta = 0$. Modulo the space of distributions $Q(x, y)$ such that $Q(x - s\theta, y - s\theta) \to 0$ when $s \to +\infty$ we have by (5.2) and (5.3)

$$Q_\theta \equiv \partial_\theta A_\theta - (\partial_\theta E_\theta) * (V \circ A_\theta) \equiv 0.$$

Since $\langle x - y, \theta\rangle \geq 0$ in the support of $Q_\theta$, it follows from Lemma 3.4 that $Q_\theta = 0$. If we set $R_\theta = V \circ A_\theta$ we have proved therefore that the following system of equations is fulfilled:

$$\partial_\theta R_\theta = R_\theta \circ (\partial_\theta E_\theta * R_\theta). \tag{5.4}$$

The same arguments work also in the case when $V$ is large and the solution $R_\theta$ has the following properties:

(5.i) The mapping $S^{n-1} \ni \theta \mapsto R_\theta \in \mathcal{D}'(\mathbf{R}^n \times \mathbf{R}^n)$ is in $C^1$.
(5.ii) If $x, y \in \operatorname{supp}(R_\theta)$ then $\langle x - y, \theta\rangle \geq 0$.
(5.iii) The mapping $\theta \mapsto e^{-\lambda\langle x-y,\theta\rangle} R_\theta(x, y)$ is continuous with values in $\mathcal{V}^0$ when $\lambda$ is large.

The following result tells us that there is a one-to-one correspondence between solutions to (5.4), (5.i)–(5.iii) and local potentials in the class $\mathcal{V}^0$.

**Theorem 5.1.** *Let $\{A_{\theta,V}\}$ be the family of intertwining operators that belongs to $V \in \mathcal{V}^0_{\mathrm{loc}}$. Then $R_\theta = V \circ A_{\theta,V}$ satisfies (5.4) and (5.i)–(5.iii). Conversely, every solution to (5.4) with these properties is of the form $R_\theta = V \circ A_{\theta,V}$ for a uniquely determined $V \in \mathcal{V}^0_{\mathrm{loc}}$.*

*Remark.* The equation (5.4) was derived by using uniqueness results for the ultra-hyperbolic equation. Such results are also important for proving the other part of the theorem.

Let $p \in \mathbf{C}^n \setminus \mathbf{R}^n$ and let $\theta$ be the unit vector parallel to $\operatorname{Im} p$. Let $V$ be multiplication by $v$ and assume for simplicity that $V \in \mathcal{V}^0$. Set

$$A_p(x, y) = A_{p,V}(x, y) = e^{i\langle x-y,p\rangle} A_\theta(x, y).$$

If $V$ is small then $v(x)A_p(x,y)$ is a bounded measure on $\mathbf{R}^n \times \mathbf{R}^n$, and we define the generalized scattering amplitude by

$$G(\xi,p) = (\mathcal{F}(V \circ A_p))(\xi,0) = \iint e^{-i\langle x,\xi\rangle} v(x)A_p(x,y)\,dxdy$$

when $\xi \in \mathbf{R}^n$, $p \in \mathbf{C}^n \setminus \mathbf{R}^n$. (See [HN], [We] and [M3].) This function is also defined for large $V$ in a set where $|\operatorname{Im} p|$ is sufficiently large.

In order to extend the domain of the generalized scattering amplitude to a larger set we shall need a blow-up construction around the real points in $\mathbf{C}^n$. We consider the mapping

$$\mathbf{R}^n \times \overline{\mathbf{R}_+} \times S^{n-1} \ni (x,r,\theta) \mapsto x + ir\theta \in \mathbf{C}^n.$$

The restriction to $\mathbf{R}^n \times \mathbf{R}_+ \times S^{n-1}$ is a diffeomorphism onto $\mathbf{C}^n \setminus \mathbf{R}^n$, which may therefore be considered as the interior of a manifold $\widetilde{\mathbf{C}}^n$, diffeomorphic to $\mathbf{R}^n \times \overline{\mathbf{R}_+} \times S^{n-1}$. The mapping $\mathbf{C}^n \setminus \mathbf{R}^n \ni p \mapsto \theta(p) = \operatorname{Im} p/|\operatorname{Im} p|$ extends now to a smooth mapping from $\widetilde{\mathbf{C}}^n$ to $S^{n-1}$, and it is true also that $p \mapsto A_p$ extends to a continuous mapping from $\widetilde{\mathbf{C}^n}$ to the space of (possibly unbounded) measures on $\mathbf{R}^n \times \mathbf{R}^n$. It turns out that there is a continuous function $q : \widetilde{\mathbf{C}^n} \mapsto \mathbf{C}$, with $q(p) \to 1$ as $p \to \infty$, so that $G(\xi,p)q(p)$ is continuous on $\mathbf{R}^n \times \widetilde{\mathbf{C}^n}$. The zero set of $q$ is of measure 0 and its elements are called exceptional points. (See [HN], [We], [M3], [M4].) It follows that the generalized scattering amplitude is defined almost everywhere when $V \in \mathcal{V}^0_{\text{loc}}$ is arbitrary.

It is a simple consequence of the Riemann-Lebesgue lemma and the properties of $q$ that $G(\xi,p) \to \widehat{v}(\xi)$ when $p \to \infty$ in $\widetilde{\mathbf{C}}^n$. Hence $v$ can be computed from the generalized scattering amplitude. This is not at all surprising, since $vA_\theta$ is uniquely determined from the generalized scattering amplitude and $A_\theta(x,y) = \delta(x-y)$ plus terms which are more regular. (They are locally integrable.)

Most results on characterization of scattering data deal with the generalized scattering amplitude ([HN], [We]). In the case of a local potential in $\mathcal{V}^0 \cap \mathcal{V}^1$ for example it turns out that these results follow immediately from Theorem 5.1. Also the $\overline{\partial}$-equations are easily obtained from Theorem 5.1 (see [M3]).

The question that remains to be discussed now is the relation between the ordinary scattering amplitude and the generalized scattering amplitude (or, alternatively, the operators $vA_\theta$ occurring in Theorem 5.1.) The scattering amplitude is the restriction of $G(\xi,p)$ to a certain subset of $\mathbf{R}^n \times \widetilde{\mathbf{C}^n}$. In fact, when $v$ is small, then the information provided by the scattering operator is the values of $(\mathcal{F}(vA_\theta))(\xi,\eta)$ at points $(\xi,\eta)$ with $|\xi| = |\eta| > 0$ and $\eta/|\eta| = \theta$. By solving certain integral equations these functions can be extended to larger sets in $\mathbf{R}^n \times \widetilde{\mathbf{C}^n}$ so that the generalized scattering amplitude can be determined. (For further details we refer to the first section of [We] and the references given there.)

It seems to be out of reach to implement the constructions mentioned above in numerical computations. We shall finish our presentation by giving some comments on yet another method to recover the potential from the scattering matrix. This method, part of which may be found in [F2] for example, may also be viewed as a high frequency method since it relies on a singularity analysis of the distribution kernel of $A_\theta$. In order to avoid some technical difficulties we assume that $n$ is odd and that $v \in C_0^\infty$. We shall blow up $A_\theta(x, y)$ at the diagonal $x = y$ by using the transformation $\gamma$ in Section 4. By considering $\lim_{r \to +0} a_\theta(x, r, \phi)$, where $a_\theta$ is a pre-image of $A_\theta$ under the mapping $\gamma_*$, one defines a 'residue', $\operatorname{res}(A_\theta)$, which is a smooth function of $(x, \phi) \in \mathbf{R}^n \times S^{n-1}$ except for a jump discontinuity when $\langle \theta, \phi \rangle = 0$. In fact,

$$(\operatorname{res}(A_\theta))(x, \phi) = C_n Y_+(-\langle \theta, \phi \rangle) \int_{\langle z, \phi \rangle = 0} \langle \phi, \partial_z \rangle^{n-2} v(x + z) \, dz. \tag{5.5}$$

If we apply $\langle \phi, \partial_x \rangle$ to the right-hand side and integrate with respect to $\phi$ we obtain the expression

$$\begin{aligned} &C_{n,1} \int_{S^{n-1}} \left( \int_{\langle z, \phi \rangle = 0} \langle \phi, \partial_x \rangle^{n-1} v(x + z) \, dz \right) d\phi \\ &= C_{n,2} \int_{\mathbf{R}^n} |y|^{-1} \Delta^{(n-1)/2} v(x + y) \, dy = C_{n,3} v(x), \end{aligned}$$

since $|y|^{-1}$ is proportional to a fundamental solution of $\Delta_y^{(n-1)/2}$. This proves that $\operatorname{res}(A_\theta)$ determines $v$.

Since $A_\theta$ and $A_\theta^*$ are supported in opposite half-planes it is easily proved that also the residue of $A_\theta^* A_\theta$ determines $v$. Hence $v$ can be computed from $B_\theta = A_\theta^* A_\theta$. This operator commutes with the Laplacian, and in analogy with the scattering operator it can be represented as a direct integral:

$$B_\theta = \int_0^\infty B_\theta(k) \, dk,$$

where the $B_\theta(k)$ are bounded linear operators on $L^2(S^{n-1})$. It turns out now that $B(k)$ may be computed from the scattering matrix $S(k)$ by the following procedure. The direction $\theta$ introduces an ordering on $S^{n-1}$ so that $\phi' \geq \phi''$ if and only if $\langle \phi' - \phi'', \theta \rangle \geq 0$. With respect to this ordering one can talk about lower and upper triangular operators on $L^2(S^{n-1})$. The scattering matrix may be factorized into a product of a lower and an upper triangular factor, and $B(k)$ is then given by one of these factors multiplied by its adjoint. (This is true at least for $k$ outside a set of measure 0.)

*Remark.* The residue constructions mentioned here are analogous to the corresponding one-dimensional constructions. In fact the formula (5.5) is the natural generalization to higher dimensions of formula (2.6) of [M1]. Thus, by introducing the mapping $\gamma$ in Section 4, we were able to view some scattering objects in the multi-dimensional case as angular dependent one-dimensional objects of similar type.

## References


[AN] M. J. Ablowitz and A. I. Nachman, *A multidimensional inverse scattering method*, Studies in Appl. Math. **71**, 243–250.

[BC] R. Beals and R. Coifman, *Scattering, transformations spectrales, et équations d'involution non linéaires*, Séminaire Goulaouic-Meyer-Schwartz 1980-1981, exp. 22, École Polytechnique, Palaiseau.

[ER1] G. Eskin and J. Ralston, *The inverse backscattering problem in three dimensions*, Comm. Math. Phys. **124**, no. 3, 169–215.

[ER2] ———, *Inverse backscattering in two dimensions*, Comm. Math. Phys. **138**, no. 3, 451-486.

[F1] L. D. Faddeev, *Uniquenes of the solution of the inverse scattering problem*, Vestnik Leningrad Univ. Math. Mekh. Astronom. **7** (1956), 126–130. (Russian)

[F2] ———, *Inverse problems of quantum scattering theory. II*, Itogi Nauki i Tekhniki. Sovremennye Problemy Matematiki **3** (1974), 93–180 (Russian); English transl. in J. Sov. Math. **5**, 334–396.

[GS] I. M. Gel'fand and G. E. Shilov, *Generalized functions. I*, Academic Press, New York and London, 1964.

[HN] G. M. Henkin and R. G. Novikov, *The $\overline{\partial}$-equation in multi-dimensional inverse scattering*, Uspehi Mat. Nauk **42** (1987), no. 3, 93–152 (Russian); English transl. in Russian Math. Surveys **42** (1987), no. 3, 109–180.

[Ho] A. Holst, *Approximation of Schrödinger operators and scattering data using multiresolution analysis* (1994), Thesis, University of Lund and Lund Institute of Technology.

[Hö] L. Hörmander, *The analysis of linear partial differential operators*, Springer-Verlag, Berlin-Göttingen-Heidelberg-New York-Tokyo, 1983–1985.

[Is] H. Isozaki, *Multi-dimensional inverse scattering theory for Schrödinger operators*, Preprint, Department of Mathematics, Osaka University, 1995.

[Ma] V. A. Marchenko, *Sturm-Liouville operators and applications*, Operator Theory: Advances and applications, Vol. 22, Birkhäuser Verlag, Basel-Boston-Stuttgart, 1986.

[M1] A. Melin, *Operator methods for inverse scattering on the real line*, Comm. in Partial Diff. Eq. **10** (1985), 677-766.

[M2] ———, *On the use of intertwining operators in inverse scattering*, Schrödinger operators, (H. Holden and A. Jensen, eds.), Lecture Notes in Physics, 345, Springer Verlag, Berlin-Heidelberg-New York, 1989.

[M3] ———, *Inverse scattering problem for a quantum mechanical two-body system*, Analysis, algebra and computers in mathematical research (Proceedings of the twenty-first Nordic congress of mathematicians) (M. Gyllenberg and L. E. Persson, eds.), Lecture notes in pure and applied mathematics, 156, Marcel Dekker, New York-Basel-Hong Kong, 1994.

[M4] ———, *The intertwining approach to multi-dimensional scattering and inverse scattering*, In preparation.

[Ne] R. G. Newton, *Inverse Schrödinger scattering in three dimensions*, Springer-Verlag, Text and monographs in physics, Berlin-Heidelberg-New York-London-Paris-Tokyo-Hong Kong, 1989.

[We] R. Weder, *Characterization of the scattering data in multidimensional inverse scattering theory*, Inverse problems **7** (1991), 461–489.

DEPARTMENT OF MATHEMATICS, LUND INSTITUTE OF TECHNOLOGY, BOX 118, S-221 00 LUND, SWEDEN

*E-mail address*: melin@maths.lth.se

# FIBRATIONS, COMPACTIFICATIONS AND ALGEBRAS OF PSEUDODIFFERENTIAL OPERATORS

RICHARD B. MELROSE

ABSTRACT. Some recent, and some new, results on the structure of algebras of pseudodifferential operators on compact manifolds with boundary are discussed. In particular their relationship to compactifications of non-compact spaces is emphasized. It is shown how these relationships allow the methods of pseudodifferential operators ("microlocalization") to be applied to problems in scattering and spectral theory.

## EUCLIDEAN SPACE

The first algebra of pseudodifferential operators on a general compact manifold with boundary, that I will discuss here, comes directly from the very familiar case of Euclidean space. Namely consider the standard calculus of pseudodifferential operators on $\mathbb{R}^n$. Since there are several variants of this 'standard' calculus, let me be precise. An element of the space $\Psi^m(\mathbb{R}^n)$ is an operator on $\mathcal{S}(\mathbb{R}^n)$ which can be written in the form

$$(1)\qquad Au(x) = (2\pi)^{-n}\int_{\mathbb{R}^n} e^{ix\cdot\xi}a(x,\xi)\hat{u}(\xi),\ \hat{u}(\xi) = \int_{\mathbb{R}^n} e^{-iy\cdot\xi}u(y)dy$$

with the (left-reduced) symbol satisfying the estimates 'of type $1,0$'

$$|D_x^\alpha D_\xi^\beta a(x,\xi)| \le C_{\alpha,\beta}(1+|\xi|)^{m-|\beta|}\ \forall\ \alpha,\beta\in \mathbb{N}_0^n.$$

These spaces form a filtered algebra, but I wish to consider the subalgebra of elements with symbols with complete, 1-step, asymptotic expansions in both variables. Rather than do this in the usual way, let me consider the compactification of Euclidean space to a ball.

**Radial compactification.** Let $z$ denote the variable in $\mathbb{R}^n$. Consider the map of 'stereographic projection' type

$$(2)\qquad \mathrm{SP}:\mathbb{R}^n \ni z \longmapsto Z=(Z_0, zZ_0)\in\mathbb{R}^{n+1},\ Z_0=(1+|z|^2)^{-\frac{1}{2}}.$$

The range of this map is precisely the open upper half sphere

$$(\mathbb{S}^n_+)^\circ = \mathbb{S}^n_+ \setminus \partial\mathbb{S}^n_+,\ \mathbb{S}^n_+ = \{(Z_0, Z_1, \dots, Z_n) \in \mathbb{R}^{n+1}; Z_0 \geq 0,\\ Z_0^2 + Z_1^2 + \cdots + Z_n^2 = 1\},\ \partial\mathbb{S}^n_+ = \mathbb{S}^{n-1}.$$

Thus, $\mathbb{S}^n_+$ is a compact manifold with boundary and $\mathbb{R}^n$ is identified with its interior. The function $a$ can therefore be considered as defined on the interior of the product, $\mathbb{S}^n_+ \times \mathbb{S}^n_+$, which is a compact manifold with corner. Let $\rho_\partial = (1 + |x|^2)^{-\frac{1}{2}}$ be the function '$Z_0$' on the first factor and $\rho_\sigma = (1 + |\xi|^2)^{-\frac{1}{2}}$ be the same function on the second factor. Then by '$a$ has a complete 1-step asymptotic expansion' I simply mean that

$$a = (\mathrm{SP} \times \mathrm{SP})^* a',\ \rho_\sigma^m a' \in \mathcal{C}^\infty(\mathbb{S}^n_+ \times \mathbb{S}^n_+). \tag{3}$$

The space of smooth functions $\mathcal{C}^\infty(\mathbb{S}^n_+ \times \mathbb{S}^n_+)$ is just the restriction of $\mathcal{C}^\infty(\mathbb{R}^{n+1} \times \mathbb{R}^{n+1})$ to $\mathbb{S}^n_+ \times \mathbb{S}^n_+$.

I will come back to this 'radial' compactification of Euclidean space at several points later, for the moment it allows me to define the space, which I shall denote $\Psi^m_{\mathrm{cl}}(\mathbb{R}^n)$, although I shall fairly soon discard this (objectionable) notation. The significance of the map SP is that it identifies the space of symbols of order 0 on $\mathbb{R}^n$ with complete, 1-step, asymptotic expansions with $\mathcal{C}^\infty(\mathbb{S}^n_+)$. Thus $A \in \Psi^m_{\mathrm{cl}}(\mathbb{R}^n)$ if it can be written in the form (1) with left-reduced symbol satisfying (3). I should point out that, even at the level of estimates, the assumption (3) improves the 'type 1,0' hypothesis to

$$|D_x^\alpha D_\xi^\beta a(x,\xi)| \leq C_{\alpha,\beta}(1 + |x|)^{-|\alpha|}(1 + |\xi|)^{m-|\beta|}\ \forall\ \alpha, \beta \in \mathbb{N}_0^n.$$

Operators defined by symbols satisfying these estimates were considered by Shubin in [13].

**Fredholm properties.** Amongst the elements of $\Psi^m_{\mathrm{cl}}(\mathbb{R}^n)$ are the translation invariant differential operators. In particular the flat Laplacian, $\Delta = D_1^2 + D_2^2 + \cdots + D_n^2 \in \Psi^2_{\mathrm{cl}}(\mathbb{R}^n)$. Multiplication by a bounded symbol on $\mathbb{R}^n$, $V$, with complete, 1-step, asymptotic expansion is also in the algebra, so it is quite natural to expect these pseudodifferential operators to be helpful in discussing an operator such as $\Delta + V$, which is one of the central objects treated in scattering theory. Notice that such a $V$ may be (very) long range in the sense that it does not decay at infinity at all. Let me give a simple example of the relationship between this algebra and the spectral theory of $\Delta + V$. The elements of $\Psi^m_{\mathrm{cl}}(\mathbb{R}^n)$ are bounded from the standard Sobolev space $H^m(\mathbb{R}^n)$ to $L^2(\mathbb{R}^n)$. The Fredholm elements have a simple characterization.

**Proposition 1.** *An element $A \in \Psi^m_{\mathrm{cl}}(\mathbb{R}^n)$ is Fredholm as a map from $H^m(\mathbb{R}^n)$ to $L^2(\mathbb{R}^n)$ if and only if it is 'fully elliptic' in the sense that the function $\rho_\sigma^m a'$ in (3) is non-zero at every point of the boundary of $\mathbb{S}^n_+ \times \mathbb{S}^n_+$.*

The boundary of $\mathbb{S}^n_+ \times \mathbb{S}^n_+$ consists of the two manifolds with boundary (boundary hypersurfaces) at $\rho_\sigma = 0$ and $\rho_\partial = 0$; both are diffeormorphic to $\mathbb{S}^n_+ \times \mathbb{S}^{n-1}$. The first condition, that $\rho_\sigma^m a'$, be non-zero on $\{\rho_\sigma = 0\}$ is just the usual notion of ellipticity (uniformly in $x$). The second condition, that the same function be non-vanishing on $\{\rho_\partial = 0\}$, is not a condition on the principal symbol at all, but a condition on the 'full symbol at infinity'. Using the composition properties of the calculus it is straightforward to translate Proposition 1 to a (well-known) statement about the spectral theory of $\Delta + V$.

**Proposition 2.** *If $V \in \mathcal{C}^\infty(\mathbb{R}^n)$ is real-valued and of the form $\mathrm{SP}^*(V')$ for $V' \in \mathcal{C}^\infty(\mathbb{S}^n_+)$ then $\Delta + V$ is self-adjoint with domain $H^2(\mathbb{R}^n)$, has continuous spectrum in the interval $[b, \infty)$, where $b = \inf(V' \restriction \mathbb{S}^{n-1})$ and only discrete spectrum of finite multiplicity in the complement.*

In fact the resolvent $(\Delta + V - \lambda)^{-1}$ is a meromorphic function on $\mathbb{C} \setminus [b, \infty)$ with values in $\Psi^{-2}_{\mathrm{cl}}(\mathbb{R}^n)$.

Let me 'interpret' the properties of $\Psi^m_{\mathrm{cl}}(\mathbb{R}^n)$ in the direction I wish to go. First, the constant coefficient vector fields on $\mathbb{R}^n$ can be transferred to $\mathbb{S}^n_+$ by SP. They become smooth vector fields on $\mathbb{S}^n_+$ and $\Psi^m_{\mathrm{cl}}(\mathbb{R}^n)$ can be thought of as the 'microlocalization' of the Lie algebra, $\mathcal{V}_{\mathrm{sc}}(\mathbb{S}^n_+)$, of vector fields on $\mathbb{S}^n_+$ which is the span, with coefficients from $\mathcal{C}^\infty(\mathbb{S}^n_+)$, of the images of the constant coefficient vector fields from $\mathbb{R}^n$. The subscript 'sc' stands for 'scattering' for the reasons outlined above. The next idea is to replace $\mathbb{S}^n_+$ by an arbitrary compact manifold with boundary.

## Scattering calculus

First one can work out exactly what $\mathcal{V}_{\mathrm{sc}}(\mathbb{S}^n_+)$ is, in terms of the $\mathcal{C}^\infty$ structure of $\mathbb{S}^n_+$ as a compact manifold with boundary. Let $X$ be a general compact manifold with boundary. The space $\mathcal{V}(X)$ of *all* smooth vector fields on $X$ is then defined. Here (and below), by 'smooth' I always mean smooth up to the boundary. Now, $\mathcal{V}(X)$ is a Lie algebra. It has a natural subalgebra consisting of the elements, $V \in \mathcal{V}(X)$, which are tangent to the boundary, meaning that if $f \in \mathcal{C}^\infty(X)$ vanishes at $\partial X$ then so does $Vf$. I shall denote this subalgebra as $\mathcal{V}_{\mathrm{b}}(X)$ (the 'b' just stands for boundary). Let $\rho \in \mathcal{C}^\infty(X)$ be a boundary defining function, i.e. $\rho > 0$ on $X \setminus \partial X$, $\rho = 0$ on $\partial X$ and $d\rho \neq 0$ on $\partial X$. Then we *define*

$$\mathcal{V}_{\mathrm{sc}}(X) = \rho \mathcal{V}_{\mathrm{b}}(X) = \{V \in \mathcal{V}(X); V = \rho W \text{ with } W \in \mathcal{V}_{\mathrm{b}}(X)\}. \tag{4}$$

Of course this definition would be improper were it not for the fact that it is indeed consistent with the notation above for $\mathcal{V}_{\text{sc}}(\mathbb{S}^n_+)$, defined by compactification.

**Localization.** Having generalized $\mathcal{V}_{\text{sc}}(X)$ in this way it is perhaps natural to look for a generalization of the algebra $\Psi^*_{\text{cl}}(\mathbb{R}^n)$. Observe that the map SP identifies the Schwartz space $\mathcal{S}(\mathbb{R}^n)$ with $\dot{\mathcal{C}}^\infty(\mathbb{S}^n_+) \subset \mathcal{C}^\infty(\mathbb{S}^n_+)$, precisely the space of smooth functions vanishing to infinite order at the boundary, i.e. for any compact manifold with boundary

$$\dot{\mathcal{C}}^\infty(X) = \bigcap_{k\in\mathbb{N}} \rho^k \mathcal{C}^\infty(X).$$

Thus if $A \in \Psi^*_{\text{cl}}(\mathbb{R}^n)$ then $A(\mathrm{SP}^* u) = \mathrm{SP}^*(A'u)$ defines an operator, $A'$ on $\dot{\mathcal{C}}^\infty(\mathbb{S}^n_+)$. I shall denote this image algebra by $\Psi^*_{\text{sc}}(\mathbb{S}^n_+)$ (and drop the previous notation on $\mathbb{R}^n$ as anticipated).

This generalization is quite straightforward; it is carried out in [6]. First it can be seen that $\Psi^*_{\text{sc}}(\mathbb{S}^n_+)$ is invariant under conjugation by $F^*$ for any diffeomorphism $F$ of $\mathbb{S}^n_+$. Secondly the Schwartz kernel of an element of $\Psi^{-\infty}_{\text{sc}}(\mathbb{S}^n_+)$ can be seen to be smooth on $\mathbb{S}^n_+ \times \mathbb{S}^n_+$ and to vanish to infinite order at the boundary. Indeed this is a characterization

$$\Psi^{-\infty}_{\text{sc}}(\mathbb{S}^n_+) = \dot{\mathcal{C}}^\infty(\mathbb{S}^n_+ \times \mathbb{S}^n_+; \Omega_R) \tag{5}$$

(where $\Omega_R$ is the density bundle pulled up from the right factor). We *define* $\Psi^{-\infty}_{\text{sc}}(X)$ by replacing $\mathbb{S}^n_+$ by $X$ everywhere in (5). Any point in $X$ has a neighbourhood which can be smoothly identified with an open set in $\mathbb{S}^n_+$. If $F : O \longrightarrow O'$ is such a smooth diffeomorphism and $\phi \in \mathcal{C}^\infty(X)$ has support in $O$, so $\phi = F^*\phi'$ where $\phi'$ has compact support in $O'$ then any $A \in \Psi^m_{\text{sc}}(\mathbb{S}^n_+)$ defines an operator on $\dot{\mathcal{C}}^\infty(X)$ by $u \longmapsto F^*(\phi' A(G^*(\phi u)))$, $G = F^{-1}$. Then $\Psi^m_{\text{sc}}(X)$ may be defined as the span of $\Psi^{-\infty}_{\text{sc}}(X)$ and these localized operators. By working over sets where a given pair of vector bundles, $E, F$, is trivial the space of operators, $\Psi^m_{\text{sc}}(X; E, F)$ ($=\Psi^m_{\text{sc}}(X; E)$ if $E = F$) from sections of $E$ to sections of $F$ can be defined in a similar way. There is a more elegant definition, which I shall not discuss in detail here. It arises by generalizing the direct definition of pseudodifferential operators on a compact manifold without boundary: namely that they are precisely those operators which have Schwartz kernels which are densities conormal to the diagonal. In the corresponding definition of the scattering calculus the product $X \times X$ must first be blown up in an appropriate manner. This is discussed in the Appendix to [6].

The basic properties of $\Psi^*_{\text{sc}}(X)$ can be obtained from those of the special case $\Psi^*_{\text{sc}}(\mathbb{S}^n_+)$. In particular the spaces form a filtered algebra

$$\Psi^m_{\text{sc}}(X; F, G) \circ \Psi^{m'}_{\text{sc}}(X; E, F) \subset \Psi^{m+m'}_{\text{sc}}(X; E, G).$$

To describe the analogue of Proposition 1 in this more general setting it is convenient first to describe the symbol map in an invariant way. The compactification, (2), of $\mathbb{R}^n$ is invariant under linear transformations. That is, if $L$ is an invertible linear transformation of $\mathbb{R}^n$ then there is a smooth diffeomorphism, $L'$, of $\mathbb{S}^n_+$ such that $L' \circ \mathrm{SP} = \mathrm{SP} \circ L$. This means that the compactification makes sense for any real vector space, $U$, of finite dimension. As a set this 'radial compactification' is

$$\bar{U} = U \sqcup \left((U \setminus \{0\})/\mathbb{R}^+\right) \cong \mathbb{S}^n_+, \ \dim U = n.$$

The smooth lifting of linear isomorphisms shows that the identification with $\mathbb{S}^n_+$ given by any basis of $U$, identifying $U$ with $\mathbb{R}^n$, gives $\bar{U}$ a $\mathcal{C}^\infty$ structure. In fact the smoothness of this operation in parameters means that the fibre-by-fibre compactification of a vector bundle over a manifold $X$ defines a smooth ball bundle over $X$.

**Structure algebra.** Consider again the basic 'structure algebra' $\mathcal{V}_{\mathrm{sc}}(X)$, for a compact manifold with boundary $X$. If $p \in \partial X$ then there are local coordinates $x, y_1, \ldots, y_{n-1}$, based at $p$, where $x$ is a local boundary defining function, and in terms of any such coordinates an element $V \in \mathcal{V}_{\mathrm{sc}}(X)$ is locally of the form

$$V = a(x,y)x^2\frac{\partial}{\partial x} + \sum_{j=1}^{n-1} b_j(x,y)x\frac{\partial}{\partial y_j}$$

with the coefficients $\mathcal{C}^\infty$ functions in the local coordinates. Conversely if $V \in \mathcal{V}(X)$ is of this form in a covering of the boundary by coordinate systems then it is an element of $\mathcal{V}_{\mathrm{sc}}(X)$. This implies that there is a smooth vector bundle, ${}^{\mathrm{sc}}TX$ over $X$, with coordinate basis $x^2\partial/\partial x$, $x\partial/\partial y_j$, $j = 1, \ldots, n-1$, such that

$$\mathcal{V}_{\mathrm{sc}}(X) = \mathcal{C}^\infty(X; {}^{\mathrm{sc}}TX)$$

just as is the case for $\mathcal{V}(X)$ itself. The fibre of ${}^{\mathrm{sc}}TX$ at any point $p \in X$ can be defined as an abstract vector space by

$${}^{\mathrm{sc}}T_pX = \mathcal{V}_{\mathrm{sc}}(X)/\mathcal{I}_p(X) \cdot \mathcal{V}_{\mathrm{sc}}(X)$$

where $\mathcal{I}_p(X) \subset \mathcal{C}^\infty(X)$ is the ideal of functions vanishing at $p$ and $\mathcal{I}_p(X) \cdot \mathcal{V}_{\mathrm{sc}}(X)$ is the span of the products. If $p \in \partial X$ then $\mathcal{I}_p(X) \cdot \mathcal{V}_{\mathrm{sc}}(X)$ is a Lie ideal in $\mathcal{V}_{\mathrm{sc}}(X)$ with commutative quotient, i.e. ${}^{\mathrm{sc}}T_pX$ in that case is naturally an Abelian Lie algebra.

**Symbol map.** Combining these two constructions, let ${}^{\mathrm{sc}}\bar{T}X$ be the fibre-by-fibre compactification of ${}^{\mathrm{sc}}TX$. Since $X$ is a compact manifold with boundary, ${}^{\mathrm{sc}}\bar{T}X$ is a compact manifold with corners of codimension two. In case $X = \mathbb{S}^n_+$, ${}^{\mathrm{sc}}\bar{T}X = \mathbb{S}^n_+ \times \mathbb{S}^n_+$ is precisely the space on which the symbols were defined. As in that case ${}^{\mathrm{sc}}\bar{T}X$ has two types of boundary hypersurfaces, that coming from the compactification of the fibres and the part of the bundle over the boundary. Let $\rho_\sigma$ be a defining function for the first part and $\rho_\partial$ for the second (which may not be connected). The symbol space for operators of order $m$ is then a quotient

$$S^{[m]}_{\mathrm{sc}}(X;E) = \frac{\rho_\sigma^{-m}\mathcal{C}^\infty({}^{\mathrm{sc}}\bar{T}X;\pi^*\hom E)}{\rho_\sigma^{-m+1}\mathcal{C}^\infty({}^{\mathrm{sc}}\bar{T}X;\pi^*\hom E) + \rho_\sigma^{-m}\rho_\partial\mathcal{C}^\infty({}^{\mathrm{sc}}\bar{T}X;\pi^*\hom E)}, \quad \pi : {}^{\mathrm{sc}}\bar{T}X \longrightarrow X.$$

Thus, if $m = 0$, an element of $S^{[0]}_{\mathrm{sc}}(X;E)$ is just the restriction to the boundary of ${}^{\mathrm{sc}}\bar{T}X$ of a smooth section of $\pi^*\hom E$. The symbol map, $\sigma_{m,\mathrm{sc}}$, is multiplicative, and gives a short exact sequence

$$0 \longrightarrow \rho\Psi^{m-1}_{\mathrm{sc}}(X;E) \hookrightarrow \Psi^m_{\mathrm{sc}}(X;E) \stackrel{\sigma_{m,\mathrm{sc}}}{\longrightarrow} S^{[m]}_{\mathrm{sc}}(X;E) \longrightarrow 0. \tag{6}$$

The Sobolev spaces $H^m_{\mathrm{sc}}(X;E)$ can be defined by local identification with $\mathbb{S}^n_+$ (and hence ultimately by local identification with $H^m(\mathbb{R}^n)$ in conic neighbourhoods of infinity) and the elements $A \in \Psi^m_{\mathrm{sc}}(X;E)$ define bounded maps

$$A \in \Psi^m_{\mathrm{sc}}(X;E) \Longrightarrow A : H^m_{\mathrm{sc}}(X;E) \longrightarrow L^2_{\mathrm{sc}}(X;E). \tag{7}$$

Here $L^2_{\mathrm{sc}}(X)$ is the metric space of square-integrable functions, which reduces exactly to $L^2(\mathbb{R}^n)$ (rather than $L^2(\mathbb{S}^n_+)$) in the case of Euclidean space. Furthermore the first space in (6), the null space of $\sigma_{m,\mathrm{sc}}$, is precisely the subset of $\Psi^m_{\mathrm{sc}}(X;E)$ defining compact maps in (7).

Routine symbolic constructions then allow us to obtain a direct generalization of Proposition 1.

**Proposition 3.** *An element $A \in \Psi^m_{\mathrm{sc}}(X;E)$ is Fredholm as a map from $H^m_{\mathrm{sc}}(X;E)$ to $L^2_{\mathrm{sc}}(X;E)$ if and only if it is 'fully elliptic' in the sense that the symbol $\sigma_{m,\mathrm{sc}}(A)$ has an inverse in $S^{[-m]}_{\mathrm{sc}}(X;E)$.*

To get an analogue of Proposition 2 we first need a generalization of the flat Laplacian to an arbitrary compact manifold with boundary. This is provided by the choice of a 'scattering metric' which makes the space look asymptotically Euclidean at the boundary. Near the boundary of $\mathbb{S}^n_+$, viewed as the compactification of $\mathbb{R}^n$, the function $x = 1/r$ is a boundary defining function so that $\mathbb{S}^n_+ \cong [0,1)_x \times \mathbb{S}^{n-1}$ near $\partial\mathbb{S}^n_+$. In terms of this decomposition the usual formula for the Euclidean metric in polar coordinates becomes

$$g_{\mathrm{Eu}} = \frac{dx^2}{x^4} + \frac{|d\omega|^2}{x^2}$$

where $|d\omega|^2$ is the standard metric on the sphere. This generalizes to an arbitrary compact manifold with boundary by taking a boundary defining function $\rho \in \mathcal{C}^\infty(X)$ and considering any Riemann metric on the interior which is of an analogous form near the boundary

$$g_{\mathrm{sc}} = \frac{d\rho^2}{\rho^4} + \frac{h'}{\rho^2} \text{ near } \partial X. \tag{8}$$

Here $h'$ is a symmetric 2-cotensor which is smooth up to the boundary and such that $h = h' \restriction \partial X$ is positive definite, and so defines a metric on $\partial X$. Clearly such a metric exists on any compact manifold with boundary.

**Proposition 4.** *The Laplacian, $\Delta$ of a scattering metric, in the sense of* (8), *(acting on any form bundle) has spectrum* $[0,\infty)$ *consisting of continuous spectrum of infinite multiplicity.*

**Generalized eigenfunctions.** In fact a little more work is required to show that the spectrum is continuous of infinite multiplicity but its location can be deduced directly from Proposition 3 since the resolvent $(\Delta-\sigma)^{-1}$ can be seen to be a holomorphic function on $\mathbb{C}\setminus[0,\infty)$ with values in $\Psi^{-2}_{\mathrm{sc}}(X)$ and having no bounded analytic continuation across any point of $[0,\infty)$. To get stronger results on the structure of the spectrum it is natural to use ideas from scattering theory combined with microlocal analysis. Similar approaches have been made to the analysis of rather general constant coefficient operators and for long-range potentials, see [1] and [4; Chapter XXX]. The first step is to discuss the weak limit as $\operatorname{Im}\lambda \uparrow 0$ near $\lambda' > 0$ of the resolvent written in the from $R(\lambda) = (\Delta-\lambda^2)^{-1}$, with the resolvent domain taken to be $\operatorname{Im}\lambda < 0$.

**Lemma 1** ([6]). *For a scattering Laplacian on a compact manifold with boundary and $f \in \dot{\mathcal{C}}^\infty(X)$, as $\lambda \to \lambda' > 0$, from $\operatorname{Im}\lambda < 0$, $R(\lambda)f$ converges in $\rho^{-\frac12-\epsilon}L^2_{\mathrm{sc}}(X)$, for any $\epsilon > 0$, to a function of the form*

$$R(\lambda')f = e^{-i\lambda/\rho}\rho^{\frac12(n-1)}g,\ g \in \mathcal{C}^\infty(X).$$

The existence of this limit also depends on the use of the form of the unique continuation theorem of Aronszajn-Cordes type given by Hörmander in [4] since this shows that there can be no point spectrum in $(0,\infty)$.

From this lemma it can be seen by direct construction that for each $0 \neq \lambda \in \mathbb{R}$ and $f \in \mathcal{C}^\infty(\partial X)$ there is a unique eigenfunction $u$, satisfying $(\Delta - \lambda^2)u = 0$, of the form

$$u = e^{-i\lambda/\rho}\rho^{\frac{1}{2}(n-1)}f' + e^{i\lambda/\rho}\rho^{\frac{1}{2}(n-1)}f'' \qquad \text{with } f', f'' \in \mathcal{C}^\infty(X) \text{ and } f' \restriction \partial X = f.$$

The resulting map $\mathcal{C}^\infty(\partial X) \ni f \longmapsto u \in \mathcal{C}^{-\infty}(X)$ is the *Poisson operator* for the problem and its leading part at the boundary,

$$A(\lambda) : \mathcal{C}^\infty(\partial X) \ni f \longmapsto f'' \restriction \partial X \in \mathcal{C}^\infty(\partial X)$$

is the absolute scattering operator for the metric. In the case of Euclidean space, as a scattering metric on $\mathbb{S}^n_+$, it can be seen by direct computation that $A(\lambda)f(\omega) = i^{\frac{1}{2}(n-1)}f(-\omega)$. In recent work with M. Zworski the analogous result in the general case has been obtained.

**Theorem** ([12]). *The scattering matrix for a scattering metric, at any non-zero real frequency $\lambda$, is a Fourier integral operator of order 0 associated to geodesic flow, for the metric $h$, on $\partial X$ at distance $\pi$.*

**Microlocalization.** Both Lemma 1 and this theorem are proved by the techniques of microlocal analysis. The quotient algebra $\Psi^*_{\mathrm{sc}}(X)/\rho^\infty\Psi^{-\infty}_{\mathrm{sc}}(X)$ is localized at the boundary of ${}^{\mathrm{sc}}\bar{T}^*X$, and hence a notion of wavefront set $\mathrm{WF}_{\mathrm{sc}}(u) \subset \partial{}^{\mathrm{sc}}\bar{T}^*X$, for extendible distributions $u \in \mathcal{C}^{-\infty}(X)$, can be defined essentially as in the case of a compact manifold without boundary. The precise limiting absorption principle, Lemma 1, is then seen to be a uniform version, in this context, of Hörmander's theorem on the propagation of singularities. The more precise result in the theorem is obtained by constructing a parametrix for the Poisson operator as a Legendre distribution, which is the analogue in this setting of a Lagrangian distribution. Indeed, except for the fact that the Legendre manifold involved is singular, the notion of Legendre distribution in [12] is reduced to the usual notion of a Lagrangian distribution by Fourier transformation.

## Fibred cusp calculus

The Lie algebra $\mathcal{V}_{\mathrm{sc}}(X)$, defined in (4) can also be described in the following way

$$\mathcal{V}_{\mathrm{sc}}(X) = \left\{V \in \mathcal{V}(X); V\rho \in \rho^2\mathcal{C}^\infty(X) \text{ and } V \text{ vanishes at } \partial X\right\}. \tag{9}$$

In this form the definition can be generalized by replacing the vanishing condition at the boundary by tangency to the fibres of a fibration; vanishing then corresponds to the 'trivial' fibration with each fibre a point. Thus let $\phi : \partial X \longrightarrow Y$ be a fibration of the boundary. Choose a boundary defining function $\rho \in \mathcal{C}^\infty(X)$ and set

$$(10)\quad \mathcal{V}_\Phi(X) = \{V \in \mathcal{V}(X); V\rho \in \rho^2\mathcal{C}^\infty(X) \text{ and } V \text{ is tangent to the fibres of } \phi \text{ at } \partial X\}.$$

Notice that the first condition certainly implies that $V \in \mathcal{V}_{\mathrm{b}}(X)$. The other extreme case, of $\phi : \partial X \longrightarrow \{\mathrm{pt}\}$ the one-fibre fibration, is called the 'cusp' algebra.

**Model case.** As opposed to the case of the scattering Lie algebra the definition (10) does in fact depend on the choice of defining function; the subscript $\Phi$ is supposed to indicate that it depends on more than $\phi$. The model case analogous to $\mathbb{R}^n$ for the scattering calculus is $\mathbb{R}^p \times F$ where the compact manifold without boundary $F$ of dimension $k$, $n = p + k$, is a typical fibre of $\phi$. A calculus of operators on this model space can be obtained by localizing on $F$ and considering the operators

$$(11)\quad \begin{aligned} Au(x,y) &= (2\pi)^{-n} \int_{\mathbb{R}^n \times \mathbb{R}^k} e^{ix\cdot\xi + (y-y')\cdot\eta} a(x,y,\xi,\eta)\hat{u}(\xi,y')dy'd\eta d\xi, \\ \hat{u}(\xi,y') &= \int_{\mathbb{R}^p} e^{-ix'\cdot\xi} u(x',y')dx', \ n = p+k \end{aligned}$$

where the amplitude is of the form

$$(12)\qquad a = (\mathrm{SP}_p \times \mathrm{SP}_n)^* a',\ \rho_\sigma^m a' \in \mathcal{C}^\infty(\mathbb{S}^p_+ \times \mathbb{S}^n_+ \times O).$$

Here, $\mathrm{SP}_p$ is radial compactification in $\mathbb{R}^p$ and $\mathrm{SP}_n$ is radial compactification in the product $\mathbb{R}^n = \mathbb{R}^p_\xi \times \mathbb{R}^k_\eta$ to $\mathbb{S}^n_+$, $\rho_\sigma$ is a boundary defining function in this factor and $O$ is a coordinate patch in $F$. It is again relatively easy to show that these local kernels near the diagonal, together with the smooth kernels vanishing to infinite order on both boundary faces, sum to give an algebra of operators on $\dot{\mathcal{C}}^\infty(\mathbb{S}^p_+ \times F)$ after radial compactification of $\mathbb{R}^p$ to $\mathbb{S}^p_+$.

Although, as already noted in the case of the scattering calculus, there is a more elegant formulation of this definition directly in terms of the Schwartz kernels I shall not describe this approach here; the detailed discussion can be found in [6].

**Suspended algebras.** The main formal difference between the scattering and cusp calculi, with intermediate behaviour for a general fibration of the boundary, is that in the latter case the operators 'remain global' at the boundary. The definition in (11) can be modified to define pseudodifferential operators on $F$ which depend symbolically on variables $\xi \in \mathbb{R}^p$ and smoothly on parameters $z \in S \subset \mathbb{R}^p$ $S$ open, by dropping the $\xi$ integrals and considering

$$Au(z,\xi,y) = (2\pi)^{-k} \int_{\mathbb{R}^{2k}} e^{(y-y')\cdot\eta} a(z,y,\xi,\eta)\hat{u}(\xi,y')dy'd\eta,$$

where $a$ is an amplitude as in (12), with the $x$ variables replaced by $z \in S$. These local kernels near the diagonal in $F \times F$ can be extended to global kernels by summing over a partition of unity and admitting global smooth kernels in $z, \xi, y, y'$ which are rapidly decreasing, with all derivatives in $\xi$. Again this gives an algebra of operators, parametrized by $(z,\xi) \in S \times \mathbb{R}^p$. These algebras are invariant under general diffeomorphisms of $F$, diffeomorphic change of the parameters in $S$ and linear transformations in the variables $\xi$. This allows the definition to be globalized to the following geometric arrangement. Let $Y$, replacing $S$, be a compact manifold without boundary. Let $V$ be a vector bundle over $Y$, the fibre of which replaces $\mathbb{R}^p$ and let $\phi : X' \longrightarrow Y$ be a fibration with typical fibre $F$. The fibre product $G = V \times_Y X'$ then reduces locally to the product $S \times \mathbb{R}^p \times F$ considered above and the local definition of the algebra extends to define a global algebra which we shall denote $\Psi^m_{\mathrm{sus}(V)-\phi}(G)$. Notice that its elements act on each fibre of $\phi$ and only depend parametrically on the other variables.

**Normal operator.** Each of the Lie algebras $\mathcal{V}_\Phi(X)$ corresponds to a vector bundle over $X$, denoted ${}^\Phi TX$, with the property that $\mathcal{V}_\Phi(X) = \mathcal{C}^\infty(X; {}^\Phi TX)$. This bundle comes equipped with a natural vector bundle map $\iota_\Phi : {}^\Phi TX \longrightarrow TX$ into the standard tangent bundle; it is an isomorphism over the interior and over the boundary has range of rank $k$, the fibre dimension, in $T\partial X$. Let ${}^\Phi N\partial X$ be the bundle over $\partial X$ consisting of the null spaces of $\iota_\Phi$. Now, this bundle is naturally trivial over each of the fibres of $\phi$, i.e. it is actually the lift under $\phi$ of a bundle ${}^\phi NY$ over $Y$. Let $G = {}^\phi N^*Y \times_Y \partial X$ be the fibre product, so the discussion above applies. The fibred cusp algebra then has two symbol maps

$$\begin{aligned}
&\sigma_{m,\Phi} : \Psi^m_\Phi(X;E) \longrightarrow S^{[m]}_\sigma({}^\Phi\bar{T}^*X;\pi^*\hom E) \text{ and}\\
&N : \Psi^m_\Phi(X;E) \longrightarrow \Psi^m_{\mathrm{sus}({}^\phi N^*Y)-\phi}(G;\pi^*\hom E) \text{ where}\\
&S^{[m]}_\sigma({}^\Phi\bar{T}^*X;\pi^*\hom E)\\
&\quad = \rho_\sigma^{-m}\mathcal{C}^\infty({}^\Phi\bar{T}^*X;\pi^*\hom E)/\rho_\sigma^{-m+1}\mathcal{C}^\infty({}^\Phi\bar{T}^*X;\pi^*\hom E).
\end{aligned} \tag{13}$$

In case the fibres of $\phi$ are not points, i.e. other than for the scattering calculus, the image space of the second map is a non-commutative algebra.

Both maps are surjective homomorphisms, but their direct sum is not surjective. Rather the compatibility condition between the two images is that the symbol of the 'normal operator' $N(A)$ is the restriction to the boundary of the symbol $\sigma_{m,\Phi}(A)$. The subspace so defined of the direct sum is an algebra and the null space of the joint symbol map is exactly the space $\rho\Psi_{\Phi}^{m-1}(X;E)$.

In this wider setting we can again obtain a complete characterization of the Fredholm properties of these operators acting on the corresponding Sobolev spaces. The latter are defined as before from the basic $L^2$ space. Notice that the algebra $\mathcal{V}_{\Phi}(X)$ fixes a positive measure, up to a smooth positive multiple, on $X$ by the condition that it pair smoothly and positively with the element $|v_1 \wedge \cdots \wedge v_n|$ for any local smooth basis. Let $L^2_{\Phi}(X)$ be defined with respect to this measure and set

$$H_{\Phi}^{m}(X;E) = \Psi_{\Phi}^{-m}(X;E) \cdot L^2(X;E).$$

**Lemma 2.** *There is an element $A_m \in \Psi_{\Phi}^{m}(X;E)$ for each $m$ and any vector bundle $E$ over $X$ such that*

$$H_{\Phi}^{m}(X;E) = A_{-m} L^2_{\Phi}(X;E).$$

From this and the properties of the algebra it is easy to deduce that elements of the calculus define bounded operators

$$A : H_{\Phi}^{m}(X;E) \longrightarrow L^2_{\Phi}(X;E),\ A \in \Psi_{\Phi}^{m}(X;E). \tag{14}$$

**Proposition 5.** *An element $A \in \Psi_{\Phi}^{m}(X;E)$ is Fredholm as a map (14) if and only if it is totally elliptic in the sense that its symbol $\sigma_{m,\Phi}(A)$ and normal operator $N(A)$ have inverses, respectively, in $S_{\sigma}^{[-m]}({}^{\Phi}T^*X;\pi^*\hom E)$ and $\Psi^{-m}_{\mathrm{sus}({}^{\phi}N^*Y)-\phi}(G;E)$.*

**Fibred cusp metrics.** As before this leads to an effective description of the spectrum of the Laplacian associated to a metric from the appropriate class. Given the defining function $\rho$ and the fibration of the boundary, $\phi$, it is always possible to find a Riemann metric on the interior of $X$ which takes the following form near the boundary

$$g = \frac{d\rho^2}{\rho^4} + \frac{\tilde{\phi}^* g_Y}{\rho^2} + h'. \tag{15}$$

Here $\tilde{\phi} : U \longrightarrow Y$ is an extension of the fibration of $\partial X$ to a neighbourhood of $\partial X$ given by the choice of a product decomposition near the boundary $U \sim [0,\epsilon)_{\rho} \times \partial X$; $g_Y$ is a metric on $Y$ (possibly depending on $\rho$) and $h'$ is a smooth 2-cotensor on $X$ near the boundary which restricts to define a metric on each fibre over the boundary. Using the fact that such a 'fibred cusp metric' defines a non-degenerate fibre metric on the bundle ${}^{\Phi}TX$ it is straightforward to show that the associated Laplacian is an elliptic element of $\Psi^2_\Phi(X)$.

From Proposition 5 it follows directly that the spectrum of the Laplacian for such a metric is $[0,\infty)$. There are 'thresholds' at which the multiplicity of the spectrum changes; these are given by the eigenvalues of the Laplacian on the fibres of $\phi$. Since these may vary with the point in $Y$, which labels the fibres, the general structure of the spectrum is somewhat complicated. Since the null space of the fibre Laplacians consists always of the constants alone, the spectrum *is* relatively simple near 0. It has a parametrization much as in the extreme case of the scattering metric discussed above.

**Proposition 6.** *For the Laplacian, acting on functions, of a fibred cusp metric* (15), *for each* $0 \neq \lambda \in \mathbb{R}$ *with* $\lambda^2$ *smaller than the infimum over* $Y$ *of the smallest positive eigenvalue of the fibre Laplacians, there is for each* $f \in \mathcal{C}^\infty(Y)$ *a unique generalized eigenfunction of the form*

$$(16)\quad u = e^{i\lambda/\rho}\rho^{\frac{1}{2}q}f' + e^{-i\lambda/\rho}\rho^{\frac{1}{2}q}f'',\ q = \dim Y,\qquad f', f'' \in \mathcal{C}^\infty(X),\ f' \restriction \partial X = \phi^* f.$$

*The closure of this set in* $\mathcal{C}^{-\infty}(X)$ *consists of all the eigenfunctions in this space.*

As in the scattering case this can be proved by the use of a variant of Hörmander's theorem on the propagation of singularities for the wavefront set corresponding to the fibred cusp calculus. This parametrization of the generalized eigenspaces, for small positive eigenvalues, also leads to an absolute scattering matrix, $A(\lambda) : f \longmapsto f'' \restriction \partial X$. In fact $f'' \restriction \partial X = \phi^* g$ for $g \in \mathcal{C}^\infty(Y)$ so the scattering operator acts on $\mathcal{C}^\infty(Y)$.

**Conjecture.** *For small* $\lambda$ *(such that* $\lambda^2$ *is less than the infimum of the positive fibre eigenvalues) the absolute scattering matrix for a fibred cusp metric is a Fourier integral operator on* $Y$ *associated to geodesic flow, at time* $\pi$, *for the metric* $g_Y$ *on* $Y$.

## Other calculi

I will finish this brief overview of some of the properties of 'cusp-type' algebras of pseudodifferential operators on a compact manifold with boundary with some comments on the relationship to other calculi and further likely developments. For simplicity I shall only consider operators on functions; however operators on sections of vector bundles behave similarly.

**Cusp and b-pseudodifferential operators.** First let me mention that the cusp calculus, which in terms of the fibred cusp structure discussed above is the opposite extreme to the scattering calculus, is closely related to the b-calculus introduced in [7] and discussed in detail in [4; Chapter XVII] and [8]. This relationship can be described, succinctly, as follows.

The algebra $\Psi_{\mathrm{b}}^*(X)$ is defined for any compact manifold with boundary $X$. One can change the $\mathcal{C}^\infty$ structure near the boundary of $X$ by introducing as a new boundary defining function

$$\rho_{\log} = \frac{1}{\log \frac{1}{\rho}} \tag{17}$$

where $\rho \in \mathcal{C}^\infty(X)$ is a boundary defining function such that $0 \le \rho < 1$ on $X$. The space $\mathcal{C}^\infty(X_{\log})$ of $\mathcal{C}^\infty$ functions for the new $\mathcal{C}^\infty$ structure (on the same set $X$) consists of function of the form $f(\rho_{\log}, f_1, \dots, f_N)$ where $f$ is $\mathcal{C}^\infty$ on $\mathbb{R}^{N+1}$ and the $f_i \in \mathcal{C}^\infty(X)$ are real-valued. Since this includes $\mathcal{C}^\infty(X)$ itself, the identity map is smooth $\iota_{\log} : X_{\log} \longrightarrow X$. Over the interior this is a diffeomorphism, so any operator on $X$ of the form $A : \mathcal{C}_c^\infty(X^\circ) \longrightarrow \mathcal{C}^\infty(X)$, $X^\circ = X \setminus \partial X$, lifts to an operator on $X_{\log}$. In this sense the space $\Psi_{\mathrm{b}}^m(X)$ lifts into the cusp algebra $\Psi_{\mathrm{cusp}}^m(X_{\log})$ defined with respect to $\rho_{\log}$, which is independent of the choice of $\rho$ modulo $\rho_{\log}^2\mathcal{C}^\infty(X_{\log})$, i.e. up to a term vanishing to second order at the boundary of $X_{\log}$. The difference between the two calculi is then only that the coefficients are more general in $\Psi_{\mathrm{cusp}}^*(X_{\log})$ than in $\Psi_{\mathrm{b}}^*(X)$, namely $\mathcal{C}^\infty$ functions on $X_{\log}$ in the first case rather than $\mathcal{C}^\infty$ functions on $X$ in the second.

With this identification, the structure of the scattering matrix and trace formula for the Laplacian of a b-metric carried out by T. Christiansen [2], using results from [8], can be extended to the more general case of cusp metrics.

**Index problem.** Various problems for cusp and b-pseudodifferential operators are therefore essentially the same. For example, the open problem of finding an explicit index formula for Fredholm pseudodifferential operators (characterized in Proposition 5) is the same in the two calculi. This index problem for pseudodifferential operators is, as yet, unsolved for all the fibred cusp algebras except the scattering algebra (see [10] for that case). The eta homomorphism discussed in [11] can be expected to play a rôle. It is defined on the boundary algebra in (13) for the cusp algebra. The extension of the construction of [11] to the boundary algebras for the other fibred cusp algebras (except the commutative case of the scattering algebra) would therefore be of interest.

**Quadratic scattering calculus.** The relationship between the scattering and the cusp calculi is rather more complicated than that between the cusp and b-calculi. However, it can be seen directly from (4) that there is such a relationship. Namely, the scattering algebra is obtained by quantization of a subalgebra of the algebra from which the b-pseudodifferential operators are obtained. One can consider smaller subalgebras of a similar nature, for example

$$\mathcal{V}_{\mathrm{qsc}}(X) = \rho^2\mathcal{V}_{\mathrm{b}} \subset \mathcal{V}_{\mathrm{sc}}(X) = \rho\mathcal{V}_{\mathrm{b}}(X) \subset \mathcal{V}_{\mathrm{b}}(X). \tag{18}$$

This 'quadratic scattering algebra' is therefore also well defined on any compact manifold with boundary. It has been analyzed in some detail by J. Wunsch, [16], who has used it to give a theorem on the propagation of singularities (in the sense of growth) for the Schrödinger operator. Again the quadratic scattering algebra is only slightly more general than the scattering algebra (as discussed above). Namely if $X$ is any compact manifold with boundary then instead of $\rho_{\log}$ in (17) one can get a new $\mathcal{C}^\infty$ structure by introducing

$$\rho_{\frac{1}{2}} = \rho^{\frac{1}{2}} \tag{19}$$

as a new boundary defining function. The resulting $\mathcal{C}^\infty$ structure turns the same set $X$ into a new manifold with boundary $X_{\frac{1}{2}}$ (diffeomorphic to the original one) with $\iota_{\frac{1}{2}} : X_{\frac{1}{2}} \longrightarrow X$ a $\mathcal{C}^\infty$ map which is a diffeomorphism over the interior. Apart from a small question of the smoothness of coefficients, the scattering calculus on $X$ then lifts to $X_{\frac{1}{2}}$ to be the 'quadratic scattering calculus' on $X_{\frac{1}{2}}$. That this can be reasonably associated to the (time dependent) Schrödinger operator $D_t + \Delta$ can be seen by introduction of the appropriate singular coordinates. As Wunsch shows, the propagation statements of [3] (see also [5]) can be interpreted in this way as a theorem on the propagation of wavefront sets.

**N-body problem.** Another analytic problem to which a microlocal approach similar to that described above is applicable is the quantum N-body problem. Consider $N$ particles with positions $x_j \in \mathbb{R}^n$ for $j = 1, \ldots, N$. For simplicity suppose that the potentials describing the relative interactions are real-valued functions $V_{ij} \in \mathcal{C}_c^\infty(\mathbb{R}^n)$. Then the energy states of the quantum system are described by the spectrum of the operator

$$\sum_{j=1}^{N} \Delta_j + \sum_{i>j=1}^{N} V_{ij}(x_i - x_j)$$

as an operator on $\mathbb{R}^{nN}$, where each $\Delta_j$ is the Laplacian in the variable $x_j \in \mathbb{R}^n$. The span of the variables on which the potentials depend is a hyperplane of dimension $n(N-1)$. The operator is translation-invariant in the extra 'centre of mass' coordinate which therefore, after Fourier transformation in those variables, can be absorbed into the spectral variable. Thus it is traditional to consider the 'reduced operator' which is of the form

$$B = \Delta_x + \sum_p V_p(L_p x), \ x \in \mathbb{R}^M, \ M = n(N-1),$$

with $L_j : \mathbb{R}^M \longrightarrow \mathbb{R}^n$ surjective linear maps which can be assumed to have no joint null space. The spectrum of such operators has been much discussed and the 'asymptotic completeness' has been shown by Sigal and Soffer; see [14] and references therein.

The methods used in this and many related papers do not easily yield explicit information on the structure of the resolvent of $B$. It is natural to associate to the operator $B$ a calculus of pseudodifferential operators. This calculus has not yet been discussed in detail, but it arises from the following compactification of $\mathbb{R}^M$. First consider the radial compactification to $\mathbb{S}^M_+$ of $\mathbb{R}^M$. The Laplacian lifts to an element of the scattering calculus, as discussed above. However the interaction potentials do not lift to be smooth on $\mathbb{S}^M_+$. Each of the linear maps $L_p$ has as null space a linear subspace of $\mathbb{R}^M$ which defines a spherical submanifold of the boundary of $\mathbb{S}^M_+$. The appropriate configuration space for the N-body problem is $\mathbb{S}^M_+$ with all these subspaces, and all their intersections, blown up in increasing order of dimension. One can expect that the generalized eigenfunctions of $B$ are relatively simple functions on this space. The extra boundary faces introduced at infinity are the 'scattering channels'.

The three-body problem is particularly simple since the (three) channels at infinity are disjoint. A. Vasy, see [15], has recently shown how the propagation properties of the scattering wavefront set can be used to construct a precise parametrix for the resolvent, and the Poisson operator. This is done using a finite iteration of unreduced two-body resolvents and Poisson operators. Since all these operators can be analyzed in detail, this allows rather precise information about the generalized eigenfunctions to be obtained. For example it follows that the absolute scattering matrix in this case is a Fourier integral operator associated to a singular canonical relation which consists of the broken geodesic flow on the boundary $\mathbb{S}^{M-1}$ at distance $\pi$. The 'breaks' occur at the channels at infinity. A. Hassell had previously shown that the scattering matrix in the unreduced two-body problem has a similar structure.

## References


1. S. Agmon and L. Hörmander, *Asymptotic properties of solutions of differential equations with simple characteristics*, J. Analyse Math. **30** (1976), 1-38.
2. T. Christiansen, *Scattering theory for manifolds with asymptotically cylindrical ends*, J. Funct. Anal. **131** (1995), 499-530.

3. W. Craig, T. Kappeler and W. Strauss, *Microlocal dispersive smoothing for the Schrödinger equation*, Preprint.
4. L. Hörmander, *The Analysis of Linear Partial Differential Operators*, vol. 3, 4, Springer-Verlag, Berlin, Heidelberg, New York, Tokyo, 1985.
5. L. Kapitanski and Y. Safarov, *Dispersive smoothing for Schrödinger equations*, Preprint, 1995.
6. R. Mazzeo and R. B. Melrose, *Pseudodifferential operators on manifolds with fibred boundaries*, In preparation.
7. R. B. Melrose, *Transformation of boundary problems*, Acta Math. **147** (1981), 149-236.
8. ———, *The Atiyah-Patodi-Singer Index Theorem*, A K Peters, Wellesley, Mass, 1993.
9. ———, *Spectral and scattering theory for the Laplacian on asymptotically Euclidian spaces*, Spectral and scattering theory (M. Ikawa, ed.), Marcel Dekker, 1994.
10. ———, *Geometric scattering theory*, Cambridge University Press, 1995.
11. ———, *The eta invariant and families of pseudodifferential operators*, Math. Res. Letters (to appear).
12. R. B. Melrose and M. Zworski, *Scattering metrics and geodesic flow at infinity*, Invent. Math. (1996) (to appear).
13. M. A. Shubin, *Pseudodifferential operators on* $\mathbb{R}^n$, Soviet Math. Dokl. **12** (1971), 147-151.
14. I. M. Sigal and A. Soffer, *Asymptotic completeness of* $N$*-particle long-range scattering*, J. Amer. Math. Soc. **7** (1994), 307-334.
15. A. Vasy, *Structure of the resolvent for three-body potentials*, In preparation.
16. J. Wunsch, *Propagation of singularities and growth for Schrödinger operators*, In preparation.

Department of Mathematics, MIT, Cambridge, MA 02139, USA
*E-mail address*: rbm@math.mit.edu

# DEGREE THEORY BEYOND CONTINUOUS MAPS

LOUIS NIRENBERG

**Abstract.** This is a report of joint work with H. Brezis to appear in *Selecta Mathematica.* Classical degree theory for maps from a smooth oriented manifold $X$, to another, $Y$, of the same dimension $n$, is defined for continuous maps. In recent years, in connection with certain variational problems, it has become of interest to extend degree theory to certain families of maps which need not be continuous. For simplicity, we consider only $X, Y$ compact without boundary, $Y$ connected. We also suppose that $Y$ is smoothly embedded in some $\mathbb{R}^N$. Then the degree of a map $u$ at a point $y$, an integer, is independent of the point $y$, and may be written as $\deg(u, X, Y)$.

If $u$ is smooth, and $X, Y$ are Riemannian manifolds, then

$$\deg(u, X, Y) = \frac{1}{\text{Vol}(Y)} \int_X \det J_u(x)\, d\sigma(x) \tag{1}$$

where $d\sigma$ is the volume element on $X$ and $J_u$ is the Jacobian of the map $u$ (computed using geodesic normal coordinates at $x$ and at $u(x)$). The integral makes sense if the map is in the Sobolev class $W^{1,n}(X, Y)$, i.e., as a map from $X$ to $\mathbb{R}^N$, $u$ has first derivatives in $L^n$, and $u(x) \in Y$ a.e. Then, the fact that the right hand side is an integer is proved by approximating $u$ in $W^{1,n}(X, Y)$ by smooth maps.

In this talk we extend degree theory, via approximation by smooth maps, to a class of maps including $W^{1,n}(X, Y)$ and $C^0(X, Y)$. The class of maps is the completion of smooth maps in the BMO (bounded mean oscillation) topology. Recall that for $u \in L^1(X, \mathbb{R}^N)$, the BMO norm of $u$ is

$$\|u\|_{\text{BMO}} = \sup_{\substack{x \in X \\ \varepsilon \leq \varepsilon_0}} ⨍_{B_\varepsilon(x)} \left| u - ⨍_{B_\varepsilon(x)} u \right|.$$

Here $B_\varepsilon(x)$ is a geodesic ball in $X$ centred at $x$ of radius $\varepsilon \leq \varepsilon_0$, the injectivity radius of $X$, and the slash mean averaging the integral. The set of maps with finite BMO norm forms a Banach space — modulo adding constants. The completion of smooth (or continuous) maps in this topology forms a subspace of BMO called VMO, first introduced by D. Sarason. He characterized $u \in$ VMO by the property (hence the name VMO):

$$\lim_{\varepsilon\to 0} \fint_{B_\varepsilon(x)} \left|u - \fint_{B_\varepsilon(x)} u\right| \to 0 \text{ uniformly in } x \in X.$$

For maps from $S^1$ to $S^1$, L. Boutet de Monvel and O. Gabber had, earlier, suggested the space VMO for defining degree.

We prove that the degree for maps $u \in \mathrm{VMO}(X,Y)$ obtained by approximation, has the usual properties of a degree. One may also define degree for $u \in \mathrm{BMO}(X,Y)$ provided the distance from $u$ to $C^{\mathrm{C}}(X,Y)$, measured in the BMO norm is small.

COURANT INSTITUTE, NEW YORK UNIVERSITY, NEW YORK, N. Y. 10012, U. S. A.
*E-mail address*: nirenl@MATH12.CIMS.NYU.EDU

# ON THE WEYL FORMULA FOR OBSTACLES

DIDIER ROBERT

**Abstract.** In this paper a trace perturbation formula for scattering by obstacles is proved. The main application of this formula concerns high energy asymptotics for scattering phases (or Krein spectral shift functions). In particular we prove a Weyl type asymptotic for obstacle problems with arbitrary geometry, in any dimension, extending a result proved by R. Melrose in odd dimension.

For example, let us consider in the Euclidean plane $\mathbb{R}^2$ a compact obstacle $K$ and the Laplace-Dirichlet problem in the domain $\Omega := \mathbb{R}^2 \backslash K$, denoted by $P_\Omega$. Then the scattering phase $s(\lambda)$ for the pair $(P_\Omega, P_{\mathbb{R}^2})$ satisfies

$$s(\lambda) = \frac{\text{Area}(K)}{4\pi}\lambda + o(\lambda) \quad \text{as } \lambda \to +\infty,$$

without any smoothness assumption on $K$. If the boundary of $K$ is smooth then we have the remainder term estimate:

$$s(\lambda) = \frac{\text{Area}(K)}{4\pi}\lambda + O(\lambda^{1/2}) \quad \text{as } \lambda \to +\infty.$$

More generally, we prove that similar results hold for a large class of perturbations of elliptic operators at infinity which may be degenerate in a bounded set.

**1. Introduction.** First of all, let us recall the well-known Weyl formula [46] for an open bounded set $\Omega \subset \mathbb{R}^n$. Let $\{\lambda_{\Omega,j}\}_{j\in\mathbb{N}}$ be the nondecreasing sequence of eigenvalues for the Laplace-Dirichlet problem in $\Omega$:

$$-\triangle\phi_j = \lambda_{\Omega,j}\phi_j; \quad \phi_j|_{\partial\Omega} = 0,$$

and let us introduce the eigenvalue counting function:

$$N(\lambda) := \text{Card}\{j \in \mathbb{N};\ \lambda_{\Omega,j} \le \lambda\}.$$

The Weyl asymptotic formula is ([1, 13, 15, 21])

$$N(\lambda) = (2\pi)^{-n}\omega_n \operatorname{Vol}(\Omega)\, \lambda^{n/2} + R_n(\lambda) \tag{1}$$

with $R_n(\lambda) = o(\lambda^{n/2})$ for arbitrary bounded domains $\Omega$ and $R_n(\lambda) = O(\lambda^{(n-1)/2})$ for smoothly bounded domains; $\omega_n$ is the Euclidean volume of the unit ball in $\mathbb{R}^n$.

Let us now consider a compact obstacle $K \subset \mathbb{R}^n$ and the Laplace-Dirichlet problem in $\Omega := \mathbb{R}^n \backslash K$. It is known (M. G. Krein [18]) that there exists an analogue of the eigenvalue counting function $N(\lambda)$, called the spectral shift function or the scattering phase, and which is denoted here by $s(\lambda)$. A naïve definition of $s(\lambda)$ could be the following (we shall prove later that this definition is correct).

Let us denote by $B_R$ the open ball in $\mathbb{R}^n$ of radius $R > 0$ with center at the origin and $\Omega_R := \Omega \cap B_R$. Taking $R_0$ large enough we have $K \subset B_{R_0}$.

**Proposition 1.1.** *For every $f$ in the Schwartz space $\mathcal{S}(\mathbb{R})$, the following limit exists*

$$\lim_{R \to +\infty} \Big\{ \sum_{j \in \mathbb{N}} f(\lambda_{\Omega_R, j}) - f(\lambda_{B_R, j}) \Big\} = L(f). \tag{2}$$

*Moreover there is a unique function $s : \mathbb{R} \to \mathbb{R}$, right continuous in $\mathbb{R}$, with polynomial growth at $+\infty$ and $s(\lambda) = 0$ for $\lambda < 0$, such that, for every $f$ in $\mathcal{S}(\mathbb{R})$ we have*

$$L(f) = \int_{-\infty}^{+\infty} s(\lambda)\, f'(\lambda) d\lambda. \tag{3}$$

Our goal here is to prove an asymptotic formula for $s(\lambda)$ as $\lambda \nearrow +\infty$. In particular we get

**Theorem 1.2.** *For every $n \geq 1$ we have*

$$s(\lambda) = (2\pi)^{-n} \omega_n \operatorname{Vol}(K)\, \lambda^{n/2} + R_n(\lambda) \tag{4}$$

*with:*

$$R_n(\lambda) = o(\lambda^{n/2}) \tag{5}$$

*for arbitrary compact obstacles $K$, and:*

$$R_n(\lambda) = O(\lambda^{(n-1)/2}) \tag{6}$$

*for smooth compact obstacles. Furthermore:*

$$R_n(\lambda) = \frac{\omega_{n-1}}{4(2\pi)^{(n-1)}} \operatorname{Vol}(\partial K) \lambda^{(n-1)/2} + o(\lambda^{(n-1)/2}) \tag{7}$$

*if the set of billiard trajectories in $\Omega$ has measure 0 in the cotangent space.*

Let us mention some points of the history concerning Weyl asymptotics for obstacles. One of the first papers concerning asymptotics for $s(\lambda)$ is the pioneering work of Buslaev [5] (1975). But it seems that no complete proof appeared. Then follows Jensen-Kato [16] (1978) where (5) is proved assuming the obstacle star-shaped. Majda-Ralston [20] (1978) get (7) for smooth strictly convex obstacles. For smooth and nontrapping obstacles Petkov-Popov [28] (1982) get complete asymptotics for $s(\lambda)$ as $\lambda \nearrow +\infty$, for $n \geq 3$. Nevertheless, the problem of asymptotics of $s(\lambda)$ for smooth but arbitrarily shaped domains remained an open problem till 1988. It was not known before that $s(\lambda)$ has polynomial growth. In [22] Melrose gives the first general result for arbitrarily shaped obstacles in odd dimension. At that point we have to emphasize that an essential difference between the functions $N(\lambda)$ and $s(\lambda)$ concerns monotonicity. $N(\lambda)$ is obviously nondecreasing but it is not known if $s(\lambda)$ is monotone for any obstacle (it is true for star-shaped domains [16]). However monotonicity is important to apply Tauberian theorems. For example Grubb [11] (1984) proves complete asymptotics for the Stieltjès transform of $s(\lambda)$, $\sigma_m(\lambda) = \int \frac{s(\mu)}{\mu^m+\lambda} d\mu$, for $m > \frac{n}{2} + 1$. But this is not sufficient to prove (4). In [22] Melrose solves the problem by using his estimate on the growth of the scattering poles in odd dimensional domains and splitting $s(\lambda)$ into two terms such that one is monotone, the other is a symbol, hence the Tauberian argument of Hörmander-Levitan [13] can be applied.

In very recent work, Smilansky [42] and Eckmann-Pillet [9] dicuss connections between the eigenvalues of the interior problem and the scattering matrix for the exterior problem, for planar obstacles.

Here, we shall present a different approach, essentially by establishing a suitable perturbation trace formula and splitting $s(\lambda)$ into several monotone pieces. Some results contained in this paper were announced in [34, 35]. Here we shall give complete proofs.

**2. Assumptions and general results in Obstacle Scattering.** Our general frame work is the following. Let $K$ be a compact obstacle, $K \subset \mathbb{R}^n$ and set $\Omega = \mathbb{R}^n \backslash K$. Let us introduce the operators $e_\Omega$, extension by 0 outside $\Omega$, and $r_\Omega$, the restriction operator to $\Omega$. We have clearly $r_\Omega^* = e_\Omega$ ($A^*$ denotes the adjoint of the operator $A$), and $r_\Omega\, e_\Omega = \mathbb{I}$. Let us also introduce the notation

$$\langle x \rangle := (1 + |x|^2)^{1/2};$$

$$B_R = \{x \in \mathbb{R}^n;\ |x| < R\}; \quad B_R^c = \mathbb{R}^n \backslash B_R; \quad \Omega_R := \Omega \cap B_R.$$

We shall consider differential operators of order $2m$,

$$p(x, D) = \sum_{|\alpha| \leq 2m} a_\alpha(x) D^\alpha, \quad \text{where } D := -i\nabla.$$

We assume that $a_\alpha \in L^1_{\mathrm{loc}}(\Omega)$ and that the following assumptions hold:
$(H_1)$ $\exists R_0 > 0$ such that $K \subset B_{R_0}$ and $a_\alpha \in C^\infty(B^c_{R_0})$.
$(H_2)$ $\exists R_0 > 0$, $\rho > n$ and an elliptic homogeneous polynomial $p_0(\xi) = \sum_{|\alpha|=2m} a_{0,\alpha}\xi^\alpha$, $a_{0,\alpha} \in \mathbb{C}$, such that:

$$(8) \qquad |\partial_x^\beta (a_\alpha(x) - a_{0,\alpha})| \leq c_{\alpha,\beta} \langle x \rangle^{-\rho-|\beta|}, \ \forall x, \ |x| > R_0,$$

where $a_{0,\alpha} = 0$ for $|\alpha| < 2m$, and
$(H_3)$ $p(x, D)$ is symmetric on $C_0^\infty(\Omega)$.
Let us consider a self-adjoint extension $P$ of $p(x, D)$ in $L^2(\Omega)$ with a domain $D(P)$. Assume furthermore:
$(H_4)$ $\zeta D(P) \subset D(P)$ $\forall \zeta \in C_0^\infty(\mathbb{R}^n)$ with $\zeta \equiv 1$ in a neighborhood of $K$.
$(H_5)$ $\exists \mu > 0$ such that $\forall \zeta \in C_0^\infty(\mathbb{R}^n)$, the product $\zeta \cdot (P + i)^{-1}$ is in the Schatten operator class $S^\mu$ in $L^2(\Omega)$.
For definitions and properties of Schatten classes $S^\mu$ we refer to [10]. In the whole paper $\|\cdots\|_\mu$ denotes the operator norm in the Schatten class $S^\mu$. The limiting case $\mu = 1$ corresponds to the trace-norm, denoted also by $\|\cdots\|_{\mathrm{tr}}$, and $\mu = +\infty$ corresponds to the uniform operator norm, denoted by $\|\cdots\|$.
For convenience we also assume
$(H_6)$ $P$ is semibounded below in $L^2(\Omega)$.

*Remark 2.1.* Ellipticity for $P$ is assumed only at infinity. In particular $P$ may be degenerate on $\partial\Omega$ (see section 4) and its coefficients may have integrable singularities at finite distance.

*Remark 2.2.* A similar framework was considered in [41] to study scattering resonances.

Let us denote by $P_0$ the self-adjoint extension of $p_0(D)$ in $L^2(\mathbb{R}^n)$ ($D(P_0) = H^{2m}(\mathbb{R}^n)$). As usual, to compare the Hamiltonians $P$ and $P_0$ we use the identification operator $r_\Omega$ defined above.

In what follows, $C$ is a generic constant, independent on the point $z \in \mathbb{C}$.

The following proposition is a preliminary result to define the spectral shift function for the pair $(P_0, P)$.

**Proposition 2.3.** *Assume that $(H_1)$ to $(H_5)$ hold and that*

$$N \geq N_1 > \max\{N_0, \frac{n}{2m+1}\}.$$

*Then* $\forall z \in \mathbb{C}\backslash\mathbb{R}$ *we have the following uniform estimate in trace-norm in* $L^2(\mathbb{R}^n)$*:*

$$\|e_\Omega(P-z)^{-N}r_\Omega - (P_0-z)^{-N}\|_{\mathrm{tr}} \leq C\frac{\langle z\rangle^{N_1}}{|\operatorname{Im} z|^N}. \tag{9}$$

*Proof.* We will use standard estimates for elliptic operators and classical arguments from perturbation theory. Let us consider a cut-off $\zeta \in C_0^\infty(\mathbb{R}^n)$, $\zeta \equiv 1$ on $B_{R_0}$, $R_0$ being chosen so that $K \subset B_{R_0}$. At first, using classical estimates and assumptions $(H_1)$ to $(H_5)$, we get, for $N \geq 1$,

$$\|(P+i)^{-N}(P-z)^N\| \leq C\langle z\rangle^N, \tag{10}$$

$$\|(P+i)^N(P-z)^{-N}\| \leq C\Big(\frac{\langle z\rangle}{|\operatorname{Im}(z)|}\Big)^N, \tag{11}$$

and for $N > \max\{N_{\mathrm{C}}, \frac{n}{2m}\}$,

$$\|\zeta(P_0+i)^{-N}\|_{\mathrm{tr}} \leq C, \tag{12}$$

$$\|\zeta\, e_\Omega(P-z)^{-N}r_\Omega\|_{\mathrm{tr}} + \|\zeta\,(P_0-z)^{-N}\|_{\mathrm{tr}} \leq C\Big(\frac{\langle z\rangle}{|\operatorname{Im}(z)|}\Big)^N. \tag{13}$$

So we have to estimate the contribution far from the boundary of $K$. Let us introduce the notation

$$\triangle_{\zeta,N}(z) := (1-\zeta)(P-z)^{-N}r_\Omega - (1-\zeta)(P_0-z)^{-N}, \tag{14}$$

$$q(x,D) := p(x,D) - p_0(D) = \sum_{|\alpha|\leq 2m} q_\alpha(x)D^\alpha. \tag{15}$$

From $(H_2)$ we know that $|D^\beta q_\alpha(x)| \leq C\langle x\rangle^{-\rho-|\beta|}$ (with $\rho > n$). For $N = 1$ we get by an easy computation

$$\triangle_{\zeta,1} = (P-z)^{-1}\left([P,\zeta] - (1-\zeta)q(x,D)\right)(P_0-z)^{-1} + (P-z)^{-1}[P,\zeta](P-z)^{-1}. \tag{16}$$

Taking the derivative in (16) $(N-1)$ times in $z$ we get

$$\triangle_{\zeta,N}(z) = \sum_{k+l=N} c_{k,l}(P-z)^{-k}\left([P,\zeta] - (1-\zeta)q(x,D)\right)(P_0-z)^{-l} + \sum_{k+l=N} d_{k,l}(P-z)^{-k}[P,\zeta](P-z)^{-l}. \tag{17}$$

Let $\tilde{\zeta} \in C_0^\infty(\mathbb{R}^n)$ be another cut-off function, such that $\zeta\tilde{\zeta} \equiv 1$ and $\tilde{\zeta} \equiv 1$ on $B_{R_0}$. We have easily

(18) $$\|(P-z)^{-k}(1-\zeta)q(x,D)(P_0-z)^{-l}\|_{\mathrm{tr}}$$
$$\leq \|(P-z)^{-k}(1-\zeta)(P_0-z)^{k}\|\|(P_0-z)^{-k}(1-\tilde{\zeta})q(x,D)(P_0-z)^{-l}\|_{\mathrm{tr}}.$$

Using the estimates (10), (12) we can estimate each term in the product and the conclusion follows from the following classical trace-class estimate in $L^2(\mathbb{R}^n)$:

(19) $$\|(P_0+i)^{-k}(1-\tilde{\zeta})q(x,D)(P_0+i)^{-l}\|_{\mathrm{tr}} \leq C, \quad \text{if } k+l \geq \frac{n}{2m}+1.$$

□

**Corollary 2.4.** *$f \to \mathrm{tr}\,(e_\Omega f(P)r_\Omega - f(P_0))$ is a Schwartz temperate distribution on $\mathbb{R}$.*

*Proof.* We use the Cauchy and Mellin representation formulas, like in [12], for the functional on self-adjoint operators. We refer to [36] for details. It is convenient to assume that $(H_6)$ holds. Then $P_0 \geq c_0\mathbb{I}$, $P \geq c_0\mathbb{I}$, for some $c_0 > 0$. Let $c_1 > -c_0$ and let $\Gamma$ be an infinite complex contour surrounding the spectra of $P_0$ and $P_1$, $\Gamma \subset \{z \in \mathbb{C}, \mathrm{Re}\, z > 0\}$ (see [36]). The Cauchy formula gives, for $s \in \mathbb{C}$, $\mathrm{Re}\, s > 0$, $s \notin \mathbb{N}$,

(20)
$$(P+c_1)^{-s} = \frac{(N-1)!}{(s-1)(s-2)\cdots(s-N+1)}\int_\Gamma z^{-s+N-1}(P+c_1-z)^{-N}dz.$$

Choosing for $\Gamma$ two straight lines outside a compact set, with angles $\pm\theta$ and choosing $\theta = |\,\mathrm{Im}(s)|^{-1}$, we get, for $\sigma \notin \mathbb{N}$, $\sigma > N_1+1$, $\mathrm{Re}\, s = \sigma$, (see [12])

(21) $$\|e_\Omega(P+c_1)^{-s}r_\Omega - (P_0+c_1)^{-s}\|_{\mathrm{tr}} \leq C.$$

Let us consider $f \in \mathcal{S}(\mathbb{R})$ (the usual Schwartz space of fast decreasing functions). Then using the inverse Mellin transform, it follows from (20) that we have, with $C$ independent on $f$,

(22) $$\|e_\Omega f(P)r_\Omega - f(P_0)\|_{\mathrm{tr}} \leq C\int_{\mathbb{R}} \langle\lambda\rangle^{\sigma-1}\Big(\sum_{0\leq j\leq 2}\langle\lambda\rangle^j|f^{(j)}(\lambda)|\Big)\,d\lambda.$$

□

We shall improve this estimate in Section 3. Now we are going to prove the first part of Proposition 1.1. Let us consider the realization $P_R$ of $P$ in $\Omega_R := \Omega \cap B_R$ by putting Dirichlet condition on $\partial B_R$ (which implies that $D(P_R) = D(P) \cap H_0^m(B_R)$). The analogous notation is used for $P_{0,R}$. By the functional calculus techniques used above, the result will follow from the following lemmas.

Let us consider the family of cut-off functions $\zeta_R(x) := \zeta(x/R)$ where $\zeta \in C_0^\infty(\mathbb{R}^n)$, $\zeta(x) = 1$ for $|x| \leq 1/2$, $\zeta(x) = 0$ for $|x| \geq 1$.

**Lemma 2.5.** *There exists $N_2 > 0$ such that for every $\zeta \in C_0^\infty(\mathbb{R}^n)$ and every integer $N > N_2$, there exists $C > 0$ such that, for every $R > R_1$ we have*

$$\|\zeta\left(e_{\Omega_R}(P_R - z)^{-N} r_{\Omega_R} - (P - z)^{-N}\right)\zeta\|_{\mathrm{tr}} \le \frac{C}{R}\frac{\langle z\rangle^{N_2}}{|\operatorname{Im} z|^N}. \tag{23}$$

*We also have the analogous result for $(P_0,\ P_{0,R})$.*

*Proof.* We have the formula

$$\zeta_R(P_R - z)^{-1} - (P - z)^{-1} = (P - z)^{-1}[P, \zeta_R](P_R - z)^{-1}, \tag{24}$$

hence, using that $\zeta\zeta_R \equiv 1$ for $R > 2$, we have

$$\zeta\left(e_\Omega(P_R - z)^{-1} r_\Omega - (P - z)^{-1}\right)\zeta = \zeta(P - z)^{-1}[P, \zeta_R](P_R - z)^{-1}\zeta. \tag{25}$$

Taking derivatives in $z$ and using the assumption $(H_2)$, we get the lemma. □

Let us introduce the notation

$$\triangle_N(R, z) := e_{\Omega_R}(P_R - z)^{-N} r_{\Omega_R} - e_{B_R}(P_{0,R} - z)^{-N} r_{B_R} \tag{26}$$

and $\triangle_N(+\infty, z) := \triangle_N(z)$. Then we have

**Lemma 2.6.** *With the above notation, there exist $N_2$ and $\varepsilon \in ]0,1]$ such that, for every integer $N > N_2$, there exists $C > 0$ such that for every $z \in \mathbb{C}\backslash\mathbb{R}$ we have*

$$\|(1 - \zeta)\zeta_{R/4}\left(\triangle_N(R, z) - \triangle_N(z)\right)\zeta_{R/4}\|_{\mathrm{tr}} \le \frac{C}{R^\varepsilon}\frac{\langle z\rangle^{N_2}}{|\operatorname{Im} z|^N}. \tag{27}$$

*Proof.* Let us start with the usual perturbation formulae

$$\begin{aligned}(1 - \zeta)&\big((P_{0,R} - z)^{-1} - (P_R - z)^{-1}\big) \\ &= (P_R - z)^{-1}((1 - \zeta)Q + [\zeta, P])(P_{0,R} - z)^{-1},\end{aligned} \tag{28}$$

$$\begin{aligned}(1 - \zeta)&\big((P_0 - z)^{-1} - (P - z)^{-1}\big) \\ &= (P - z)^{-1}((1 - \zeta)Q + [\zeta, P])(P_0 - z)^{-1},\end{aligned} \tag{29}$$

and

$$(P_{0,R} - z)^{-1}\zeta_R - \zeta_R(P_0 - z)^{-1} = (P_{0,R} - z)^{-1}[P_0, \zeta_R](P_0 - z)^{-1}, \tag{30}$$

$$(P_R - z)^{-1}\zeta_R - \zeta_R(P - z)^{-1} = (P_R - z)^{-1}[P, \zeta_R](P - z)^{-1}. \tag{31}$$

For $C > 0$, large enough, we clearly have

$$\text{dist}\{\text{supp}(\nabla\zeta_R), \text{supp}(\zeta_{R/4})\} \geq \frac{R}{C}, \tag{32}$$

$$\zeta_{R/4}\zeta_R \equiv \zeta_{R/4}. \tag{33}$$

After some standard computations we achieve the proof of the lemma by taking derivatives in $z$ and using that the Schwartz kernel of pseudodifferential operators is fast decreasing outside the diagonal. This gives the following

**Lemma 2.7.** *There exist $n_0$, $n_1 \in \mathbb{N}$ such that for every $M > 0$ and every multiindex $\alpha \neq 0$, there exists $C > 0$ such that*

$$\|[D^\alpha, \zeta_R](P - z)^{-1}\zeta_{R/4}\|_{\text{tr}} \leq \frac{C}{R^M}\frac{\langle z\rangle^{n_0}}{|\operatorname{Im}(z)|^{M+n_1}}. \tag{34}$$

The proof of this lemma is standard: for $P = P_0$ it is obtained by integration by parts and for general $P$ by construction of a parametrix (see [36]).

**Lemma 2.8.** *With the notation above there exist $N_2$ and $\varepsilon \in ]0,1]$ such that, for every integer $N > N_2$, there exists $C > 0$ such that for every $z \in \mathbb{C}\backslash\mathbb{R}$ we have*

$$\|(1-\zeta^2_{R/4})(\triangle_N(R,z) - \triangle_N(z))(1-\zeta^2_{R/9})\|_{\text{tr}} \leq \frac{C}{R^\varepsilon}\frac{\langle z\rangle^{N_2}}{|\operatorname{Im} z|^N}. \tag{35}$$

*Proof.* Let us remark that we have $(1-\zeta^2_{R/4})(1-\zeta^2_{R/9}) = (1-\zeta^2_{R/4})$. Thus we can apply formulae (28), (30) with $\zeta = \zeta^2_{R/4}$ and use (34) as in the proof of Lemma 2.6. □

*End of the proof of* (2) *of Proposition 1.1.* We use the same functional calculus method as in the proof of Corollary 2.4 and the above lemmas. Let us denote, for $s \in \mathbb{C}$,

$$F(R,s) := e_{\Omega_R}(P_R + c_1)^{-s} r_{\Omega_R} - e_{B_R}(P_{0,R} + c_1)^{-s} r_{B_R} \tag{36}$$

and $F(+\infty, s) := F(s)$. As in the proof of Corollary 2.4, using Lemmas 2.5, 2.6 and 2.8, we get, for $\sigma \notin \mathbb{N}$, $\sigma > N_1 + 1$, $\operatorname{Re} s = \sigma$, (see [12])

$$\|F(R,s) - F(s)\|_{\text{tr}} \leq \frac{C}{R^\varepsilon}. \tag{37}$$

Let us consider $f \in \mathcal{S}(\mathbb{R})$. Then there exists $C_f$ such that for all $R \geq R_1$ we have

$$
\begin{aligned}
(38)\quad & |\mathrm{tr}\,(e_{\Omega_R} f(P_R) r_{\Omega_R} - e_{B_R} f(P_{0,R}) r_{B_R}) - \\
& \qquad \mathrm{tr}\,(e_\Omega f(P) r_\Omega - f(P_0))| \leq \frac{C_f}{R^\varepsilon},
\end{aligned}
$$

hence we get the formula (2). □

*Remark 2.9.* A result similar to (2) is given in [38] for Sturm-Liouville problems on the real line.

Until now we have introduced the scattering phase without referring to the usual scattering theory (time dependent or time independent) in the framework presented in $(H_1)$ to $(H_6)$. Under these assumptions we can use the trace-class perturbation theory of Birman-Krein (see [45] for a self contained presentation). So from the above proposition and [45] we get easily

**Corollary 2.10.** (i) *The wave operators:*

$$
W_\pm := \lim_{t \to \pm\infty} \exp(itP) r_\Omega \exp(-itP_0) \tag{39}
$$

*exist and are complete.*
(ii) *The scattering matrix, $S(\lambda)$, is a trace-class perturbation of the identity in $L^2(\Sigma)$, where $\Sigma = \{\xi \in \mathbb{R}^n, p_0(\xi) = 1\}$ with the Euclidean measure.*
(iii) *There exists a Borel mesurable function $s : \mathbb{R} \to \mathbb{R}$ and $N > 0$ such that*

$$
\int_{\mathbb{R}} \frac{|s(\lambda)|}{(1 + |\lambda|)^N} d\lambda < +\infty, \tag{40}
$$

$$
\mathrm{tr}\,(e_\Omega f(P) r_\Omega - f(P_0)) = \int_{\mathbb{R}} s(\lambda) f'(\lambda) d\lambda, \tag{41}
$$

$$
\det(S(\lambda)) = \exp(-2i\pi s(\lambda)) \quad \text{a.e. } on \ ]0, +\infty[. \tag{42}
$$

*Remark 2.11.* Usually the scattering phase for obstacles is defined by the wave equation and the scattering matrix associated with the wave group [22, 19, 30]. Let us denote by $\xi(\lambda)$ the scattering phase defined in this way. We have a direct connection with the scattering phase $s(\lambda)$ defined above through the Schrödinger group: $s(\lambda) = \gamma\xi(\sqrt{\lambda})$ where $\gamma$ is a numerical factor.

**3. A Perturbation Trace Formula.** Now we state the main technical tool of this paper from which we shall deduce a Weyl type formula for obstacles. Let us introduce the notations: $Q = P - P_0$, $A = \frac{1}{2}(x \cdot D + D \cdot x)$ ($A$ is the generator of the dilation group of $\mathbb{R}^n$). Since $p_0$ is homogeneous of degree $2m$, we have, by the Euler identity:

$$\frac{1}{i}[A, p_0(D)] = 2m\, p_0(D). \tag{43}$$

Let us fix $\zeta$ as in $(H_1)$, $(H_2)$, with $R_0$ large enough. Then we have:

**Theorem 3.1.** *Under the assumptions $(H_1)$ to $(H_5)$, for every $f \in C_0^\infty(\mathbb{R}\backslash\{0\})$ we have the trace formula:*

$$\int f'(\lambda)\, s(\lambda) d\lambda = \mathrm{tr}\,(\zeta^2 f(P)) + \mathrm{tr}\Big(\frac{A[\zeta^2, P]}{2im} P^{-1} f(P)\Big) + +\mathrm{tr}\Big((Q - \frac{[A, Q]}{2im})(1-\zeta^2)P^{-1} f(P)\Big). \tag{44}$$

*Proof.* By definition, we have:

$$\int \frac{d}{d\lambda}(\lambda f(\lambda))\, s(\lambda)\, d\lambda = \mathrm{tr}\,(e_\Omega P f(P) r_\Omega - P_0 f(P_0)). \tag{45}$$

We have first, using (43),

$$P = \frac{1}{2im}[A, P] + Q - \frac{1}{2im}[A, Q], \tag{46}$$

so we get:

$$\begin{aligned} &\mathrm{tr}\,(e_\Omega J P(1-\zeta^2) f(P) r_\Omega P_0(1-\zeta^2) f(P_0)) \\ &= \mathrm{tr}\Big((Q - \frac{1}{2im}[A, Q])(1-\zeta^2)\, f(P)\Big) - \\ &\mathrm{tr}\Big(\frac{1}{2im}[A, P](1-\zeta^2) f(P) r_\Omega - \frac{1}{2im}[A, P_0](1-\zeta^2) f(P_0)\Big). \end{aligned} \tag{47}$$

The first term of the right hand side is in the trace class by the assumption $(H_2)$. The second term also. We want to compute this term more explicitly. At first let us compute formally. Using cyclicity of traces:

$$\begin{aligned} \mathrm{tr}\,([A, P](1-\zeta^2) f(P)) &= \mathrm{tr}\,((AP - PA)(1-\zeta^2) f(P)) \\ &= \mathrm{tr}\,((AP(1-\zeta^2) - A(1-\zeta^2)P) f(P)) \\ &= \mathrm{tr}\,(A[\zeta^2, P]\, f(P)). \end{aligned} \tag{48}$$

The same computation is valid for $P_0$, taking $\zeta \equiv 0$, which gives:

$$\mathrm{tr}\,(P_0 f(P_0)) = 0.$$

Now we are going to give a rigorous proof for all these formal computations.

Let us consider the differential operators with coefficients of compact support: $A_R = \chi(\frac{x}{R})\, A\, \chi(\frac{x}{R})$ where $\chi \in C_0^\infty(\mathbb{R}^n)$ satisfies: $\chi(x) = 1$ for $|x| \le 1$ and $\chi(x) = 0$ for $|x| \ge 2$. To compute some limits of traces we shall use the following dominated convergence lemma for a family of operators.

**Lemma 3.2.** [10] *Let $A \in S^1(H)$, be a trace class operator in the Hilbert space $H$ and $\{B_j\}_{j\geq 1}$ a family of norm-bounded operators in $H$ such that $\|B_j\| \leq C$, $\forall j \geq 1$ and $B_j$ converge weakly to $0$ as $j \to +\infty$. Then we have*

$$\lim_{j\to 0} \|B_j A\|_{\mathrm{tr}} = 0, \quad \textit{hence} \lim_{j\to 0} \mathrm{tr}(B_j A) = 0.$$

**Lemma 3.3.** *With the notation above we have*

$$\lim_{R\nearrow +\infty} \mathrm{tr}\left(A_R[\zeta^2, P]\, f(P)\right) = \mathrm{tr}\left(A\, [\zeta^2, P]\, f(P)\right).$$

*Proof.* We have

$$A_R[\zeta^2, P]\, f(P) = \chi^2 A\, [\zeta^2, P]\, f(P) + \chi_R[A, \chi_R][\zeta^2, P] f(P)$$

and

$$[A, \chi_R] = \frac{1}{R} x \cdot \nabla\chi\left(\frac{x}{R}\right).$$

But $[\zeta^2, P]f(P)$ is clearly of trace class and $\chi_R^2$, $[A, \chi_R]$, are uniformly bounded and converge weakly to 0 in $L^2(\Omega)$. Then the conclusion follows from Lemma 3.2. □

**Lemma 3.4.** *With the notation above we have*

$$\begin{aligned}(49)\quad \lim_{R\nearrow +\infty} &\mathrm{tr}\left([A_R, P](1-\zeta^2)f(P)r_\Omega - [A_R, P_0](1-\zeta^2)f(P_0)\right) \\ &= \mathrm{tr}\left([A, P](1-\zeta^2)f(P)r_\Omega - [A, P_0](1-\zeta^2)f(P_0)\right).\end{aligned}$$

*Proof.* We first compute the commutator

$$\begin{aligned}[A_R, P] &= \chi_R[A, P]\chi_R + \chi_R A[\chi_R, P] + [\chi_R, P]A\chi_R \\ &= \chi_R^2[A, P] + b_R(P),\end{aligned} \tag{50}$$

where $b_R(P)$ is a differential operator of order $\leq 2m$, with coefficients depending linearly on $P$, bounded in $L^\infty$-norm, and supported in the set $\{R/C \leq |x| \leq CR\}$ for $C > 0$ large enough. So Lemma 3.4 will follow from the next lemma:

**Lemma 3.5.** *We have*

$$\lim_{R\nearrow +\infty} \|b_R(P)(1-\zeta^2)f(P)r_\Omega - b_R(P_0)(1-\zeta^2)f(P_0)\|_{\mathrm{tr}} = 0. \tag{51}$$

*Proof.* We have, by the triangle inequality,

$$(52)\quad \|b_R(P)(1-\zeta^2)f(P)r_\Omega - b_R(P_0)(1-\zeta^2)f(P_0)\|_{\rm tr}$$
$$\leq \|(b_R(P)-b_R(P_0))(1-\zeta^2)f(P)\|_{\rm tr}+\|b_R(P_0)(1-\zeta^2)\big(f(P)r_\Omega-f(P_0)\big)\|_{\rm tr}.$$

Using $(H_2)$ we see that $b_R(P)-b_R(P_0)$ is a differential operator with coefficients having $L^\infty$-norm dominated by $CR^{-\rho}$. So we write down

$$(53)\quad (b_R(P)-b_R(P_0))(1-\zeta^2)f(P)$$
$$= (b_R(P)-b_R(P_0))\langle x\rangle^\rho(P+i)^{-1}(P+i)\langle x\rangle^{-\rho}(1-\zeta^2)f(P).$$

We can see easily that $(b_R(P)-b_R(P_0))\langle x\rangle^\rho(P+i)^{-1}$ is uniformly bounded, converges weakly to 0 as $R\nearrow+\infty$, and $(P+i)\langle x\rangle^{-\rho}(1-\zeta^2)f(P)$ is a trace class operator. Applying Lemma 3.2 we get

$$(54)\quad \lim_{R\nearrow+\infty}\|(b_R(P)-b_R(P_0))(1-\zeta^2)f(P)\|_{\rm tr}=0.$$

We have also

$$(55)\quad b_R(P_0)(1-\zeta^2)\big(f(P)r_\Omega-f(P_0)\big)$$
$$= b_R(P_0)(P_0+i)^{-1}(P+i)(1-\zeta^2)(f(P)r_\Omega-f(P_0))$$
$$- b_R(P_0)(P_0+i)^{-1}Q(1-\zeta^2)f(P),$$

where $b_R(P_0)(P_0+i)^{-1}$ is uniformly bounded, converges weakly to 0 as $R\nearrow+\infty$, and $Q(1-\zeta^2)f(P)$, $(P+i)(1-\zeta^2)(f(P)r_\Omega-f(P_0))$ are trace class operators. Applying again Lemma 3.2 we get

$$(56)\quad \lim_{R\nearrow+\infty}\|b_R(P_0)(1-\zeta^2)\big(f(P)r_\Omega-f(P_0)\big)\|_{\rm tr}=0.$$

Then (54), (56) give (51) and this achieves the proof of (44). □

We shall deduce now a useful pointwise representation formula for $s(\lambda)$. Let us introduce $E_P(\lambda_0,\lambda)$ $(\lambda>\lambda_0>0)$, the spectral projector of $P$ on $]\lambda_0,\lambda]$, and $\chi_0\in C^\infty(\mathbb{R})$ a cut-off function, $\chi_0(\mu)=1$ for $\mu>\lambda_0/2$ and $\chi_0(\mu)=0$ if $\mu<\lambda_0/3$. Then we have

**Corollary 3.6.** *With the notations above, for all $\lambda>\lambda_0$, we have:*

$$(57)\quad s(\lambda)-s(\lambda_0)=-\mathrm{tr}\,(\zeta E_P(\lambda_0,\lambda)\zeta)-\mathrm{tr}\Big(\frac{\chi_0(P)}{P}\frac{[A,\zeta^2]}{2im}E_P(\lambda_0,\lambda)\Big)$$
$$-\,\mathrm{tr}\Big(\frac{\chi_0(P)}{P}(Q-\frac{1}{2im}[A,Q])(1-\zeta^2)E_P(\lambda_0,\lambda)\Big).$$

*In particular $s$ is right continuous on $\mathbb{R}$.*

*Proof.* It is easy to check that each operator involved in the trace operations in the RHS of formula (57) is trace-class. Moreover using general properties of the spectral projectors $E_P(\lambda_0,\lambda)$ we can see also that each of the three terms is locally of bounded variation and right continuous in $\lambda$. □

*Remark 3.7.* In many cases it can be proved that $s(\lambda)$ is $C^\infty$ or real analytic on $]0,+\infty[$.

**4. The Weyl formula for Obstacles.** Using the formula (57) we can split $s(\lambda)$ into three terms:

(1) $s_{bd}(\lambda) = \operatorname{tr}(\zeta E_P(\lambda_0,\lambda)\zeta)$; this term contains the contribution of the obstacle. Its behaviour when $\lambda \nearrow +\infty$ depends of the nature of the boundary condition on $\partial K$ and can be studied almost in the same way as for bounded domains.

(2) $s_b(\lambda) = \operatorname{tr}(b(x,D)E_P(\lambda_0,\lambda))$ where $b(x,D) = \frac{\chi_0(P)}{P}\frac{[A,\zeta^2]}{2im}$ is a classical pseudodifferential operator of order 0, compactly supported in $\Omega$.

(3) $s_\sigma(\lambda) = \operatorname{tr}(\sigma(x,D)E_P(\lambda_0,\lambda))$, where

$$\sigma(x,D) = \frac{\chi_0(P)}{P}(Q - \frac{1}{2im}[A,Q])(1-\zeta^2)$$

is also a classical pseudodifferential operator of order 0, with a symbol supported very far away from the obstacle, and with a behaviour in the variable $x$ like $O(\langle x\rangle^{-\rho})$, $\rho > n$.

It is clear that $s_{bd}(\lambda)$ is a nondecreasing function when $\lambda > \lambda_0$. Concerning $s_b(\lambda)$ and $s_\sigma(\lambda)$, let us remark that we can assume that the operators $b(x,D)$ and $\sigma(x,D)$ are self-adjoint and can be split into positive and negative parts with the same properties

$$b(x,D) = b^+(x,D) - b^-(x,D), \tag{58}$$

$$\sigma(x,D) = \langle x\rangle^{-\rho/2}\sigma^+(x,D)\langle x\rangle^{-\rho/2} - \langle x\rangle^{-\rho/2}\sigma^-(x,D)\langle x\rangle^{-\rho/2}, \tag{59}$$

so that $b^\pm(x,D)$, $\sigma^\pm(x,D)$ are classical, nonnegative, pseudifferential operators of order 0, with the same properties of support respectively as $b$ and $\sigma$.

Then $s_b(\lambda)$ and $s_\sigma(\lambda)$ are particular examples of local spectral densities defined as $s_w(\lambda) = \operatorname{tr}(w(x,D)E_P(\lambda_0,\lambda))$ where $w(x,D)$ is a *nonnegative*, classical pseudodifferential operator of order 0, supported outside the obstacle and decreasing fast enough in $x$. In particular $s_w(\lambda)$ is a nondecreasing function, so Tauberian techniques can be used to prove asymptotics as $\lambda \nearrow +\infty$ ([13, 15, 31, 32]).

More generally, let us consider any nonnegative classical pseudodifferential operator $w(x,D)$ of order 0, such that $w(x,\xi) = 0$ if $|x| > R_0$, $w(x,\xi) \asymp \sum_{j\geq 0} w_j(x,\xi)$, where $w_j(x,\xi)$ is homogeneous in $\xi$ of degree $-j$, outside a small ball, and satisfies, for some $\delta > n$,

$$|\partial_x^\alpha w_j(x,\xi)| \leq \langle x\rangle^{-\delta-|\alpha|}, \quad \text{for } x \in \mathbb{R}^n,\ |\xi| = 1, \tag{60}$$

and for all $N \geq 1$ we have

$$|\partial_x^\alpha\partial_\xi^\beta(w(x,\xi) - \sum_{0\leq j\leq N-1} w_j(x,\xi))| \leq C\langle x\rangle^{-\rho-|\alpha|}\langle\xi\rangle^{-N-|\beta|}. \tag{61}$$

Now the contribution to the Weyl formula for $s(\lambda)$ outside the boundary and outside the singularities of the coefficients of $P$ (if any) is given by

**Theorem 4.1.** *Let us assume that* $(H_1)$, $(H_2)$ *with some* $\rho > 0$ *and* (60) *hold. Then we have:*
(i)

$$s_w(\lambda) = \gamma_{w,0}\lambda^{\frac{n}{2m}} + O(\lambda^{\frac{n-1}{2m}}). \tag{62}$$

(ii) *Furthermore, there exists* $R_1 > 0$ *large enough, such that if* $w(x,\xi) = 0$ *for* $|x| > R_1$, *then we have the two term asymptotics*

$$s_w(\lambda) = \gamma_{w,0}\lambda^{\frac{n}{2m}} + \gamma_{w,1}\lambda^{\frac{n-1}{2m}} + o(\lambda^{\frac{n-1}{2m}}). \tag{63}$$

(iii) *Moreover, if the following asymptotic limiting absorption principal holds:*
*there exists* $\mu > 0$, $S > 0$, $k > 0$ *such that*

$$\|\langle x\rangle^{-\mu}(P-\lambda-i0)^{-1}\langle x\rangle^{-\mu}\| = O(e^{S\lambda^k}),\ \lambda \nearrow +\infty, \tag{64}$$

*then there exists* $R_1 > 0$ *large enough such that if* $w(x,\xi) = 0$ *for* $|x| > R_1$ *then we have the complete asymptotic expansion*

$$s_w(\lambda) \asymp \sum_{j\geq 0}\gamma_{w,j}\lambda^{\frac{n-j}{2m}}, \qquad \mathrm{mod}\ \ O(\lambda^{-\infty}). \tag{65}$$

*Proof.* Let us first assume that there exists $R_2 > R_0$ such that $w(x,\xi) = 0$ for $|x| < R_2$. Then (62) follows from Hörmander's Theorem [13, 14] on local asymptotics for the spectral function of any elliptic scalar operator. To prove (63) we shall show that a method already used in [33] for perturbations of elliptic operators in $\mathbb{R}^n$ can be adapted to obstacles. Let us consider $L = (P + c_1)^{\frac{1}{2m}}$ ($c_1$ large enough, see the proof of Corollary 2.4). In $B^c_{R_0}$, $L$ is a classical elliptic pseudodifferential operator of order 1. To apply the constructions of [33, 32] we transform our problem into a semiclassical problem. For that let us first remark that

$$s_w(\lambda) = \mathrm{tr}(w(x,D)E_L(\mu)), \ \text{ where } \mu = (\lambda + c_1)^{\frac{1}{2m}}.$$

Let us introduce the small parameter $h = \mu^{-1}$. Using the homogeneity in the $\xi$ variable, we have clearly $E_L(\mu) = E_{hL}$ and $hL = l(h,x,hD)$ where $l(h)$ is a $h$-asymptotic symbol, $l(h) \asymp \sum_{j\geq 0} h^j l_j$, in particular $l_0(x,\xi) = p_{2m}^{\frac{1}{2m}}(x,\xi)$. In the same way we define $L_0 = (P_0 + c_1)^{\frac{1}{2m}}$ and $L_0(h)$. We have also $w(x,D) = w(h,x,hD)$, $w(h)$ being also a $h$-asymptotic symbol, $w(h) \asymp \sum_{j\geq 0} h^j w_j$, $w_0$ is the principal symbol of $w(x,D)$, $w_j(x,\xi) = 0$ for $|x| \leq R_1$.

Let us introduce the Schrödinger propagators

$$U(t,h) := \exp\Big(-\frac{it}{h}L(h)\Big), \quad U_0(t,h) := \exp\Big(-\frac{it}{h}L_0(h)\Big).$$

Then choosing $R_1$ large enough, like in Theorem 4.10 of [33], we can construct parametrices for $U(t,h)$ for all $t \geq 0$ in outgoing areas and for all $t \leq 0$ in incoming areas. Let us formulate the result (for a more accurate statement and definitions see [33]):
For every $N$, $M$, there exist

$$a^{(N)}(h) = \sum_{0\leq j\leq N} h^j a_j \quad \text{and } b^{(N)}(h) = \sum_{0\leq j\leq N} h^j b_j$$

such that we have

$$\chi(x,hD) = J(\varphi, a^{(N)}(h))J(\varphi, b^{(N)}(h))^* + h^N\chi_N(h,x,hD), \tag{66}$$

$$\begin{aligned} U(t,h)\chi(x,hD) = {} & J(\varphi, a^{(N)}(h))U_0(t,h)J(\varphi, b^{(N)}(h))^* \\ & + h^N U(t,h)\chi_N(h,x,D) + h^{N+1}R_N(t,h), \end{aligned} \tag{67}$$

where $\mathrm{supp}(\chi) \subset \Gamma_3^+$, $\mathrm{supp}(a_j) \cup \mathrm{supp}(b_j) \subset \Gamma_0^+$. Here $\Gamma_j^+$ are outgoing areas such that $\Gamma_3^+ \Subset \Gamma_0^+$, and $J(\varphi, a)$ denotes a Fourier integral operator associated to the phase function $\varphi_\pm$ defined in [33], Proposition 4.1. $R_N(t,h)$ is given by

$$R_N(t,h) = \int_0^{+\infty} U(t-s,h)J(\varphi, r_{N+1}(h))U_0(s,h)J(\varphi, b^{(N)}(h))^* ds.$$

It is not very difficult to see, by revisiting the proof of Theorem 4.10 in [33], that (67) holds also after adding a perturbation by a compact obstacle. Intuitively, this is clear, because the classical flow defined by the Hamiltonian $l_0$ transforms outgoing areas into outgoing areas for $t \geq 0$ (replace outgoing by incoming for $t \leq 0$) (see [33], Lemma 4.2), so that the obstacle stays far away.

So we have to consider

$$s_w(h,\mu) := \mathrm{tr}\,\big(w(h,x,hD)\theta(L(h)-1)E_{L(h)}(-\infty,\mu)\big)$$

for $\mu \in [1/2, 3/2])$ where $\theta \in C_0^\infty(]-3/4, 3/4[)$. We can study $s_w(h,\mu)$ as in [37, 32], using a partition of unity, localization and cyclicity of trace. Let us sketch the main ideas. We introduce, for $M \geq 1$, $\eta \in C_0^\infty(]-T_0, T_0[)$, $\eta = 1$ on $[-\frac{T_0}{2}, \frac{T_0}{2}]$, $\eta$ even, $\eta_M(t) := \eta(th^M)$,

$$\tau_{\eta,M}(h,\mu) = \operatorname{tr}\Big(w(h;x,D)\theta(L(h)-1)\int_{\mathbb{R}} \eta(th^M)U(t,h)\exp\left(i\frac{t\mu}{h}\right)\frac{dt}{2\pi}\Big).$$

This is an approximation of $\tau_w(h,\mu) := \frac{d}{d\mu}s_w(h,\mu)$,

$$\tau_{\eta,M}(h,\mu) = \tau_w(h) * \frac{\hat{\eta}_M}{2\pi}(\mu).$$

Let us denote $G(t,h) := \operatorname{tr}\left(w(h;x,D)\theta(L(h)-1)U(t,h)\right)$. As in [32], using (67) and the nonstationary phase theorem, we can prove easily

**Lemma 4.2.** *There exists $T_1 > 0$ such that for every $N \geq 1$, $M \geq 1$, we have $G(t,h) = O(h^N)$, uniformly in $T_1 \leq |t| \leq h^{-M}$.*

*Proof of Theorem 4.1-(ii).* Because the support of $\eta_M$ in $t$ is arbitrarily large, the result follows from the Tauberian theorem of [29]. □

*Proof of Theorem 4.1-(iii).* As in [33], we can use (67) to get a more explicit representation formula for $\tau_w(h)$ with the boundary value of the resolvents $(L(h)-\mu\pm 0)^{-1}$ and $(L_0(h)-\mu\pm 0)^{-1}$ in outgoing and incoming areas. So by a complex deformation argument, using the property (64), as in [32], we get the complete asymptotics (65). □

*Remark 4.3.* The assumption (64) seems very weak but we do not know whether it is always satisfied.

Now we are left with the contribution of a neighborhood of the boundary, $s_{bd}(\lambda)$, which will be considered below.

*4.1. Nonsmooth Obstacles.* We shall use here the heat kernel method. For simplicity, let us consider first the Laplace-Dirichlet problem with an arbitrary compact obstacle $K \subset \mathbb{R}^n$ and $P = -\Delta$. Let denote by $H_P(t;x,y)$ the integral kernel of the heat semi-group, $\exp(-tP)$. We have the following two properties (see [7] p. 52),

$$(4\pi t)^{n/2} H_P(t;x,x) \leq 1, \ \ \forall x \in \Omega, \ \forall t > 0, \tag{68}$$

$$\lim_{t\searrow 0}(4\pi t)^{n/2} H_P(t;x,x) = 1, \ \ \forall x \in \Omega. \tag{69}$$

So, by the Lebesgue dominated convergence theorem, we get

$$\lim_{t\searrow 0}(4\pi t)^{n/2}\operatorname{tr}(\zeta\exp(-tP)\zeta) = \int_\Omega \zeta^2(x)dx, \tag{70}$$

hence, using the Karamata Tauberian theorem, we have

$$s_{bd}(\lambda) = \lambda^{n/2}(2\pi)^{-n}\omega_n \int_\Omega \zeta^2(x)dx + o(\lambda^{n/2}). \tag{71}$$

Now we have to compute the term

$$s_\bullet(\lambda) := \operatorname{tr}\Big(\frac{\chi_0(P)}{2iP}A[\zeta^2, P]E_P(\lambda_0, \lambda)\Big). \tag{72}$$

The principal symbol of $b(x, D) =: \frac{\chi_0(P)}{2iP}A[\zeta^2, P]$ is clearly $b_0(x, \xi) = (x \cdot \xi)(\nabla\zeta^2 \cdot \xi)/|\xi|^2$, and we have classically

$$s_b(\lambda) = (2\pi)^{-n}\lambda^{n/2} \int_{\Omega\times B_1} b_0(x, \xi)dxd\xi + o(\lambda^{n/2}). \tag{73}$$

Using integration by parts we get easily

$$\int_{\Omega\times B_1} b_0(x, \xi)dxd\xi = -\omega_n \int_{\mathbb{R}^n} \zeta^2(x)\, dx. \tag{74}$$

Now, using (57) and that $\zeta = 1$ on $K$, we get (5), (6) of Theorem 1.2. □

The result above can be extended in several directions. For example, we can replace $-\Delta$ by any strictly elliptic second order differential operator ([7] p. 73), and Dirichlet by Neumann boundary conditions if the boundary is a Lipschitz manifold ([7] p. 77), so we get formulas analogous to (5), (6).

For higher order operators we have, for example:

**Proposition 4.4.** *Let $P = \Delta^2$ with Dirichlet boundary condition. Then for $1 \le n \le 3$, we have*

$$s(\lambda) = \lambda^{n/4}(2\pi)^{-n}\omega_n \mathrm{Vol}(K) + o(\lambda^{n/4}). \tag{75}$$

*Proof.* As in the second order case, we have a uniform estimate for the heat kernel $H_P(t, x, y)$ of $\exp(-tP)$ (Davies [8]). For $1 \le n \le 3$, there exists $d_n > 0$ such that

$$t^{n/4}H_\Omega(t, x, x) \le d_n, \ \ \forall x \in \Omega, \ \forall t \in ]0, 1], \tag{76}$$

so we can conclude as above, by the dominated convergence theorem. □

*Remark 4.5.* For the asymptotic distribution of eigenvalues in a nonsmooth environment the minimax principle and the Courant variational method is a powerful technique [3, 24]. Unfortunately this method does not work for the continuous spectrum considered here.

*4.2. Smooth Elliptic Boundary Value Problems.* We assume in this section that $K$ is a smooth manifold with boundary. For simplicity, let us formulate the results for second order operators (see the remark below for more general cases)

$$p(x,D) = \sum_{1\le j,k\le n} a_{j,k}(x)D_jD_k + \sum_{1\le j\le n} b_j(x)D_j - b_0(x). \tag{77}$$

We assume that the general assumptions $(H_1)$, $(H_2)$ are satisfied with $p_0(D) = -\Delta$ and that $a_{j,k}, b_j \in C^\infty(\overline{\Omega})$, $p(x,D)$ being uniformly elliptic in $\overline{\Omega}$. We assume also that the realization $P$ of $p(x,D)$ is defined by the Dirichlet, Neumann or Robin boundary condition.

**Theorem 4.6.** *Under the conditions above we have*

$$s(\lambda) = (2\pi)^{-n}\omega_n\overline{\mathrm{Vol}}(K)\lambda^{n/2} + O(\lambda^{(n-1)/2});\ \lambda \nearrow +\infty. \tag{78}$$

*Moreover, if the set of periodic generalized rays for the principal symbol of $p(x,D)$ in $T^*(\Omega\backslash 0)$ is of measure 0, then we have a two term asymptotic:*

$$s(\lambda) = (2\pi)^{-n}\omega_n\overline{\mathrm{Vol}}(K)\lambda^{n/2} + \gamma_{bd}\lambda^{(n-1)/2} + o(\lambda^{(n-1)/2}), \tag{79}$$

*where the volume coefficient is defined by:*

$$\overline{\mathrm{Vol}}(K) := \mathrm{Vol}(K) + \int_{\mathbb{R}^n\backslash K}\left((\det[a(x)])^{-1/2} - 1\right)dx,\ a(x) := \{a_{jk}(x)\}_{\substack{1\le j\le n\\ 1\le k\le n}}.$$

*In particular if $p(x,D) = p_0(D)$ in a neighborhood of $K$ we have*

$$\gamma_{bd} = \pm\frac{\omega_{n-1}}{4(2\pi)^{(n-1)}}\mathrm{Vol}(\partial K)$$

*where $(+)$ is for the Dirichlet and $(-)$ for the Robin boundary condition.*

*Proof.* Using (57) and Theorem 4.1, it is clear that the only term we have to consider is the contribution of the boundary $s_{bd}(\lambda) = \mathrm{tr}(\zeta^2 E_P(\lambda))$. Hence Ivrii's results [15], which are proved locally, can easily be applied to this term. □

*Remark 4.7.* By the same method used in the proof of Theorem 4.6, the results obtained by Vassiliev [43, 44] concerning the Weyl asymptotic formula for the eigenvalue counting function for higher order elliptic boundary problems can also be easily extended to the scattering phase $s(\lambda)$, using (57). In the same way the results obtained by Safarov [39] concerning quasi-Weyl asymptotics of the eigenvalue counting function can also be easily extended to the scattering phase $s(\lambda)$.

*4.3. Elliptic Degenerate Operators.* Let us consider the second order operators:

$$p(x, D) = -\sum \partial_j(\varphi a_{jk}\partial_k)$$

where the matrix $a_{jk}$ satisfies the assumptions:

(A.1) $a_{jk} \in C^\infty(\bar{\Omega})$ and there exists $E > 0$ such that

$$\sum a_{jk}(x)\xi_j\xi_k \geq E|\xi|^2,\ \forall x \in \bar{\Omega};\ \forall \xi \in \mathbb{R}^n.$$

(A.2) $\exists \rho > n$ such that $\forall \alpha$; $\forall (j,k)$ we have

$$|D^\alpha(a_{jk}(x) - \delta_{jk})| \leq C_\alpha \langle x\rangle^{-\rho-|\alpha|}.$$

(A.3) $\varphi : \mathbb{R}^n \to [0, +\infty[$ is a smooth function such that:

$$\begin{aligned} &\partial K = \{x : \varphi(x) = 0\};\ \varphi(x) > 0 \iff x \in \Omega, \\ &x \in \partial K \implies \nabla\varphi \neq 0, \\ &\exists \rho > n \text{ such that } |D^\alpha(\varphi(x) - 1)| \leq C_\alpha \langle x\rangle^{-\rho-|\alpha|}. \end{aligned} \tag{80}$$

For example we could have: $\varphi(x) = (|x|^{2k} - 1)/(|x|^{2k} + 1)$ with $k \in \mathbb{N}$ and $k > \frac{n}{2}$.

Let us consider the realization $P$ of $p(x, D)$ generated by the variational form $b[u, v]$ in the Hilbert space:

$$H_1^{1/2}(\Omega) := \{u \in \mathcal{D}'(\Omega);\ \varphi^{1/2}\partial^\alpha u \in L^2(\Omega);\ |\alpha| \leq 1\},$$

where

$$b[u, v] = \int_\Omega \varphi(x) \sum_{1\leq j,k\leq n} a_{jk}(x)D_ju(x)\overline{D_kv(x)}\,dx.$$

It is known that for every $\zeta \in C_0^\infty(\mathbb{R}^n)$ and for $k > n$, $\zeta(P + i)^{-k}$ is a trace class operator in $L^2(\Omega)$ [27].

The Weyl formula for degenerate elliptic operators on bounded domains was proved in different contexts by several authors: Baouendi-Goulaouic [2], Nordin [26], Pham The Lai [27], Bolley-Camus-Pham The Lai [4], Menikoff-Sjöstrand [23]. Using their methods and our trace formula we can get the asymptotics for the scattering phase of the pair $(P_0, P)$.

**Theorem 4.8.** *We have the asymptotic formulae*

$$\text{for } n > 2;\ s(\lambda) = d_n\lambda^{n-1} + o(\lambda^{n-1}), \tag{81}$$

$$\text{for } n = 2;\ s(\lambda) = d_2\lambda\log(\lambda) + o(\lambda\log(\lambda)), \tag{82}$$

*with*

$$d_2 = \frac{1}{4}\int_{\partial K}\frac{d\sigma(x)}{(\det[a(x)])^{1/2}}, \tag{83}$$

$$d_n = c_n\int_{\partial K}\frac{< a(x)\nu_x,\nu_x >^{1-n/2}}{|\nabla\varphi(x)|^{n-1}(\det[a(x)])^{1/2}}\,d\sigma(x), \quad n > 2, \tag{84}$$

*where $d\sigma$ is the Euclidean induced measure on $\partial K$, $\nu_x$ is the exterior normal to the obstacle $K$, $c_n = (2\pi)^{1-n}\omega_{n-1}\sum_{j\geq 1}(2j-1)^{1-n}$.*

*Proof.* As above we have to compute asymptotics for $s_{bd}(\lambda) = \mathrm{tr}(\zeta^2 E_P(\lambda))$. It is convenient here to use the resolvent method as in [27]. It is not clear for us how to use the variational principle of Courant like in [3, 24, 26].

Let us fix $m > n$ and consider the Stieltjès transform

$$\mathcal{T}(z) = \mathrm{tr}(\zeta^2(P^m - z)^{-1}), \quad z \in \mathbb{C}\backslash\mathbb{R}.$$

Then the methods used in [27] can be applied to give asymptotics for $\mathcal{T}(z)$ as $z \to -\infty$. The Tauberian theorem of Hardy-Littlewood gives the behaviour of $s_{bd}(\lambda^{\frac{1}{m}})$ which is a nondecreasing function. □

## REFERENCES

1. S. Agmon, *Asymptotic formulas with remainder estimates of eigenvalues of elliptic operators*, Arch. Rational Mech. Anal. **28** (1968), 165–183.
2. M. S. Baouendi and C. Goulaouic, *Régularité et théorie spectrale pour une classe d'opérateurs elliptiques dégénérés*, Arch. Rational Mech. Anal. **34** (1969), 361–379.
3. M. S. Birman and M. Z. Solomyak, *Leading term in the asymptotic spectral formula for "nonsmooth" elliptic problems*, Funct. Anal. Appl. **4** (1970), 265–275.
4. P. Bolley, J. Camus and Pham The Lai, *Noyau, résolvente et valeurs propres d'une classe d'opérateurs elliptiques et dégénérés*, Springer Lecture Notes in Math. N° 660 "Équations aux dérivées partielle", Proceedings, Saint Jean de Monts (1977).
5. V. S. Buslaev, *On the asymptotic behaviour of the spectral characteristics of exterior problems for the Schrödinger operator*, Math. USSR-Izv. **9** No 1 (1975), 139–223.
6. M. S. Birman and M. G. Krein, *On the theory of wave operators and scattering operators*, Dokl. Akad. Nauk SSSR **144** (1962), 475–478.
7. E. B. Davies, *Heat kernels and spectral theory*, Cambridge University Press, 1989.
8. ———, *Long time asymptotics of fourth order parabolic equations*, preprint 1994.
9. J. P. Eckmann and C. A. Pillet, *Spectral Duality for Planar Billiards*, Comm. Math. Phys. (1995).
10. I. C. Gohberg, M. G. Krein, *Introduction à la théorie des opérateurs non autoadjoints*, Dunod, 1972.
11. G. Grubb, *Remarks on trace estimates for exterior boundary problems*, Comm. in Partial Diff. Eq. **9** no 3 (1984), 231–270.
12. B. Helffer and D. Robert, *Calcul fonctionnel par la transformée de Mellin et opérateurs admissibles*, J. Funct. Anal. **53** no 3 (1983), 246–268.
13. L. Hörmander, *The spectral function of an elliptic operator*, Acta Math. **121** (1968), 193–218.

14. ———, *The Analysis of Linear Partial Differential Operators I to IV*, Springer Verlag, 1983–1985.
15. V. Ivrii, *On the second term in the spectral asymptotics for the Laplace-Beltrami operator on a manifold with boundary*, Funct. Anal. Appl. **14** (1980), 98–106.
16. A. Jensen and T. Kato, *Asymptotic behaviour of the scattering phase for exterior domains*, Comm. in Partial Diff. Eq. **3** no 12 (1978), 1165–1195.
17. M. Kac, *Can we hear the shape of a drum?*, Amer. Math. Monthly **73** (1966), 1–23.
18. M. G. Krein, *Perturbation determinants and a formula for the traces of unitary and self adjoint operators*, Soviet Math. Dokl. **3** (1962), 707–710.
19. P. Lax and R. Phillips, *Scattering theory*, Academic Press, 1967.
20. A. Majda and J. Ralston, *An analogue of Weyl's formula for unbounded domains. I*, Duke Math. J. **45** (1978), 183–196; II, **45** (1978), 513–536; III. An epilogue, **46** (1979), 725–731.
21. R. Melrose, *Weyl's conjecture for manifolds with concave boundary*, Proc. Symp. on Pure Math. A.M.S (1980), 257–273.
22. ———, *Weyl asymptotics for the phase in obstacle scattering*, Comm. in Partial Diff. Eq. **13** no 11 (1988), 1431–1439.
23. A. Menikoff and J. Sjöstrand, *On the eigenvalues of a class of hypoelliptic operators*, Math. Ann. 235 (1978), 55–84.
24. G. Métivier, *Valeurs propres de problèmes aux limites irréguliers*, Bull. Soc. Math. France, mémoire **51-52** (1977), 125–219.
25. E. Mourre, *Opérateurs conjugués et propriétés de propagation*, Comm. Math. Phys. **91** (1981), 279–300.
26. C. Nordin, *The asymptotic distribution of the eigenvalues of a degenerate elliptic operator*, Ark. Mat. **10** (1972), 3–21.
27. Pham The Lai, *Comportement asymptotique du noyau de la résolvente et des valeurs propres d'une classe d'opérateurs elliptiques dégénérés non necessairement auto-adjoint*, J. Math. Pures Appl. **55** (1976), 1–42.
28. V. Petkov and G. Popov, *Asymptotic behaviour of the scattering phase for non trapping obstacles*, Ann. Inst. Fourier, Grenoble **32** no 3 (1982), 111-149.
29. V. Petkov and D. Robert, *Asymptotique semiclassique du spectre d'hamiltoniens quantiques et trajectoires périodiques*, Comm in Partial Diff. Eq. **10** (1985), 365–390.
30. M. Reed and B. Simon, *Scattering theory*, Academic Press, 1979.
31. D. Robert, *Asymptotique à grande énergie de la phase de diffusion pour un potentiel*, Asymptotic Analysis **3** (1991), 301–320.
32. ———, *Asymptotique de la phase de diffusion à haute énergie pour des perturbations du second ordre du Laplacien*, Ann. Éc. Norm. Sup **25** (1992), 107–134.
33. ———, *Relative time delay for perturbations of elliptic operators and semiclassical asymptotics*, J. Funct. Anal. **126** no 1 (1994), 36–82.
34. ———, *Sur la formule de Weyl pour des ouverts non bornés*, C. R. Acad. Sci. Paris **319** Série I (1994), 29–34.
35. ———, *A trace formula for obstacles problems and applications*, Mathematical results in quantum mechanics (M. Demuth, P. Exner, H. Neidhardt and V. Zagrebnov, eds.), Birkhäuser Verlag, Basel, 1994, pp. 283–292, Operator Theory; Advances and applications Vol. 70.
36. ———, *Autour de l'Approximation Semi-Classique*, Birkhäuser, Basel, Progress in Math. **68**.
37. D. Robert and H. Tamura, *Semiclassical asymptotics for local spectral densities and time delay problems in scattering process*, J. Funct. Anal. **80** (1988), 124–147.

38. D. Robert and V. Sordoni, *Trace formulas and Dirichlet-Neumann problems with variable boundary*, Preprint, June 1995.
39. Yu. G. Safarov, *Exact asymptotics of the spectrum of a boundary value problem and periodic billiards*, Math. USSR-Izv. **33** no 3 (1989), 553–573.
40. R. Seeley, *A sharp asymptotic remainder estimate for the eigenvalues of the Laplacian in a domain of* $\mathbb{R}^3$, Adv. Math. **29** (1978), 244–269.
41. J. Sjöstrand and M. Zworski, *Complex scaling and the distribution of scattering poles*, J. Amer. Math. Soc. **4** no 4 (1991), 729–769.
42. U. Smilansky, *Semiclassical quantization of chaotic billiards*, In Chaos and quantum chaos (W. D. Heiss, ed.), Springer-Verlag, Berlin, Lecture Notes in Physics **411** (1993), 57–120.
43. D. G. Vassil'ev, *Two-term asymptotics of the spectrum of a boundary value problem for inside reflection of a general type*, Funct. Anal. Appl. **18** (1984), 1–13.
44. ———, *Asymptotics of the spectrum of a boundary value problem*, Trans. Moscow Math. Soc. (1987), 167–237.
45. D. Yafaev, *Mathematical Scattering Theory. General Theory*, Amer. Math. Soc., Rhode Island, 1992, Vol. 105.
46. H. Weyl, *Das asymptotische Verteilungsgesetz der eigenwerte linearer partieller Differentialgleichungen*, Math. Ann. **71** (1911), 441-469.

DEPARTMENT OF MATHEMATICS, CNRS-URA 758, UNIVERSITY OF NANTES, 2, RUE DE LA HOUSSINIÈRE, F-44072 NANTES CÉDEX 03, FRANCE
*E-mail address*: robert@math.univ-nantes.fr

# ON THE ASYMPTOTIC COMPLETENESS FOR PARTICLES IN CONSTANT ELECTROMAGNETIC FIELDS

ERIK SKIBSTED

**Abstract.** We prove asymptotic completeness for systems of two or three particles in combined constant electric and magnetic fields under the assumption of pairwise short-range interaction

**1. Introduction and results.** We consider a system of $N$ 3-dimensional particles labelled $1, \ldots, N$ with masses, charges, positions and momenta denoted by $m_i$, $q_i$, $x_i$, and $p_i$, respectively. The interaction consists of two parts. One part (the external part) is due to the presence of an electric field $\mathcal{E}$ and a magnetic field $\mathcal{B}$, both fields being constant and nonzero. The other part (the internal interaction) is given as a sum of real pair potentials, $v_{ij}$, assumed to be in $C^2$ and to obey the following decay conditions:

$$\text{For some } \sigma > \tfrac{1}{2} : \partial^\alpha v_{ij}(y) = O\left(|y|^{-\frac{2-|\alpha|}{2}\sigma}\right); \quad |\alpha| \leq 2. \tag{1.1}$$

The Hamilton action on $L^2(\mathbf{R}^{3N})$ reads

$$H = \sum_{i=1}^{N} \left( \frac{1}{2m_i} \left( p_i - \frac{1}{2} q_i \mathcal{B} \times x_i \right)^2 - q_i \mathcal{E} \cdot x_i \right) + \sum_{1 \leq i < j \leq N} v_{ij}(x_i - x_j). \tag{1.2}$$

The aim of this paper is to describe the large time behaviour of the evolution group generated by $H$ for $N = 2$ and $N = 3$. It is natural to conjecture that our results should generalize to arbitrary $N$. One can also pose the problem in the following more general setting in which we in fact prefer to state our results for $N \leq 3$.

Consider a finite dimensional real vector space $X$ equipped with a metric (i.e. a Euclidean space). Clearly the metric is identified with a linear mapping $G : X \to X'$ from $X$ to its dual space by the identity $\langle x, y \rangle = x \cdot Gy$; $x, y \in X$. Suppose $E \in X'$, and $A : X \to X'$ is a linear antisymmetric mapping (i.e. $x \cdot Ay = -y \cdot Ax$). Consider a "generalized potential" $V(x) = \sum_a V_a(x^a)$, where the index $a$ runs over a finite family of subspaces of $X$, denoted $X^a$, on which the associated "pair potential" $V_a$ is defined and obeys the conditions (1.1). The Hamiltonian $H$, defined as the closure of its natural action on $C_0^\infty(X) \subset L^2(X)$, reads

$$H = H(E) = (p - Ax)^2 - E \cdot x + V(x), \tag{1.3}$$

where $p$ refers to the momentum operator.

The physical Hamiltonian given by (1.2) can be put on this form (see (1.13) and (1.16) at the end of this section). We shall need a symmetry that in the physical case holds by translation invariance with respect to vectors representing particles all positioned at the same point. In the general case we decompose (orthogonally)

$$E = E_\perp \oplus E_\parallel \in X' = \operatorname{Ran} A \oplus G \operatorname{Ker} A$$

and pick $\alpha \in X$ such that

$$E_\perp = 2A\alpha \quad \text{and } \alpha \cdot E_\parallel = 0. \tag{1.4}$$

We shall assume

$$V(x + s\alpha) = V(x); \quad s \in \mathbf{R}. \tag{1.5}$$

We can now describe the first step of a procedure for simplifying the evolution group generated by $H(E)$. Consider the unitary operator

$$U_1(t) = e^{-it|\alpha/2|^2} e^{-i(tA\alpha + G\alpha/2)\cdot x} e^{it\alpha\, p}. \tag{1.6}$$

A straightforward computation using (1.5) shows that

$$e^{-itH(E)} = U_1(t)^* e^{-itH(E_\parallel)} U_1(0). \tag{1.7}$$

The next step is the difficult one. We want to get rid of the part of the potential that goes to zero in the direction of $G^{-1}E_\parallel$. So we introduce

$$\begin{aligned} V^{\tilde a}(x) &= \sum_{(G^{-1}E_\parallel)^a = 0} V_a(x^a), \\ H_{\tilde a}(E_\parallel) &= (p - Ax)^2 - E_\parallel \cdot x + V^{\tilde a}(x). \end{aligned} \tag{1.8}$$

We would like to prove that the wave operators

$$\text{s-}\lim_{t\to+\infty} e^{itH(E_\parallel)} e^{-itH_{\tilde a}(E_\parallel)}, \quad \text{s-}\lim_{t\to+\infty} e^{itH_{\tilde a}(E_\parallel)} e^{-itH(E_\parallel)} \tag{1.9}$$

exist.

The main part of this paper consists in showing the existence of these limits in the physical framework for $N \leq 3$. We shall rely heavily on the methods of [HMS1], [HMS2]. We believe that the wave operators also exist for $N > 3$ but in this case there are additional technical difficulties.

Suppose the existence of (1.9), then the third step consists in simplifying $e^{-itH_{\tilde{a}}(E_{\|})}$ as follows. Consider the unitary operator

$$U_2(t) = e^{it^3|E_{\|}|^2/3}e^{it^2G^{-1}E_{\|}\cdot p}e^{-itE_{\|}\cdot x}, \tag{1.10}$$

and the Hamiltonian

$$H_{\tilde{a}} = H_{\tilde{a}}(0) = (p - Ax)^2 + V^{\tilde{a}}(x).$$

A straightforward computation using the fact that

$$AG^{-1}E_{\|} = 0 \quad \text{and} \quad V^{\tilde{a}}(x + sG^{-1}E_{\|}) = V^{\tilde{a}}(x); \quad s \in \mathbf{R}, \tag{1.11}$$

shows the Avron-Herbst type formula

$$e^{-itH_{\tilde{a}}(E_{\|})} = U_2(t)^*e^{-itH_{\tilde{a}}}. \tag{1.12}$$

We have proved

**Theorem 1.1.** *Suppose the existence of* (1.9). *Then the wave operator*

$$s-\lim_{t\to+\infty} e^{itH(E)}(U_2(t)U_1(t))^*e^{-itH_{\tilde{a}}}$$

*exists, and it is onto.*

A last step could be to analyze the group $e^{-itH_{\tilde{a}}}$. We shall not elaborate on this except for remarking that for the physical model (in fact for arbitrary $N$) we could use the recent asymptotic completeness results of Gérard and Laba [GL1], [GL2] for particles in a constant magnetic field to obtain a full description in terms of free objects (with no internal potential at all) under additional conditions on the pair potentials. The operator $H_{\tilde{a}}$ is then a tensor sum of cluster Hamiltonians for particles in a constant magnetic field and having the same charge/mass ratio.

Asymptotic completeness for particles in a constant electric field was recently established rather generally [AT], [HMS1], [HMS2]. The present paper seems to be the first that deals with asymptotic completeness for particles in combined constant electric and magnetic fields.

We end this section by a certain restatement of the existence of (1.9) in the physical model and by some notation.

We recall the following standard notation in the $N$-body theory. Let $a = (C_1, \ldots, C_{\#a})$ be a (cluster) decomposition of the $N$ particles. Then we define

$$X_a = \{x = (x_1, \ldots, x_N) \in \mathbf{R}^{3N} |\, x_i = x_j \text{ if } i,j \in C \text{ for some } C \in a\},$$

and $X^a$ as its orthogonal complement in $X = \mathbf{R}^{3N}$ with respect to the metric

$$\langle x, y\rangle = \sum_{j=1}^{N} 2m_j x_j \cdot y_j. \tag{1.13}$$

The one-cluster decomposition is denoted by $a_{\max}$. Clearly

$$X^{a_{\max}} = \Big\{x = (x_1, \ldots, x_N) \in \mathbf{R}^{3N} |\, \sum_{i=1}^{N} m_i x_i = 0\Big\}.$$

We introduce coordinates $(R_1, R_2, R_3)$ for $X_{a_{\max}}$ by choosing the basis $\{\bar{c}_1, \bar{c}_2, \bar{c}_3\} \subset X_{a_{\max}}$,

$$\begin{aligned} \bar{c}_1 &= (\bar{e}_1, \ldots, \bar{e}_1); \quad \bar{e}_1 = (1,0,0),\\ \bar{c}_2 &= (\bar{e}_2, \ldots, \bar{e}_2); \quad \bar{e}_2 = (0,1,0),\\ \bar{c}_3 &= (\bar{e}_3, \ldots, \bar{e}_3); \quad \bar{e}_3 = (0,0,1), \end{aligned} \tag{1.14}$$

and then writing

$$x_{a_{\max}} = \sum_{j=1}^{3} R_j \bar{c}_j. \tag{1.15}$$

We introduce $E \in X' (= \mathbf{R}^{3N})$, $A : X \mapsto X'$ and $\alpha \in X$ by

$$\begin{aligned} E &= (q_1 \mathcal{E}, \ldots, q_N \mathcal{E}),\\ Ax &= \left(\tfrac{1}{2} q_1 \mathcal{B} \times x_1, \ldots, \tfrac{1}{2} q_N \mathcal{B} \times x_N\right),\\ \alpha &= \left(\mathcal{E} \times \frac{\mathcal{B}}{|\mathcal{B}|^2}, \ldots, \mathcal{E} \times \frac{\mathcal{B}}{|\mathcal{B}|^2}\right). \end{aligned} \tag{1.16}$$

Then the Hamiltonian (1.2) is on the form (1.3), and the conditions (1.4) and (1.5) hold.

In order to prove the existence of (1.9) we can assume by a rotation that $\mathcal{B} = (0,0,2b)$. Furthermore, by redefining the charges and the electric field $\mathcal{E} = (\mathcal{E}_1, \mathcal{E}_2, \mathcal{E}_3)$ we can assume that $b = 1$. Hence we can write

$$E_{\|} = (q_1 \mathcal{E}_3 \bar{e}_3, \ldots, q_N \mathcal{E}_3 \bar{e}_3), \quad Ax = (q_1 J x_1, \ldots, q_N J x_N), \tag{1.17}$$

where $J = \{j_{lk}\}_{l,k\leq 3}$ is the $3\times 3$ matrix given by $j_{12} = -1$, $j_{21} = 1$, $j_{lk} = 0$ otherwise.

We use the notation $x = x^a \oplus x_a \in X^a \oplus X_a$, $\xi = \xi^a \oplus \xi_a \in (X^a)' \oplus (X_a)'$, and similarly for the momentum operator $p = (p^a, p_a)$. The composed maps $X \ni x \mapsto (Ax)_a \in (X_a)'$ and $X \ni x \mapsto (Ax)^a \in (X^a)'$ are denoted by $A_a$ and $A^a$ respectively.

The decomposition

$$\mathbf{R}^{3N} = Y \times \mathbf{R}_{R_3}; \quad Y = X^{a_{\max}} \times \mathbf{R}^2_{(R_1,R_2)},$$

simplifies the operators $H(E_\|)$ and $H_{\tilde{a}}(E_\|)$ (involved in (1.9)), namely in the sense that

$$\begin{aligned} H(E_\|) &= H(E_\|^{a_{\max}}, Y) \otimes I + I \otimes h, \\ H_{\tilde{a}}(E_\|) &= H_{\tilde{a}}(E_\|^{a_{\max}}, Y) \otimes I + I \otimes h, \end{aligned} \tag{1.18}$$

on the corresponding tensor product of $L^2$-spaces, $L^2(X) = L^2(Y) \otimes L^2(\mathbf{R}_{R_3})$. The operator $h$ is a free Stark operator on $L^2(\mathbf{R}_{R_3})$. In Section 2 we shall analyze the basic form (up to unitary equivalence) of the operators on the right hand sides that act on $L^2(Y)$.

The remaining part of this paper concerns a proof of

**Theorem 1.2.** *Suppose* (1.1), *and* $N = 2$ *or* $N = 3$. *Then there exist the strong limits*

$$s-\lim_{t\to+\infty} e^{itH(E_\|^{a_{\max}},Y)} e^{-itH_{\tilde{a}}(E_\|^{a_{\max}},Y)}$$

*and*

$$s-\lim_{t\to+\infty} e^{itH_{\tilde{a}}(E_\|^{a_{\max}},Y)} e^{-itH(E_\|^{a_{\max}},Y)}$$

*on* $L^2(Y)$.

We notice that Theorem 1.2 is nontrivial only if $E_\|^{a_{\max}} \neq 0$, that is $\mathcal{E}_3 \neq 0$ and $q_i/m_i \neq q_j/m_j$ for some pair $i, j \leq N$.

Although Theorems 1.1 and 1.2 only deal with the limit $t \to +\infty$ there are similar results for $t \to -\infty$.

We remark that (1.1) is reasonably phrased as a short-range condition. Notice also that the decay condition for $|\alpha| = 1$ follows from the other conditions. If we instead of (1.1) impose for some $\sigma > 1$, $\partial^\alpha v_{ij}(y) = O(|y|^{-\sigma})$; $|\alpha| = 1$, and the same boundedness condition for $|\alpha| = 2$ as before, then this class constitutes a long-range class for which our results generalize. We need only to modify the above wave operators by a certain time-dependent but purely multiplicative phase factor, cf. [HMS2]. The proof is more complicated, cf. [HMS2], and shall not be discussed here. (We remark though that the condition $\sigma > 1/2$ in (1.1) is used crucially only at one point in this paper namely at the end of Section 8. Up to this point $\sigma > 0$ would suffice.) Moreover we shall not discuss inclusion of local singularities. The Coulomb singularity was delicate for the Stark problem [HMS1, Sect. 6].

In the first part of Section 2 we introduce certain reduced Hamiltonians. In the second part we study the corresponding group in the free case for $N = 2$. We use the explicit formulas obtained to establish analogues of [HMS1, Lemma 3.3], Lemmas 3.1 and 3.2. The proof of the latter results, given in Section 3, is rather complicated (as opposed to the one of [HMS1, Lemma 3.3]). The results lead to an important intermediate 2-body result stated as Proposition 4.1. It can be viewed as an analogue of [HMS1, Lemma 3.6]. In the remaining sections, Sections 5–8, we follow [HMS1] and [HMS2] rather closely.

*Acknowledgement.* We thank Ira Herbst for making us aware at an early stage of this work of a classical analogue to (1.7), cf. for example [C, pp. 329–330]. This classical result led us naturally to (1.7).

**2. Unitary equivalence and free dynamics.** We want to describe the operators $H(E_{\|}^{a_{\max}}, Y)$ and $H_{\tilde{a}}(E_{\|}^{a_{\max}}, Y)$ in the tensor decomposition (1.18) and to establish formulas for the corresponding group in the free case for $N = 2$. We shall rely on the papers [AHS1], [AHS2] and [GL1]. The subscript 0 indicates that the potential (the internal interaction) is put equal to zero. For example we shall denote by $H_0(E_{\|})$ the operator $(p - Ax)^2 - E_{\|} \cdot x$ on $L^2(X)$. For notational convenience let $a = a_{\max}$ throughout this section.

A computation shows that

$$\begin{aligned} &e^{-ix^a \cdot Ax_a} H_0(E_{\|}) e^{ix^a \cdot Ax_a} \\ &= (p^a - A^a x^a)^2 + (p_a - 2A_a x^a - A_a x_a)^2 - E_{\|} \cdot x. \end{aligned}$$

Combined with (1.15) this gives

$$\begin{aligned} &e^{-ix^a \cdot Ax_a} H_0(E_{\|}^a, Y) e^{ix^a \cdot Ax_a} \\ &= (p^a - A^a x^a)^2 + (p_{R_1} \bar{c}_1^* + p_{R_2} \bar{c}_2^* - 2A_a x^a - A_a x_a)^2 - E_{\|}^a \cdot x^a, \end{aligned} \tag{2.1}$$

where $p_{R_j}$ denotes the momentum operator with respect to the variable $R_j$ and

$$\bar{c}_j^* = \Big(\frac{m_1}{M}\bar{e}_j, \ldots, \frac{m_N}{M}\bar{e}_j\Big); \quad M = \sum_{i=1}^N m_i, \quad j = 1, 2.$$

A further simplification depends on whether the total charge $Q = \sum_{j=1}^N q_j$ is zero or nonzero (cf. [AHS2] and [GL1]).

In the first case, the neutral case, the term $A_a x_a$ is zero. If $F$ denotes the Fourier transform with respect to $R = (R_1, R_2)$ we thus obtain

$$\begin{aligned} &Fe^{-ix^a\cdot Ax_a} H_0(E_{\|}^a, Y) e^{ix^a\cdot Ax_a} F^{-1} \\ &= (p^a - A^a x^a)^2 + (k_1\bar{c}_1^* + k_2\bar{c}_2^* - 2A_a x^a)^2 - E_{\|}^a \cdot x^a, \end{aligned} \tag{2.2}$$

on $L^2(Y) = L^2(X^a)\otimes L^2(\mathbf{R}_k^2)$; $k = (k_1, k_2)$. Clearly this operator is a direct integral with respect to the variable $k$, called the pseudo momentum, and with the fiber operator given by

$$H_0(E_{\|}^a, k, X^a) = (p^a - A^a x^a)^2 + (k_1\bar{c}_1^* + k_2\bar{c}_2^* - 2A_2 x^a)^2 - E_{\|}^a \cdot x^a. \tag{2.3}$$

Since $Fe^{-ix^a\cdot Ax_a}$ commutes with the variable $x^a$ similar formulas hold for the fiber operators $H(E_{\|}^a, k, X^a)$ and $H_{\tilde{a}}(E_{\|}^a, k, X^a)$ of $H(E_{\|}^{a_{\max}}, Y)$ and $H_{\tilde{a}}(E_{\|}^{a_{\max}}, Y)$, respectively.

For $N = 2$ a further simplification is provided by the unitary operator

$$V_k = e^{-i\mu r\cdot\tilde{k}};\ r = x_1 - x_2,\ \tilde{k} = (k_1, k_2, 0),\ \mu = \frac{m_2^2 - m_1^2}{2M^2}. \tag{2.4}$$

We introduce the translation operator $T_\kappa$ on $L^2(X^a)$ $(= L^2(\mathbf{R}_r^3))$ given by

$$(T_\kappa f)(r) = f(r - \kappa); \quad \kappa = \frac{1}{2q_1} J\tilde{k}. \tag{2.5}$$

Here we assume (of course) that $q_1 = -q_2 \neq 0$. Now a computation shows that in fact

$$T_\kappa V_k H_{\text{C}}(E_{\|}^a, k, X^a) V_k^{-1} T_\kappa^{-1} = H_0(E_{\|}^a, 0, X^a).$$

We shall later use the following unitary operator on $L^2(Y)$ for $N = 2$:

$$U_n = V_k F e^{-ix^a\cdot Ax_a}. \tag{2.6}$$

Here we just notice that it commutes with the variable $r$ and that it takes $H_0(E_{\|}^a, Y)$ and $H(E_{\|}^a, Y)$ to fibered operators (direct integrals).

We now turn to the charged case $Q \neq 0$. Consider the unitary operator

$$U_c = e^{-ip_{R_1}p_{R_2}/2Q} e^{-iQR_1R_2} e^{-ix^a\cdot Ax_a} \tag{2.7}$$

on $L^2(Y)$. A computation using (2.1) shows that

$$\begin{aligned} &U_c H_0(E_{\|}^a, Y) U_c^{-1} \\ &= (p^a - A^a x^a)^2 + (2QR_2\bar{c}_1^* + p_{R_2}\bar{c}_2^* - 2A_a x^a)^2 - E_{\|}^a \cdot x^a \\ &= (p^a - A^a x^a)^2 - E_{\|}^a \cdot x^a + T; \end{aligned} \tag{2.8}$$

$$T = \frac{1}{2M}\left\{4(QR_2 - \bar{c}_1\cdot Ax^a)^2 + (p_{R_2} - 2\bar{c}_2\cdot Ax^a)^2\right\}.$$

Hence if we write $L^2(Y) = L^2(\tilde{X}^a) \otimes L^2(\mathbf{R}_{R_1})$, $\tilde{X}^a = X^a \times \mathbf{R}_{R_2}$, we have

$$(2.9)\qquad \begin{aligned} U_c H_0(E_\parallel^a, Y) U_c^{-1} &= H_0(E_\parallel^a, \tilde{X}^a) \otimes I; \\ H_0(E_\parallel^a, \tilde{X}^a) &= (p^a - A^a x^a)^2 - E_\parallel^a \cdot x^a + T. \end{aligned}$$

We notice that $T$ is independent of $x^a$ if all particles have the same mass/charge ratio. In particular if $N = 2$ and $q_1/m_1 = q_2/m_2$, then we can decompose

$$(2.10)\qquad H_0(E_\parallel^a, \tilde{X}^a) = H_0(E_\parallel^a, X^a) \otimes I + I \otimes T$$

on $L^2(X^a) \otimes L^2(\mathbf{R}_{R_2})$.

We also notice that $U_c$ commutes with the variable $x^a$. So there are formulas similar to (2.9) and (2.10) for $H(E_\parallel^a, Y)$ and $H_{\tilde{a}}(E_\parallel^a, Y)$.

*Free dynamics.* In the free case there is an obvious additional simplification. If we define

$$X_\perp^a = \{x = (x_1, \ldots, x_N) \in X^a \,|\, x_j = ((x_j)_1, (x_j)_2, 0);\ j = 1, \ldots, N\}$$

and

$$Y_\perp = X_\perp^a \times \mathbf{R}_R^2,$$

then we can decompose

$$(2.11)\qquad H_0(E_\parallel^a, Y) = H_0(Y_\perp) \otimes I + I \otimes h$$

on

$$(2.12)\qquad L^2(Y) = L^2(Y_\perp) \otimes L^2(X^a \ominus X_\perp^a).$$

The operator $h$ is a free Stark operator on $L^2(X^a \ominus X_\perp^a)$.

We notice that all unitaries mentioned in the first part of this section respect the decomposition (2.12) in the sense that they act on the first factor on the right hand side. Since we shall here aim at explicit formulas in the case $N = 2$ for $Ue^{-itH_0(E_\parallel^a, Y)}U^{-1}$; $U = U_n, U_c$, it suffices to examine $Ue^{-itH_0(Y_\perp)}U^{-1}$; $U = U_n, U_c$. Then the contribution from the free Stark operator just enters as an additional explicit factor (cf. (1.12)).

In the rest of this section $N = 2$. Abusing the notation (2.4) slightly we choose coordinates $r = (r_1, r_2)$ for $X_\perp^a$ specified as the first two coordinates of the vector $x_1 - x_2$. The coordinates for $Y_\perp$ are then $(r_1, r_2, R_1, R_2)$. First we compute a formula for the kernel of $e^{-itH_0(Y_\perp)}$ in terms of these coordinates. It will be assumed that $q_j \neq 0$; $j = 1, 2$. Let

$$(2.13)\qquad K_1 = \frac{q_1}{m_1},\quad K_2 = \frac{q_2}{m_2}.$$

We shall use the following formula for a charged 2-dimensional particle in a constant magnetic field cf. [AHS1, (3.5)] (corrected for a sign error). For $u, v \in \mathbf{R}^2$ we define

$$u \times v = u_1 v_2 - u_2 v_1; \quad u = (u_1, u_2),\ v = (v_1, v_2).$$

Let $H = \frac{1}{2m}\{(p_{u_1} + qu_2)^2 + (p_{u_2} - qu_1)^2)\}$ on $L^2(\mathbf{R}^2_u)$. Then (assuming $\sin(qt/m) \neq 0$)

$$\begin{aligned} e^{-itH}(u, u') &= \frac{-i\frac{q}{2\pi}}{\sin(\frac{qt}{m})} e^{i\{A_1 + A_2\}}; \\ A_1 = \frac{q}{2}\cot(\frac{q}{m}t)(u - u')^2, &\quad A_2 = -qu \times u'. \end{aligned} \tag{2.14}$$

By using (2.14) for each of the two particles and then changing coordinates we obtain the following kernel

$$\begin{aligned} e^{-itH_0(Y_\perp)}(r, R; r', R') &= \frac{-q_1 q_2}{4\pi^2 \sin(K_1 t)\sin(K_2 t)} e^{i\{B_1 + \cdots + B_4\}}; \\ B_1 &= \frac{q_1}{2}\cot(K_1 t)\left(R + \frac{m_2}{M}r - R' - \frac{m_2}{M}r'\right)^2, \\ B_2 &= \frac{q_2}{2}\cot(K_2 t)\left(R - \frac{m_1}{M}r - R' + \frac{m_1}{M}r'\right)^2, \\ B_3 &= -q_1\left(R + \frac{m_2}{M}r\right) \times \left(R' + \frac{m_2}{M}r'\right), \\ B_4 &= -q_2\left(R - \frac{m_1}{M}r\right) \times \left(R' - \frac{m'}{M}r'\right), \end{aligned} \tag{2.15}$$

under the assumption that

$$\sin(K_1 t), \sin(K_2 t) \neq 0. \tag{2.16}$$

Since

$$x^a \cdot Ax_a = \theta R \times r; \quad \theta = (K_1 - K_2)\frac{m_1 m_2}{M}, \tag{2.17}$$

the kernel of the group generated by $e^{-ix^a \cdot Ax_a} H_0(Y_\perp) e^{ix^a \cdot Ax_a}$ is given by the expression on the right hand side of (2.15) upon adding to the exponent the term $B_5$ given by

$$B_5 = -\theta R \times r + \theta R' \times r'. \tag{2.18}$$

To proceed from here we distinguish between the neutral and the charged case, cf. (2.6) and (2.7).

In the neutral case a small computation shows that the exponential takes the form

$$e^{i\{B_1+\cdots+B_5\}} = e^{ip_t(R-R',r,r')}.$$

Hence conjugation by $F$ (given as in (2.6)) amounts to computing the (two-dimensional) Fourier transform of the function $p_t$ with respect to the variable $R-R'$. (This is a matter of computing the integral of a complex Gaussian function, cf. [H, Theorem 7.6.1].) The operator then becomes multiplicative with respect to the dual variable $k=(k_1,k_2)$. It remains to conjugate by $V_k$ given by (2.4), which amounts to multiplying the kernel by the factor $e^{-i\mu(r-r')\cdot k}$, where $\mu$ is given by (2.4). A simplification of the resulting expression then gives the following result. We shall here (and later) use the notation

$$a(t)=\frac{q_1}{2}\cot(K_1t)+\frac{q_2}{2}\cot(K_2t), \tag{2.19}$$

and

$$C_1=K_2-K_1,\ C_2=K_2+K_1,\ \kappa=\frac{1}{2q_1}(-k_2,k_1), \tag{2.20}$$

the latter being a slight abuse of the notation (2.5).

**Lemma 2.1.** *Suppose $q:=q_1=-q_2\neq 0$. Then writing*

$$U_nH_0(Y_\perp)U_n^{-1}=\int_{\mathbf{R}^2}\oplus\, dk\, H_{0n}(k,X_\perp^a) \tag{2.20}$$

*we have, under the assumption*

$$\sin(K_2t),\ \sin(K_1t),\ \sin(C_1t)\neq 0, \tag{2.21}$$

*the following kernel:*

$$e^{-itH_{0n}(k,X_\perp^a)}(r;r')=\frac{iq}{2\pi\sin(C_1t)}e^{ip_t(r+\kappa,r'+\kappa)};$$

$$p_t(r,r')=-\frac{q}{2}\begin{pmatrix} r\\ r'\end{pmatrix}\cdot B(t)\begin{pmatrix} r\\ r'\end{pmatrix}:$$

*where $B(t)$ is the symmetric $4\times4$ matrix given (in terms of $2\times2$ matrices) as*

$$\sin(C_1t)B(t)=\begin{pmatrix}\cos(C_1t)I_2 & -D(t)\\ -D(t)^{\mathrm{T}} & \cos(C_1t)I_2\end{pmatrix}$$

$$D(t)=\begin{pmatrix}\cos(C_2t) & \sin(C_2t)\\ -\sin(C_2t) & \cos(C_2t)\end{pmatrix}.$$

In Section 3 we shall use Lemma 2.1 exploiting that under the conditions of the lemma

$$|C_1| > |C_2|. \tag{2.22}$$

In the charged case we introduce

$$B_6 = -QR_1R_2 + QR_1'R_2'. \tag{2.23}$$

Up to a time-dependent factor the kernel of

$$Ue^{-itH_0(Y_\perp)}U^{-1}; \quad U = e^{-iQR_1R_2}e^{-ix^a\cdot Ax_a},$$

is then on the form

$$e^{i\{B_1+\cdots+B_6\}} = e^{ip_t(R_1-R_1',R_2,R_2',r,r')}, \tag{2.24}$$

where $B_1,\ldots,B_6$ are given by (2.15), (2.18), (2.23). It remains to conjugate by the unitary operator $e^{-ip_{R_1}p_{R_2}/2Q}$. This can be done explicitly by computing the (partial) Fourier transform of the function $p_t$ on the right hand side of (2.24) with respect to the variable $R_1 - R_1'$ and then evaluating at zero. We obtain the following result using in addition to previous notation, cf. (2.17) and (2.19),

$$b(t) = (K_1\cot(K_1t) - K_2\cot(K_2t))\frac{m_1m_2}{M}. \tag{2.25}$$

**Lemma 2.2.** *Suppose $Q$, $q_1$, $q_2 \neq 0$. Then writing $\tilde{X}_\perp^a = X_\perp^a \times \mathbf{R}_{R_2}$, $L^2(Y_\perp) = L^2(\tilde{X}_\perp^a)\otimes L^2(\mathbf{R}_{R_1})$ and*

$$U_cH_0(Y_\perp)U_c^{-1} = H_0(\tilde{X}_\perp^a)\otimes I$$

*we have, under the assumption*

$$\sin(K_2t),\ \sin(K_1t),\ a(t)\neq 0, \tag{2.26}$$

*the following kernel:*

$$e^{-itH_0(\tilde{X}_\perp^a)}(r,R_2;r',R_2') = f(t)e^{ip_t(r,R_2,r',R_2')},$$

where

$$\begin{aligned} f(t) =& \frac{-\pi^{-3/2}q_1q_2e^{i\frac{\pi}{4}\operatorname{sgn}(a(t))}}{4|a(t)|^{1/2}\sin(K_1t)\sin(K_2t)},\\ p_t(r,R_2,r',R_2') =& \frac{-1}{4a(t)}\left(b(t)(r_1-r_1') - \theta(r_2+r_2') - Q(R_2+R_2')\right)^2\\ &+ \frac{1}{2M^2}\left(q_1m_2^2\cot(K_1t) + q_2m_1^2\cot(K_2t)\right)(r_1-r_1')^2\\ &+ \frac{q_1}{2}\cot(K_1t)\left(R_2 + \frac{m_2}{M}r_2 - R_2' - \frac{m_2}{M}r_2'\right)^2\\ &+ \frac{q_2}{2}\cot(K_2t)\left(R_2 - \frac{m_1}{M}r_2 - R_2' + \frac{m_1}{M}r_2'\right)^2\\ &+ \theta(R_2-R_2')(r_1'+r_1) - \frac{q_1m_2^2+q_2m_1^2}{M^2}r\times r'. \end{aligned}$$

*Remark 2.3.* If only one of the charges is nonzero the expression on the right hand side of the formula in Lemma 2.2 does not make sense. However the limit does: For example if $q_1 = 0$ and $q_2 \neq 0$ we get a well defined expression by letting $q_1 \to 0$ in this expression. As is easily seen from the derivation of the above formula we obtain in this way the correct formula for the case $q_1 = 0$ and $q_2 \neq 0$.

**3. Free resolvent estimates.** In this section we shall use Lemmas 2.1 – 2.2 and Remark 2.3 to obtain resolvent estimates in the free case for $N = 2$. The estimates will play a crucial role in Section 4. We shall use the same notation and conventions as in Section 2.

For $\rho > 0$ the notation $F(\cdot < \rho)$ stands for the characteristic function of the interval $(-\infty, \rho)$. Let $F(\cdot \geq \rho) = 1 - F(\cdot < \rho)$.

**Lemma 3.1.** *Suppose $N = 2$, $q_1 \neq 0$ and $Q = 0$. We represent*

$$U_n H_0(E_{\|}^a, Y) U_n^{-1} = \int_{\mathbf{R}^2} \oplus\, dk\, H_{0n}(E_{\|}^a, k, X^a).$$

*Then for all $\rho \in \mathbf{R}$ and $\mu \in \mathbf{R} \setminus \{0\}$*

$$\|(H_{0n}(E_{\|}^a, k, X^a) - \lambda - i\mu)^{-1} F(|x^a| < \rho)\| \to 0 \quad \text{as } |k| + |\lambda| \to \infty. \tag{3.1}$$

If $q_1 = q_2 = 0$, then (3.1) is false. (In this case it can be viewed as a statement for the 3-dimensional Laplace operator.)

**Lemma 3.2.** *Suppose $N = 2$, $Q \neq 0$ and $K_1 \neq K_2$ (i.e. $q_1/m_1 \neq q_2/m_2$). We represent (cf. (2.9))*

$$U_c H_0(E_{\|}^a, Y) U_c^{-1} = H_0(E_{\|}^a, \tilde{X}^a) \otimes I.$$

*Then for all $\rho \in \mathbf{R}$ and $\mu \in \mathbf{R} \setminus \{0\}$*

$$\|(H_0(E_{\|}^a, \tilde{X}^a) - \lambda - i\mu)^{-1} F(|x^c| < \rho)\| \to 0 \quad \text{as } |\lambda| \to \infty. \tag{3.2}$$

If $K_1 = K_2$, then (3.2) is false, cf. (2.10).

For the sake of presentation we shall first reduce the above estimates (3.1) and (3.2) to somewhat similar statements that do not involve the electric field as well as the third component of the internal position vector. Then we shall prove these separately.

Clearly by the first resolvent equation we can assume that $\mu = 1$. By squaring the norms it is then enough (by another application of the first resolvent equation) to estimate the norm of $F(|x^a| < \rho)(H - \lambda - i)^{-1} F(|x^a| < \rho)$ for the relevant inputs of $H$. We represent

$$(H - \lambda - i)^{-1} = i \int_0^\infty e^{it(\lambda+i)} e^{-itH}\, dt. \tag{3.3}$$

Next we notice that the group $e^{-itH}$ factorizes in both cases corresponding to the tensor decompositions (2.11) and (2.12). For example, in the case of Lemma 3.1

$$e^{-itH_{0n}(E^a_{\|},k,X^a)} = e^{-itH_{0n}(k,X^a_{\perp})} \otimes e^{-ith},$$

where the first factor is given explicitly in Lemma 2.1 and the other is given by the Avron-Herbst formula [HMS1, (3.1)] (see also (1.12)). We shall show that

$$\sup_{T>0} \left\| \int_0^T F(|r| < \rho) e^{it(\lambda+i)} e^{-itH_{0n}(k,X^a_{\perp})} F(|r| < \rho)\, dt \right\| \to 0 \quad \text{as } |k| + |\lambda| \to \infty. \tag{3.4}$$

Here $r = (r_1, r_2)$ denotes the coordinates for $X^a_{\perp}$ as in the second part of Section 2. Given the statement (3.4) we can easily prove (3.1) by an integration by parts on the right hand of (3.3) and applying [HMS1, (3.1)]. Similarly (3.2) will follow if we can prove

$$\sup_{T>0} \left\| \int_0^T F(|r| < \rho) e^{it(\lambda+i)} e^{-itH_0(\tilde{X}^a_{\perp})} F(|r| < \rho)\, dt \right\| \to 0 \quad \text{as } |\lambda| \to \infty. \tag{3.5}$$

We now proceed to the proof of (3.4) and (3.5).

*Proof of* (3.4). We shall use Lemma 2.1. Let $A_{\lambda,\kappa}(t)$ denote the integrand in (3.4). First we shall show

$$\sup_{T>0,\, \lambda\in\mathbf{R}} \left\| \int_0^T A_{\lambda,\kappa}(t)\, dt \right\| \to 0 \quad \text{as } |k| \to \infty. \tag{3.6}$$

We notice that $\|A_{\lambda,\kappa}(t)\| \le e^{-t}$ and that (2.21) holds except for $t$ in a discrete set. Hence to prove (3.6) it suffices to prove that for a fixed interval $[a,b] \subset (0,\infty)$ that does not intersect this discrete set

$$\sup_{T\in[a,b],\, \lambda\in\mathbf{R}} \left\| \int_a^T A_{\lambda,\kappa}(t)\, dt \right\| \to 0 \quad \text{as } |k| \to \infty. \tag{3.7}$$

By squaring the norm we need to bound $\int_S A_{\lambda,\kappa}(s)^* A_{\lambda,\kappa}\, ds\, dt$ for sets $S$ that we can assume do not intersect the diagonal, i.e. for

$$S = S(T,\varepsilon) = [a,T]^2 \cap \{(s,t) \in \mathbf{R}^2 \,|\, \varepsilon \le |s-t|\}; \quad 0 < \varepsilon.$$

To do this we assume Lemma 2.1 (for each factor) to obtain the expression

$$f(s,t,\lambda) F(|r| < \rho) F(|r''| < \rho) \int_{|r'|<\rho} e^{i(-p_s(r'+\kappa, r+\kappa) + p_t(r'+\kappa, r''+\kappa))}\, dr' \tag{3.8}$$

for the kernel $(A_{\lambda,\kappa}(s)^* A_{\lambda,\kappa}(t))(r, r'')$. Here

$$|f(s,t,\lambda)| \leq C$$

for a constant $C$ independent of $\lambda \in \mathbf{R}$, $T \in [a,b]$, and $(s,t) \in S(T,\varepsilon)$. We expand the integrand in (3.8) into a product of exponentials. The factor depending on both $r'$ and $\kappa$ reads

$$e^{iqr'\cdot\{\cdot\}_1}, \quad \text{where } \{\cdot\}_1 = \left\{(B(s) - B(t))\begin{pmatrix}\kappa\\ \kappa\end{pmatrix}\right\}_1.$$

Here $\{\cdot\}_1$ refers do the first component of the input vector in $\mathbf{R}^2 \times \mathbf{R}^2$. We want to integrate by parts in (3.8) using the formula

$$e^{iqr'\cdot\{\cdot\}_1} = -\frac{i}{q}\frac{\{\cdot\}_1}{|\{\cdot\}_1|^2}\cdot\nabla_{r'} e^{iqr'\cdot\{\cdot\}_1}.$$

This can be done if the denominator is nonzero. To estimate it we shall use the formula

$$\text{(3.9)}\quad \{\cdot\}_1 = \int_t^s \left\{B'(u)\begin{pmatrix}\kappa\\ \kappa\end{pmatrix}\right\}_1 du = \int_t^s \frac{1}{\sin^2(C_1 u)}(-C_1 I_2 + E(u))\kappa\, du,$$

obtained by computing the derivative

$$B'(u) = \frac{1}{\sin^2(C_1 u)}\begin{pmatrix} -C_1 I_2 & E(u) \\ E(u)^{\mathrm{T}} & -C_1 I_2\end{pmatrix},$$

where

$$E(u) = C_1 \cos(C_1 u) D(u) + C_2 \sin(C_1 u)\tilde{D}(u),$$
$$\tilde{D}(u) = \begin{pmatrix} \sin(C_2 u) & -\cos(C_2 u) \\ \cos(C_2 u) & \sin(C_2 u)\end{pmatrix}.$$

A computation shows that

$$\|E(u)\| = |C_1|\left(1 - \left(1 - \left|\frac{C_2}{C_1}\right|^2\right)\sin^2(C_1 u)\right)^{\frac{1}{2}}.$$

Hence by (2.22) and (3.9) we can estimate

$$\text{(3.10)}\qquad |\{\cdot\}_1| \geq \left|\int_t^s \frac{C_1\delta}{\sin^2(C_1 u)}|\kappa|\, ds\right| \geq |C_1\delta(s-t)\kappa|,$$

where $\delta$ is a positive number depending only on $[a,b]$ (and $C_1$, $C_2$).

Now we can perform one integration by parts in (3.8). By (3.10) this implies a bound of the form

$$|(A_{\lambda,\kappa}(s)^* A_{\lambda,\kappa}(t))(r,r'')| \le \frac{C}{|(s-t)\kappa|} F(|r|<\rho) F(|r''|<\rho),$$

which upon integration on $S$ gives

$$\left\| \int_{S(T,\varepsilon)} A_{\lambda,\kappa}(s)^* A_{\lambda,\kappa}(t)\, ds\, dt \right\| \le \frac{\tilde{C}}{|\kappa|}; \quad \tilde{C} = \tilde{C}(a,b,\varepsilon).$$

Clearly the right hand side $\to 0$ as $|k| \to \infty$. We conclude (3.7) and hence (3.6).

To complete the proof of (3.4) it suffices to bound

$$\sup_{T>0,\, \kappa\in\mathcal{K}} \left\| \int_0^T A_{\lambda,\kappa}(t)\, dt \right\| \to 0 \quad \text{as } |\lambda| \to \infty, \tag{3.11}$$

where $\mathcal{K} \subset \mathbf{R}^2$ is a fixed compact set. For that we use Lemma 2.1 again. It suffice to look at the interval given by replacing $[0,T]$ by a finite union of compact intervals on which (2.21) holds. Then we proceed as in the proof of [HMS1, Lemma 3.3] writing $e^{it\lambda} = -\frac{i}{\lambda}\frac{d}{dt}e^{it\lambda}$ and integrate by parts.

We have proved (3.11) and hence (3.4).

*Proof of* (3.5). First we assume that $q_1, q_2 \neq 0$ which allows us to apply Lemma 2.2. At the end of the proof we shall briefly comment on the case where one of the charges is zero.

We shall follow a somewhat similar procedure as that of the proof of (3.4). While the previous problem involved estimates with respect to the parameter $k$, the present one concerns the complication coming from the fact that the characteristic weight functions are only partial weights. These functions are independent of the variable $R_2$. For notational convenience we shall use the notation $y$ for this variable. We shall somehow introduce a localization with respect to $y$ (giving a full weight) before we perform the integration by parts with respect to the time variable.

Clearly we can replace the function $F(|r|<\rho)$ in (3.5) by a real nonnegative function $\chi(r) \in C_0^\infty(\mathbf{R}_r^2)$. Using Lemma 2.2 the kernel of the integrand then reads

$$A_\lambda(t)(r,y;r',y') = f(t)e^{it(\lambda+i)}\chi(r)e^{ip_t(r,y,r',y')}\chi(r'),$$

on the set given by the conditions (2.26), i.e. outside a discrete set. As before we need to bound the operator $\int_S A_\lambda(s)^* A_\lambda(t)\, ds\, dt$ for sufficiently nice sets $S$. Up to a null set these should cover a neighbourhood of the diagonal. In the following we let $S$ be a small rectangle near (but not including) a fixed point $(\bar{t},\bar{t})$ such that (2.26) holds with $t=\bar{t}$. We shall assume that $S$ does not intersect the diagonal nor the zero set of a certain (nonzero) real analytic function $\tilde{g}(s,t)$ (see (3.15) below). On such a set $S$ we compute the kernel

$$\begin{aligned}(A_\lambda(s)^*A_\lambda(t))(r,y;r'',y'')\\ = \overline{f(s)}f(t)e^{i(t-s)\lambda}e^{-(t+s)}\chi(r)\chi(r'')\int dr'\chi(r')^2\int dy' e^{ip_{s,t}(\cdot)};\end{aligned} \tag{3.12}$$

$$p_{s,t}(\cdot)=p_{s,t}(r',y',r,y,r'',y'')=p_t(r',y',r'',y'')-p_s(r',y',r,y).$$

Let

$$g(s,t)=\frac{\partial^2}{\partial y'^2}p_{s,t}(\cdot)=(a(t)-a(s))\left(2+\frac{Q^2}{2a(t)a(s)}\right). \tag{3.13}$$

Then we conclude from the fact that $\frac{d}{du}a(u)<0$

$$g(s,t)\neq 0;\quad (s,t)\in S. \tag{3.14}$$

Now we can compute the first integral on the right hand side of (3.12)

$$\begin{aligned}\int dy' e^{ip_{s,t}(\cdot)} &= \sqrt{\frac{2\pi}{|g(s,t)|}}e^{i\pi\,\mathrm{sgn}(g(s,t))/4}e^{iq_{s,t}(\cdot)};\\ q_{s,t}(\cdot)&=q_{s,t}(r',r,y,r'',y'')\\ &=-\frac{\left(\frac{\partial}{\partial y'}p_{s,t}\right)^2(r',0,r,y,r'',y'')}{2g(s,t)}+p_{s,t}(r',0,r,y,r'',y'').\end{aligned}$$

We insert this expression. The idea is now to integrate by parts in the remaining integral thereby producing decay with respect to the variables $y$ and $y''$. To examine this we introduce the determinant

$$\tilde{g}(s,t)=\det(\nabla_{(y,y'')}\nabla_{r'}^T q_{s,t}(\cdot)). \tag{3.15}$$

We shall need

$$\tilde{g}(s,t)\neq 0. \tag{3.16}$$

Suppose for a moment that this condition holds. Then we write

$$e^{iq_{s,t}(\cdot)}=\left(\frac{1}{1+|\nabla_{r'}q_{s,t}(\cdot)|^2}-i\frac{\nabla_{r'}q_{s,t}(\cdot)}{1+|\nabla_{r'}q_{s,t}(\cdot)|^2}\cdot\nabla_{r'}\right)^4 e^{iq_{s,t}(\cdot)}$$

and perform four integrations by parts with respect to the variable $r'$. The resulting function has the property that $|(y, y'')|^4$ is bounded on its support. Now we write $e^{it\lambda} = -\frac{i}{\lambda}\frac{d}{dt}e^{it\lambda}$ and perform one integration by parts with respect to the variable $t$ in the integral over $S$. This gives decay in $\lambda$ at the expense of growing factors in $y$ and $y'$ of at most order two. As a conclusion we obtain an expression for the kernel of the operator $\int_S A_\lambda(s)^* A_\lambda(t)\,ds\,dt$ that is pointwise bounded by

$$\frac{C}{|\lambda|}\chi(r)\chi(r'')(|y|^2 + |y''|^2 + 1)^{-1}. \tag{3.17}$$

Clearly (3.17) implies the estimate $O(|\lambda|^{-1})$ of the above operator, which in turn proves (3.5).

It remains to examine (3.16). More specifically we introduce the function

$$h(s,t) = Q^{-2}g(s,t)^2\tilde{g}(s,t) = \det(D(s,t));$$
$$D(s,t) = Q^{-1}g(s,t)\nabla_{(y,y'')}\nabla_{r'}^T q_{s,t}(\cdot),$$

and notice that it is naturally defined (by analytic extension) on the connected subset

$$\{z \in \mathbf{C} |\ \sin(K_2 z),\ \sin(K_1 z),\ a(z) \neq 0\}^2$$

of $\mathbf{C}^2$ (cf. (2.26)). We shall show that the function $h$ is not the zero function. The zero set of the restriction of $h$ to the corresponding real domain will then have zero relative measure by a result for real analytic functions. This will justify the above arguments and thus complete the proof of (3.5).

To show that $h$ is not the zero function we need to calculate the entries of the matrix. Straightforward (but rather tedious) computations give the following result. We use the notation (2.17), (2.19), (2.25), (3.13), and set $D(s,t) = \{d_{ij}\}_{i,j\leq 2}$ where

$$\begin{aligned}
d_{11} &= \left(\frac{b(s)}{a(s)} - \frac{b(t)}{a(t)}\right)\left(\frac{Q^2}{4a(s)} + a(s)\right) + g(s,t)\left(\frac{\theta}{Q} - \frac{b(s)}{2a(s)}\right),\\
d_{12} &= -\left(\frac{b(s)}{a(s)} - \frac{b(t)}{a(t)}\right)\left(\frac{Q^2}{4a(t)} + a(t)\right) - g(s,t)\left(\frac{\theta}{Q} - \frac{b(t)}{2a(t)}\right),\\
d_{21} &= \left\{\frac{\theta}{2}\left(\frac{1}{a(t)} - \frac{1}{a(s)}\right) - \frac{b(t) - b(s)}{Q}\right\}\left(\frac{Q^2}{2a(s)} + 2a(s)\right)\\
&\quad + g(s,t)\left(\frac{b(s)}{Q} + \frac{\theta}{2a(s)}\right),\\
d_{22} &= \left\{-\frac{\theta}{2}\left(\frac{1}{a(t)} - \frac{1}{a(s)}\right) + \frac{b(t) - b(s)}{Q}\right\}\left(\frac{Q^2}{2a(t)} + 2a(t)\right)\\
&\quad - g(s,t)\left(\frac{b(t)}{Q} + \frac{\theta}{2a(t)}\right).
\end{aligned} \tag{3.18}$$

We claim that for any real $t$ obeying (2.26)

$$\lim_{s\to 0} s^2 h(s,t) = -\frac{M^2}{Qa(t)}(b(t)^2+\theta^2). \tag{3.19}$$

Since the right hand side is nonzero by the assumption $K_1 \neq K_2$ this will complete the proof.

We notice that

$$\lim_{s\to 0} sa(s) = \frac{M}{2}, \quad \lim_{s\to 0} b(s) = 0$$

which used in (3.18) gives

$$\begin{aligned}
\lim_{s\to 0} sd_{11} &= -M\left(\frac{b(t)}{2a(t)} + \frac{\theta}{Q}\right),\\
\lim_{s\to 0} sd_{12} &= -M\left(\frac{b(t)}{2a(t)} - \frac{\theta}{Q}\right),\\
\lim_{s\to 0} sd_{21} &= M\left(\frac{\theta}{2a(t)} - \frac{b(t)}{Q}\right),\\
\lim_{s\to 0} sd_{22} &= M\left(\frac{\theta}{2a(t)} + \frac{b(t)}{Q}\right).
\end{aligned}$$

We have thus computed the matrix $\lim_{s\to 0} sD(s,t)$. It remains to show that its determinant is given by the formula (3.19). But this is easily obtained by adding the second column to the first and $-\frac{1}{2}$ times the first (new column) to the second before expanding the determinant.

We have completed the proof of (3.5) under the assumption $q_1, q_2 \neq 0$.

If one of the charges is zero we can not use Lemma 2.2 directly. However, as pointed out in Remark 2.3, the kernel can be obtained as the limit of the one in Lemma 2.2. Using that expression we can obtain (3.5) in this case also by mimicking the above proof.

**4. Uniform squeezing.** We shall consider the 2-body problem assuming $E_{\|}^a \neq 0$, where $a$ as in the previous section refers to $a_{\max}$. Explicitly this condition means that $\mathcal{E}_3 \neq 0$ and $K_1 \neq K_2$. It implies absence of eigenvalues for the reduced Hamiltonians, a fact we shall use in this section. Under the conditions of Lemma 3.1 these operators are the fiber operators appearing on the right hand side of the identity

$$U_n H(E_{\|}^a, Y) U_n^{-1} = \int_{\mathbf{R}^2} \oplus\, dk\; H_n(E_{\|}^a, k, X^a).$$

Under the conditions of Lemma 3.2 the reduced Hamiltonian is given as the first factor on the right hand side of the identity

$$U_c H(E_{\|}^a, Y) U_c^{-1} = H(E_{\|}^a, \tilde{X}^a) \otimes I.$$

The fact that the above reduced Hamiltonians do not have eigenvalues (under the condition (1.1) on the pair potentials), i.e.

$$\sigma_{pp}(H_n(E^a_\|, k, X^a)) = \emptyset, \quad \sigma_{pp}(H(E^a_\|, \tilde{X}^a)) = \emptyset, \tag{4.1}$$

follows by the same basic method as the one explained in the next section giving similar results for $N = 3$. The proof for $N = 2$ is somewhat simpler though. As a consequence we shall not comment further on the proof.

For $\varepsilon > 0$ the notation $\eta_\varepsilon$ stands for any function $\eta \in C_0^\infty(\mathbf{R})$ obeying $0 \leq \eta \leq 1$ and $\eta(t) = 1$ for $|t| \leq \varepsilon$, $\eta(t) = 0$ for $|t| > 2\varepsilon$.

**Proposition 4.1.** *Suppose $N = 2$ and $E^a_\| \neq 0$. Then for any $\rho > 0$*

$$\|\eta_\varepsilon(H(E_\|) - \lambda)F(|x^a| < \rho)\| \to 0 \quad \text{for } \varepsilon \to 0, \tag{4.2}$$

*uniformly in $\lambda \in \mathbf{R}$.*

*Proof.* First we consider the neutral case. We need to prove that

$$\|\eta_\varepsilon(H_n(E^a_\|, k, X^a) - \lambda)F(|x| < \rho)\| \to 0 \quad \text{for } \varepsilon \to 0,$$

uniformly in $k \in \mathbf{R}^2$ and $\lambda \in \mathbf{R}$. For that we consider arbitrarily given sequences $(k_m) \subset \mathbf{R}^2$ and $(\lambda_m) \subset \mathbf{R}$. We need to show

$$\sup_{m\in\mathbf{N}} \|\eta_\varepsilon(H_n(E^a_\|, k_m, X^a) - \lambda_m)F(|x^a| < \rho)\| \to 0 \quad \text{for } \varepsilon \to 0. \tag{4.3}$$

By inserting

$$\eta_\varepsilon(H) = \{\eta_\varepsilon(H)(H - i)\}(H - i)^{-1}; \quad H = H_n(E^a_\|, k_m, X^a) - \lambda_m,$$

and applying the second resolvent equation and Lemma 3.1 we see that it can be assumed that the above sequences are bounded. Under this condition we shall prove (4.3) by the way of contradiction. Suppose (4.3) is false. Then upon changing the previous sequences we can assume that for some $\mu > 0$ there is a sequence $\varepsilon_m \to 0$ such that

$$\|\eta_{\varepsilon_m}(H_n(E^a_\|, k_m, X^a) - \lambda_m)F(|x^a| < \rho)\| > \mu, \tag{4.4}$$

and moreover

$$k_m \to k, \quad \lambda_m \to \lambda.$$

For each $k_m$ we define $\kappa_m$ and $T_{\kappa_m}$ by (2.5). Let

$$H(k_m) = T_{\kappa_m} H_n(E^a_\|, k_m, X^a) T_{\kappa_m}^{-1}.$$

We introduce $H(k)$ similarly. Then

$$H(k_m) = H_0(E_\|^a, 0, X^a) + v_{12}(r - \kappa_m), \tag{4.5}$$

and similarly for $H(k)$. Moreover upon enlarging $\rho$, (4.4) implies

$$\|\eta_{\varepsilon_m}(H(k_m) - \lambda_m)F(|x^a| < \rho)\| > \mu. \tag{4.6}$$

We shall find a contradiction to (4.6) by using that $\lambda$ is not an eigenvalue of $H(k)$ (cf. (4.1)). We notice that this fact and compactness (cf. (5.3)) give

$$\|\eta_{\varepsilon'}(H(k) - \lambda)F(|x^a| < \rho)\| \to 0 \quad \text{for } \varepsilon' \to 0.$$

Fix $\varepsilon' > 0$ such that the norm is $< \mu/2$. Then we write for $m$ large enough

$$\begin{aligned}&\eta_{\varepsilon_m}(H(k_m) - \lambda_m)F(|x^a| < \rho)\\ &= \eta_{\varepsilon_m}(\cdot)\eta_{\varepsilon'}(H(k) - \lambda)F(\cdot) + \eta_{\varepsilon_m}(\cdot)\{\eta_{\varepsilon'}(H(k_m) - \lambda) - \eta_{\varepsilon'}(H(k) - \lambda)\}F(\cdot).\end{aligned}$$

The norm of the first term is $< \mu/2$ (for all $m$). The norm of the second term is $< \mu/2$ for $m$ large enough, since

$$\|\eta_{\varepsilon'}(H(k_m) - \lambda) - \eta_{\varepsilon'}(H(k) - \lambda)\| \to 0 \quad \text{for } m \to \infty.$$

This statement follows readily from an application of the Stone-Weierstrass theorem, the second resolvent equation and (4.5). Notice that (1.1) implies that

$$\|v_{12}(r - \kappa_m) - v_{12}(r - \kappa)\| \le C|\kappa_m - \kappa|.$$

Clearly we have found a contradiction to (4.6) for large $m$. The proof of (4.2) in the neutral case is complete.

In the charged case we need to show that

$$\|\eta_\varepsilon(H(E_\|^a, \tilde{X}^a) - \lambda)F(|x^a| < \rho)\| \to 0 \quad \text{for } \varepsilon \to 0, \tag{4.7}$$

uniformly in $\lambda \in \mathbf{R}$. We can proceed in a similar (but simpler) way. By Lemma 3.2 it is enough to estimate the operator uniformly in $\lambda$ belonging to a compact set. Then we use (4.1) and the fact that $\eta_1(H(E_\|^a, \tilde{X}^a) - \lambda)F(|x^a| < \rho)$ is compact (cf. (5.3)) to complete the proof of (4.7).

**5. Absence of eigenvalues.** We shall consider the 3-body problem assuming $E_\|^{a_{\max}} \neq 0$, (1.1) and that (4.1) holds for 2-body subsystems satisfying $E_\|^a \neq 0$. We shall aim at proving absence of eigenvalues for the 3-body reduced Hamiltonians. As in the previous section these are defined differently according to $Q$ being zero or nonzero.

In the neutral case the reduced Hamiltonians are fiber operators. For any fiber $k$ we shall use the definition (2.3) in the free case. With potential the reduced Hamiltonian $H(E_{\|}^{a_{\max}}, k, X^{a_{\max}})$ is given by just adding the (internal) potential to this expression.

In the free charged case the reduced Hamiltonian is given as the first factor on the right hand side of the identity (2.9). With internal potential we just add it to this expression and obtain a (reduced) Hamiltonian which we shall denote by $H(E_{\|}^{a_{\max}}, \tilde{X}^{a_{\max}})$.

We shall abbreviate the notation for the above reduced Hamiltonians by simply writing $H^r$. Although the expressions are different, and, in fact, the space the operator acts on differs for the neutral and charged cases, there are common features that make it convenient to treat them together. Of course, the pseudo momentum $k$ for the neutral case is suppressed in the above notation. Throughout this section (and in fact the rest of the paper), it is arbitrarily given (for the neutral case).

For $x^{a_{\max}} = (x_1, \dots, x_N) \in X^{a_{\max}}$ we let $x_{\|}^{a_{\max}} = ((x_1)_3\bar{e}_3, \dots, (x_N)_3\bar{e}_3)$ where $\bar{e}_3$ is given by (1.14). We abbreviate

$$z = x_{\|}^{a_{\max}}, \quad \langle z\rangle = (1+|z|^2)^{1/2}, \quad \hat{z} = \frac{z}{\langle z\rangle}.$$

Similar notation $\langle A\rangle = (1+|A|^2)^{1/2}$ applies to numbers and selfadjoint operators.

We shall modify the arguments in [HMS1, Sect. 4]. For that we need generalizations of [HMS1, Lemmas 3.1 and 3.2] concerning domain questions. This is accomplished by the same method of proof, namely by simple quadratic estimates. The keypoint is the following form of the Hamiltonian (obvious from (2.3) and (2.9)):

$$H^r + E_{\|}^{a_{\max}} \cdot z - V(x) = \sum_j D_j^2, \tag{5.1}$$

where the right hand side is a finite sum of squares of differential operators with real affine symbols. Then the above mentioned method gives:

**Lemma 5.1.** *The domain $\mathcal{D}(H^r) \subset H^2_{\mathrm{loc}}$, the local Sobolev space of order 2. Suppose $0 \le h \in C^\infty(X^{a_{\max}})$ ($C^\infty(\tilde{X}^{a_{\max}})$ in the charged case) and $\psi \in \mathcal{D}(H^r)$ are given such that $hH^r\psi$, $\langle E_{\|}^{a_{\max}} \cdot z\rangle^{1/2} h\psi$, $(\partial_k h)\psi \in L^2(X^{a_{\max}})$ (or $L^2(\tilde{X}^{a_{\max}})$) for any component $\partial_k h$ of $\nabla h$. Then for any of the factors in* (5.1)

$$D_j h\psi,\ hD_j\psi \in L^2(X^{a_{\max}}) \quad (\text{or } L^2(\tilde{X}^{a_{\max}})).$$

Clearly Lemma 5.1 implies

$$D_j\langle E_{\|}^{a_{\max}} \cdot z\rangle^{-1/2}(H^r - i)^{-1},\ \langle E_{\|}^{a_{\max}} \cdot z\rangle^{-1/2} D_j (H^r - i)^{-1} \tag{5.2}$$

are bounded for any of the factors $D_j$ in (5.1).

Another useful result, readily obtained by a perturbation argument under use of (5.2), reads: For any $\rho > 0$

$$F(|x^{a_{\max}}| < \rho)(H^r - i)^{-1} \quad \text{is compact.} \tag{5.3}$$

We can now to a large extent mimic [HMS1, Sect. 4]. The heart of the matter is certain estimates which here shall be established under use of Proposition 4.1. We consider the function

$$f(x) = f_\kappa(x) = \kappa\langle z\rangle + E_{\|}^{a_{\max}} \cdot z$$

with $\kappa \gg 1$. As the notation indicates $f : X^{a_{\max}}$ (or $\tilde{X}^{a_{\max}}$) $\to \mathbf{R}_+$. For some smooth functions $\theta : \mathbf{R}_+ \mapsto \mathbf{R}$ to be specified below we put

$$A_f = (\nabla f \cdot p + p \cdot \nabla f)/2,$$
$$\theta = \theta(f),\ \theta' = \left(\frac{d}{dt}\theta\right)(f),\ \theta'' = \left(\frac{d^2}{dt^2}\theta\right)(f),$$

and compute the commutator

$$\begin{aligned} i[H^r, e^\theta A_f e^\theta] &= e^\theta \{T_1 + \cdots + T_4\} e^\theta; \\ T_1 &= 4A_f\theta' A_f, \\ T_2 &= 2pf^{(2)}p, \\ T_3 &= (E_{\|}^{a_{\max}} - \nabla V) \cdot \nabla f, \\ T_4 &= -\tfrac{1}{2}\Delta^2 f + \nabla f \cdot \nabla(((\theta')^2 - \theta'')|\nabla f|^2). \end{aligned} \tag{5.4}$$

Moreover

$$(H^r - \lambda - ((\theta')^2 - \theta'')|\nabla f|^2 + 2i\theta' A_f)e^\theta = e^\theta (H^r - \lambda). \tag{5.5}$$

These formulas will be used for $\theta$ obeying

$$\frac{d}{dt}\theta(t) \geq 0, \quad \left(\frac{d^n}{dt^n}\right)\theta(t) = O(t^{1-n}) \text{ for } t \to \infty;\ n \geq 0.$$

Hence

$$T_1 \geq 0,\ T_4 = O(\langle z\rangle^{-1}) \quad \text{for } |z| \to \infty. \tag{5.6}$$

A computation shows also that

$$T_2 \geq 0.$$

To deal with the term $T_3$ we notice that

$$\nabla V \cdot \nabla\langle z\rangle = o(1).$$

Here and below $o(1)$ means $o(1)$ for $|z| \to \infty$. Hence

$$T_3 = \kappa E_{\|}^{a_{\max}} \cdot \hat{z} + \varepsilon_0 - \sum_{a:\, E_{\|}^{a} \neq 0} \nabla V_a(x^a) \cdot E_{\|}^{a_{\max}} + o(1); \quad \varepsilon_0 = |E_{\|}^{a_{\max}}|^2.$$

For the first term we have

$$\begin{aligned} E_{\|}^{a_{\max}} \cdot \hat{z} &= \mathrm{Re}(\langle z\rangle^{-1}\{V(x) - H^r + \sum_j D_j^* D_j\}) \\ &= \sum_j D_j^* \langle z\rangle^{-1} D_j + \left(\frac{V(x)}{\langle z\rangle} - \left(\Delta \frac{1}{2\langle z\rangle}\right)\right) - \mathrm{Re}(\langle z\rangle^{-1} H^r) \\ &\geq - \mathrm{Re}(\langle z\rangle^{-1} H^r) + o(1). \end{aligned}$$

Putting this together we obtain

$$T_2 + T_3 \geq \varepsilon_0 - \sum_{a:\, E_{\|}^{a_{\max}} \neq 0} \nabla V_a(x^a) \cdot E_{\|}^{a_{\max}} - \mathrm{Re}\left(\frac{\kappa}{\langle z\rangle} H^r\right) + o(1). \tag{5.7}$$

Finally, by (5.4), (5.6) and (5.7),

$$i[H^r, e^{\theta} A_f e^{\theta}] = e^{\theta}\{\ldots\}e^{\theta};$$

$$\{\ldots\} \geq T_1 + \varepsilon_0 - \sum_{b:\, E_{\|}^{b} \neq 0} \nabla V_b(x^b) \cdot E_{\|}^{a_{\max}} - \mathrm{Re}\left(\frac{\kappa}{\langle z\rangle} H^r\right) + o(1). \tag{5.8}$$

In the following applications we consider some further specified functions $\theta$ depending on parameters. The estimate (5.8) will be uniform with respect to these parameters. As above we suppress the dependence of $\kappa$ in the notation for $f = f_\kappa$ and $\theta = \theta_\kappa(f_\kappa)$.

**Lemma 5.2.** *Let $\lambda \in \mathbf{R}$, $s > 0$ and*

$$\theta(t) = s \log\left(t(1 + t/\mu)^{-1}\right); \quad \mu > 1.$$

*Then for all sufficiently large $\kappa$ there exist constants $C, \rho > 0$ such that*

$$\begin{aligned} &\forall\, \mu > 1,\ \phi \in C_0^\infty(X^{a_{\max}}) \text{ (or } C_0^\infty(\tilde{X}^{a_{\max}})) : \\ &\langle \phi, i[(H^r - \lambda), e^{\theta} A_f e^{\theta}]\phi\rangle + C\|e^{\theta}(H^r - \lambda)\phi\|^2 \\ &\geq \frac{1}{C}\|e^{\theta}\phi\|^2 - C\|F(|z| < \rho)e^{\theta}\phi)\|^2. \end{aligned}$$

**Lemma 5.3.** *Let $\lambda \in \mathbf{R}$, $\alpha_0 \geq 0$ and*

$$\theta(t) = \mu \log\left(1 + \frac{\gamma t}{\kappa\mu}\right) + \frac{\alpha}{\kappa}t; \quad \mu > 1, \gamma > 0, \alpha \geq 0.$$

*Then for a (small) $\delta > 0$ and all sufficiently large $\kappa$ there exist constants $C, \rho > 0$ such that*

$$\begin{aligned} &\forall\, \mu > 1, \gamma + |\alpha - \alpha_0| \leq \delta,\ \phi \in C_0^\infty(X^{c_{\max}})\ (\text{or } C_0^\infty(\tilde{X}^{a_{\max}})): \\ &\quad \langle \phi, i[(H^r - \lambda), e^\theta A_f e^\theta]\phi\rangle + C\|e^\theta(H^r - \lambda)\phi\|^2 \\ &\quad \geq \frac{1}{C}\|e^\theta\phi\|^2 - C\|F(|z| < \rho)e^\theta\phi\|^2. \end{aligned}$$

*Proof of Lemmas 5.2 and 5.3.* We compute

$$\theta'(t) \leq \frac{s}{t},\ |\theta''(t)| \leq 2\frac{s}{t^2},\ |\theta'''(t)| \leq \frac{8s}{t^3} \tag{5.9}$$

in case of Lemma 5.2, while

$$|\theta'(t) - \frac{\alpha}{\kappa}| \leq \frac{\gamma}{\kappa},\ |\theta''(t)| \leq \frac{\gamma}{\kappa t},\ |\theta'''(t)| \leq \frac{2\gamma}{\kappa t^2} \tag{5.10}$$

in case of Lemma 5.3.

Thus by inspection we obtain for $h = T_4$ given by (5.4) that

$$|h(x)| \leq C\langle z\rangle^{-3}; \quad C = C_\kappa \text{ independent of } \mu, \tag{5.11}$$

under the condition (5.9), while

$$|h(x)| \leq C\langle z\rangle^{-1}; \quad C = C_\kappa \text{ independent of } \mu, \gamma, \alpha, \tag{5.12}$$

under the conditions (5.10) and $\gamma + |\alpha - \alpha_0| \leq 1$.

By (5.11) and (5.12) the fifth term on the right hand side of (5.8) is $o(1)$ uniformly with respect to the parameters. Similarly all statements below will be uniform with respect to the parameter $\mu$ in case of Lemma 5.2 and $\mu$, $\gamma$, $\alpha$ in case of Lemma 5.3 (constrained as in the lemmas), and in addition with respect to $\phi \in C_0^\infty$. For notational convenience we put $\alpha_0 = 0$ in case of Lemma 4.1, introduce $\bar{\lambda} = \lambda + \alpha_0^2$ and abbreviate the expectation $\langle e^\theta\phi, Te^\theta\phi\rangle = \langle T\rangle$ for any linear operator $T$.

As for the fourth term on the right hand side of (5.8) we obtain from (5.9) and (5.10) that

$$-\operatorname{Re}\left(\langle z\rangle^{-1}H^r\right) = -\operatorname{Re}\left(\langle z\rangle^{-1}\{H^r - \lambda - ((\theta')^2 - \theta'')|\nabla f|^2\}\right) + O\left(\langle z\rangle^{-1}\right).$$

Hence by (5.5), the Cauchy-Schwarz inequality and the same bounds again, it follows that for any $\phi \in C_0^\infty$

$$
\begin{aligned}
&-\langle \mathrm{Re}(\langle z\rangle^{-1}H^r)\rangle \\
(5.13)\quad &= -\mathrm{Re}\langle e^\theta\phi, \langle z\rangle^{-1}e^\theta(H^r-\lambda)\phi\rangle + \mathrm{Re}\langle 2i\langle z\rangle^{-1}\theta' A_f\rangle + \langle O(\langle z\rangle^{-1})\rangle \\
&\geq -\|e^\theta(H^r-\lambda)\phi\|^2 + \langle i[\theta'/\langle z\rangle, A_f]\rangle + \langle O(\langle z\rangle^{-1})\rangle \\
&\geq -\|e^\theta(H^r-\lambda)\phi\|^2 + \langle O(\langle z\rangle^{-1})\rangle.
\end{aligned}
$$

As for the third term on the right hand side of (5.8) we obtain by choosing $\rho$ large enough that it is estimated by

$$
\begin{aligned}
(5.14)\quad &-\sum_{b:\, E_{\|}^b \neq 0} h_b(x) \geq -\sum_{b:\, E_{\|}^b\neq 0} F(|x^b|<\rho)h_b(x) - \varepsilon_0/6; \\
&h_b(x) = \nabla V_b(x^b)\cdot E_{\|}^{a_{\max}}.
\end{aligned}
$$

We look closer at the first term on the right hand side of (5.14). The idea is to write

$$
\begin{aligned}
(5.15)\quad &F(|x^b|<\rho) = \eta_\varepsilon(H^r-\bar\lambda)F(|x^b|<\rho) + \mathcal{E}_b; \\
&\mathcal{E}_b = (H^r-\bar\lambda)\zeta_\varepsilon(H^r-\bar\lambda)F(|x^b|<\rho), \quad \zeta_\varepsilon(t) = t^{-1}(1-\eta_\varepsilon(t)),
\end{aligned}
$$

and choose $\varepsilon$ small to treat the first term by Proposition 4.1. The appearance of functions $h_b$ as factors is a minor complication since these are bounded. It is useful to introduce a family $\{j_a\}$, $a \neq a_{\max}$, of functions on $X^{a_{\max}}$ each one being smooth and homogeneous of degree zero outside a compact set. Moreover we assume that

$$
(5.16)\quad \sum_a j_a(x^{a_{\max}}) = 1 \quad \text{for } |x^{a_{\max}}| \text{ large,}
$$

and for any $c \not\subset a$

$$
(5.17)\quad |x^c| \geq \varepsilon|x^{a_{\max}}| \text{ on } \operatorname{supp} j_a; \quad \varepsilon = \varepsilon(a,c) > 0.
$$

The used ordering of the cluster is standard: $c \subset a \iff X^c \subset X^a$.

We claim that

$$
(5.18)\quad \eta_\varepsilon(H^r-\bar\lambda)F(|x^b|<\rho) = F(|x^b|<\rho)\eta_\varepsilon(H_b^r-\bar\lambda)j_b + K; \quad K \text{ compact.}
$$

Here $H_b^r$ is given as $H^r$ but with $V(x)$ replaced by $V_b(x^b)$.

By using (5.16) on the left hand side and applying (5.3) and (5.17) the statement (5.18) follows if we can prove that $\eta_\varepsilon(H^r-\bar\lambda)j_b - j_b\eta_\varepsilon(H_b^r-\bar\lambda)$

is compact. As for the latter it is enough, by the representation formula [HMS2, (3.6)], to show that $(H^r-\zeta)^{-1}j_b - j_b(H^r-\zeta)^{-1}$ is compact for $\zeta \in \mathbf{C}\setminus\mathbf{R}$. To do this we compute

$$\begin{aligned}(H^r-\zeta)^{-1}j_b - j_b(H^r_b-\zeta)^{-1} &= (H^r-\zeta)^{-1}(A_{b,1}+A_{b,2})(H^r_b-\zeta)^{-1};\\ A_{b,1} = -I_b j_b, &\quad I_b = V(x)-V_b(x^b),\\ A_{b,2} = -i\{D\cdot\nabla j_b + \nabla j_b\cdot D\}, &\quad D = p^{a_{\max}} - A^{a_{\max}}x^{a_{\max}}.\end{aligned}$$

The contribution from $A_{b,1}$ is compact by (5.3) and (5.17). The contribution from $A_{b,2}$ is readily seen to be compact by (5.2) and(5.3).

Now we look at the first term on the right hand side of (5.18). We claim that the norm goes to zero as $\varepsilon \to 0$. To see this we let $H(E^b_{\|})$ be the full 2-body Hamiltonian given by the pair $b$ and the indicated electric field (cf. (1.2), (1.3)). Then by a conjugation we can estimate

$$\|F(|x^b|<\rho)\eta_\varepsilon(H^r_b-\bar\lambda)\| \le \sup_{\lambda\in\mathbf{R}}\|\eta_\varepsilon(H(E^b_{\|})-\lambda)F(|x^b|<\rho)\|. \tag{5.19}$$

By Proposition 4.1 the right hand side goes to zero as $\varepsilon \to 0$. We conclude that for $\varepsilon$ small enough

$$-\operatorname{Re}\sum_{b:\,E^b_{\|}\neq 0}\eta_\varepsilon(H^r-\bar\lambda)F(|x^b|<\rho)h_b \ge -\frac{\varepsilon_0}{6}I - K;\quad K \text{ compact.} \tag{5.20}$$

On the other hand for fixed $\varepsilon$ we can estimate

$$\begin{aligned}|\langle\mathcal{E}_b h_b\rangle| &\le C\|(H^r-\bar\lambda)e^\theta\phi\|\,\|e^\theta\phi\| \le C'\|(H^r-\bar\lambda)e^\theta\varphi\|^2 + \varepsilon'\|e^\theta\phi\|^2\\ &\le 2C'\|(H^r-\lambda-((\theta')^2-\theta'')|\nabla f|^2+2i\theta' A_f)e^\theta\phi\|^2\\ &\quad + 2C'\|(\{\alpha_0^2-((\theta')^2-\theta'')|\nabla f|^2\}+2i\theta' A_f)e^\theta\phi\|^2 + \varepsilon'\|e^\theta\phi\|^2;\\ &\qquad \varepsilon'>0,\ C'=C'_{\varepsilon,\varepsilon'}.\end{aligned}$$

As for the first term we use (5.5). To estimate the second term we notice the bound

$$\theta' \le C\kappa^{-1}.$$

Moreover

$$(\alpha_0)^2 - ((\theta')^2-\theta'')|\nabla f|^2 = O(\langle z\rangle^{-1})$$

under the condition (5.9), while

$$|(\alpha_0)^2 - ((\theta')^2-\theta'')|\nabla f|^2| \le C(\langle z\rangle^{-1}+\kappa^{-1}+\delta)$$

under the condition (5.10). In both cases $C$ is independent of $\kappa$ and the parameters (constrained to $\gamma + |\alpha - \alpha_0| \leq \delta$ in case of Lemma 5.3). We apply the above estimates for $\kappa$ large, and in case of Lemma 5.3 for $\delta$ small. (Notice the bound $\|2i\theta' A_f e^\theta \phi\|^2 \leq \sup \theta' \langle T_1 \rangle$.) As a result

$$\begin{aligned} &-\Big\langle \operatorname{Re} \sum_{b:\, E_{\|}^b \neq 0} \mathcal{E}_b h_b \Big\rangle \\ &\geq -\frac{\varepsilon_0}{6}\|e^\theta \phi\|^2 - C\|e^\theta (H^r - \lambda)\phi\|^2 - \langle T_1/2 \rangle + \big\langle O(\langle z \rangle^{-2})\big\rangle. \end{aligned} \tag{5.21}$$

By combining (5.8), (5.13), (5.14), (5.20) and (5.21) we finally obtain that for a constant $C > 0$ and compact $K$ (both independent of $\mu$, and in case of Lemma 5.3 also of $\gamma$ and $\alpha$)

$$\begin{aligned} &\langle \phi, i[(H^r - \lambda), e^\theta A_f e^\theta]\phi\rangle + C\|e^\theta (H^r - \lambda)\phi\|^2 \\ &\geq \frac{\varepsilon_0}{2}\|e^\theta \phi\|^2 + \langle T_1/2 \rangle - \|K e^\theta \phi\|^2 + \langle \tilde{h} \rangle; \quad \tilde{h} = o(1). \end{aligned} \tag{5.22}$$

It remains to treat the last two terms on the right hand side of (5.22). For that we pick $C'$, $\rho' > 1$ independent of the parameters such that

$$\tilde{h} \geq -C' F(|z| < \rho') - \frac{\varepsilon_0}{6}, \tag{5.23}$$

cf. (5.11) and (5.12). We pick $\rho'' > 1$ such that

$$2\|K F(|z| \geq \rho'')\|^2 \leq \frac{\varepsilon_0}{6}. \tag{5.24}$$

Let $\rho = \max(\rho', \rho'')$. Then we obtain from (5.22) – (5.24) that

$$\begin{aligned} &\langle \phi, i[(H^r - \lambda), e^\theta A_f e^\theta]\phi\rangle + C\|e^\theta (H^r - \lambda)\phi\|^2 \\ &\geq \frac{\varepsilon_0}{6}\|e^\theta \phi\|^2 - \big(2\|K F(|z| < \rho'')\|^2 + C'\big)\,\|F(|z| < \rho) e^\theta \phi\|^2. \end{aligned} \tag{5.25}$$

Clearly the lemmas follow from (5.25) by choosing the constant $C$ in these statements sufficiently large.

The previous estimates lead to

**Lemma 5.4.** *Suppose $(H^r - \lambda)\psi = 0; \lambda \in \mathbf{R}$, $\psi \in \mathcal{D}(H^r)$. Then*

$$\forall \alpha \geq 0 : e^{\alpha\langle z \rangle}\psi(x) \in L^2(X^{a_{\max}}) \ (\text{or } L^2(\tilde{X}^{a_{\max}})).$$

*Proof.* First we prove

$$\forall s \geq 0 : \langle z \rangle^s \psi(x) \in L^2. \tag{5.26}$$

For that we use (5.2) and Lemma 5.2 to prove by an approximation argument that

$$C^2\|F(|z| < \rho)e^{\theta}\psi\|^2 \geq \|e^{\theta}\psi\|^2 \tag{5.27}$$

where $C$, $\rho$ and $\theta$ are as in Lemma 5.2. If (5.26) does not hold we obtain a contradiction by letting $\mu \to \infty$ in (5.27).

Now let

$$\alpha_0 = \sup\{\alpha \geq 0 |\, e^{\alpha\langle z\rangle}\psi(x) \in L^2\}.$$

We need to show that $\alpha_0 = \infty$. Suppose this statement is false.

If $\alpha_0 = 0$ we come to a contraction by using (5.26), Lemma 5.1 and Lemma 5.3 with $\alpha = \alpha_0$. It is obtained by letting $\mu \to \infty$ in an estimate similar to (5.27).

If $\alpha_0 > 0$ the contradiction arises by the following use of Lemma 5.3. Let $\delta > 0$ be given in accordance with the lemma. Then we choose any $\alpha < \alpha_0$ such that $\alpha_0 < \alpha + \delta/2$ and put $\gamma = \delta/2$. It remains to specify $\kappa$. We choose it so large that, in addition to the estimates in the lemma, the conditions

$$\alpha(1 + \kappa^{-1}|E_{\|}^{a_{\max}}|) < \alpha_0 \text{ and } \alpha_0 < (\alpha + \gamma)(1 - \kappa^{-1}|E_{\|}^{a_{\max}}|)$$

hold. Now we proceed as above, that is by first using Lemma 5.1 to replace $\phi$ in the estimate by $\psi$ and then letting $\mu \to \infty$.

**Proposition 5.5.** *Suppose $N = 3$ and $E_{\|}^{a_{\max}} \neq 0$. Let $H^r$ be an abbreviation for any reduced Hamiltonian as defined in the beginning of this section. Then*

$$\sigma_{pp}(H^r) = \emptyset.$$

*Proof.* We compute for any $\phi \in C_0^\infty(X)$:

$$\begin{aligned} &\|e^{\alpha E_{\|}^{a_{\max}}\cdot z}\{(H^r - \lambda) - V\}\phi\|^2 \\ &= \|(H_0^r - \lambda - \alpha^2)|E_{\|}^{a_{\max}}|^2)e^{\alpha E_{\|}^{a_{\max}}\cdot z}\phi\|^2 + 4\alpha^2\|E_{\|}^{a_{\max}}\cdot pe^{\alpha E_{\|}^{a_{\max}}\cdot z}\phi\|^2 \\ &\quad +2\alpha|E_{\|}^{a_{\max}}|^2\|e^{\alpha E_{\|}^{a_{\max}}\cdot z}\phi\|^2; \quad \lambda \in \mathbf{R},\, \alpha \geq 0. \end{aligned}$$

In particular

$$\|e^{\alpha E_{\|}^{a_{\max}}\cdot z}\{(H^r - \lambda) - V\}\phi\|^2 \geq 2\alpha|E_{\|}^{a_{\max}}|^2\|e^{\alpha E_{\|}^{a_{\max}}\cdot z}\phi\|^2.$$

Because of Lemmas 5.1 and 5.4 we can apply this inequality to $\phi = \psi$ if $(H^r - \lambda)\psi = 0$. Since $\alpha$ is arbitrary, clearly we thus obtain $\psi = 0$.

**6. Limiting absorption.** By using various statements of Section 5 we can show

**Proposition 6.1.** *Suppose $N = 3$ and $E_{\|}^{a_{\max}} \neq 0$. Let $H^r$ be an abbreviation for any reduced Hamiltonian as defined in the beginning of Section 5. Then the singular continuous spectrum $\sigma_{sc}(H^r)$ is empty. In fact, for any $\delta > \frac{1}{4}$ the limits (in the space of bounded operators on the corresponding $L^2$-space)*

$$\lim_{\operatorname{Im}\zeta\downarrow(\uparrow)0} \langle E_{\|}^{a_{\max}} \cdot z\rangle^{-\delta}(H^r - \zeta)^{-1}\langle E_{\|}^{a_{\max}} \cdot z\rangle^{-\delta}$$

*exist and are attained uniformly for* $\operatorname{Re}\zeta$ *in compact sets.*

*Proof.* As done in [HMS1] we shall use the Mourre method [M] (see also [PSS]) with the conjugate operator $A = E_{\|}^{a_{\max}} \cdot p$. We claim that for any $\lambda \in \mathbf{R}$ and $\rho < |E_{\|}^{a_{\max}}|^2$ we can pick $\varepsilon > 0$ small enough such that

$$\eta_\varepsilon(H^r - \lambda)i[H^r, A]\eta_\varepsilon(H^r - \lambda) \geq \rho\eta_\varepsilon(H^r - \lambda)^2. \tag{6.1}$$

To show (6.1) we insert the factor $\eta_{\varepsilon'}(H^r - \lambda)$; $\varepsilon' > 2\varepsilon$, to the left of the commutator. The latter is given by $|E_{\|}^{a_{\max}}|^2 - \nabla V \cdot E_{\|}^{a_{\max}}$. So it is enough to show that

$$\|\eta_{\varepsilon'}(H^r - \lambda)\nabla V_b(x^b) \cdot E_{\|}^{a_{\max}}\| \to 0 \quad \text{for } \varepsilon' \to 0\,;\ E_{\|}^b \neq 0. \tag{6.2}$$

We readily obtain (6.2) by combining (5.18), (5.19) and Propositions 4.1 and 5.5.

The statement (6.1) in conjunction with the Mourre theory gives the proposition with the weight $\langle E_{\|}^{a_{\max}} \cdot z\rangle^{-\delta}$ replaced by $\langle A\rangle^{-2\delta}$. The proof is completed by using the fact that $\langle A\rangle^{2\delta}(H^r - i)^{-1}\langle E_{\|}^{a_{\max}} \cdot z\rangle^{-\delta}$ is bounded if $\frac{1}{4} < \delta \leq \frac{1}{2}$, cf. (5.2) and a simple interpolation.

*Remark 6.2.* A simplified version of the above proof works for $N = 2$, giving an analogous result.

**7. Local smoothness.** In this section we introduce two local smoothness results that are useful for our completeness problem. We give time-dependent versions. The first is a corollary of Proposition 6.1.

**Proposition 7.1.** *Under the conditions of Proposition 6.1 the operator $\langle E_{\|}^{a_{\max}} \cdot z\rangle^{-\delta}$ is locally $H^r$-smooth for any $\delta > 1/4$; that is for any $f \in C_0^\infty(\mathbf{R})$ there exists $C > 0$ such that*

$$\int_{-\infty}^{\infty} \|\langle E_{\|}^{a_{\max}} \cdot z\rangle^{-\delta} \exp(-itH^r)f(H^r)\psi\|^2\, dt \leq C\|\psi\|^2,$$

*for all* $\psi \in L^2(X^{a_{\max}})$ *(or* $L^2(\tilde{X}^{a_{\max}})$*).*

To state the next result we introduce the function

$$\tilde{q}_0 = |E_{\|}^{a_{\max}}|^{1/2}(1 - \omega \cdot \hat{x}^{a_{\max}})^{1/2}; \quad \omega = |E_{\|}^{a_{\max}}|^{-1} E_{\|}^{a_{\max}},\ \hat{x}^{a_{\max}} = \frac{x^{a_{\max}}}{\langle x^{a_{\max}} \rangle}.$$

It is considered as a function on $X^{a_{\max}}$ in the neutral case and naturally as a function on $\tilde{X}^{a_{\max}}$ in the charged case.

**Proposition 7.2.** *Under the conditions of Proposition 6.1 the operator* $\tilde{q}_0 \langle x^{a_{\max}} \rangle^{-1/4}$ *is locally* $H^r$*-smooth.*

In the remaining part of this section we will prove Proposition 7.2 by mimicking [HMS2, Sect. 3] under use of Propositions 4.1, 5.5 and 7.1. The presentation of [HMS2] is more detailed.

Let

$$Q_k = \{\text{real } q \in C^\infty(X^{a_{\max}}) |\ \partial^\alpha q(x^{a_{\max}}) = O(\langle x^{a_{\max}} \rangle^{k - |\alpha|})\}; \quad k \in \mathbf{R}.$$

The elements of $Q_k$ will be denoted by $q_k$. In the charged case they are naturally considered as functions on $\tilde{X}^{a_{\max}}$. We introduce $j_1, j_2 \in Q_0$ with the following properties

$$\begin{gathered} j_i \text{ are homogeneous of degree 0 for } |x^{a_{\max}}| > 1, \\ 0 \le j_i \le 1,\ j_2 j_1 = j_1, \\ j_i = 1 \text{ in a neighbourhood of } G^{-1}\omega,\ \operatorname{supp}(j_i) \subset X^{a_{\max}} \setminus \bigcup_{b:\ E_{\|}^b \neq 0} X_b. \end{gathered} \tag{7.1}$$

We consider the following $H^r$-bounded operator

$$A = \langle x^{a_{\max}} \rangle^{-1/4} \omega \cdot p \langle x^{a_{\max}} \rangle^{-1/4}.$$

Clearly Proposition 7.2 will follow from Proposition 7.1 and the following estimate.

**Lemma 7.3.** *For any compact set* $\mathcal{K} \subset \mathbf{R}$ *there exists* $\varepsilon > 0$ *such that for all real* $f \in C_0^\infty((\lambda - \varepsilon, \lambda + \varepsilon))$ *with* $\lambda \in \mathcal{K}$

$$\begin{aligned} f(H^r) i[H^r, A] f(H^r) \ge & \tfrac{1}{2} f(H^r) \langle x^{a_{\max}} \rangle^{-1/4} \tilde{q}_0^2 \langle x^{a_{\max}} \rangle^{-1/4} f(H^r) \\ & - f(H^r) \langle x^{a_{\max}} \rangle^{-3/8} B \langle x^{a_{\max}} \rangle^{-3/8} f(H^r), \end{aligned} \tag{7.2}$$

*where* $B$ *is bounded.*

*Proof.* We compute the commutator form

$$\begin{aligned} & i[H^r, A] = T_1 + T_2 + T_3; \\ & T_1 = |E_{\|}^{a_{\max}}| \langle x^{a_{\max}} \rangle^{-1/2}, \\ & T_2 = -\langle x^{a_{\max}} \rangle^{-1/2} \nabla V \cdot \omega, \\ & T_3 = 2 \operatorname{Re} \left( i \langle x^{a_{\max}} \rangle^{-1/4} \omega \cdot p [(p^{a_{\max}} - A^{a_{\max}} x^{a_{\max}})^2, \langle x^{a_{\max}} \rangle^{-1/4}] \right). \end{aligned} \tag{7.3}$$

The latter commutator is on the form

$$i[(p^{a_{\max}} - A^{a_{\max}} x^{a_{\max}})^2, \langle x^{a_{\max}}\rangle^{-1/4}] = -\tfrac{1}{2}\langle x^{a_{\max}}\rangle^{-5/4} D_0 + iq_{-9/4};$$
$$D_0 = \hat{x}^{a_{\max}} \cdot (p^{a_{\max}} - A^{a_{\max}} x^{a_{\max}}).$$

By using this in the expression $T_3$ we obtain after symmetrizing

$$T_3 = -\langle x^{a_{\max}}\rangle^{-3/4} \operatorname{Re}(\omega \cdot p D_0)\langle x^{a_{\max}}\rangle^{-3/4} + q_{-7/2}.$$

In conjunction with the identity

$$\omega \cdot p = \omega \cdot (p^{a_{\max}} - A^{a_{\max}} x^{a_{\max}})$$

and the estimate

$$(p^{a_{\max}} - A^{a_{\max}} x^{a_{\max}})^2 \leq H^r + E_{\parallel}^{a_{\max}} \cdot z - V$$

we can thus bound

$$T_3 \geq -\omega \cdot \hat{x}^{a_{\max}} |E_{\parallel}^{a_{\max}}| \langle x^{a_{\max}}\rangle^{-1/2}$$
$$+\langle x^{a_{\max}}\rangle^{-3/4}(V - H^r)\langle x^{a_{\max}}\rangle^{-3/4} + q_{-7/2}.$$

Putting together we get

$$T_1 + T_3 \geq \langle x^{a_{\max}}\rangle^{-1/4} \tilde{q}_0^2 \langle x^{a_{\max}}\rangle^{-1/4}$$
$$+\langle x^{a_{\max}}\rangle^{-3/4}(V - H^r)\langle x^{a_{\max}}\rangle^{-3/4} + q_{-7/2}. \tag{7.4}$$

For given compact set $\mathcal{K} \subset \mathbf{R}$ we pick real $g \in C_0^\infty(\mathbf{R})$ equal to one on a neighbourhood of $\mathcal{K}$. Then $\operatorname{Re}\left(\langle x^{a_{\max}}\rangle^{-3/4}(V - H^r)\langle x^{a_{\max}}\rangle^{-3/4} g(H^r)\right)$ as well as the term $q_{-7/2}$ are on the form $\langle x^{a_{\max}}\rangle^{-3/8} B \langle x^{a_{\max}}\rangle^{-3/8}$ as required on the right hand side of (7.2). This can be seen by commuting $g(H^r)$ through the second factor $\langle x^{a_{\max}}\rangle^{-3/4}$ and noticing for the commutator that

$$(V - H^r)[g(H^r), \langle x^{a_{\max}}\rangle^{-3/4}]\langle x^{a_{\max}}\rangle^{3/8} \quad \text{is bounded.} \tag{7.5}$$

The statement (7.5) can be proved along the line of the proof of [HMS2, (3.7)] using (5.2).

It remains to deal with the term $T_2$. We need to find $\varepsilon > 0$ such that for all real $f \in C_0^\infty((\lambda - \varepsilon, \lambda + \varepsilon))$ with $\lambda \in \mathcal{K}$

$$f(H^r) T_2 f(H^r) \geq -\tfrac{1}{2} f(H^r)\langle x^{a_{\max}}\rangle^{-1/4} \tilde{q}_0^2 \langle x^{a_{\max}}\rangle^{-1/4} f(H^r)$$
$$- f(H^r)\langle x^{a_{\max}}\rangle^{-3/8} B \langle x^{a_{\max}}\rangle^{-3/8} f(H^r), \tag{7.6}$$

where $B$ is bounded.

To prove (7.6) we write $T_2$ as a sum of terms by expanding the potential. We treat the contribution from each term, $h_b := -\langle x^{a_{\max}}\rangle^{-1/2}\nabla V_b(x^b)\cdot\omega$, in the same way. Clearly we can assume $E_{\|}^b \neq 0$. We split

$$(7.7)\qquad \begin{aligned} h_b &= q_{-1/4}h_{b,1}q_{-1/4} + h_{b,2};\\ q_{-1/4} &= \langle x^{a_{\max}}\rangle^{-1/4}\tilde{q}_0(1-j_1),\\ h_{b,1} &= -(1-j_2)(\tilde{q}_0)^{-2}\nabla V_b(x^b)\cdot\omega,\\ h_{b,2} &= h_b j_2, \end{aligned}$$

where $j_1$ and $j_2$ are given by (7.1). To bound the contribution from the first term we multiply the first factor $f(H^r)$ by $\eta_\varepsilon(H^r-\lambda)$, take the real part and decompose

(7.8)
$$\begin{aligned} f(H^r)q_{-1/4}h_{b,1}q_{-1/4}f(H^r) &= f\,\mathrm{Re}(\eta_\varepsilon q_{-1/4}h_{b,1}q_{-1/4})f = f(A_{b,1}+A_{b,2})f;\\ A_{b,1} &= q_{-1/4}\,\mathrm{Re}(\eta_\varepsilon h_{b,1})q_{-1/4},\\ A_{b,2} &= \mathrm{Re}([\eta_\varepsilon, q_{-1/4}]h_{b,1}q_{-1/4}). \end{aligned}$$

As for the term $A_{b,1}$ we notice, cf. (6.2), that

$$\|\eta_\varepsilon(H^r-\lambda)h_{b,1}\| \le \tfrac{1}{6} \quad \text{for } \varepsilon \text{ sufficiently small,}$$

uniformly in $\lambda\in\mathcal{K}$. Here we used the proof of (6.2). Hence there exists (a small) $\varepsilon>0$ such that

(7.9)
$$\sum_{b:\,E_{\|}^b\neq 0} f(H^r)A_{b,1}f(H^r) \ge -\tfrac{1}{2}f(H^r)\langle x^{a_{\max}}\rangle^{-1/4}\tilde{q}_0^2\langle x^{a_{\max}}\rangle^{-1/4}f(H^r),$$

uniformly in $\lambda\in\mathcal{K}$.

Since one can show (cf. (5.2) and [HMS2, Sect. 3]) that

$$(7.10)\qquad \langle x^{a_{\max}}\rangle^{1/2}A_{b,2}\langle x^{a_{\max}}\rangle^{1/2} \quad \text{is bounded,}$$

obviously the term $A_{b,2}$ on the right hand side of (7.8) is on the form $\langle x^{a_{\max}}\rangle^{-3/8}B\langle x^{a_{\max}}\rangle^{-3/8}$.

By (7.9) it remains to show that the contribution from the term $h_{b,2}$ on the right hand side of (7.7) also has this form. For that we notice that (1.1) and the support property of the function $j_2$, (7.1), imply that $h_{b,2}=O(\langle x^{a_{\max}}\rangle^{-(1+\sigma)/2})$. Since by assumption $\sigma>1/2$ we are done.

*Remark 7.4.* Clearly the statement of Proposition 7.1 also holds for $N=2$, cf. Remark 6.2. As for Proposition 7.2 we notice that a simplified version of the above proof works for $N=2$, giving an analogous result.

**8. Proof of Theorem 1.2.** We shall give a proof of Theorem 1.2 for $N = 3$ using Propositions 7.1 and 7.2. The result for $N = 2$ can be proved similarly, cf. Remark 7.4. We shall use the notation $H^r (= H^r_{a_{\max}})$ of Section 5 for reduced Hamiltonians. If the full potential $V$ is replaced by $V^{\tilde{a}}$ defined in (1.8) we use the notation $H^r_{\tilde{a}}$. Clearly Propositions 7.1 and 7.2 also hold for any reduced Hamiltonian $H^r_{\tilde{a}}$. It suffices to show the existence of the limits

$$\text{(8.1)} \qquad \text{s}-\lim_{t\to+\infty} e^{itH^r}e^{-itH^r_{\tilde{a}}}, \quad \text{s}-\lim_{t\to+\infty} e^{itH^r_{\tilde{a}}}e^{-itH^r}.$$

Of course, the limits are taken in $L^2(X^{a_{\max}})$ in the neutral case and in $L^2(\tilde{X}^{a_{\max}})$ in the charged case. In the former case the pseudo momentum is fixed and the same for $H^r$ and $H^r_{\tilde{a}}$. We shall only prove the existence of the first limit in (8.1) since the existence of the other limit can be proved similarly.

Let $j_i$ be the smooth functions obeying (7.1).

**Lemma 8.1.** *We have*

$$\text{(8.2)} \qquad s-\lim_{t\to+\infty}(1-j_2)e^{-itH^r_{\tilde{a}}} = 0.$$

We notice the similarity of (8.2) and Proposition 7.2. The result follows from the latter proposition in the same way as an analogous result [HMS2, Proposition 4.2] was proved in that paper. The basic scheme is (was) first to prove the existence of

$$W = \text{s}-\lim_{t\to+\infty} e^{itH^r_{\tilde{a}}}(1-j_2)e^{-itH^r_{\tilde{a}}},$$

and then to use Proposition 7.2 to establish $W = 0$. We refer to [HMS2] for further details noticing though that the existence part follows from the proof of Lemma 8.2 to be given below.

**Lemma 8.2.** *Let*

$$W(t) = e^{itH^r} j_2 e^{-itH^r_{\tilde{a}}}.$$

*Then there exists*

$$\text{(8.3)} \qquad s-\lim_{t\to+\infty} W(t).$$

*Proof.* We follow [HMS2, Sect. 5]. Let $\psi = f(H^r_{\tilde{a}})\psi_0$, where $f \in C_0^\infty(\mathbf{R})$ and $\psi_0 \in L^2(X^{a_{\max}})$ (or $L^2(\tilde{X}^{a_{\max}})$). We pick $g \in C_0^\infty(\mathbf{R})$ such that $gf = f$ and notice that

$$\text{(8.4)} \qquad \lim_{t\to+\infty}(I - g(H^r))W(t)\psi = 0.$$

Here we have used that $g(H^r)j_2 - j_2 g(H^r_{\tilde{a}})$ is compact, cf. the proof of (5.18), and that

$$\mathrm{w}-\lim_{t\to+\infty} e^{-itH^r_{\tilde{a}}} = 0,$$

cf. Proposition 6.1.

By (8.4) we need to show the existence of the limit of

$$\begin{aligned} g(H^r)W(t)\psi &= g(H^r)j_2\psi \\ &+ \int_0^t g(H^r)e^{isH^r} i(H^r j_2 - j_2 H^r_{\tilde{a}})g(H^r_{\tilde{a}})e^{-isH^r_{\tilde{a}}}\psi\, ds. \end{aligned} \tag{8.5}$$

For that we compute

$$\begin{aligned} i(H^r j_2 - j_2 H^r_{\tilde{a}})g(H^r_{\tilde{a}}) &= T_1 + T_2; \\ T_1 &= 2\nabla j_2 \cdot (p^{a_{\max}} - A^{a_{\max}} x^{a_{\max}})g(H^r_{\tilde{a}}), \\ T_2 &= i(-\Delta j_2 + (V - V^{\tilde{a}})j_2)g(H^r_{\tilde{a}}). \end{aligned}$$

With $q_{-1/4}$ given as in (7.7) the first term can be written as

$$\begin{aligned} T_1 = q_{-1/4}^2 B &= q_{-1/4}Bq_{-1/4} + q_{-1/4}[q_{-1/4}, B]; \\ B &= 2\langle x^{a_{\max}}\rangle^{1/2}(\tilde{q}_0)^{-2}\nabla j_2 \cdot (p^{a_{\max}} - A^{a_{\max}} x^{a_{\max}})g(H^r_{\tilde{a}}). \end{aligned} \tag{8.6}$$

By (5.2) $B$ is bounded. Hence the contribution from the first term on the right hand side of (8.6) to the integral in (8.5) can be by treated (in the standard way) by Proposition 7.2.

Since it can be shown (cf. (5.2) and [HMS2, Sect. 3]) that

$$\langle x^{a_{\max}}\rangle^{1/2}q_{-1/4}[q_{-1/4}, B]\langle x^{a_{\max}}\rangle^{1/2} \quad \text{is bounded,} \tag{8.7}$$

the contribution from the second term on the right hand side of (8.6) to the integral in (8.5) can be treated by Proposition 7.1. Since $(V - V^{\tilde{a}})j_2 = O(\langle x^{a_{\max}}\rangle^{-\sigma})$ and by assumption $\sigma > 1/2$ the contribution from the term $T_2$ can be treated similarly.

The existence of

$$\mathrm{s}-\lim_{t\to+\infty} e^{itH^r}e^{-itH^r_{\tilde{a}}},$$

is an immediate consequence of Lemmas 8.1 and 8.2.

## References

[AHS1] J. Avron, I. Herbst, B. Simon, *Schrödinger operators with magnetic fields. I. General interaction*, Duke Math. J. **45** (No 4) (1978), 847–883.

[AHS2] ———, *Separation of center of mass in homogeneous magnetic fields*, Ann. Phys. **114** (1978), 431–451.

[AT] T. Adachi, H. Tamura, *Asymptotic completeness for long-range many-particle systems with Stark effect. II*, Comm. Math. Phys. (1995) (to appear).

[C] W. B. Cheston, *Elementary theory of electric and magnetic fields*, John Wiley and Sons, New York, London, Sydney, 1964.

[GL1] C. Gérard, I. Laba, *Scattering theory for N-particle systems in constant magnetic fields*, Duke Math. J. **76** (1994), 433–465.

[GL2] ———, *Scattering theory for N-particle systems in constant magnetic fields. II. Long-range interactions*, Comm. in Partial Diff. Eq. (1995) (to appear).

[H] L. Hörmander, *The analysis of linear partial differential operators I*, Springer Verlag, Berlin, Heidelberg, New York, London, Paris, Tokyo, Hongkong, 1990.

[HMS1] I. Herbst, J. S. Möller, E. Skibsted, *Spectral analysis of N-body Stark Hamiltonians*, Comm. Math. Phys. (1995) (to appear).

[HMS2] ———, *Asymptotic completeness of N-body Stark Hamiltonians*, Comm. Math. Phys. (1995) (to appear).

[M] E. Mourre, *Absence of singular continuous spectrum for certain selfadjoint operators*, Comm. Math. Phys. **91** (1981), 391–408.

[PSS] P. Perry, I. M. Sigal, B. Simon, *Spectral analysis of N- body Schrödinger operators*, Ann. Math. **114** (1981), 519–567.

Matematisk Institut, Aarhus Universitet, Ny Munkegade, 8000 Aarhus C, Denmark.
*E-mail address*: skibsted@mi.aau.dk

# THE SIZE OF ATOMS IN HARTREE-FOCK THEORY

JAN PHILIP SOLOVEJ

ABSTRACT. The purpose of this paper is to announce the proof (within Hartree-Fock theory) of what is often called the ionization conjecture. It states that the size, maximal excess charge, and ionization energy of atoms are bounded independently of the atomic number.

**1. Introduction.** One of the main motivations for quantum mechanics was to explain the stability of the hydrogen atom. More precisely, it explained why the electron is not sitting on top of the nucleus as it would classically. The quantum theory of hydrogen is of course special because it is exactly solvable. For all the heavier atoms it is not. For heavier atoms, one has had to rely on large scale computing to determine the exact structure.

One of the great mathematical puzzles of quantum theory is to explain the experimental (as well as computational) fact that all atoms, no matter how heavy, are of about the same size as hydrogen. More precisely, the statement to be proved can be phrased as follows. *As the atomic number $Z$ (together with the number $N$ of electrons) tends to infinity all but a finite number (independent of $Z$) of electrons take up a fixed region of space, which neither increases nor decreases with $Z$. Moreover, the ionization energy, and the maximal possible excess charge are bounded independently of $Z$.*

In many respects quantum theory is too complicated for both theoretical and for numerical studies. Much of theoretical and computational chemistry has been based on the Hartree-Fock approximation that we shall describe below. The success of Hartree-Fock theory in describing the structure of atoms has been astounding.

The purpose of this paper is to announce a proof of the above statement within Hartree-Fock theory.

Over the past 15 years there has been much mathematical work on this problem. For the full quantum problem, all we have are some $Z$-dependent

Work done while the author was at Princeton University partially supported by U.S. National Science Foundation grant DMS 92-03829.

bounds [2, 5, 6, 10, 11, 12]. A few years ago I solved the problem for a simplified version of the Hartree-Fock model [13]. The main simplification in this model was its spherical symmetry. The atoms in Hartree-Fock theory, however, are in general not spherically symmetric. It may at first sound surprising that Hartree-Fock atoms are not spherical. It is, however, also true in quantum theory that atoms may have non-zero angular momentum and are therefore not invariant under rotations.

It is my intent here, not only to announce the result, but also to very briefly sketch the main steps in its proof.

**2. Hartree-Fock theory and the main result.** An atom is in quantum theory described by the following Hamiltonian acting on the space $\bigwedge^N L^2(\mathbb{R}^3;\mathbb{C}^2)$ of antisymmetric spinor valued functions.

$$H = \sum_{i=1}^{N} (-\tfrac{1}{2}\Delta_i - V(x_i)) + \sum_{1\leq i<j\leq N} |x_i - x_j|^{-1}, \quad x_1, x_2, \ldots, x_N \in \mathbb{R}^3,$$

where $V(x) = Z|x|^{-1}$ is the nuclear potential.

The Hartree-Fock (HF) theory consists of restricting attention to functions of the form $\Psi = (N!)^{-1/2}\phi_1\wedge\ldots\wedge\phi_N$, where $\phi_1,\ldots,\phi_N \in L^2(\mathbb{R}^3;\mathbb{C}^2)$ are orthonormal. It is natural to introduce the projection operator $\gamma$ onto the space spanned by the functions $\phi_1,\ldots,\phi_N$. If we let $\gamma(x,y)$ denote the ($2\times 2$-matrix valued) kernel for $\gamma$ we may write

$$(\Psi, H\Psi) = \mathrm{Tr}[(-\tfrac{1}{2}\Delta - V)\gamma] + \mathcal{D}(\gamma) - \mathcal{E}X(\gamma). \tag{1}$$

Here we have introduced the direct Coulomb energy

$$\mathcal{D}(\gamma) = D(\rho_\gamma) = \tfrac{1}{2}\iint \rho_\gamma(x)|x-y|^{-1}\rho_\gamma(y)\,dx\,dy, \tag{2}$$

where the density of $\gamma$ is given by $\rho_\gamma(x) = \mathrm{Tr}_{\mathbb{C}^2}[\gamma(x,x)] = \sum_i |\phi_i(x)|^2$, and the exchange Coulomb energy

$$\mathcal{E}X(\gamma) = \tfrac{1}{2}\iint \mathrm{Tr}_{\mathbb{C}^2}[|\gamma(x,y)|^2]|x-y|^{-1}dxdy. \tag{3}$$

It is known [7] that the functional of $\gamma$ defined by the right side of (1) has minimizers if we fix $N = \int\rho_\gamma < Z+1$. Here we shall assume that we deal with neutral or super neutral atoms, i.e., $N \geq Z$.

It is also known [4] that we may (without introducing new minimizers) consider not only projections, $\gamma$, but more general *density matrices*. By a *density matrix* we mean any operator $\gamma$ satisfying the operator inequality $0 \leq \gamma \leq \mathrm{I}$ (I is here the identity). There are in general more than one minimizer for the HF functional (as we explained earlier they need not even be spherically symmetric).

The way we shall prove the main result is to show that the density $\rho_{\mathrm{HF}} = \sum_{i=1}^N |\phi_i|^2$ of a HF minimizer is well approximated by the Thomas-Fermi (TF) density defined below. It was previously known [7, 8] that the HF density is well approximated by the TF density at distances of order $Z^{-1/3}$ from the nucleus. What we prove is, the somewhat surprising result, that the approximation is good also at distances independent of $Z$.

**2.1. Theorem (The Thomas-Fermi approximation for the potential).** *Let $\rho_{\mathrm{HF}}$ denote the density of a HF minimizer and let $\phi_{\mathrm{HF}} = Z|x|^{-1} - \rho_{\mathrm{HF}} * |x|^{-1}$ be the HF potential. There exist universal constants $\varepsilon_1, D > 0$ such that for all $x$ satisfying $0 < |x| \leq D$ we have*

$$|\phi_{\mathrm{HF}}(x) - \phi_{\mathrm{TF}}(x)| \leq |x|^{-4+\varepsilon_1} \tag{4}$$

*and*

$$\left| \int_{|y|<|x|} [\rho_{\mathrm{TF}}(y) - \rho_{\mathrm{HF}}(y)]\, |x-y|^{-1} dy \right| \leq |x|^{-4+\varepsilon_1}. \tag{5}$$

*Here $\phi_{\mathrm{TF}}$ is the Thomas-Fermi (TF) potential that is the unique (positive) solution to the TF equation*

$$\Delta\phi_{\mathrm{TF}} = 2^{7/2}3^{-1}\pi^{-1}\phi_{\mathrm{TF}}^{3/2} - 4\pi Z\delta(x) \tag{6}$$

*and the TF density is $\rho_{\mathrm{TF}} = 2^{3/2}(3\pi^2)^{-1}\phi_{\mathrm{TF}}^{3/2}$. Equivalently we have*

$$\phi_{\mathrm{TF}}(x) = Z|x|^{-1} - \rho_{\mathrm{TF}} * |x|^{-1}. \tag{7}$$

We have for simplicity chosen not to put constants in front of the right sides of (4) and (5). The constants can of course always be chosen to be one by simply sacrificing in the exponent $\varepsilon_1$ and in $D$.

The importance of the exponent $-4$ in the bounds in Theorem 2.1 is that the TF potential satisfies the following asymptotic relation first found by Sommerfeld [14] (the first mathematical study of the TF equation goes back to [3]; the most complete analysis, however, is to be found in [8]).

**2.2. Theorem (Sommerfeld asymptotics).** *The solution $\phi_{\mathrm{TF}}$ to the TF equation (6) satisfies*[1]

$$3^4 2^{-3}\pi^2|x|^{-4} - CZ^{(4-\tau)/3}|x|^{-\tau} \leq \phi_{\mathrm{TF}}(x) \leq 3^4 2^{-3}\pi^2|x|^{-4}, \tag{8}$$

*where $\tau = (1+\sqrt{73})/2 \approx 4.8$. The density $\rho_{\mathrm{TF}}$ then satisfies*

$$3^5 2^{-3}\pi|x|^{-6} - CZ^{(4-\tau)/3}|x|^{-\tau-2} \leq \rho_{\mathrm{TF}}(x) \leq 3^5 2^{-3}\pi|x|^{-6}. \tag{9}$$

There is a more general estimate of this kind which is very useful (a somewhat similar estimate can be found in [1]).

[1] We denote all constants by $C$. They are meant to be universal, i.e, explicitly computable numbers

**2.3. Theorem (TF estimate).** *Assume $\phi$ satisfies the TF differential equation*

$$\Delta\phi(x) = 2^{7/2}3^{-1}\pi^{-1}\phi(x)^{3/2} \quad \text{for } |x| > r.$$

*Let*

$$a(r) = \inf_{|x|=r}[\phi(x) - 3^4 2^{-3}\pi^2|x|^{-4}] \quad \text{and} \quad A(r) = \sup_{|x|=r}[\phi(x) - 3^4 2^{-3}\pi^2|x|^{-4}].$$

*Then*

$$a(r)r^{\eta}|x|^{-\eta} \leq \phi(x) - 3^4 2^{-3}\pi^2|x|^{-4} \leq A(r)r^{\tau}|x|^{-\tau}, \tag{10}$$

*where $\tau = (1+\sqrt{73})/2 \approx 4.8$ is the same exponent as appeared in the Sommerfeld asymptotics Theorem 2.2 and $\eta = (1+\sqrt{55})/2 \approx 4.2$.*

While the upper bound in (10) has the best possible exponent $\tau$ we have sacrificed in the exponent in the lower bound in order to simplify the expression.

**3. The Coulomb norm estimates.** One of the technical tools in the analysis is the Coulomb norm defined by $D(f)^{1/2}$, where $D(f)$ was given in (2). The usefulness of this norm lies in the following inequality first applied in the context of atomic problems in [2].

**3.1. Lemma (Coulomb norm estimate).** *If $|\nabla g| \in L^2(\mathbb{R}^3)$ and $f \in L^{6/5}(\mathbb{R}^3)$ then*

$$\left|\int fg\right| \leq \|\nabla g\|_2 D(f)^{1/2}.$$

**3.2. Corollary.** *If $\mathcal{B}$ is the ball of radius $s$ centered at $x$ then*

$$|f * |x|^{-1}| \leq C\left(s^{1/5}\|f\|_{L^{5/3}(\mathcal{B})} + s^{-1/2}D(f)^{1/2}\right). \tag{11}$$

*Let $0 < \kappa < 1/4$. Denote by $\mathcal{A}$ the annulus $\mathcal{A} = \{y : (1-2\kappa)|x| \leq |y| \leq |x|\}$. Then*

$$\left|\int_{|y|<|x|} f(y)|x-y|^{-1}dy\right| \leq C\left(\kappa^{1/15}|x|^{1/5}\|f\|_{L^{5/3}(\mathcal{A})} + \kappa^{-1}|x|^{-1/2}D(f)^{1/2}\right). \tag{12}$$

**4. Proof of the main theorem.** We shall now sketch the steps in the proof of the TF approximation, Theorem 2.1. We shall concentrate on the second estimate (5). The first estimate (4) on the HF potential is easy to prove from (5). We shall proceed in an iterative manner as described in the following lemma.

**4.1. Lemma (Iterative step).** *There exist (universal) positive constants $D$, $\varepsilon_1$, $\varepsilon_2$ and $\alpha$ such that if the TF approximation (5) holds for all $x$ satisfying $r \le |x| \le r^{(1+\varepsilon_2)^{-1}} \le D$ (assuming $D < 1$), for some $r$ with $\alpha Z^{-1/3} \le r$, it then holds for all $x$ satisfying $r \le |x| \le r^{(1+2\varepsilon_2)^{-1}}$.*

In order to use the iterative lemma to prove the TF approximation (5) for all $x$ with $0 < |x| \le D$ one must of course first show that the approximation holds for $0 < |x| \le (\alpha Z^{-1/3})^{(1+\varepsilon_2)^{-1}}$. This first step in the iterative process is very similar to (in fact, easier than) the proof of the iterative step; we shall therefore not give the details of it here. Having proved the first step, the TF approximation for all $x$ with $0 < |x| \le D$ is then a simple consequence of the iterative step since $r^{(1+\varepsilon_2)^{-1}} < r^{(1+2\varepsilon_2)^{-1}}$.

In order to prove Lemma 4.1 we shall compare the HF density $\rho_{\rm HF}$ with a density that minimizes a TF problem outside a ball.

**4.2. Definition (The outside TF density and potential).** Let $\rho_{\rm TF}^{(r)}$ be the unique density supported in $\{x : |x| > r\}$ that minimizes the **outside TF functional**

$$\mathcal{E}_r^{\rm TF}(\rho) = \tfrac{3}{10}(3\pi^2)^{2/3}\int \rho^{5/3} - \int V_r\rho + D(\rho),$$

with the exterior potential $V_r$ being the nuclear potential $V(x)$ screened by the density $\rho_{\rm HF}$ inside the ball of radius $r$, i.e.,

$$V_r(x) = Z|x|^{-1} - \int_{|y|<r} \rho_{\rm HF}(y)|x-y|^{-1}dy.$$

We also define the **outside TF potential**

$$\phi_{\rm TF}^{(r)} = V_r - \rho_{\rm TF}^{(r)} * |x|^{-1}. \tag{13}$$

We shall not here discuss the fairly standard techniques used to prove the existence of the minimizer $\rho_{\rm TF}^{(r)}$ and the fact that $\rho_{\rm TF}^{(r)} \in L^1(\mathbb{R}^3)$. Moreover, $\rho_{\rm TF}^{(r)}$ is the (unique) solution in $|x| > r$ to the **outside TF equation**

$$\tfrac{1}{2}(3\pi^2)^{2/3}\rho_{\rm TF}^{(r)}(x)^{2/3} = \left[V_r(x) - \rho_{\rm TF}^{(r)} * |x|^{-1}\right]_+, \tag{14}$$

where $[t]_+$ denotes the positive part of $t$.

In terms of the exterior TF density and potential we may for $|x| \ge r$ use (7) and (13) to rewrite

$$\int_{|y|<|x|} [\rho_{\rm TF}(y) - \rho_{\rm HF}(y)]\,|x-y|^{-1}dy = \mathcal{A}_1(x) + \mathcal{A}_2(x) + \mathcal{A}_3(x), \tag{15}$$

where

$$\mathcal{A}_1(x) = \phi_{\mathrm{TF}}^{(r)}(x) - \phi_{\mathrm{TF}}(x), \quad \mathcal{A}_2(x) = \int_{|y|>|x|} \left[\rho_{\mathrm{TF}}^{(r)}(y) - \rho_{\mathrm{TF}}(y)\right] |x-y|^{-1} dy$$

and

$$\mathcal{A}_3(x) = \int_{r<|y|<|x|} \left[\rho_{\mathrm{TF}}^{(r)}(y) - \rho_{\mathrm{HF}}(y)\right] |x-y|^{-1} dy.$$

We shall prove the iterative lemma by showing that the three terms $\mathcal{A}_1, \mathcal{A}_2$ and $\mathcal{A}_3$ each are small.

**The analysis of $\mathcal{A}_1$ and $\mathcal{A}_2$.** The most difficult part of this analysis is to show that $\phi_{\mathrm{TF}}^{(r)}(x) > 0$ for all $|x| \geq r$. Having shown this it then follows from the exterior TF equation (14) that $\phi_{\mathrm{TF}}^{(r)}$ satisfies the **outside TF differential equation**

$$\Delta \phi_{\mathrm{TF}}^{(r)}(x) = 2^{7/2} 3^{-1} \pi^{-1} \phi_{\mathrm{TF}}^{(r)}(x)^{3/2} \quad \text{for } |x| > r. \tag{16}$$

Notice that this equation does not uniquely determine $\phi_{\mathrm{TF}}^{(r)}$ since we have not specified the boundary condition at $|x| = r$. When we have established the outside TF differential equation (16) the necessary estimates on $\mathcal{A}_1$ and $\mathcal{A}_2$ will be simple consequences of the TF estimate in Theorem 2.3.

In order to prove that $\phi_{\mathrm{TF}}^{(r)}(x) > 0$ for all $|x| > r$ we begin by showing that $\rho_{\mathrm{TF}}^{(r)}$ is indeed close to $\rho_{\mathrm{TF}}$ on the set $\{x : |x| > r\}$. We shall control the difference in the Coulomb norm.

**4.3. Proposition.** *Let $\chi_+$ denote the characteristic function of the set $\{y : |y| > r\}$. Then assuming the TF approximation (5) for $|x| = r$ we have*

$$D(\rho_{\mathrm{TF}}^{(r)} - \chi_+ \rho_{\mathrm{TF}}) \leq C r^{(-7+\varepsilon_1)}.$$

*Proof.* To prove this we make a perturbation analysis of the TF functional. We introduce the perturbation potential

$$W(x) = \int_{|y|<r} [\rho_{\mathrm{TF}}(y) - \rho_{\mathrm{HF}}(y)] |x-y|^{-1} dy. \tag{17}$$

Then from (7)

$$\phi_{\mathrm{TF}}(x) + (\chi_+ \rho_{\mathrm{TF}}) * |x|^{-1} = V_r(x) - W(x). \tag{18}$$

Note that, by the assumption that the TF approximation (5) holds for $|x| = r$, we have

$$\sup_{|x|=r} |W(x)| \leq r^{-4+\varepsilon_1}. \tag{19}$$

Since $\rho_{\mathrm{TF}}^{(r)}$ is a minimizer for the functional $\mathcal{E}_r^{\mathrm{TF}}$ on the set $\{y : |y| > r\}$ we have

$$0 \le \mathcal{E}_r^{\mathrm{TF}}(\chi_+\rho_{\mathrm{TF}}) - \mathcal{E}_r^{\mathrm{TF}}(\rho_{\mathrm{TF}}^{(r)}) = \int W(\rho_{\mathrm{TF}}^{(r)} - \chi_+\rho_{\mathrm{TF}}) - D(\rho_{\mathrm{TF}}^{(r)} - \chi_+\rho_{\mathrm{TF}})$$
$$+ \int_{|y|\ge r} \Big(\Big[\tfrac{3}{10}(3\pi^2)^{2/3}\rho_{\mathrm{TF}}(y)^{5/3} - \phi_{\mathrm{TF}}(y)\rho_{\mathrm{TF}}(y)\Big]$$
$$- \Big[\tfrac{3}{10}(3\pi^2)^{2/3}\rho_{\mathrm{TF}}^{(r)}(y)^{5/3} - \phi_{\mathrm{TF}}(y)\rho_{\mathrm{TF}}^{(r)}(y)\Big]\Big)\,dy.$$

Since $\rho_{\mathrm{TF}} = 2^{3/2}(3\pi^2)^{-1}\phi_{\mathrm{TF}}^{3/2}$ we see that the expression $\frac{3}{10}(3\pi^2)^{2/3}t^{5/3} - \phi_{\mathrm{TF}}(y)t$ takes its minimal value for $t = \rho_{\mathrm{TF}}(y)$. Hence we conclude that the last integral above is negative and hence that $D(\rho_{\mathrm{TF}}^{(r)} - \chi_+\rho_{\mathrm{TF}}) \le \int W(\rho_{\mathrm{TF}}^{(r)} - \chi_+\rho_{\mathrm{TF}})$.

We must now estimate the right side of this inequality. Note first that $W$ is a harmonic function of $x$ for $|x| > r$ and that $W$ tends to zero at infinity. It is therefore easy to see that we may write $W = W_1 + W_2$ such that $W_1$ is supported on $\{x : |x| < 3r\}$ and $W_2$ is supported on $\{x : |x| > 2r\}$ and satisfy

$$\|W_1\|_\infty \le C \sup_{|x|=r} |W(x)|, \quad \|\nabla W_2\|_2 \le C r^{1/2} \sup_{|x|=r} |W(x)|.$$

It follows from the Coulomb norm estimate Lemma 3.1 that

(20)
$$D(\rho_{\mathrm{TF}}^{(r)} - \chi_+\rho_{\mathrm{TF}}) \le C \sup_{|x|=r} |W(x)| \Big[\int_{r<|y|<3r} \Big(\rho_{\mathrm{TF}}^{(r)}(y) + \chi_+\rho_{\mathrm{TF}}(y)\Big)\,dy$$
$$+ r^{1/2} D(\rho_{\mathrm{TF}}^{(r)} - \chi_+\rho_{\mathrm{TF}})^{1/2}\Big].$$

We control the integral on the right using the outside TF equation (14) to get that for $r < |x| < 3r$

$$\rho_{\mathrm{TF}}^{(r)}(x) \le C V_r(x)^{3/2} \le C r^{-6}. \tag{21}$$

We have here used (8), (9), (18), (19) and the harmonicity of $W$ to obtain

$$V_r(x) = \phi_{\mathrm{TF}}(x) + \int_{|y|>r} \rho_{\mathrm{TF}}(y)|x-y|^{-1}dy + W(x) \le C r^{-3}|x|^{-1}. \tag{22}$$

If we insert the Sommerfeld estimate (9) and the estimate (21) into (20) and then use the iterative assumption (19) we immediately conclude the result of the proposition.

**4.4. Lemma (Controlling the outside TF problem).** *Assume that the TF approximation (5) holds for $|x| = r$. There exist constants $\alpha$ and $D$ (depending only on $\varepsilon_1$) such that if $\alpha Z^{-1/3} \leq r \leq D$ we have the following Sommerfeld estimates for the outside TF problem:*

$$-Cr^{\eta-4}|y|^{-\eta} \leq \phi_{\mathrm{TF}}^{(r)}(y) - 3^4 2^{-3}\pi^2|y|^{-4} \leq Cr^{\tau-4}|y|^{-\tau} \tag{23}$$

*and*

$$-Cr^{\eta-4}|y|^{-\eta-2} \leq \rho_{\mathrm{TF}}^{(r)}(y) - 3^5 2^{-3}\pi|y|^{-6} \leq Cr^{\tau-4}|x|^{-\tau-2}. \tag{24}$$

*Proof.* The first step is to control $\phi_{\mathrm{TF}}^{(r)}(x)$ for $|x| = r$. Recall that with $W$ defined in (17) we may write

$$\phi_{\mathrm{TF}}(x) - \phi_{\mathrm{TF}}^{(r)}(x) = (\rho_{\mathrm{TF}}^{(r)} - \chi_+\rho_{\mathrm{TF}}) * |x|^{-1} - W(x). \tag{25}$$

By the estimates (9) and (21) we see that $\|\rho_{\mathrm{TF}}\|_{L^{5/3}(\mathcal{B})}$, $\|\rho_{\mathrm{TF}}^{(r)}\|_{L^{5/3}(\mathcal{B})} \leq Cs^2r^{-6}$, where $\mathcal{B}$ is a ball of radius $s < r$ centered at $x$. It therefore follows from the Coulomb norm estimate (11), Proposition 4.3 and the iterative assumption (19) that

$$|\varphi_{\mathrm{TF}}(x) - \phi_{\mathrm{TF}}^{(r)}(x)| \leq Cr^{-4+(2\varepsilon_1/5)},$$

where we made the optimal choice $s = r^{(1+(\varepsilon_1/5))}$.

From the Sommerfeld estimate (8) we see that if $\alpha Z^{-1/3} < r < D$ we have for $|x| = r$

$$\left|\phi_{\mathrm{TF}}^{(r)}(x) - 3^4 2^{-3}\pi^2 r^{-4}\right| \leq C\left(\alpha^{(4-\tau)}r^{-4} + r^{-4+(2\varepsilon_1/5)}\right).$$

We may therefore choose $\alpha$ large enough and $D$ small enough to ensure that $\phi_{\mathrm{TF}}^{(r)}(x)$ is positive for $|x| = r$. A simple harmonicity argument using the TF equation (14) then shows that $\phi_{\mathrm{TF}}^{(r)}(x)$ is positive for all $|x| \geq r$ and hence satisfies the TF differential equation (16).

The estimate (23) is now an immediate consequence of the TF estimates in Theorem 2.3. Since $\rho_{\mathrm{TF}}^{(r)} = C(\phi_{\mathrm{TF}}^{(r)})^{3/2}$ it is clear that we could have chosen $\alpha$ and $D$ to ensure also the second estimate (24).

From Lemma 4.4 and the Sommerfeld asymptotics, Theorem 2.2, we can now get the following estimates on $\mathcal{A}_1$ and $\mathcal{A}_2$.

**4.5. Corollary.** *Assume again that the TF approximation* (5) *holds for* $x$ *with* $|x| = r$. *We may choose constants* $\alpha$ *and* $D$ *(depending only on* $\varepsilon_1$*) such that if* $\alpha Z^{-1/3} \le r \le |x'| \le D$ *we have*

$$|\mathcal{A}_1(x')|\,,\ |\mathcal{A}_2(x')| \le Cr^{\gamma-4}|x'|^{-\eta}. \tag{26}$$

**The analysis of $\mathcal{A}_3$ .** We define a smooth "characteristic" function $\theta_+$ of the set $\{x : |x| \ge r\}$ and in terms of this an outside HF density matrix $\gamma_{\mathrm{HF}}^{(r)} = \theta_+\gamma_{\mathrm{HF}}\theta_+$ with density $\rho_{\mathrm{HF}}^{(r)} = \theta_+^2\rho_{\mathrm{HF}}$. We estimate $\rho_{\mathrm{HF}}^{(r)}$ by considering the contribution of $\gamma_{\mathrm{HF}}^{(r)}$ to the HF energy, i.e., we separate the inside from the outside. In order to do this we introduce the auxiliary functional

$$\mathcal{E}^A(\gamma) = \mathrm{Tr}[(-\tfrac{1}{2}\Delta - V_r)\gamma] + D(\rho_\gamma). \tag{27}$$

Like the outside TF functional this functional has the exterior potential $V_r$. Moreover, there is no exchange term in this functional. The proof of the following lemma is technically very involved. It uses heavily the famous Lieb-Thirring inequality [9].

**4.6. Lemma (The outside energy).** *Assume that the TF approximation* (5) *holds for* $r \le |x| \le 2r$. *Then*

$$\mathcal{E}^A(\gamma_{\mathrm{HF}}^{(r)}) \le \inf_\gamma\{\mathcal{E}^A(\gamma)\} + Cr^{-7+a_1},$$

*for a universal exponent* $a_1 > 0$. *Here the infimum is over all density matrices* $\gamma$ *with* $\mathrm{supp}\,\rho_\gamma \subset \{y : |y| \ge r\}$. *Moreover, we have the bound* $\int(\rho_{\mathrm{HF}}^{(r)})^{5/3} \le Cr^{-7}$.

One may ask what the connection is between the auxiliary functional $\mathcal{E}^A$ and the outside TF functional in Definition 4.2. The answer is that the connection is established using semiclassical estimates.

The semiclassical approximation to the sum of the negative eigenvalues of a Schrödinger operator $-\frac{1}{2}\Delta - V$ on $L^2(\mathbb{R}^3; \mathbb{C}^2)$ is given by

$$-2^{5/2}(15\pi^2)^{-1}\int V^{5/2}.$$

(Here we have assumed that $V \ge 0$.) Moreover, the density of the projection onto the eigenfunctions is, in semiclassics, approximated by the function $2^{3/2}(3\pi^2)^{-1}V(x)^{3/2}$.

Usually semiclassics is done in terms of an asymptotic expansion in a semiclassical parameter. We need more explicit estimates. The following two theorems are not very difficult to prove.

**4.7. Theorem (Semiclassics upper bound).** *Assume that $V \geq 0$ is in $L^{5/3}(\mathbb{R}^3)$ and $L^{3/2}(\mathbb{R}^3)$ and that $V = V_1 + V_2$, where $|\nabla V_1|, V_2 \in L^{5/2}(\mathbb{R}^3)$. For all $s > 0$ there exists a density matrix $\gamma$ in $L^2(\mathbb{R}^3; \mathbb{C}^2)$ such that*

$$\mathrm{Tr}[(-\tfrac{1}{2}\Delta - V)\gamma] \leq -2^{5/2}(15\pi^2)^{-1} \int V^{5/2} + C\left[\|V\|_{5/2}^{3/2}\left(s\|\nabla V_1\|_{5/2} + \|V_2\|_{5/2}\right) + s^{-2}\|V\|_{3/2}^{3/2}\right]$$

*and such that*

$$D\left(\rho_\gamma - 2^{3/2}(3\pi^2)^{-1}V^{3/2}\right) \leq Cs^{7/5}\|V\|_{5/2}^3.$$

**4.8. Theorem (Semiclassics lower bound).** *Assume that $V$ satisfies the assumptions of Theorem 4.7. Let $e_1, e_2, \ldots$ denote the negative eigenvalues of the operator $-\frac{1}{2}\Delta - V$ on $L^2(\mathbb{R}^3; \mathbb{C}^2)$. Then*

$$\sum_j e_j \geq -2^{5/2}(15\pi^2)^{-1}\left(\int V^{5/2}\right)\left[1 + C\|V\|_{5/2}^{-3/2}\|\nabla V_1\|_{5/2}^{2/3}\|V\|_{3/2}^{1/2} + C\|V\|_{5/2}^{-1}\|V_2\|_{5/2}\right]^{5/2}.$$

Using the semiclassical bounds we can now control the density $\rho_{\mathrm{HF}}^{(r)}$ in the Coulomb norm.

**4.9. Lemma.** *Assuming that the TF approximation (5) holds for $r \leq |x| \leq 2r$, we can find $\alpha$ and $D$ (depending only on $\varepsilon_1$) such that if $\alpha Z^{-1/3} \leq r \leq D$ we have*

$$D(\rho_{\mathrm{HF}}^{(r)} - \rho_{\mathrm{TF}}^{(r)}) \leq Cr^{-7+a_2}$$

*for a universal exponent $a_2$.*

*Proof.* We shall do semiclassics with the potential $V = \phi_{\mathrm{TF}}^{(r)}$. From Lemma 4.4 it is easy to see (choosing $\alpha$ and $D$ depending on $\varepsilon_1$) that one can write $\phi_{\mathrm{TF}}^{(r)} = V = V_1 + V_2$ in such a way that the errors in the semiclassical estimates are smaller than the leading term $\int V^{5/2} \sim r^{-7}$ for small $r$.

More precisely, we choose[2] $\gamma$ in accordance with Theorem 4.7 for $V = \phi_{\mathrm{TF}}^{(r)}$. If we recall (13) we get from Lemma 4.6 that

$$\mathcal{E}^A(\gamma_{\mathrm{HF}}) - Cr^{-7+a_1} \leq \mathcal{E}^A(\gamma) = \mathrm{Tr}[(-\tfrac{1}{2}\Delta - \phi_{\mathrm{TF}}^{(r)})\gamma] - D(\rho_{\mathrm{TF}}^{(r)}) + D(\rho_{\mathrm{TF}}^{(r)} - \rho_\gamma).$$

[2] This $\gamma$ is not really appropriate since it is not supported for $|x| \geq r$. It is, however, easy to formulate the semiclassics to incorporate boundary conditions.

Then from the semiclassical upper bound and (14) we have

$$\mathcal{E}^A(\gamma_{\mathrm{HF}}) \le -2^{5/2}(15\pi^2)^{-1}\int (\phi_{\mathrm{TF}}^{(r)})^{5/2} - D(\rho_{\mathrm{TF}}^{(r)}) + Cr^{-7+a_2}.$$

From the semiclassical lower bound we similarly obtain

$$\begin{aligned}\mathcal{E}^A(\gamma_{\mathrm{HF}}) &= \mathrm{Tr}[(-\tfrac{1}{2}\Delta - \phi_{\mathrm{TF}}^{(r)})\gamma_{\mathrm{HF}}] - D(\rho_{\mathrm{TF}}^{(r)}) + D(\rho_{\mathrm{HF}}^{(r)} - \rho_{\mathrm{TF}}^{(r)}) \\ &\ge -2^{5/2}(15\pi^2)^{-1}\int (\phi_{\mathrm{TF}}^{(r)})^{5/2} - D(\rho_{\mathrm{TF}}^{(r)}) + D(\rho_{\mathrm{HF}}^{(r)} - \rho_{\mathrm{TF}}^{(r)}) - Cr^{-7+a}.\end{aligned}$$

The last two inequalities prove the lemma.

Using the estimate just proved together with the $L^{5/3}$ bound in Lemma 4.6 we can now use the Coulomb norm estimate (12) to control the term $\mathcal{A}_3$. (For simplicity we ignore that $\theta_+$ is not really a characteristic function.)

**4.10. Corollary.** *Assume that the TF approximation* (5) *holds for $x$ with $r \le |x| \le 2r$. We may choose constants $\alpha$ and $D$ (depending only on $\varepsilon_1$) such that if $\alpha Z^{-1/3} \le r \le |x'| \le D$ we have*

$$|\mathcal{A}_3(x')| \le C(|x'|/r)^{a_3} r^{-4+a_4} \tag{28}$$

*for universal exponents $a_3, a_4 > 0$ (both smaller than 4).*

We can now conclude the iterative lemma as follows. Given $r \ge \alpha Z^{-1/3}$ such that the TF approximation (5) holds for all $x$ with $r \le |x| \le 2r$. We may then use (26) and (28) for $x'$ satisfying $r < |x'| < D < 1$. We define $\varepsilon > 0$ by $|x'|^{(1+\varepsilon)} = r$. We then get from (26) and (28) the estimate

$$|\mathcal{A}_1(x')| + |\mathcal{A}_2(x')| + |\mathcal{A}_3(x')| \le C|x'|^{-4}\left(|x'|^{\varepsilon(\eta-4)} + |x'|^{(a_4-(4+a_3-a_4)\varepsilon)}\right).$$

Choose $\varepsilon_0$ such that $\varepsilon_0(\eta-4) = a_4 - (4+a_3-a_4)\varepsilon_0$. Note that $\varepsilon_0$ is an explicitly computable number. Choose $\varepsilon_2 = \varepsilon_0/2$. We see that if $\varepsilon_2 \le \varepsilon \le 2\varepsilon_2 = \varepsilon_0$ (i.e., $r^{(1+\varepsilon_2)^{-1}} \le |x'| \le r^{(1+2\varepsilon_2)^{-1}}$) then the above error is bounded by $C|x'|^{-4+(\eta-4)\varepsilon_2}$. If we now assume that we had originally chosen $\varepsilon_1 < (\eta-4)\varepsilon_2$ we see that the error is bounded by $|x'|^{-4+\varepsilon_1}$. Here we assumed that $|x'| < D$ and that $D$ is small enough. Thus if we assume that the TF approximation holds for $r \le |x| \le r^{(1+\varepsilon_2)^{-1}}$ we now know that it holds for $r \le |x| \le r^{(1+2\varepsilon_2)^{-1}}$.

## REFERENCES


1. H. Brezis & E. H. Lieb, *Long range potentials in Thomas-Fermi theory*, Comm. Math. Phys. **65** (1979), 231–246.
2. C. Fefferman & L. A. Seco, *Asymptotic neutrality of large ions*, Comm. Math. Phys. **128** (1990), 109–130.
3. E. Hille, *On the TF equation*, Proc. Nat. Acad. Sci., U.S.A. **62** (1969), 7–10.
4. E. H. Lieb, *Variational principle for many-fermion systems*, Phys. Rev. Lett. **46** (1981), 457–459.
5. ———, *Bound on the maximum negative ionization of atoms and molecules*, Phys. Rev. A **29** (1984), 3018–3028.
6. E. H. Lieb, I. M. Sigal, B. Simon, & W. Thirring, *Asymptotic neutrality of large-Z ions*, Comm. Math. Phys. **116** (1988), 635–644.
7. E. H. Lieb & B. Simon, *The Hartree-Fock theory for Coulomb systems*, Comm. Math. Phys. **53** (1977), 185–194.
8. ———, *Thomas-Fermi theory of atoms, molecules and solids*, Adv. in Math. **23** (1977), 22-116.
9. E. H. Lieb & W. E. Thirring, *Inequalities for the moments of the eigenvalues of the Schrödinger Hamiltonian and their relation to Sobolev inequalities*, Studies in mathematical physics (E. Lieb, B. Simon, and A. S. Wightman, eds.), Princeton Univ. Press, Princeton, New Jersey, 1976, pp. 269–330.
10. M. B. Ruskai, *Absence of discrete spectrum in highly negative ions*, Comm. Math. Phys. **82** (1982), 457–469.
11. L. A. Seco, I. M. Sigal, & J. P. Solovej, *Bound on the ionization energy of large atoms*, Comm. Math. Phys. **131** (1990), 307–315.
12. I. M. Sigal, *Geometric methods in the quantum many-body problem. Nonexistence of very negative ions.*, Comm. Math. Phys. **85** (1982), 309-324.
13. J. P. Solovej, *Proof of the ionization conjecture in a reduced Hartree-Fock model*, Inv. Math. **104** (1991), 291–311.
14. A. Sommerfeld, *Asymptotische Integration der Differentialgleichung des Thomas-Fermischen Atoms*, Z. Phys. **78** (1932), 283–308.


DEPARTMENTS OF MATHEMATICAL SCIENCES, UNIVERSITY OF AARHUS, NY MUNKEGADE, DK-8000 AARHUS C, DENMARK
*E-mail address*: solovej@mi.aau.dk

# HOLOMORPHIC EXTENSION OF *CR* FUNCTIONS: A SURVEY

JEAN-MARIE TRÉPREAU

*À la mémoire de E. Andronikof*

**Abstract.** A *CR* function on a generic real submanifold of $\mathbf{C}^n$ is a solution of the induced Cauchy-Riemann system; for example, traces of holomorphic functions are *CR*. This lecture will be a survey, intended for the non-specialist, of the results obtained during the last ten years on the question of the holomorphic extendibility of *CR* functions. Emphasis will be put on the method of the complex disc, the Bishop equation, and Tumanov's theorem on the holomorphic extendibility into a wedge from minimal manifolds. The basic notions about *CR* functions will be introduced and several important results will be presented.

**1. Holomorphic extension using complex discs.** In the whole lecture, $D$ is the unit disc and $S$ is the unit circle in $\mathbf{C}$; $\mathcal{O}(\Omega)$ denotes the space of holomorphic functions on the domain $\Omega$.

Before introducing the *CR* functions, we shall recall how complex discs can be used to extend holomorphic functions holomorphically. It is well known that, given a domain $\Omega \subset \mathbf{C}^n$, it may happen, if $n \geq 2$, that every $f \in \mathcal{O}(\Omega)$ extends holomorphically through some part of $\partial\Omega$; this is by definition the case unless $\Omega$ is a domain of holomorphy. We shall use the following definition:

**Definition 1.1.** $\Omega$ has the *extension property* at $z \in \partial\Omega$ if, for every open neighborhood $U$ of $z$ there exists an open neighborhood $V \subset U$ of $z$ such that every $u \in \mathcal{O}(U \cap \Omega)$ coincides on $V \cap \Omega$ with some $\hat{u} \in \mathcal{O}(V)$.

A simple necessary condition for the extension property to hold is that there exists no germ of a holomorphic function at $z$, vanishing at $z$ but not in $\Omega$ close to $z$ (if such a function, say $u$, exists, consider $1/u$); geometrically this means that there exists no germ of a complex hypersurface passing through $z$ without intersecting $\Omega$. This condition is by no means sufficient, as shown by an example of Kohn and Nirenberg [17] of a pseudoconvex domain for which no such hypersurface exists at some of its boundary points. To study the extension property we shall use only the method of the complex disc, which we now introduce.

**Definition 1.2.** A *complex disc* in $\mathbf{C}^n$ is a continuous map $A : \bar{D} \to \mathbf{C}^n$, holomorphic on $D$. The *boundary* of $A$ is its restriction $\partial A$ to $S$.

As is usual we shall use the notation $A$ for a disc as well as for its image in $\mathbf{C}^n$.

Our basic recipe for extending functions is contained in the following elementary lemma:

**Lemma 1.3.** *If $\Omega$ is $C^1$-smooth near $z_0 \in \partial\Omega$ and, for every neighborhood $U$ of $z_0$, there exists a complex disc $A \subset U$ such that $A(0) = z_0$ and $\partial A \subset \Omega$, then $\Omega$ has the extension property at $z_0$.*

*Proof.* We may assume that $z_0 = 0$. Let $U$ be a neighborhood of $z_0$ and $u \in \mathcal{O}(U \cap \Omega)$ the function to be extended. Let $[z_0, z_a] \subset U \cap \overline{\Omega}$ be a small segment normal to $\partial\Omega$ at $z_0$ and $A$ be a complex disc as in Lemma 1.3. We may assume that $A$ is so small that $z + \partial A \subset U \cap \Omega$ for all $z$ in a neighborhood of $[z_0, z_a]$ while $z + A \subset U \cap \Omega$ for all $z$ close to $z_a$. The function

$$\hat{u}(z) = \frac{1}{2\pi i} \int_S u(z + A(\tau)) d\tau / \tau$$

is then well defined and holomorphic in a neighborhood of $[z_0, z_a]$ and by the Cauchy formula, it coincides with $u$ close to $z_a$; $\hat{u}$ extends $u$ close to $z_0$. □

In this section we shall give two simple examples of how the recipe applies; more subtle results will be given later. It may happen that a disc as in Lemma 1.3 exists but has a complicated equation; a holomorphic change of variables (which obviously preserves the extension property) may simplify matters. For simplicity, let us assume that $\Omega$ is smooth and that, after a complex affine transformation, $z_0 = 0$ and $\Omega$ has a defining function $r$ with the Taylor expansion

$$(*)_m \qquad r(z) = -\operatorname{Re} z_n + p_m(z') + o(|z_n|) + o(|z'|^m),$$

where $p_m$ is a polynomial in $z' = (z_1, \dots, z_{n-1})$ and $\bar{z}'$, homogeneous of degree $m \geq 2$. If $p_m$ is pluriharmonic, a change of holomorphic variables can be performed so that in the new coordinates a defining function is of the form $(*)_{m+1}$.

**Definition 1.4.** $\Omega$ has finite type $m \geq 2$ at 0 if, in some local holomorphic coordinates, $\Omega$ has a defining function $r$ of the form $(*)_m$ with a polynomial $p_m$ homogeneous of degree $m$ which is not pluriharmonic. $\Omega$ has infinite type at 0 if $\Omega$ does not have type $m$ at 0 for any $m \geq 2$.

The most classical example of the extension property is the following result of H. Lewy [18]:

**Theorem 1.5.** *If the Levi form of $\Omega$ at $z_0$ has a negative eigenvalue, then $\Omega$ has the extension property at $z_0$.*

*Proof.* $\Omega$ has a local definition of the form

$$\Omega := \quad -\operatorname{Re}(z_n + \sum_{i,j=1}^{n-1} a_{i,j} z_i z_j) + \sum_{i,j=1}^{n-1} c_{i,j} z_i \bar{z}_j + o(|z_n|) + o(|z'|^2) < 0.$$

An obvious change of variables reduces the first term to be $-\operatorname{Re} z_n$, and a linear change of variables then diagonalizes the second sum (it is the Levi form). By the assumption we may assume that

$$\Omega := \quad -\operatorname{Re} z_n + \sum_{i=1}^{n-1} c_i |z_i|^2 + o(|z_n|) + o(|z'|^2) < 0,$$

where $c_i \in \{-1, 0, +1\}$ and at least one of these numbers, say $c_1$, is $-1$. Then, using the recipe, the discs $A_\epsilon(\tau) = (\epsilon\tau, 0, \ldots, 0)$, $\epsilon > 0$ small, prove the theorem. □

We shall not pursue the extension property further. Even in $\mathbf{C}^2$ a simple characterization of the extension property at a given point is not known, and perhaps does not exist. (See [10] for the best results on this question.) Instead we shall consider a weaker property. Let $M$ be a real hypersurface near $z_0 \in M$ dividing locally the space into two open sets $\Omega^+$ and $\Omega^-$.

**Definition 1.6.** $M$ has the *one-sided extension property* at $z_0$ if at least one of the open sets $\Omega^\pm$ has the extension property at $z_0$.

The reason for introducing this notion is that it has a simple geometric characterization, which is not the case for the extension property. In this section we consider only the case of smooth hypersurfaces of finite type. The following result was first proved by Baouendi and Treves [6]. We give the simple proof of Rea [20].

**Theorem 1.7.** *If $M$ is a smooth real hypersurface of finite type at $z_0$, then $M$ has the one-sided extension property at $z_0$.*

The result is sharp (only) in the real analytic category:

**Corollary 1.8.** *If $M$ is a real-analytic hypersurface, the following properties are equivalent:*

(1) *$M$ has finite type at $z_0$.*
(2) *There is no germ of a complex hypersurface $\mathcal{H}$ such that $z_0 \in \mathcal{H} \subset M$.*
(3) *$M$ has the one-sided extension property at $z_0$.*

The equivalence between (1) and (2) is special to the analytic category and a simple exercise; (1) implies (3) by Theorem 1.7, and (3) obviously implies (2) since, if $\{u = 0\}$ is an equation of a complex hypersurface $\mathcal{H}$, $z_0 \in \mathcal{H} \subset M$, then $1/u$ is defined on both sides of $M$ but cannot be extended at $z_0$.

*Proof of Theorem 1.7.* We may assume $z_0 = 0$ and that $M$ has a defining function of the form $(*)_m$ with $p_m$ not pluriharmonic. A complex linear change of variables allows us to assume that $p_m(z_1, 0, \ldots, 0)$ is not harmonic. Making $z_2 = \cdots = z_{n-1} = 0$ it is sufficient to construct a "good" disc (i.e. as in Lemma 1.3 for one side of $M$) for the hypersurface in $\mathbf{C}^2$:

$$M := \quad r(z) = -\operatorname{Re} z_n + p(z_1) + o(|z_n|) + o(|z_1|^m) = 0,$$

where $p$ is a nonharmonic homogeneous polynomial of degree $m$ in $z_1, \bar{z}_1$. We shall use the following:

**Claim.** *There exists a disc $D(a,1) \subset \mathbf{C}$ with center $a$, $|a| < 1$, and radius 1, such that the harmonic polynomial $H_{S(a,1)}p$ with value $p$ on the circle $S(a,1)$ satisfies $H_{S(a,1)}p(0) \neq 0$.*

*Proof of the claim.* In case $a = 0$, the solution of the Dirichlet problem is given by the Poisson kernel

$$(H_{S(0,1)}p)(\tau) = \frac{1}{2\pi i}\int_S p(\sigma)\frac{1 - |\tau|^2}{|\sigma - \tau|^2}\frac{d\sigma}{\sigma}.$$

By translation

$$(H_{S(a,1)}p)(0) = \frac{1}{2\pi i}\int_S p(a+\sigma)\frac{1 - |a|^2}{|\sigma + a|^2}\frac{d\sigma}{\sigma}.$$

The harmonic monomials in $p$ give no contribution to this value, so we may assume $p(\sigma) = |\sigma|^2 q(\sigma)$ for some nonzero $q$. We get

$$(H_{S(a,1)}p)(0) = \frac{1 - |a|^2}{2\pi}\int_0^{2\pi} q(a + e^{it})\, dt,$$

which is identically zero for small $a$ only if $q$ is zero. □

*End of the proof of Theorem 1.7.* Choose $a$ as in the claim. The real harmonic polynomial $H_{S(a,1)}p$ has degree $m$; it is the real part of a holomorphic polynomial $h$ of degree $m$ with $\operatorname{Im} h(0) = 0$. By homogeneity, the holomorphic polynomial $h_\epsilon(\tau) = \epsilon^m h(\tau/\epsilon)$ has a real part which coincides with $p$ on the circle $S(\epsilon a, \epsilon)$. The complex disc $A_\epsilon(\tau) = (\tau, h_\epsilon(\tau) - h_\epsilon(0))$, $\tau \in D(\epsilon a, \epsilon)$, is as small as we wish, satisfies $A_\epsilon(0) = 0$, and if $\tau \in S(\epsilon a, \epsilon)$ then

$$r(A_\epsilon(\tau)) = \epsilon^m h(0) + o(\epsilon^m)$$

so that $\partial A_\epsilon$ is contained in one of the open sets defined by $M$. Hence Lemma 1.3 applies. □

**2. Complex discs attached to a hypersurface.** In Section 1 we used smoothness and a finite type condition in an essential way to prove the existence of a "good" disc, that is, one passing through the base point $z_0$ and with boundary lying in one of the open half spaces delimited by $M$. In general it might seem difficult to construct such a disc but in fact one can construct discs satisfying the much stronger requirement that their boundaries lie on $M$! Lemma 1.3 can then be applied in many cases:

**Lemma 2.1.** *If $M$ is a $C^1$-smooth real hypersurface such that for every neighborhood $U$ of $z_0 \in M$ there exists a complex disc $A \subset U$ with $\partial A \subset M$ and $A(0)$ lying on the normal line at $z_0$ to $M$ but $\neq z_0$, then $M$ has the one-sided extension property at $z_0$.*

It suffices to translate the disc a little along this normal line (choose the side with care!) in order to meet the hypothesis in Lemma 1.3. □

If $u \in L^2(S)$ is real, there exists one and only one real $v \in L^2(S)$ such that $u + iv$ is the boundary value of a holomorphic function $h \in \mathcal{O}(D)$ verifying $\operatorname{Im} h(0) = 0$; if

$$u(\tau) \sim u_0 + 2\operatorname{Re}\sum_{n>0} u_n\tau^n \quad \text{(Fourier expansion), then}$$

$$h(\tau) = u_0 + 2\sum_{n>0} u_n\tau^n \quad \text{in } D, \text{ so } v(\tau) \sim 2\operatorname{Im}\sum_{n>0} u_n\tau^n.$$

The function $v$ is called the *Hilbert transform* of $u$, and we write $v = Tu$. Clearly, $T$ induces bounded endomorphims on the Sobolev spaces. Here it will be more convenient to use the Hölder spaces. For $0 < \alpha < 1$ we denote by $C^\alpha(S)$ the space of continuous functions on $S$ satisfying a Hölder estimate $|u(\tau) - u(\sigma)| \leq C|\tau - \sigma|^\alpha$ for all $\tau, \sigma \in S$. The smallest possible $C$ is the seminorm $|u|_\alpha$. The norm $|u|_\alpha + \sup|u|$ makes $C^\alpha$ a Banach algebra. If $k \in \mathbb{N}$ and $0 < \alpha < 1$, then $C^{k,\alpha}(S)$ denotes the Banach algebra of functions of class $C^k$ whose derivatives up to order $k$ are in $C^\alpha$, with the obvious seminorm $|u|_{k,\alpha}$. The definition of $C^{k,\alpha}(\bar{D})$ is similar. It is a very classical fact that the Hilbert transform induces bounded endomorphisms of all spaces $C^{k,\alpha}(S)$ and that the Poisson operator (harmonic extension from $S$ to $D$) induces bounded operators from $C^{k,\alpha}(S)$ to $C^{k,\alpha}(\bar{D})$. A basic fact about Hölder spaces is that $C^{k,\alpha}$ is stable under composition of maps if $k \geq 1$.

We now come to the construction of discs with boundary lying on a given real hypersurface $M$ of class $C^2$. Let $z_0 \in M$ and choose complex coordinates so that $z_0 = 0$ and $M$ is defined by an equation

$$M := \quad -\operatorname{Im} z_n + h(z', \operatorname{Re} z_n) = 0,$$

with $h(0) = 0$, $dh(0) = 0$.

**Lemma 2.2.** *Let $0 < \alpha < 1$ be given. For every complex disc $A' = (A_1, \ldots, A_{n-1}) \in C^\alpha(\bar{D}, \mathbf{C}^{n-1})$ with small $C^\alpha$ norm and small $t \in \mathbf{R}$, there exists one and only one small complex disc $A \in C^\alpha(\bar{D}, \mathbf{C}^n)$ of the form $A = (A', A_n)$ with $\operatorname{Re} A_n(0) = t$ and $\partial A \subset M$.*

Thus we can lift any (small) complex disc in the complex hyperplane tangent to $M$ at $z_0$ to a complex disc with boundary in $M$, and prescribe the value at 0 of the real part of the last component. If we choose $A'(0) = 0$ and $t = 0$, we get a disc with boundary in $M$ and $A(0)$ lying on the normal line to $M$ at 0. Then Lemma 2.1 applies unless $\operatorname{Im} A_n(0) = 0$.

*Proof of Lemma 2.2.* We need only find $A_n$, which is determined by its boundary value $u + iv$ on $S$. The functions $u$ and $v$ are related by $v = h(A', u)$ on $S$, $v = Tu + \text{const}$, and the harmonic extension of $u$ has the value $t$ at the origin. The last two conditions are equivalent to $u = t - Tv$, so $u + iv$ solves the problem if and only if $v = h(A', u)$ on $S$ and $u$ is a solution of the so-called *Bishop equation:*

$$u = t - Th(A', u) \quad \text{on } S.$$

Note that this is a *functional* equation, not a point by point one. However, it is easily seen that $u \mapsto t - Th(A', u)$ is a contraction of the ball of center 0 and radius $\rho$ in $C^\alpha(S)$ if $t$, $A'$ and $\rho$ are small enough. So the equation is solved using the fixed point theorem. □

Using this construction, we can sketch the proof of the following generalization of Corollary 1.8:

**Theorem 2.3.** *A $C^2$-smooth real hypersurface $M$ has the one-sided extension property at $z_0 \in M$ if and only if there exists no germ of a complex hypersurface $\mathcal{H}$ such that $z_0 \in \mathcal{H} \subset M$.*

As already mentioned the necessity of the condition is obvious. It is true that if it is satisfied there exists a good disc as in Lemma 2.1 but we won't be able to prove this before Section 5. The original proof in [25] used instead an argument of propagation of extendibility which we only sketch here:

*Sketch of the proof.* We assume that $z_0 = 0$ and that $M$ has been prepared as in Lemma 2.2. If $A'$ is a disc in $\mathbf{C}^{n-1}$ with $A'(0) = 0$ we lift it to a disc $A$ with boundary in $M$ and $A(0)$ lying on the normal line at 0 as in Lemma 2.2. If $\operatorname{Im} A_n(0) \neq 0$, then Lemma 2.1 applies and we get the one-sided extension. So the problem remains only when all such lifted discs go through 0, i.e. satisfy $A(0) = 0$. Using the (local) envelopes of holomorphy of the two open sets delimited by $M$ and the Kontinuitätssatz, one can show that, if a disc has its boundary in $M$, goes through 0, and intersects one of these open sets, then every function holomorphic on this side of $M$ extends to the other side. This gives the result unless all lifted discs are in fact contained (not only their boundaries!) in $M$. In this way, we get plenty of complex curves contained in $M$ and going through 0. The second argument is that if $M$ contains a complex curve and $u$ is a holomorphic function on one side of $M$, such that $u$ extends across $M$ at some point of this curve, then it extends across $M$ at any point of the curve. Using this propagation argument one gets that, if one-sided extension at 0 does not hold, then all lifted discs at 0 lie in $M$ as well as all lifted discs at any point of these lifted discs, and so on. One then shows that these lifted discs generate (a germ of) a complex hypersurface in $M$ at 0, concluding the proof of the theorem. □

This proof leaves open the question of the existence of a good disc and cannot be generalized to the higher codimensional case. These questions will be handled later on.

**3. *CR* functions.** Traces of holomorphic functions on a real hypersurface in $\mathbf{C}^n$ satisfy differential equations if $n \geq 2$. For example, if $M = \{z \in \mathbf{C}^2; \operatorname{Im} z_2 = (\operatorname{Re} z_1)^2\}$ and $t_1 = \operatorname{Re} z_1$, $t_2 = \operatorname{Im} z_1$, $t_3 = \operatorname{Re} z_2$ are used as local coordinates on $M$, then the trace $u(t) = \hat{u}(t_1 + it_2, t_3 + it_1^2)$ of any holomorphic function satisfies the equation

$$(\partial/\partial t_1 + i\partial/\partial t_2 - 2it_1\partial/\partial t_3)u = 0,$$

obtained by eliminating the partial derivatives of $\hat{u}$ in the relations

$$\partial u/\partial t_1 = \partial\hat{u}/\partial z_1 + 2it_1\partial\hat{u}/\partial z_2, \quad \partial u/\partial t_2 = i\partial\hat{u}/\partial z_1, \quad \partial u/\partial t_3 = \partial\hat{u}/\partial z_2.$$

Generalizing this example, one gets the notion of the *induced Cauchy-Riemann system* whose solutions are the *CR* functions. Let $M$ be a $q$-codimensional real submanifold of $\mathbf{C}^n$, defined by

$$M := \quad r_j(z) = 0, \; j = 1, \dots, q, \tag{3.1}$$

locally near $z_0 \in M$, where the $r_j$ are $q$ real functions with $dr_1(z_0) \wedge \cdots \wedge dr_q(z_0) \neq 0$. If $\hat{u}$ is holomorphic in a neighborhood of $z_0$, its restriction $u = \hat{u}|_M$ is annihilated at $z \in M$ by every antiholomorphic vector $\sum_k a_k \partial/\partial\bar{z}_k$ tangent to $M$ at $z$, i.e. such that the coefficients $a_k$ satisfy the linear system $\sum_k a_k \partial r_j(z)/\partial\bar{z}_k = 0$, $j = 1, \dots, q$. The number of independent equations is equal to the rank of the matrix with entries $\partial r_j(z)/\partial z_k$ or equivalently to the rank of the system of holomorphic forms $\partial r_j$, $j = 1, \dots, q$, where $\partial r$ denotes the holomorphic differential $\sum_k \partial r/\partial z_k dz_k$.

**Definition 3.1.** The real $q$-codimensional manifold $M$ defined by (3.1) near $z_0 \in M$ is a *CR manifold* if $\partial r_1, \ldots, \partial r_q$ has constant rank close to $z_0$ in $M$, and a *generic manifold* if $\partial r_1(z_0) \wedge \cdots \wedge \partial r_q(z_0) \neq 0$. $M$ is called *maximally real* if it is generic of codimension (or dimension) $n$.

For example, a real hypersurface is always generic; a complex submanifold is $CR$ but is not generic unless it is $n$-dimensional; the basic example of a maximally real manifold is $\mathbf{R}^n$ in $\mathbf{C}^n$. Let us denote by $T^{1,0}\mathbf{C}^n$ (respectively $T^{0,1}\mathbf{C}^n$) the bundle of holomorphic (respectively antiholomorphic) vectors in $\mathbf{C}^n$ and $T^{1,0}M = T^{1,0}\mathbf{C}^n \cap \mathbf{C}TM$ (respectively $T^{0,1}M = T^{0,1}\mathbf{C}^n \cap \mathbf{C}TM$). If $M$ is $CR$ these are complex subbundles of $\mathbf{C}TM$. The following lemma is a direct consequence of the definitions:

**Lemma 3.2.** *The following properties are equivalent at $z_0$:*

(1) *$M$ is generic.*
(2) *$T^{0,1}_{z_0}M$ has (minimal) dimension $p = n - q$.*
(3) *There exists a maximally real $L$ with $z_0 \in L \subset M$.*
(4) *The restrictions of $z_1, \ldots, z_n$ to $M$ have independent differentials.*
(5) *One can choose local holomorphic coordinates so that $z_0 = 0$ and*

$$r_j(z) = \operatorname{Im} z_{n-q+j} + s_j(z_1, \ldots, z_{n-q}, \operatorname{Re} z_{n-q+1}, \ldots, \operatorname{Re} z_n), \quad j = 1, \ldots, q,$$

*with $s_j(0) = 0$, $ds_j(0) = 0$.*

From now on we assume that $M$ is generic of codimension $q$ and let $p = n - q$ denote the *CR dimension* of $M$, that is, the dimension of the (complex) bundle $T^{0,1}M$ of antiholomorphic vectors on $M$. It follows from its definition that $T^{0,1}M$ is the annihilator of the bundle of forms spanned by $dz_1|_M, \ldots, dz_n|_M$; it is locally generated by a system of $p$ complex vector fields $L_1, \ldots, L_p$ satisfying the Frobenius condition and such that $L_1, \ldots, L_p, \bar{L}_1, \ldots, \bar{L}_p$ are independent. Note that the system generated by the $L_j$ and their conjugates need not satisfy the Frobenius condition.

We are now in a position to define $CR$ functions:

**Definition 3.3.** A $C^1$-smooth function on $M$ is $CR$ if $Lu = 0$ for every section $L$ of $T^{0,1}M$, or equivalently if $du \wedge dz_1|_M \wedge \cdots \wedge dz_n|_M = 0$. A continuous or a locally integrable function on $M$ is $CR$ if it satisfies this condition in the weak sense.

Though our main concern will be with continuous $CR$ functions, we shall use $CR$ currents from time to time, assuming that $M$ is oriented and smooth. A function $u \in C^1(M)$ defines a current of maximal dimension $\dim M = p + n$ by $u(\phi) = \int_M u\phi$ for all test forms on $M$ of degree $n + p$. By Stokes' theorem, $u$ is $CR$ if and only if $u(d\psi) = 0$ for all test forms $\psi$ divisible by $dz_1|_M \wedge \cdots \wedge dz_n|_M$, hence the definition:

**Definition 3.4.** A current $u$ on $M$ of maximal dimension is *CR* if $u(d\psi) = 0$ for all test forms $\psi$ divisible by $dz_1|_M \wedge \cdots \wedge dz_n|_M$.

Of course restrictions of holomorphic functions to $M$ are *CR*. More generally one gets *CR* functions by taking "boundary values" of functions holomorphic in a domain whose boundary contains $M$. Let $z_0 \in M$ and $\theta_0 \in \mathbf{C}^N$, identified here with the tangent space $T_{z_0}\mathbf{C}^n$, and assume $\theta_0 \notin T_{z_0}M$. A *wedge with edge $M$ at $z_0$ in the direction $\theta_0$* is defined to be an open set of the form

$$\mathcal{W} = \{z + t\theta; z \in M \text{ close to } z_0,\ \theta \in \mathbf{C}^n \text{ close to } \theta_0,\ t > 0 \text{ close to } 0\},$$

or more precisely as the germ $\mathcal{W}_M(z_0, \theta_0)$ of open set defined by such an open set $\mathcal{W}$. It is easy to verify that, as a germ, $\mathcal{W}_M$ does not depend on the choices made in the definition of $\mathcal{W}$ and only depends on the choice of $\theta_0$ modulo $T_{z_0}M$ (and also modulo the action of $\mathbf{R}^+$). In the sequel we shall view $(z_0, \theta_0)$ as an element of the normal bundle to $M$ in $\mathbf{C}^n$. If a function is holomorphic in a wedge $\mathcal{W}_M(z_0, \theta_0)$ and continuous up to $M$, then its restriction to $M$ is a continuous *CR* function near $z_0$. When $M$ is smooth there is the following generalization, see [2]:

**Theorem 3.5.** *If $u \in \mathcal{O}(\mathcal{W}_M(z_0, \theta_0))$ has slow growth along $M$ (i.e. $|u(z)|$ is controlled by a sufficiently high power of the inverse of the distance of $z$ to $M$), then the limit $\lim_{\epsilon \to +0} u(\cdot + \epsilon\theta_0)|_M$ exists (close to $z_0$) in the sense of currents. It is a CR current, called the* boundary value *of $u$ and denoted $b_{\theta_0}(u)$. Moreover, if $u$ is fixed, $b_{\theta_0}(u) = b_{\theta_1}(u)$ provided $\theta_1$ is close enough to $\theta_0$, and $b_{\theta_0}(u) = 0$ if and only if $u = 0$.*

The last statements generalize the principle of analytic continuation.

Now we can ask the main questions considered in this lecture; is it possible to decompose (locally) every *CR* function as a sum of boundary values of holomorphic functions? Find a condition which assures that every *CR* function extends holomorphically to a wedge, i.e. is the boundary value of a single holomorphic function.

In case $M$ is a hypersurface near $z_0 \in M$, the answer to both questions is known. Note that the normal space to $M$ at $z_0$ is one dimensional, so there are two directions $\pm$ allowing to take boundary values, corresponding to the two open sets delimited by $M$. Decomposition is always possible:

**Theorem 3.6.** *If $M$ is a real hypersurface, then every CR function $u$ on $M$ can be decomposed as $u = b^+(u^+) + b^-(u^-)$ near $z_0$, where each function $u^\pm$ is holomorphic on one side of $M$, close to $z_0$.*

*Sketch of proof.* Several proofs are known of this result. We sketch a proof (see [1]) under the simplifying assumption that $M$ and $u$ are smooth. One of the vector fields $\partial/\partial\bar{z}_k$, say $\partial/\partial\bar{z}_n$, is transversal to $M$ at $z_0$. Using the formal Cauchy-Kovalewsky theorem we can solve formally the equation $\partial\hat{u}/\partial\bar{z}_n = 0$, $\hat{u} = u$ on $M$ near $z_0$, with $\hat{u}$ a formal series in the powers of $r$, a defining function of $M$, and coefficients which are functions on $M$. From the fact that $u$ satisfies the induced Cauchy-Riemann system one easily sees that $\hat{u}$ formally satisfies the Cauchy-Riemann system on $M$. Let $v$ be a smooth function in a neighborhood of $z_0$, with the same derivatives on $M$ as $\hat{u}$. Then $\omega = \bar{\partial}v$ is a smooth $\bar{\partial}$-closed form vanishing of infinite order on $M$. This is also the case for the form $\omega'$ defined as being equal to $\omega$ on one side of $M$ and 0 on the other side. Due to the local solvability of the $\bar{\partial}$ system $\omega' = \bar{\partial}w$ for some function $w$ near $z_0$. Let $\hat{u}^+ = v - w$ and $\hat{u}^- = w$. Then $\hat{u}^{\pm}$ are holomorphic on one or the other side of $M$ and $u$ is the sum of the restrictions to $M$ of $\hat{u}^+$ and $\hat{u}^-$. □

From Theorem 2.3 and Theorem 3.6 we get the following:

**Theorem 3.7.** *If $M$ is a $C^2$-smooth real hypersurface and $z_0 \in M$, then every CR function near $z_0$ on $M$ extends holomorphically to a wedge at $z_0$ if and only if there exists no germ of a complex hypersurface $\mathcal{H}$ such that $z_0 \in \mathcal{H} \subset M$.*

The answer to the questions is known also in case $M$ is maximally real. In that case there is no *CR* equation; every function is a *CR* function. Of course it cannot be expected (for example by the principle of analytic continuation!) that every function extends to a wedge. However, decomposability holds:

**Theorem 3.8.** *If $M$ is maximally real, then every function in a neighborhood of $z_0$ in $M$ can be decomposed near $z_0$ as a sum of boundary values of functions holomorphic in (germs of) wedges at $z_0$ with edge $M$.*

Several proofs are known of this fact, see [2], but all proofs involve some technicalities. The question of the different ways a function can be decomposed gives rise to the microlocal theory of analytic singularities, which goes back to the fundamental work of Sato, see [21]. Let us only recall the quite classical proof using the Fourier transform in the special but suggestive case of $M = \mathbf{R}^n$. If $u$ is (for simplicity) a smooth function on $\mathbf{R}^n$ with compact support, then $u(x) = c\int_{\mathbf{R}^n} e^{ix\cdot\xi}\hat{u}(\xi)\,d\xi$ by the Fourier inversion formula. Let us write $\mathbf{R}^n = \cup\Gamma_j$ as a finite union of closed strictly convex cones with negligible intersections $\Gamma_j \cap \Gamma_k$, $j \neq k$. Then $\hat{u}_j(z) = c\int_{\Gamma_j} e^{iz\cdot\xi}\hat{u}(\xi)\,d\xi$ defines a holomorphic function in the tube $\mathbf{R}^n + i\Gamma_j^\circ$, smooth up to the edge $\mathbf{R}^n$, where $\Gamma_j^\circ = \{y \in \mathbf{R}^n; y\cdot\xi > 0 \text{ for all } \xi \in \Gamma_j\}$ and $u$ is the sum of the traces of the $\hat{u}_j$ on $\mathbf{R}^n$. □

These two extreme cases are the only ones where decomposition holds in full generality:

**Theorem 3.9.** *There exists a 2-codimensional generic submanifold in $\mathbf{C}^3$ which admits CR functions which are not decomposable near some point.*

The idea of the construction involves results on the propagation of analytic singularities and we shall not give it here, see [26].

However, the second problem has a general answer, which generalizes Theorem 3.7. It uses the notion of minimality, which we now introduce.

Let $M$ be a generic manifold of $CR$ dimension $p$ and $z_0 \in M$. The real parts of the antiholomorphic (or holomorphic) vectors tangent to $M$ generate a subbundle $T^cM$ of $TM$ of rank $2p$. Let $\mathcal{O}(z_0)$ be the orbit of $z_0$ for this bundle, that is, the set of points in $M$ which can be reached from $z_0$ following continuous piecewise $C^1$ curves with tangent in $T^cM$. It follows from a result of Sussmann [24] (recall that $T^cM$ does not satisfy any kind of Frobenius integrability condition) that this orbit has a natural structure of a differentiable manifold for which the inclusion in $M$ is an immersion. So a sufficiently small neighborhood of $z_0$ in the orbit defines a submanifold of $M$. Replacing $M$ by a neighborhood $U$ of $z_0$ in $M$, the same construction gives a submanifold at $z_0$, whose dimension can only decrease when $U$ is shrunk. So as a germ of a submanifold at $z_0$, the orbit in $U$ is stationary for sufficiently small $U$. The resulting germ of a submanifold at $z_0$ is called the *local orbit* of $z_0$ and denoted $\mathcal{O}^{\mathrm{loc}}(z_0)$. By the construction, it is clear that the restriction of $T^cM$ to the local orbit is tangent to the local orbit, so $\mathcal{O}^{\mathrm{loc}}(z_0)$ has dimension at least $2p$. Also, viewing $\mathcal{O}^{\mathrm{loc}}(z_0)$ as a submanifold of $\mathbf{C}^n$ we see that it is a $CR$ manifold with the same $CR$ dimension as $M$ since $T^{0,1}\mathcal{O}^{\mathrm{loc}}(z_0) = T^{0,1}M|_{\mathcal{O}^{\mathrm{loc}}(z_0)}$. Note that it is not generic unless it has the same dimension as $M$, i.e. unless it is a neighborhood of $z_0$ in $M$. Conversely, if $N \subset M$ is a germ of a $CR$ manifold at $z_0$ with the same $CR$ dimension as $M$, it is clear that the local orbit of $z_0$ is contained in $N$.

**Definition 3.10.** $M$ is minimal at $z_0$ if the local orbit of $z_0$ is a neighborhood of $z_0$ in $M$, or equivalently if there does not exist a germ of a $CR$ manifold $N$ with the same $CR$ dimension as $M$ such that $z_0 \in N \subset M$ and $\dim N < \dim M$.

For example, if $M$ is maximally real, then $T^cM = 0$ so the orbits are single points and $M$ is not minimal. If $M$ is a hypersurface, it has dimension $2n-1$ while $T^cM$ has rank $2n-2$. The local orbit can have dimension $2n-1$ ($M$ is then minimal at $z_0$) or $2n-2$ ($\mathcal{O}^{\mathrm{loc}}(z_0)$ is then a germ of a complex hypersurface at $z_0$).

**Theorem 3.11.** *Let $M$ be a generic submanifold of $\mathbf{C}^n$ and $z_0 \in M$. Then $M$ has the* wedge extension property *at $z_0$, which means that every germ of a CR function near $z_0$ on $M$ extends holomorphically to a wedge at $z_0$, if and only if $M$ is minimal at $z_0$.*

The sufficiency of the minimality is due to Tumanov [29] and is the main theme of this lecture. We shall sketch the proof later. The necessity was proved by Baouendi and Rothschild [3].

*Proof of the necessity.* We follow Hanges and Treves [14]. For the sake of simplicity we assume that $M$ is smooth and we construct a nonextendible $CR$ current, not a function. Let $N$ be a proper submanifold of $M$ at $z_0$, with the same $CR$ dimension $p$ as $M$ but with dimension $p + r < p + n = \dim M$. The pullback to $N$ of the system of forms $dz_1, \ldots, dz_n$ has constant rank $r$, and we may assume that $dz_1, \ldots, dz_r$ are independent on $N$. We introduce $p$ functions $t_1, \ldots, t_p$ on $M$ such that (the restrictions of) $dz_1, \ldots, dz_n, dt_1, \ldots, dt_p$ are independent on $M$ and $dz_1, \ldots, dz_r, dt_1, \ldots, dt_p$ are independent on $N$. The current $u$ defined by

$$u(\psi dz_1 \wedge \cdots \wedge dz_n \wedge dt_1 \wedge \cdots \wedge dt_p) = \int_N \psi\, dz_1 \wedge \cdots \wedge dz_r \wedge dt_1 \wedge \cdots \wedge dt_p,$$

where $\psi \in C_0^\infty(M)$, is a $CR$ current since

$$u(dz_1 \wedge \cdots \wedge dz_n \wedge d\phi) = \int_N dz_1 \wedge \cdots \wedge dz_r \wedge d\phi$$

due to the fact that $dz_1, \ldots, dz_n$ has constant rank $r$ on $N$; this is zero by Stokes' theorem. This $CR$ current cannot be a boundary value close to $z_0$, since the support is (locally) a proper submanifold of $M$, which is forbidden for the boundary values of holomorphic functions by Theorem 3.5. □

We now prove the approximation theorem of Baouendi and Treves [5] which is most important when applying the method of the complex disc to $CR$ functions.

A classical proof of the Weierstrass approximation theorem uses the Gaussian kernel as follows: if $u$ is a continuous function on $\mathbf{R}^n$ with compact support, then

$$T^\epsilon_{\mathbf{R}^n} u(z) = (\epsilon\pi)^{-n/2} \int_{\mathbf{R}^n} e^{-(z-x')^2/\epsilon} u(x')\, dx'$$

is an entire function whose restriction to $\mathbf{R}^n$ converges uniformly to $u$ as $\epsilon$ goes to zero. As a corollary one gets that $u$ is locally a uniform limit of a sequence of polynomials. The same proof applies when $\mathbf{R}^n$ is replaced by any maximally real manifold close to $\mathbf{R}^n$. To be precise, let $L$ be a maximally real submanifold in $\mathbf{C}^n$ and $z_0 \in L$. We may perform a complex linear change of coordinates such that $L$ is defined near $z_0$ by an equation of the form $\operatorname{Im} z = \Phi(\operatorname{Re} z)$ with $d\Phi(\operatorname{Re} z_0) = 0$. (It is sufficient for what follows that $d\Phi(z_0)$ be small.) If $u$ is a continuous function on $L$ (defined in a fixed neighborhood of $z_0$), we define

$$T_L^\epsilon u(z) = (\epsilon\pi)^{-n/2} \int_L e^{-(z-z')^2/\epsilon} u(z')\chi(z')\, dz'$$

where $\chi$ is a cutoff function with small support near $z_0$.

**Theorem 3.12.** *$T_L^\epsilon u$ is an entire function and its restriction to $L$ converges uniformly to $u$ in a (controlled) neighborhood of $z_0$.*

*Proof.* It is important to notice that, due to the fact that $d\Phi$ is small, $\operatorname{Re}(z-z')^2 \geq \text{const}\,|z-z'|^2$ for $z, z' \in L$ close enough to $z_0$ (hence the choice of $\chi$) so that the exponential factor in the integral is uniformly controlled for $z \in L$ close to $z_0$ and $z' \in L \cap \operatorname{supp}\chi$. Writing $z = x + i\Phi(x)$, $z' = x' + i\Phi(x')$ and performing the change of integration variable $x' = x + \sqrt{\epsilon}t$ one gets

$$\begin{aligned} &T_L^\epsilon u(z) \\ &= \frac{1}{\pi^{\frac{1}{2}n}} \int_{\mathbf{R}^n} e^{-((I+i\Phi'(x))t+o(1))^2} (u\chi)(z+O(\sqrt{\epsilon}))\det(I+i\Phi'(x)+O(\sqrt{\epsilon}))\, dt, \end{aligned}$$

which by the Lebesgue convergence theorem goes to

$$u(z)\frac{1}{\pi^{n/2}} \int_{\mathbf{R}^n} e^{-((I+i\Phi'(x))t)^2} \det(I + i\Phi'(x))\, dt = u(z). \quad \square$$

Now assume that $M$ is a generic manifold near $z_0 \in M$. Let $L_0$ be a maximally real manifold in $M$ through $z_0$ and make a linear transformation so that $\mathbf{R}^n$ becomes tangent to $L_0$ at $z_0$. The previous result applies: $T_{L_0}^\epsilon u$ is an entire function and its restriction to $L_0$ tends to $u$ for all continuous functions on $L_0$. In fact more is true if $u$ is *CR* on $M$. If $z \in M$ lies in a sufficiently small neighborhood of $z_0$ (in which $\chi \equiv 1$) we may deform $L_0$ near $z_0$ to a maximally real manifold $L_z$ through $z$, satisfying the condition in Theorem 3.12. Noticing that the integrated form, in the definition of $T_{L_0}^\epsilon u$, is closed where the deformation takes place (because $u$ is *CR*!) we obtain by Stokes' theorem that $T_{L_0}^\epsilon u(z) = T_{L_z}^\epsilon u(z)$ which converges to $u(z)$ by the previous theorem. We have obtained:

**Theorem 3.13.** *If $u$ is a continuous CR function on $M$, then the restriction to $M$ of $T_{L_0}^\epsilon u$ converges uniformly to $u$ in a (controlled) neighborhood of $z_0$.*

In this lecture we shall use the following important consequence of this theorem:

**Theorem 3.14.** *If $M$ is a generic manifold and $z_0 \in M$, then there exists a neighborhood $U$ of $z_0$ in $M$ such that every continuous CR function on $M$ is the uniform limit in $U$ of a sequence of holomorphic polynomials.*

Theorem 3.13 has also important consequences concerning the "unique propagation" of support for *CR* functions. In fact, we see from its proof that if a *CR* function vanishes on a maximally real submanifold $L$, it vanishes on a controlled neighborhood of $L$. From this Holmgren type theorem one gets by standard arguments, see [28]:

**Theorem 3.15.** *The support of a CR function on $M$ is a union of orbits of $T^cM$.*

**4. The method of the complex disc.** As in the case of a hypersurface we shall prove extension results by using complex discs with boundary on a given generic manifold $M$.

**Definition 4.1.** The complex disc $A : \bar{D} \to \mathbf{C}^n$ is *attached to* $M$ if $\partial A \subset M$. It is *attached to* $M$ *at* $z_0$ if moreover $A(1) = z_0$.

The second part of the definition will be discussed later. In the applications Lemma 2.1 is replaced by the following one:

**Lemma 4.2.** *If for every neighborhood $U$ of $z_0$ the union of all complex discs $A \subset U$ attached to $M$ contains a wedge $\mathcal{W}$ at $z_0$, then every continuous CR function on $M$ extends (close to $z_0$) to a function holomorphic on $\mathcal{W}$, continuous up to the edge $M$.*

*Proof.* Let $u$ be a continuous *CR* function. By the approximation theorem 3.14 we may assume that $u$ is on $M \cap U$ the uniform limit of a sequence of holomorphic polynomials $P_n$. By the maximum principle, if $A$ is a complex disc attached to $M \cap U$, then

$$|P_n \circ A(\tau)| \leq \sup_{\sigma \in S} |P_n \circ A(\sigma)|, \quad \text{for all } \tau \in D,$$

so the sequence $P_n$ is a Cauchy sequence on every complex disc attached to $M \cap U$. Thus the sequence converges uniformly on the union of all such discs towards a continuous function, holomorphic in the interior of this union. The lemma follows. □

The problem is now to construct such discs and evaluate the set they generate. The construction is similar to the case of a hypersurface. To simplify the presentation, we shall assume that $M$ is smooth.

We change our notation: $Z = (w, z)$ denotes the variable in $\mathbf{C}^n$, with $w \in \mathbf{C}^p$, $z \in \mathbf{C}^q$, $p + q = n$. After a complex linear transformation, we may assume that $z_0 = 0$ and that $M$ has a local equation of the form

$$M := \quad r(Z) = (r_1(Z), \dots, r_q(Z)) = 0, \tag{4.1}$$

where $r$ is a smooth function $\mathbf{C}^n \to \mathbf{R}^q$ defined near 0 with (variables as subscript indicate derivatives) $r_w(0) = 0$ and $r_z(0) = \mathrm{I}$, the $q \times q$ identity matrix. We shall work with discs of class $C^{1,\alpha}$, for some fixed $\alpha$ with $0 < \alpha < 1$. We modify the construction of Section 2 a little by requiring that the boundary of the disc goes through the base point 0, while in Section 2 we required that the interior of the disc should meet the normal space to the base point. We shall denote by $C_0^{1,\alpha}$ the space of functions $A$ of class $C^{1,\alpha}$ with $A(1) = 0$ and

$$\mathcal{O}_0^{1,\alpha}(\bar{D}) = \mathcal{O}(D) \cap C_0^{1,\alpha}(\bar{D}).$$

In what follows, "small" always refers to the $|\ |_{1,\alpha}$ norm. Finally, if $u$ is a $C^{1,\alpha}$ function on $S$, then $T_1 u$ will denote the modified Hilbert transform $T_1 u = Tu - (Tu)(1)$, so that $T_1 u(1) = 0$.

The function $r$ induces a smooth map, still called $r$:

$$r : \mathcal{O}_0^{1,\alpha}(\bar{D}, \mathbf{C}^n) \to C_0^{1,\alpha}(S, \mathbf{R}^q), \quad r(A) = r(A(\cdot))|_S. \tag{4.2}$$

The discs attached to $M$ at 0 are thus defined by the equation $r(A) = 0$. This is indeed an equation in the sense of differential geometry:

**Lemma 4.3.** *The map (4.2) is a submersion at the trivial disc $A \equiv 0$. It follows that the set $\mathcal{W}_0^{1,\alpha}$ of all small discs of class $C^{1,\alpha}$ attached to $M$ at 0 is a Banach submanifold of $\mathcal{O}_0^{1,\alpha}(\bar{D}, \mathbf{C}^n)$, with equation $r = 0$. The tangent space at $A \equiv 0$ is isomorphic to $\mathcal{O}_0^{1,\alpha}(\bar{D}, \mathbf{C}^p)$.*

In particular $\mathcal{W}_0^{1,\alpha}$ is pretty large, unless of course $p = 0$, i.e. $M$ is maximally real.

*Proof.* Let us write $\widehat{A} = (\hat{w}, \hat{z})$ for an element of $\mathcal{O}_0^{1,\alpha}(\bar{D}, \mathbf{C}^n)$, considered as a tangent vector at $A$. We think of $\hat{w}$ and $\hat{z}$ as column vectors of functions and of the partial derivatives of $r$ as matrices, accordingly. The differential of the map $r$ is given by

$$r'(A)\,\widehat{A} = dr(A)\,\widehat{A}|_S = 2\,\mathrm{Re}\,\partial r(A)\,\widehat{A}|_S = 2\,\mathrm{Re}\,(r_w(A)\,\hat{w} + r_z(A)\,\hat{z})|_S. \tag{4.3}$$

In particular $r'(0)\,\widehat{A} = 2\,\mathrm{Re}\,\hat{z}|_S$. This shows that $r'$ is onto, hence $r$ is a submersion, since the equation $2\,\mathrm{Re}\,\hat{z}|_S = \hat{\psi} \in C_0^{1,\alpha}(S, \mathbf{R}^q)$ is solved by taking $\hat{z}$ to be the holomorphic extension to $D$ of $(\mathrm{I} + iT_1)\hat{\psi}/2$ (letting $T_1$ act componentwise). Also $r'(0)\,\widehat{A} = 0$ if and only if $\widehat{A} = (\hat{w}, 0) \in \mathcal{O}_0^{1,\alpha}(\bar{D}, \mathbf{C}^p) \times 0$. $\square$

If $A$ is a complex disc attached to $M$ at 0, then the tangent vector to $\partial A$ at $A(1) = 0$ is given by

$$\mathcal{G}(A) = \frac{d}{dt} A(e^{it})|_{t=0}, \tag{4.4}$$

while, since $A$ is holomorphic, the tangent vector to the parametrized curve $A(s)$, $s \leq 1$, is $i\mathcal{G}(A)$.

The main step in the proof of Theorem 3.11 will be to show that if $M$ is minimal at 0, then the map $\mathcal{G} : \mathcal{W}_0^{1,\alpha} \to T_0M$ defined by (4.4) is a submersion. In fact, we can use the following variant of Lemma 4.2 to extend $CR$ functions:

**Lemma 4.4.** *If $A_0$ is a small disc attached to $M$ at 0 and the map $\mathcal{G}$ is a submersion at $A_0$, then every continuous $CR$ function on $M$ extends holomorphically to a wedge at 0.*

*Sketch of proof.* We want to show (see Lemma 4.2) that the union $\mathcal{V}$ of all discs attached to $M \cap U$, $U$ a small complex neighborhood of 0, contains a wedge of edge $M$ at 0. If $A$ is such a disc, attached to 0, it contains a pencil of parametric curves (taking the image of a sector in $D$ at 1) issuing from 0 with tangent vectors $i\mathcal{G}(A) + \epsilon\mathcal{G}(A)$ at 0, with $\epsilon > 0$ small. If $\mathcal{G}$ is a submersion at $A_0$, then $\mathcal{G}(A)$, as $A$ varies in a small neighborhood of $A_0$, covers a neighborhood of $\mathcal{G}(A_0)$ in $T_0(M)$ so the associated pencil of curves fills an open cone with vertex 0 and axis $i\mathcal{G}(A_0)$, intersected with a small neighborhood of 0. In fact, everything is controlled, so applying the same construction to points in $M$ close to 0, we actually obtain a wedge. □

By (4.3) the tangent space to $\mathcal{W}_0^{1,\alpha}$ at a small disc $A$ is the subspace of $\mathcal{O}_0^{1,\alpha}(\bar{D}, \mathbf{C}^n)$ defined by the equation

$$2\,\mathrm{Re}\,(r_w(A)\,\hat{w} + r_z(A)\,\hat{z})|_S = 0. \tag{4.5}$$

This defines $\hat{z}$ as an implicit function of $\hat{w}$, but since this equation involves multiplications by functions and the Hilbert transform, it is not clear *a priori* what the solution looks like. It turns out that the solution also involves only multiplication operators and the Hilbert transform. The explicit form of the solution which we shall now exhibit will be crucial in Section 5. We begin with a lemma:

**Lemma 4.5.** *Let $a$ be a $q \times q$ matrix with $C^{1,\alpha}(S)$ entries and $a(1) = \mathrm{I}$. If $a$ is close enough to $\mathrm{I}$, there exist matrices $g, h$ with the same properties, such that $g$ is real, $h$ extends holomorphically to $D$, and $g \times a = h$.*

*Proof.* It is sufficient to show that the bilinear map $(g, h) \mapsto g \times h$, acting on the obvious spaces of matrices is a submersion at $\mathrm{I} \times \mathrm{I}$. Differentiating, one is reduced to showing that any $q \times q$ matrix $\hat{a}$ with $C^{1,\alpha}$ entries and $a(1) = 0$ can be written $\hat{a} = \hat{g} + \hat{h}$ where $\hat{g}$, $\hat{h}$ have the same properties, $\hat{g}$ is real and $\hat{h}$ extends holomorphically to $D$. This is the classical additive Riemann-Hilbert problem, solved componentwise using the Hilbert tranform. □

We are now able to parametrize the tangent space at any small disc. In fact, let $a = r_z(A)$ be close to I and $c = r_w(A)$ be close to 0. Let $g$, $h$

be associated to $a$ as in the lemma. The equation (4.5) is solved in the following way: Multiply both members by the real matrix $g$, so that the *holomorphically extendible* vector $ga\hat{z} = h\hat{z}$ has its real part determined on $S$ by the equation, which gives

$$\hat{z} = -h^{-1}(\mathrm{I}+iT_1)\,\mathrm{Re}\,gc\hat{w}, \quad \hat{w} \in \mathcal{O}_0^{1,\alpha}(\bar{D}, \mathbf{C}^p). \tag{4.6}$$

**5. The analysis of the Bishop equation.** We go on with the notation and assumptions at the end of the previous section. $M$ is a $q$-codimensional generic manifold in $\mathbf{C}^n = \mathbf{C}^p_w \times \mathbf{C}^q_z$ defined near 0 by the equation (4.1). For $Z \in \mathbf{C}^n$ we identify the real tangent space $T_Z\mathbf{C}^n$ with $\mathbf{C}^n$, $\widehat{Z}$ being identified with the derivation $u \mapsto du(Z+t\widehat{Z})/dt|_{t=0}$. If $Z \in M$, we identify the tangent space $T_ZM$ with a subspace of $\mathbf{C}^n$, with the equation

$$T_ZM := \quad \mathrm{Re}\ \partial r(Z)\,\widehat{Z} = 0$$

in matrix notation. The space $T_Z^cM$, introduced in Section 3, is defined by the equation:

$$T_Z^cM := \quad \partial r(Z)\,\widehat{Z} = 0.$$

Let $T^*\mathbf{C}^n$ denote the bundle of holomorphic forms $\lambda = \sum_{i=1}^n a_i dz_i$; note that this is a *complex space* isomorphic to $\mathbf{C}^n \times \mathbf{C}^n$. By taking the imaginary part of $\lambda$, one can use $T^*\mathbf{C}^n$ as a (real) cotangent bundle. This allows one to identify the conormal bundle to $M$ in $\mathbf{C}^n$ with a real submanifold of $T^*\mathbf{C}^n$:

**Definition 5.1.** $\Sigma M$ is the $\mathbf{R}$-subbundle of $T^*\mathbf{C}^n|_M$ defined by its fiber $\Sigma_Z M = \{\lambda; \mathrm{Im}\,\lambda\,\widehat{Z} = 0 \text{ for all } \widehat{Z} \in T_ZM\}$.

So, if $\lambda \in \Sigma_Z M$, then $\lambda$ is real on $T_ZM$, and since it is complex linear, it vanishes on $T_Z^cM$. The converse is true, and this gives a description of this space using the defining functions of $M$:

$$\Sigma_Z M = \{\lambda \in T_Z^*\mathbf{C}^n; \lambda = i\sum_{k=1}^q \theta_k \partial r_k(Z),\ \theta_k \in \mathbf{R}\},$$

or in matrix notation $\lambda = i\,{}^t\theta\,\partial r(Z)$ for a column vector $\theta \in \mathbf{R}^q$.

It may happen that a complex disc attached to $M$ can be lifted to a nontrivial complex disc in $T^*\mathbf{C}^n$ attached to $\Sigma M$:

**Definition 5.2.** Let $A$ be attached to $M$ and $V_A$ denote the real vector space of complex discs $\tau \mapsto (A(\tau), \lambda(\tau))$ attached to $\Sigma M$ upon $A$, which means that $\lambda(\tau) \in \Sigma_{A(\tau)}M$ for all $\tau \in S$. The dimension of $V_A(\tau) = \{\lambda(\tau); (A, \lambda) \in V_A\}$ is called the *defect* of the disc $A$ at $\tau$.

**Lemma 5.3.** *If $A$ is a small disc attached to $M$, then the defect of $A$ at $\tau$ does not depend on $\tau \in S$. It is equal to the dimension of $V_A$ and called the defect of the disc $A$.*

*Proof.* $V_A$ is the space of the forms $i^t\theta(\tau)\,\partial r \circ A(\tau)$ on $S$ where $\theta$ is a column $q$-vector of real functions of class $C^{1,\alpha}$ such that $i^t\theta(\tau)\partial r(A)(\tau)$ extends holomorphically to the disc $D$. Recall our notation from the end of Section 4: we write $a = r_z(A)$ and $c = r_w(A)$. Thus ${}^t\theta\, c$ and ${}^t\theta\, a$ extend holomorphically to $D$. Writing $a = g^{-1}h$ as in Lemma 4.5 we obtain that ${}^t\theta g^{-1}$ extends holomorphically. As it is real it is constant, thus ${}^t\theta(\tau) = {}^t\theta(1)g(\tau)$ while the second condition means that ${}^t\theta(1)gc$ extends holomorphically to $D$. Finally $V_A$ is the set of vector valued functions of the form $i^t\theta g\partial r(A)$ with $\theta \in \mathbf{R}^q$ such that ${}^t\theta gc$ extends holomorphically. As $g$ is invertible, the lemma follows. □

Taking $dw_1, \dots, dw_p, \partial r_1, \dots, \partial r_q$ as a basis for complex linear forms on $\mathbf{C}^n$, we see that every *real* form can be written $\lambda = \mathrm{Re}\,({}^t\alpha\, dw + {}^t\beta\, \partial r)$ for column vectors $\alpha \in \mathbf{C}^p$, $\beta \in \mathbf{C}^q$. As already noticed, $\partial r$ is purely imaginary on $TM$, so when computing on $M$, we may write any real form as $\lambda = \mathrm{Re}\,{}^t\alpha dw + i^t\theta\partial r$ for column vectors $\alpha \in \mathbf{C}^p$ and $\theta \in \mathbf{R}^q$.

We consider the maps $\mathcal{F}_{\tau_0} : \mathcal{W}_0^{1,\alpha} \to M$ and $\mathcal{G} : \mathcal{W}_0^{1,\alpha} \to T_0M$ defined by

$$\mathcal{F}_{\tau_0}(A) = A(\tau_0), \quad \mathcal{G}(A) = dA(e^{it})/dt|_{t=0}.$$

The introduction of the notion of the defect is justified by the following result:

**Proposition 5.4.** *For any small disc $A \in \mathcal{W}_0^{1,\alpha}$ and $\tau_0 \in S \setminus \{1\}$ one has*

(1) $\mathcal{F}'_{\tau_0}(A)\,T_A\mathcal{W}_0^{1,\alpha} = V_A(\tau_0)^\perp \subset T_{A(\tau_0)}M$.

(2) $\mathcal{G}'(A)\,T_A\mathcal{W}_0^{1,\alpha} = V_A(1)^\perp \subset T_0M$.

In this statement orthogonality refers to the duality between vectors and forms in $TM$.

*Proof.* We shall prove (1), the computation for (2) is similar. $\mathcal{F}_{\tau_0}$ is the restriction of the linear map of Banach spaces $A \mapsto A(\tau_0)$, so $\mathcal{F}'_{\tau_0}(A)\,\widehat{A} = \widehat{A}(\tau_0)$. First, if $\lambda \in V_A(\tau_0)$, we may write $\lambda = \lambda(\tau_0)$ where $\lambda(\cdot)$ is holomorphic on $D$ and $\mathrm{Im}\,\lambda(\tau)\,\widehat{A}(\tau) = 0$ for $\tau \in S$. So $\lambda\,\widehat{A}$ is real on $S$ and holomorhically extendible to $D$; it must be constant $= \lambda(1)\,\widehat{A}(1) = 0$. This shows that $V_A(\tau_0)$ is contained in the orthogonal of the range of $\mathcal{F}'_{\tau_0}$ for all $\tau_0 \in S$.

Conversely, let $\lambda$ be a real form on the tangent space to $M$ at $A(\tau_0)$, vanishing on the range of $\mathcal{F}'_{\tau_0}$. Writing $\lambda = \mathrm{Re}\,{}^t\alpha\, dw + i^t\theta\, \partial r$ we get

$$\mathrm{Re}\,{}^t\alpha\hat{w}(\tau_0) + i^t\theta(c(\tau_0)\hat{w}(\tau_0) + a(\tau_0)\hat{z}(\tau_0)) = 0 \tag{5.1}$$

for all $(\hat{w}, \hat{z})$ in the tangent space to $\mathcal{W}_0^{1,\alpha}$ at $A$. We use the parametrization (4.6) of the tangent space. We first restrict attention to complex discs parametrized by $C^{1,\alpha}$ discs $\hat{w}$ satisfying $\hat{w}(\tau_0) = 0$; for those discs the previous equation reduces to

$$i^t\theta a(\tau_0)\hat{z}(\tau_0) = {}^t\theta a(\tau_0)h(\tau_0)^{-1}(T_1(\operatorname{Re} gc\hat{w}))(\tau_0) = 0.$$

As this is true for $i\hat{w}$ as well, we obtain

$$T_1({}^t\mu\hat{w})(\tau_0) = 0, \quad {}^t\mu(\tau) = {}^t\theta g(\tau_0)^{-1}g(\tau)c(\tau).$$

Now, from its definition, one gets easily the following integral formula for the modified Hilbert transform:

$$T_1u(\tau_0) = \text{principal value } \Big(\frac{1}{\pi}\int_S \frac{u(\tau)(1-\tau_0)}{(\tau-1)(\tau-\tau_0)}\, d\tau\Big).$$

Choosing $\hat{w}(\tau) = (\tau-1)(\tau-\tau_0)(1-\tau_0)^{-1}\tau^k c$, $k \in \mathbf{N}$, $c \in \mathbf{C}^p$, we find (moment condition) that ${}^t\mu$ extends holomorphically to $D$. This means precisely that $i^t\theta\, \partial r(A(\tau_0))$ belongs to $V_{A(\tau_0)}$. Now we drop the assumption $\hat{w}(\tau_0) = 0$. By the first part of the proof we know that the second term in (5.1) vanishes, so the equation (5.1) reduces to $\operatorname{Re}{}^t\alpha\hat{w}(\tau_0) \equiv 0$, so $\alpha = 0$ and (1) is proved. □

We are now able to sketch the proof of Theorem 3.11.

**Theorem 5.5.** *If $M$ is a $C^2$-smooth generic manifold, minimal at $z_0$, then every continuous CR function on $M$ extends locally to a wedge at $z_0$.*

This result, with the same proof, extends to a more general situation, see Tumanov [30]; if the local orbit of $z_0$ has dimension $2CR$-$\dim M + r$, then *CR* functions on $M$ extend as *CR* functions in a wedge with edge $M$ in a generic manifold of *CR* dimension $p+r$. The minimal case corresponds to $r = \operatorname{codim} M$.

We shall sketch the proof under the hypothesis that $M$ is smooth, but the proof in the general case only involves unimportant improvements. We follow Tumanov [29] as we already did since the beginning of this section, apart from some terminology taken from [4].

*Proof.* Let $A_0$ be a small disc attached to $M$ at 0, and choose $A_0$ of minimal defect. From the definitions one easily sees that the disc $A_0(\tau^2)$ has the same properties (it has the same defect as $A_0$), so we may assume that $A_0(-1) = 0$. The map $\mathcal{F} = \mathcal{F}_{-1} : \mathcal{W}_0^{1,\alpha} \to M$ is a smooth map in a neighborhood of the disc $A_0$, with rank $\dim M - \operatorname{defect}(A_0)$ at $A_0$ by Proposition 5.4. By our choice of $A_0$, the rank is constant close to $A_0$, so that the (local) range of $\mathcal{F}$ is a smooth submanifold $N$ of $M$ going through $\mathcal{F}(A_0) = 0$. By the same proposition, the tangent space to $N$ at $\mathcal{F}(A) = A(-1)$ is the orthogonal of $V_A(1)$ which contains $T^c_{A(-1)}(M)$. This means that $N$ is a $CR$ submanifold of $M$ with the same $CR$ dimension as $M$. Since $M$ is minimal, $N = M$ locally, so $\mathcal{F}$ is a submersion, hence the map $\mathcal{G}$ is also a submersion by Proposition 5.4. Due to Lemma 4.4 the theorem is proved. □

Note that if, in the proof, the disc $A_0$ is transversal to $M$ at 0, there is by the proof extendibility in the direction $i\mathcal{G}(A_0)$ or in a family of directions converging to $i\mathcal{G}(A_0)$. If it is not transversal, then $CR$ functions extend to a whole complex neighborhood of 0.

**6. Further remarks.** The Bishop equation was first considered by Bishop [9]. In this lecture it appears in a local context. For example we saw that there exists no *small* nonconstant complex disc attached to a maximally real manifold. The situation is quite different in a global context; there are many complex discs attached to the compact totally real torus $S \times S$ in $\mathbf{C}^2$. In this global situation, the analysis of the Bishop equation in the neighborhood of a given disc may allow studying the set of nearby discs. In fact, differentiating the equation leads to the multiplicative Riemann-Hilbert problem, the simplest model of an elliptic system on a compact manifold (the circle!). This gives rise to finite dimensional spaces of discs attached to a maximally real manifold. This was first used by Bedford and Gaveau [7] to compute, using the work of Bishop, the envelope of holomorphy of certain 2-spheres in $\mathbf{C}^2$, see also [8], [11], [12]. In the more general context of almost complex manifolds, the global Bishop equation for maximally real manifolds is one of the tools used by Gromov in his impressive paper [13] including a proof of Arnold's conjecture that there exists no exact compact Lagrangean submanifold in $\mathbf{C}^n$ for the standard symplectic structure.

Due to lack of space we have not presented the elements of the microlocal study of singularities of $CR$ functions. The analytic wavefront set (or microsupport) of a hyperfunction $u$ in $\mathbf{R}^n$ was first introduced by Sato. It is a closed conic subset of the conormal bundle $\Sigma\mathbf{R}^n$ of $\mathbf{R}^n$ in $\mathbf{C}^n$ which characterizes geometrically the different ways $u$ can be decomposed as a sum of boundary values of holomorphic functions. Another complete approach to the analytic wavefront set, using general oscillatory integrals with complex phase, was introduced by Sjöstrand [22]. The case of general generic manifolds is treated by Baouendi, Chang and Treves [2], see also [23]. The main idea is to define the (hypo)analytic wavefront set of a $CR$ function at $z_0$ as the wavefront set at $z_0$ of any of its traces on a maximally real manifold through $z_0$. This notion has many applications (for example a Holmgren type theorem at the microlocal level, unique continuation of singularities, and so on). Wedge extendibility of a $CR$ function is characterized by the fact that the fiber of the wavefront set is contained in an open half-space. However, a drawback of this notion is that the wavefront set does not characterize the property for a function to extend as a *CR* function on a bigger generic manifold. It does not seem that such problems occur in the general theory of PDE.

Theorem 5.5 can of course be interpreted as a result of semi-microlocal analytic hypoellipticity; if $M$ is minimal at $z_0$ the fiber of the (hypo)analytic wavefront set of any *CR* function $u$ is contained in an open half-space. Note however that the proof is local and not microlocal, that is, $u$ is a local, not just microlocal, solution of the induced Cauchy-Riemann system, so the theorem does not mean that the system is microlocally analytic hypoelliptic outside an open half-space.

As already mentioned, even in $\mathbf{C}^2$, a simple characterization of the extension property at a given point of the boundary of a domain is not known, and perhaps does not exist. To have a chance to get a neat statement, one might be tempted to try to characterize the stronger property that the extension property holds at every boundary point. For example, it is an easy consequence of Theorem 2.3 that a smooth domain $\Omega \subset \mathbf{C}^2$ has the extension property at every point of an open set $M$ of its boundary if and only if there exists no germ of a complex hypersurface (curve) intersecting $M$ without intersecting $\Omega$. Unfortunately, no such result holds in higher dimension, see [27].

Propagation of microlocal singularities of *CR* functions is studied in [26], using the FBI transform and the Bishop equations as tools. The main result is an analogue of a previous result of Hanges and Sjöstrand for generic manifolds. The main applications are the construction of examples of generic manifolds for which decomposability does not hold, and the fact that, in nonminimal generic manifolds, the property for a *CR* function to be extendible to a wedge propagates along the orbits of the *CR* structure, as is the case for the support (see Theorem 3.15). As for the extension problem, a drawback of using the analytic wavefront set is that it does not characterize *CR* extendibility. Propagation is studied in [31] using only complex discs. Propagation of wedge or *CR* extendibility can also be proved using deformation, see [16]. Another application is that if $M$ is not assumed to be minimal at $z_0$, but only *globally* minimal, in the sense that the *global* orbit of $z_0$ is a neighborhood of $z_0$, extendibility to a wedge at $z_0$ still holds for *globally* defined *CR* functions, see Jöricke [16], Merker [19].

## References


1. A. Andreotti and C. D. Hill, *E. E. Levi convexity and the Hans Lewy problem*, Ann. Scuola Norm. Sup. Pisa **26** (1972), 325–363.
2. M. S. Baouendi, C. H. Chang and F. Treves, *Microlocal hypo-analyticity and extension of CR functions*, J. Diff. Geom. **18** (1983), 331–391.

3. M. S. Baouendi and L. P. Rothschild, *Cauchy-Riemann functions on manifolds of higher codimension in complex space*, Invent. Math. **101** (1990), 45–56.
4. M. S. Baouendi, L. P. Rothschild and J.-M. Trépreau, *On the geometry of analytic discs attached to real manifolds*, J. Diff. Geom. **39** (1994), 379–405.
5. M. S. Baouendi and F. Treves, *A property of the functions and distributions annihilated by a locally integrable system of complex vector fields*, Ann. of Math. **113** (1981), 387–421.
6. ———, *About the holomorphic extension of CR functions on real hypersurfaces in complex space*, Duke Math. J. **51** (1984), 77–107.
7. E. Bedford and B. Gaveau, *Envelopes of holomorphy of certain two-spheres in* $\mathbf{C}^2$, Amer. J. Math. **105** (1983), 975–1009.
8. E. Bedford and Klingenberg, *On the envelope of holomorphy of a 2-sphere in* $\mathbf{C}^2$, J. Amer. Math. Soc. **4** (1991), 623–646.
9. E. Bishop, *Differentiable manifolds in complex Euclidean space*, Duke Math. J. **32** (1965), 1–22.
10. J. E. Fornæss and C. Rea, *Local holomorphic extendibility and non-extendibility of CR-functions on smooth boundaries*, Ann. Scuola Norm. Sup. Pisa **12** (1985), 491–502.
11. F. Forstneric, *Analytic discs with boundary in a maximal totally real submanifold of* $\mathbf{C}^n$, Ann. Inst. Fourier, Grenoble **37** (1987), 1–44.
12. J. Globevnik, *Perturbation of analytic discs along maximal real submanifolds of* $\mathbf{C}^n$, Math. Z. **217** (1994), 287–316.
13. M. Gromov, *Pseudo-holomorphic curves in symplectic manifolds*, Invent. Math. **82** (1985), 307–347.
14. N. Hanges and F. Treves, *On the local holomorphic extension of CR functions*, Astérisque **217** (1993), 119–137, Colloque Analyse Complexe et Géométrie, Marseille 1992.
15. L. Hörmander, *An introduction to complex analysis in several variables*, North-Holland Publ. Co., Amsterdam, London, 1973.
16. B. Jöricke, *Deformation of CR manifolds, minimal points and CR manifolds with the microlocal extension property*, J. Geom. Anal. (to appear).
17. J. J. Kohn and L. Nirenberg, *A pseudoconvex domain not admitting a holomorphic support function*, Math. Ann. **201** (1973), 265–268.
18. H. Lewy, *On the local character of the solution of an atypical differential equation in three variables and a related problem for regular functions of two complex variables*, Ann. of Math. **64** (1956), 514–522.
19. J. Merker, *Global minimality of generic manifolds and holomorphic extendibility of CR functions*, Int. Math. Res. Not. **8** (1994), 329–342.
20. C. Rea, *Prolongement holomorphe des fonctions CR, conditions suffisantes*, Notes aux C. R. Acad. Sci. Paris **297** série 1 (1983), 163–165.
21. M. Sato, T. Kawai and M. Kashiwara, *Hyperfunctions and pseudodifferential equations*, Springer Lecture Notes in Math. **287** (1973), 265–529.
22. J. Sjöstrand, *Singularités analytiques microlocales*, Astérisque **95** (1982), 1–166.
23. ———, *The FBI-transform for CR submanifolds of* $\mathbf{C}^n$, Prépublications Math. d'Orsay, Université Paris-Sud 1982.
24. H. J. Sussmann, *Orbits of families of vector fields and integrability of distributions*, Trans. Amer. Math. Soc. **180** (1973), 171–187.
25. J.-M. Trépreau, *Sur le prolongement holomorphe des functions CR définies sur une hypersurface réelle de classe* $C^2$ *dans* $\mathbf{C}^n$, Invent. Math. **83** (1986), 583–592.
26. ———, *Sur la propagation des singularités dans les variétés CR*, Bull. Soc. Math. France **118** (1990), 403–450.

27. ———, *On the extension of holomorphic functions across a real hypersurface*, Math. Z. **211** (1992), 93–103.

28. F. Treves, *Approximation and representation of functions and distributions annihilated by a system of complex vector fields*, Ecole Polytechnique, Centre de Math., Palaiseau 1981.

29. A. E. Tumanov, *Extension of CR functions into a wedge from a manifold of finite type*, Mat. Sbornik **136** (1988), 129–140.

30. ———, *Extension of CR-functions into a wedge*, Mat. Sbornik **181** (1990), 951–964.

31. ———, *Connections and propagation of analyticity for CR functions*, Duke Math. J. **73** (1994), 1–24.

Université de Paris VI, Institut de Mathématiques, tour 45–46, 5ème étage, 4, Place Jussieu, F-75252 PARIS Cedex 05, France

*E-mail address*: jmt@ccr.jussieu.fr

# LOCAL SOLVABILITY IN A CLASS OF OVERDETERMINED SYSTEMS OF LINEAR PDE

François Treves

**1. The case of a single vector field.** The simplest linear partial differential equations whose *local solvability* is not automatic are those defined by complex, smooth, nowhere vanishing vector fields in regions of the plane. Let $L = A(x_1, x_2)\partial/\partial x_1 + B(x_1, x_2)\partial/\partial x_2$ be such a vector field, defined in a domain $\Omega \subset \mathbf{R}^2$. The (inhomogeneous) equation

$$Lu = f \tag{1}$$

is said to be *locally solvable at a point* $O \in \Omega$ if there is an open neighborhood $U \subset \Omega$ of $O$ such that, given any $f \in C^\infty(\Omega)$, there is a distribution $u \in \mathscr{D}'(U)$ that satisfies (1) in the open set $U$. One can vary this definition by asking that $u$ be a $C^\infty$ function, in which case one may talk of local solvability in $C^\infty$; or by asking that there be a distribution solution $u$ in $U$, for each $f \in \mathscr{D}'(\Omega)$. A moment of thought will also show that there is no loss of generality if we restrict our attention to right hand sides $f$ whose support is compact and contained in $U$.

If we write $L = X + iY$ with $X$ and $Y$ real vector fields, and possibly replace $L$ by $iL$, we may assume that $X$ does not vanish in a full neighborhood of $O$ in $\Omega$. There are coordinates $t, x$ in a (smaller) neighborhood of $O$ such that $X = \partial/\partial t$, and therefore $L = \partial/\partial t + ia(t,x)\partial/\partial t + ib(t,x)\partial/\partial x$. Below we shall always assume that $x$ and $t$ vanish at $O$. With this expression of $L$ a necessary and sufficient condition for the local solvability of (1) at $O$ can be stated as follows

(2) *There are positive numbers $\delta$ and $\varepsilon$ such that, whatever $x \in (-\delta, \delta)$, the function $(-\varepsilon, \varepsilon) \ni t \mapsto b(t,x) \in \mathbf{R}$ does not change sign.*

(Neither the proof of the sufficiency of (2) for the local solvability of (1), nor that of its necessity are easy; (2) is a version of the property usually

The research of F. Treves described in this note has been supported by various grants of the Division of Mathematical Sciences of the National Science Foundation called ($\mathscr{P}$); see [NIRENBERG-TREVES, 1963] and Ch. XXVI, [HÖRMANDER, 1985].)

We observe that $(1+ia)^{-1}L = \partial/\partial t + \lambda\partial/\partial x$ with $\lambda = ib/(1+ia)$; the fact that (1) is locally solvable in $C^\infty$ entails that the special equation

$$w_t + \lambda w_x = -\lambda_x \tag{3}$$

has a $C^\infty$ solution $w(t,x)$. The function

$$Z(t,x) = \int_0^x e^{w(t,y)}\,dy - \int_0^t \lambda(s,0)e^{w(s,0)}\,ds \tag{4}$$

satisfies the homogeneous equation $LZ = 0$ and is such that $Z_x \neq 0$. Possibly after replacing $Z$ by $iZ$ we may assume that $\operatorname{Re} Z_x \neq 0$, which allows us to take $\operatorname{Re} Z$ as a new $x$ coordinate in an open neighborhood of $O$ (which we may take to be $\Omega$). Thus in the new coordinates $Z = x + i\Phi(t,x)$ and the expression of $L$ is $\partial/\partial t + i\Phi_x\partial/\partial t - i\Phi_t\partial/\partial x$. Condition (2) reads

(5) *There are positive numbers $\delta$ and $\varepsilon$ such that, whatever $x \in (-\delta,\delta)$, the function $(-\varepsilon,\varepsilon) \ni t \mapsto \Phi(t,x) \in \mathbf{R}$ is monotone.*

The following restatement of (5) is significant for what is to come:

(6) *There is an open neighborhood of $O$ in $\Omega$ in which the pre-image of any point of $\mathbf{C}$ under the map $Z$ is connected.*

**2. The problem for $n$ vector fields.** If one wants to investigate the local solvability of overdetermined systems of linear PDE it is reasonable to begin by looking at the natural generalization of the situation in §1: namely at systems of PDE defined by $n$ complex, smooth vector fields $L_1,\dots,L_n$ in a region $\Omega$ of $\mathbf{R}^{n+1}$, with $n \geq 1$ arbitrary. The nonvanishing of a single vector field is generalized by the hypothesis

(7) $L_1,\dots,L_n$ *are linearly independent at every point of $\Omega$.*

Furthermore, in order to generalize Condition (6) we must hypothesize the existence of a function $Z$, defined and $C^\infty$ in a neighborhood of $O$ (which we take to be $\Omega$) such that

(8) $L_1Z = \cdots = L_nZ = 0$ *and $dZ \neq 0$ at every point of $\Omega$.*

Combined, (7) and (8) entail

$$[L_j, L_k] = L_jL_k - L_kL_j \in \operatorname{Span}(L_1,\dots,L_n). \tag{9}$$

(9) is often referred to as the *Frobenius condition*; it expresses the *formal integrability* of the system $L_1, \dots, L_n$. The existence of the "first integral" $Z$ in (8) expresses its *local integrability* (at $O$). There are classes in which formal integrability $\Longrightarrow$ local integrability, such as: when the vector fields $L_j$ are *real* (Frobenius' theorem); when their coefficients are of class $C^\omega$ (Frobenius' theorem in the analytic category); when the system $L_1, \dots, L_n$ is *elliptic*, meaning that the vector fields $L_1, \dots, L_n$ and their complex conjugates $\bar{L}_1, \dots, \bar{L}_n$ span the complex tangent bundle to $\mathbf{R}^{n+1}$ at every point of $\Omega$ (Newlander-Nirenberg theorem). There are also examples of systems of vector fields that are formally integrable but nowhere locally integrable. On these questions we refer the reader to Chapters VI, VII of [TREVES, 1992].

Reasoning as before we select the coordinates $t_1, \dots, t_n, x$ and the function $Z$ in (8) in such a way that $Z = x + i\Phi(t,x)$. Since the properties we propose to study are not affected by a nonsingular linear substitution of $\{L_1, \dots, L_n\}$ and since the span of $L_1, \dots, L_n$ is characterized by Property (8), there is no loss of generality in assuming that

$$L_j = \partial/\partial t_j - i(\partial\Phi/\partial t_j)L_0, \quad 1 \le j \le n, \tag{10}$$

with $L_0 = (1 + i\Phi_x)^{-1}\partial/\partial x$. With these choices we have

$$[L_j, L_k] \equiv 0, \quad 0 \le j < k \le n. \tag{11}$$

We propose to study the local solvability of the system of first-order linear PDE,

$$L_j u = f_j, \quad j = 1, \dots, n, \tag{12}$$

under the compatibility conditions ensuing from (11):

$$L_j f_k = L_k f_j, \quad 1 \le j < k \le n. \tag{13}$$

This problem is a generalization of the equation $\overrightarrow{\mathrm{grad}u} = \vec{f}$ under the compatibility condition $\mathrm{curl}\,\vec{f} \equiv 0$. But we can also generalize the equation $\mathrm{curl}\,u = f$, as well as the higher order equations. A convenient manner of writing down those generalizations is through the use of the following type of differential forms:

$$f = \sum_{|J|=q} f_J(t,x)\, dt_J, \tag{14}$$

where $J = (j_1, \dots, j_q)$, $1 \le j_1 < \cdots < j_q \le n$; $dt_J = dt_{j_1} \wedge \cdots \wedge dt_{j_q}$. Most often the coefficients $f_J$ will be smooth functions in an open neighborhood

$U$ of $O$ in $(t,x)$-space $\mathbf{R}^{n+1}$, in which case we write $f \in C^\infty(U;\bigwedge^{0,q})$; they could also be distributions, in which case we write $f \in \mathscr{D}'(U;\bigwedge^{0,q})$ (or hyperfunctions, as we shall indicate below). In all cases we define the operator

$$Lf = \sum_{|J|=q}\sum_{k=1}^{n} L_k f_J(t,x)\, dt_k \wedge dt_J.$$

Since the $L_j$ commute we have $L^2 = 0$. We define $\lambda = -i(1+i\Phi_x)^{-1}d_t\Phi \in C^\infty(U;\bigwedge^{0,1})$; then

$$Lf = d_t f + \lambda \wedge \partial_x f;$$

and we see that $L^2 = 0$ is equivalent to

$$L\lambda = 0. \tag{15}$$

We can consider the differential complex

$$L : C^\infty(U;\textstyle\bigwedge^{0,q-1}) \to C^\infty(U;\bigwedge^{0,q}), \quad q = 1,2,\dots. \tag{16}$$

If we allow the coefficients to be distributions we must deal with the differential complex

$$L : \mathscr{D}'(U;\textstyle\bigwedge^{0,q-1}) \to \mathscr{D}'(U;\bigwedge^{0,q}), \quad q = 1,2,\dots. \tag{17}$$

The local solvability (or *local exactness*) in the complex (16), at the $q$ level, at the point $O$, can be defined as follows:

$(*)_q$ *To every open neighborhood $U$ of $O$ there is another open neighborhood $V \subset U$ such that $\forall f \in C^\infty(U;\bigwedge^{0,q})$ such that $Lf = 0$, $\exists u \in C^\infty(V;\bigwedge^{0,q-1})$ such that $Lu = f$ in $V$.*

A variant of $(*)_q$ generalizes the definition of local solvability of the single equation (1) at the beginning of §1:

$(*')_q$ *To every open neighborhood $U$ of $O$ there is another open neighborhood $V \subset U$ such that $\forall f \in C^\infty(U;\bigwedge^{0,q})$ such that $Lf = 0$, $\exists u \in \mathscr{D}'(V;\bigwedge^{0,q-1})$ such that $Lu = f$ in $V$.*

Before proceeding we wish to make two remarks:

I. When $q = n$ any form $f \in C^\infty(U;\bigwedge^{0,q})$ is $L$-closed; and therefore so is $\chi f$ whatever $\chi \in C_c^\infty(U)$. Of course this is not any more true when $q < n$; in this case we cannot assume, in Properties $(*)_q$ or $(*')_q$, that the support of $f$ is compact.

II. Property $(*)_1$ entails the existence, in a suitably small open neighborhood of $O$, of a first integral $Z$. This is proved by essentially the same argument as in the case $n = 1$. One must use the fact, easy to derive from (15), that

$$L\lambda_x = 0. \tag{18}$$

Then the analogue of (3), the equation $Lw = -\lambda_x$, will have a $C^\infty$ solution $w$ in a neighborhood of $O$ and we may define $Z$ as in the case $n = 1$, namely by Formula (4).

**3. Fibres of the structure. The approximate Poincaré Lemma.** Our aim is to find a property of the vector fields $L_1, \ldots, L_n$ equivalent to $(*)_q$. Of course we would prefer a statement that is clearly invariant — invariant under coordinate changes as well as nonsingular linear substitutions of $L_1, \ldots, L_n$. The latter expresses the fact that we are really dealing with a vector subbundle $\mathscr{V}$ of the complex(ified) tangent bundle of the base manifold; whereas the invariance under coordinate changes suggests that we think of the base manifold, not as $\mathbf{R}^{n+1}$, but as an arbitrary $C^\infty$ manifold $\mathscr{M}$ of dimension $n+1$. The rank, or fibre dimension, of the vector bundle $\mathscr{V}$ is equal to $n$. Moreover $\mathscr{V}$ is formally integrable: the commutation bracket of two smooth sections of $\mathscr{V}$ is a section of $\mathscr{V}$. Actually we are reasoning under the stronger hypothesis of local integrability: namely that the orthogonal of $\mathscr{V}$ in the complex cotangent bundle $\mathbf{C}T\mathscr{M}$ (for the duality between tangent and cotangent vectors) is locally spanned by an exact differential ($dZ$ in a neighborhood of $O$).

In connection with these questions notice that it is not at all obvious that Condition (6) is independent of the choice of the first integral $Z$. At this stage we know that the invariance of (6) is a consequence of the equivalence of (6) with the local solvability of (1). But in fact there is a direct approach to the invariance of (6) through the following consequence of the *Approximation Formula* ([BAOUENDI-TREVES, 1981]; see also Section II.2, [TREVES, 1992]):

**Lemma 1.** *To each open neighborhood $U$ of $O$ there is another open neighborhood $V \subset U$ of $O$ such that any $C^\infty$ solution of the system of homogeneous equations $L_jh = 0$ $(j = 1, \ldots, n)$ in $U$ is the limit in $C^\infty(V)$ of a sequence of polynomials $P(Z)$.*

Thus (locally) the solution $h$ is constant on any set on which $Z = \text{const.}$ (on any *fibre* of $Z$, as we say from now on). It follows that the *germs of sets* at $O$ defined by the fibres of $Z$ are independent of the choice of $Z$. The partition of the germ of manifold $(\mathscr{M}, O)$ into those germs of fibres is an invariant of the locally integrable structure defined by the vector bundle $\mathscr{V}$.

We may view Lemma 1 as a particular case (for zero-forms) of the following *Approximate Poincaré Lemma* (see Section II.6, [TREVES, 1992]):

**Lemma 2.** *To each open neighborhood $U$ of $O$ there is another open neighborhood $V \subset U$ of $O$ such that any differential form $f \in C^\infty(U; \bigwedge^{0,q})$ such that $Lf \equiv 0$ in $U$ is the limit in $C^\infty(V; \bigwedge^{0,q})$ of a sequence of forms $(d_t u)(Z; t, dt) = L\big(u(Z; t, dt)\big)$, where the coefficients of*

$$u(z; t, dt) = \sum_{|J|=q-1} u_J(t, z)\, dt_J$$

*depend smoothly on $t$ (say, in $\mathbf{R}^n$) and polynomially on the complex variable $z$.*

Lemma 2 has the following consequence, crucial for the forthcoming argument:

**Corollary.** *Let $U$ and $V$ be as in Lemma 2, and let $f \in C^\infty(U; \bigwedge^{0,q})$ be a differential form such that $Lf \equiv 0$ in $U$. Then the pullback of $f$ to any smooth submanifold $\mathscr{X}$ of any fibre of $Z$ in $V$ is exact.*

Indeed, it follows from Lemma 2 that the pullback of $f$ to $\mathscr{X}$ is orthogonal to all cycles in $\mathscr{X}$ and thus exact, by De Rham's theorem (Theorem 17, p. 113 [DE RHAM, 1955]).

In the next paragraph we sketch the outline of a proof of the local solvability at $O$ in the complex (16), at the level $q$. The proof is based on the above Corollary and makes use of the concept of fibres of the locally integrable structure ensuing from Lemma 1.

**4. Outline of a proof and conjecture.** Let $U$ and $V$ be as in Lemma 2, and let $f \in C^\infty(U; \bigwedge^{0,q})$ be a differential form such that $Lf \equiv 0$ in $U$; the pullback of $f$ to any *regular* fibre

$$F(V, z_0) = \{(t, x) \in V; x = x_0, \Phi(t, x) = y_0\}, \quad z_0 = x_0 + iy_0$$

is exact ("regular" means that $d_t\Phi$ does not vanish at any point of $F(V, z_0)$; in other words, $z_0$ is a noncritical value of $Z$).

We are now going to take a very sanguine view of our capabilities. To begin with we shall posit that $f$ can be integrated on each regular fibre in a manner that varies smoothly with the fibre. More precisely, we shall posit that there is $u \in C^\infty(V; \bigwedge^{0,q-1})$ such that $d_t\Phi \wedge (d_t u - f) \equiv 0$, i.e., that $\lambda \wedge (Lu - f) \equiv 0$. This means that

$$Lu - f = \lambda \wedge g \tag{19}$$

in a neighborhood of each regular fibre $F(V, z_0)$, for some smooth $(q-1)$-form $g$ which a priori depends on $z_0$. Our second working hypothesis will be that $g \in C^\infty(V; \bigwedge^{0,q-1})$ can be taken to be the same for all $z_0$. In other words we shall assume that (19) is valid in $V$. By letting $L$ act on both sides of (19) and taking (15) into account we see that

$$d_t\Phi \wedge d_t g \equiv 0. \tag{20}$$

First we look at the case $q = 1$. Here (20) means that $g$ is *locally constant* on each regular fibre, and therefore on each fibre $F(V, z)$. We make now the hypothesis that generalizes naturally (6): there is an open neighborhood $V' \subset V$ of $O$ such that, for each $z$, $F(V', z)$ is contained in a single connected component of $F(V, z)$; whence we conclude that $g$ is constant on $F(V', z)$. After further contracting $V'$ about $O$ this allows us to write $g(t, x) = \tilde{g}(z)$ with $\tilde{g}$ a continuous function in the compact set $Z(\overline{V}')$ ($\overline{V}'$: the closure of $V'$). We define $\tilde{g}(z) = 0$ if $z \notin Z(\overline{V}')$ and solve the equation $2\partial\tilde{v}/\partial\bar{z} = \tilde{g}$ in $\mathbf{R}^2$. If $v(t, x) = \tilde{v}(Z(t, x))$ we have

$$Lv = \big((\partial\tilde{v}/\partial\bar{z}) \circ Z\big) L\overline{Z} = \tfrac{1}{2}(\tilde{g} \circ Z) L(Z + \overline{Z}) = g\lambda,$$

whence $L(u - v) = f$ in $V'$ by (19).

Next we look at the case $q \geq 2$. Here we suppose that (possibly after contracting $V$ about $O$) we can find $v \in C^\infty(V; \bigwedge^{0,q-2})$ satisfying $d_t\Phi \wedge (d_t v - g) = 0$. This is equivalent to

$$\lambda \wedge (Lv - g) = -L(\lambda \wedge v) - \lambda \wedge g = 0,$$

whence, by (19),

$$L(u + \lambda \wedge v) = f \quad \text{in } V.$$

The difficulties that would beset us in trying to make the preceding proof rigorous are obvious:

1) Singular fibres occur. Actually the local solvability problem is interesting only to the extent that $d_t\Phi = 0$ at the central point $O$. Otherwise the structure is *elliptic* in a neighborhood of $O$ and in this case local solvability at $O$ is true at all levels (see Section VI.7, [Treves, 1992]).
2) Even knowing that the pullback of $f$ to regular fibres is exact it is not clear how to solve $d_t\Phi \wedge (d_t u - f) = 0$ or $d_t\Phi \wedge (d_t v - g) = 0$ with smooth $u$ and $v$.
3) More critically it is not even clear that one can find a smooth form $v$ such that $d_t\Phi \wedge (d_t v - g) = 0$ on a *single* regular fibre.

In order to respond to the objection **3** one must know, before anything else, that the cohomology of the regular fibres vanishes in degree $q-1$. There is a "model" case whose treatment is possible thanks to partial Fourier transform with respect to $x$, and for which a complete answer to the local solvability problem can be obtained. The answer is consistent with our guess.

*The Basic Hyperquadric Example:* $\Phi(t,x) = \Phi(t) = t_1^2 + \cdots + t_\nu^2 - t_{\nu+1}^2 - \cdots - t_n^2$. We can take the basic vector fields to be the "Mizohata" vector fields,

$$\partial/\partial t_j - 2i\varepsilon_j t_j \partial/\partial x, \tag{21}$$

with $\varepsilon_j = +1$ if $j \leq \nu$, $\varepsilon_j = -1$ if $j > \nu$. The fibres can be identified to the hyperquadrics

$$t_1^2 + \cdots + t_\nu^2 - t_{\nu+1}^2 - \cdots - t_n^2 = c. \tag{22}$$

It is possible to know precisely for which values of $\nu$ and $q$ local solvability holds:

(23) *There is no local solvability at the origin at the levels $q = \nu$ and $q = n - \nu$ (provided $q \geq 1$); local solvability holds at all other levels.*

Now compare (23) with the nonvanishing of the *singular homology* of the hyperquadric (22). Let us write $t' = (t_1, \ldots, t_\nu)$, $t'' = (t_{\nu+1}, \ldots, t_n)$. If $c > 0$, the map $t \mapsto ((|t''|^2 + c)^{-\frac{1}{2}} t', t'')$ is a diffeomorphism of (22) onto $\mathbf{S}^{\nu-1} \times \mathbf{R}^{n-\nu}$. It follows that the reduced homology of the hyperquadric (22) does not vanish in dimension $\nu - 1$. When $c < 0$ it does not vanish in dimension $n - \nu - 1$.

The time has come to state precisely the homological condition on the fibres of the structure. It must reflect the possibility that the equation $Lu = f$ be solved in a smaller neighborhood than the one in which the right hand side $f$ is defined and satisfies $Lf = 0$:

$(**)_q$ *To every open neighborhood $U$ of $O$ there is another open neighborhood $V \subset U$ of $O$ such that the natural map $H_q(F(V, z_0)) \to H_q(F(U, z_0))$ vanishes.*

(When $q = 0$, $H_0$ must be replaced by $\widetilde{H}_0$, the reduced homology.) The meaning of $(**)_q$ is the following:
when $q = 0$, $F(V, z_0)$ is contained in a single connected component of $F(U, z_0)$;
when $q \geq 1$, every $q$-cycle in $F(V, z_0)$ bounds in $F(U, z_0)$;

when $q = n-1$ and $n \geq 2$, no connected component of $F(V, z_0)$ is compact ($U$ is irrelevant).

When $n = 1$, there is only one condition: $F(V, z_0)$ must be connected, i.e., $F(V, z_0) = \{x_0\} \times [\eta, \eta']$ with $-\varepsilon < \eta \leq \eta' < \varepsilon$ in the notation of (2). This is precisely Condition (6). Thus our conjecture is that, for every $q = 1, \ldots, n$,

$$(24)_q \qquad (*)_q \iff (*')_q \iff (**)_{q-1}.$$

**5. Systems of real-analytic vector fields.** The equivalence $(24)_q$ was conjectured for all $q$, $1 \leq q \leq n$, in [TREVES, 1983]. Proofs were given only in the case $q = 1$, under the hypothesis that the coefficients of the basic vector fields $L_1, \ldots, L_n$ are real-analytic (but complex-valued!). Actually a somewhat weaker result was proved: on one hand, $(*')_1 \implies (**)_0$; on the other hand, $(*)_1$ follows if one assumes $(**)_0$ to be valid not only at the central point $O$ but at every point in a neighborhood of $O$. The proof of the latter entailment was different from the outline in §4; it relied on line integration of one-forms along the fibres. In the same work it was also shown that if $(**)_0$ fails to hold then there are $C^\infty$ perturbations of the basic vector fields $L_j^\# = L_j + R_j$, $(j = 1, \ldots, n)$, that, while formally integrable, are not locally integrable, i.e., no first integral $Z$ (with $dZ$ nowhere zero) exists in a full neighborhood of $O$.

Still under the analyticity hypothesis the equivalence $(24)_n$ is proved in [CORDARO-HOUNIE, 1990]. The implication $(*')_n \implies (**)_{n-1}$ is established by the now standard Hörmander method of violating the inequality that ensues from $(*')_n$. The proof that $(**)_{n-1} \implies (*)_n$ differs from the outline in §4 and from the method in [TREVES, 1983]: it uses duality and the Hahn-Banach theorem; it exploits the fact that, in the analytic case, when $(**)_{n-1}$ holds, every point $t$ in an open ball $\mathbf{B}$ centered at 0 can be joined to the boundary of $\mathbf{B}$ by a piecewise analytic curve on which $\Phi(t, x) = \text{const.}$ ($x$ is fixed).

The recent work [CHANILLO-TREVES, 1995] adheres to the guidelines of §4 to produce a proof that $(**)_{q-1} \implies (*)_q$ for all $q = 1, \ldots, n$, still under the hypothesis that the first integral $Z$ is of class $C^\omega$. It finds it convenient to let the point $t$ vary on an $n$-dimensional sphere $\mathbf{S}$ rather than in an open ball $\mathbf{B} \subset \mathbf{R}^n$; this has the advantage that, in such a setup, all the fibres can be regarded as *compact* real-analytic subvarieties of the sphere $\mathbf{S}$. This is achieved as follows. In applying $(**)_{q-1}$ one selects an open ball $\mathbf{B}$ in $t$-space and an open interval $\mathscr{I}$ in the $x$-line, both centered at 0, such that $\mathbf{B} \times \mathscr{I} \Subset V$. One deals with $L$-closed forms $f \in C^\infty(U; \bigwedge^q)$ and whose pullbacks to every smooth submanifold of every fibre $F(V, z)$ are therefore exact. Since by (11) $L$ and $L_0$ commute, $Lf \equiv 0 \implies LL_0^k f \equiv 0$, where $L_0 = (1 + i\Phi_x)^{-1}\partial/\partial x$ acts on the coefficients, and we may also assume that the pullback of $L_0^k f$ to every smooth submanifold of every fibre $F(V, z)$ is exact, for $k = 1, 2, \ldots$. We take then $\mathbf{S}$ to be the sphere in $\mathbf{R}^{n+1}$ whose projection into the equatorial hyperplane $\mathbf{R}^n$ is equal to $\overline{\mathbf{B}}$; we pull back all forms from $V$ to $\mathbf{S} \times \mathscr{I}$ by means of the natural projection $\mathbf{S} \times \mathscr{I} \to \overline{\mathbf{B}} \times \mathscr{I}$.

In this setup the proof in [CHANILLO-TREVES, 1995, 2] exploits the subanalyticity of the set of singular fibres, and its stratification, to overcome the obstacles **2** and **3** of §4. The two main procedures used are:

**I.** *Peeling.* Thanks to the exactness of the pullbacks of the forms $L_0^k f$ ($k = 0, 1, \ldots$) to each stratum of each singular fibre $F(V, z)$ ($\subset \mathbf{S}$), and by reasoning inductively on the dimension of the strata, one can solve the congruence $Lu \cong f$ modulo forms that vanish to infinite order on every singular fibre. This reduces the problem to solving the equation $Lu = f$ for right hand sides $f$ that have this property (and are $L$-closed). Then the form $g$ in (19) also vanishes to infinite order on every singular fibre.

**II.** *Integration of exact forms on the regular fibres according to Hodge theory.* In other words, in integrating on each regular fibre $F(V, z)$ the pullbacks of the forms $f$ and $g$ one selects the antiderivative that is orthogonal to all closed forms. This produces a form which varies smoothly with $z$ in the open set of noncritical values of the function $Z$. Then one applies the main result in [CHANILLO-TREVES, 1995, 1] to show that the lowest (nonzero) eigenvalue of the Hodge Laplacian on $F(V, z)$ is bounded from below by a power of the distance from $z$ to the critical set of $Z$. Since the right hand side $f$ vanishes to infinite order on every singular fibre, Hodge's theory provides us with a solution of the exterior derivative equation on each regular fibre equal to the pullback of a smooth form in the whole domain $\mathbf{S} \times \mathscr{I}$.

**6. The situation for systems of $C^\infty$ vector fields. Solvability in hyperfunctions.** In this paragraph we look at the case of a first integral $Z$ merely $C^\infty$, not $C^\omega$. At the date of this writing the known results are the following:

The implication $(*')q \implies (**)_{q-1}$ is established for every $q = 1, \ldots, n$, in [CORDARO-TREVES, 1991]. Once again this is done by violating the Hörmander inequality which is a consequence of $(*')_q$.

The implication $(**)_{q-1} \implies (*)_q$ has been established only for $q = 1$, in [MENDOZA-TREVES, 1991]. The key to the proof lies in representing functions and distributions with respect to $x$ as jumps of holomorphic functions defined above and below the curve

$$y = \Phi(t, x) \tag{25}$$

in the complex plane. This is done explicitly by means of a variant of the Fourier transform with respect to $x$. Here the functions and distributions depend smoothly on $t$, and so do the holomorphic functions. If the latter blow up as we approach the curve (25) (from above of from below) they do so "temperedly". The solution of $Lu = f \in C^\infty(U; \bigwedge^1)$ is carried out in $(t, z)$-space, in each region, above and below the curve (25) respectively. The integration with respect to $t$ must not disrupt the tempered growth as $z$ tends to the curve (25). This is achieved by partitioning the domain of $t$-integration into cubes whose diameter tends to zero as a power of $1/|\xi|$, and whose number is bounded by a power of $|\xi|$. ($\xi$ is the variable on the Fourier transform side.)

The argument in [MENDOZA-TREVES, 1991] would be greatly simplified if we could dispense with the need to preserve tempered growth as $z$ tends to the curve (25). This is possible within the framework of *hyperfunction* theory. For each $t$ we can consider the hyperfunctions on the curve (25), and it is not difficult to give a precise definition of what we mean by saying that such a hyperfunction in $U$ depends smoothly on $t$. These objects form a linear space $\mathscr{B}(U, C_t^\infty)$. However such a definition is problematic, in that it depends on the choice of the coordinates $t_j$ as well as of the first integral $Z$. But the crucial fact (proved in the monograph [CORDARO-TREVES, 1994] for any number of $C^\infty$ functions $Z_1, \dots, Z_m$ such that $dZ_1 \wedge \cdots \wedge dZ_m \neq 0$) is the invariance of the subspace $\mathfrak{Sol}(U, C_t^\infty)$ of $\mathscr{B}(U, C_t^\infty)$ consisting of the "hyperfunction solutions", that is to say, of those hyperfunctions $f$ (depending smoothly on $t$) such that $d_t f \equiv 0$: $\mathfrak{Sol}(U, C_t^\infty)$ is independent of the choice of the coordinates $t_j$ and remains the same if we substitute $H(Z)$ for $Z$ with $H$ holomorphic and $H' \neq 0$. The important point here is that *the equation $d_t f \equiv 0$ is equivalent to the equation $Lf \equiv 0$ when $f$ is a distribution.* We also wish to point out that there is even a more direct way of looking at the concept of hyperfunction solution. We can view the map $(t, x) \mapsto (t, Z(t, x))$ as valued in $\mathbf{C}^{n+1}$ (in $\mathbf{C}^{n+m}$ when there are $m$ first integrals $Z_j$). The image of $U$ is a totally real submanifold of $\mathbf{C}^{n+1}$ of real dimension $n + 1$, on which hyperfunctions are well defined. Call then $\mathfrak{Sol}(U)$ the linear space of those hyperfunctions $f$ that satisfy the equation $\partial_t f = 0$; there is a "canonical" isomorphism $\mathfrak{Sol}(U) \cong \mathfrak{Sol}(U, C_t^\infty)$.

It is now natural to build a differential complex on the previous notions. Call $\mathscr{B}(U, C_t^\infty; \bigwedge^q)$ the space of differential "forms" (14) whose coefficients $f_J(t, x)$ belong to $\mathscr{B}(U, C_t^\infty)$. The differential complex

$$(26) \qquad d_t : \mathscr{B}(U, C_t^\infty; \textstyle\bigwedge^{q-1}) \to \mathscr{B}(U, C_t^\infty; \bigwedge^q), \quad q = 1, 2, \dots,$$

specializes to (17) if we "restrict" it to the subspaces of *currents* (14), i.e., forms (14) whose coefficients are distributions in $U$. Again it must be noted that the spaces $\mathscr{B}(U, C_t^\infty; \bigwedge^q)$ are not "invariant". But it turns out that the cohomology spaces $\mathfrak{Sol}^{(q)}(U, C_t^\infty)$ of the differential complex (26) are. And they are naturally isomorphic to the corresponding spaces $\mathfrak{Sol}^{(q)}(U)$, i.e., when the coefficients in (14) are allowed to be arbitrary hyperfunctions in $U$ (in the sense indicated above).

In this language (and roughly speaking) the solvability in hyperfunctions at the level $q$ is equated to the vanishing of the $q^{\text{th}}$ cohomology space $\mathfrak{Sol}^{(q)}(U, C_t^\infty)$. More precisely, the local solvability in hyperfunctions, at the point $O$, at the level $q$, of the system of equations defined by the vector fields $L_1, \ldots, L_n$, can be stated as follows:

$(*'')_q$ *To every open neighborhood $U$ of $O$ there is another open neighborhood $V \subset U$ such that the natural restriction map*

$$\mathfrak{r}_V^U : \mathfrak{Sol}^{(q)}(U) \to \mathfrak{Sol}^{(q)}(V)$$

*vanishes.*

In Section IV.9 of [CORDARO-TREVES, 1994] it is shown that $(**)_{q-1} \Longrightarrow (*'')_q$; and in [CORDARO-TREVES, 1995] it is shown that $(*'')_q \Longrightarrow (**)_{q-1}$. In both cases, $q$ is any integer, $1 \leq q \leq n$. The proofs rely heavily on homological algebra.

## REFERENCES


Baouendi, M. S. and Treves, F., *A property of the functions and distributions annihilated by a locally integrable system of complex vector fields*, Ann. of Math. **113** (1981), 387–421.

Chanillo, S. and Treves, F., *On the lowest eigenvalue of the Hodge Laplacian* (1995) (to appear).

______, *Local exactness in a class of differential complexes* (1995) (to appear).

Cordaro, P. D. and Hounie, J., *On local solvability of underdetermined systems of vector fields*, Amer. J. Math. **112** (1990), 243–270.

Cordaro, P. D. and Treves, F., *Homology and cohomology in hypo-analytic structures of the hypersurface type*, J. Geometric Analysis **1** (1991), 39–70.

______, *Hyperfunctions on hypo-analytic manifolds*, Princeton University Press, Princeton, N. J., 1994, Annals of Math. Studies **136**.

______, *Necessary and sufficient conditions for the local solvability in hyperfunctions of a class of systems of complex vector fields*, Inv. Math. **120** (1995), 339–360.

De Rham, G., *Variétés différentiables*, Hermann, Paris, 1955.

Hörmander, L., *The analysis of linear partial differential operators IV*, Springer Verlag, Berlin, Heidelberg, 1985.

Mendoza, G. and Treves, F., *On the local solvability in locally integrable structures of corank one*, Duke Math. J. **82** (1991), 355–377.

Nirenberg, L. and Treves, F., *Solvability of a first-order linear partial differential equation*, Comm. Pure Appl. Math. **16** (1963), 331–351.

Treves, F., *On the local solvability and local integrability of systems of vector fields*, Acta Math. **151** (1983), 1–38.

______, *Hypo-analytic structures. Local Theory*, Princeton University Press, Princeton, N. J., 1992.



DEPARTMENT OF MATHEMATICS, RUTGERS UNIVERSITY, NEW BRUNSWICK, N. J. 08903, USA

*E-mail address*: treves@math.rutgers.edu

SHORT ABSTRACTS

## STAR PRODUCTS ASSOCIATED TO A HOMOGENEOUS POISSON BRACKET OF CONSTANT RANK

L. Boutet de Monvel

A cone is a manifold with a free action of the multiplicative group $\mathbf{R}_+$. A star product on a cone is an associative algebra law $(f,g) \to f * g$ on the space of formal series $\sum_{k \leq k_0} f_k$ of homogeneous functions of degree $k$, an integer $\to -\infty$, for which 1 is unit: $f * (g * h) = (f * g) * h$ , $1 * f = f * 1 = f$ , the law being given by a formal bilinear differential operator $f * g = \sum_{k \geq 0} L_k(f,g)$ where for each $k$, $L_k$ is a bilinear differential operator homogeneous of degree $-k$. Then $L_0(f,g) = fg$ is the usual product law, and $\{f,g\} = L_1(f,g) - L_1(g,f)$ is a Poisson bracket on X, homogeneous of degree $-1$. Using Fedosov's method we show that any homogeneous Poisson bracket of constant rank is associated to some star-product.

Institut de Mathématiques, U.E.R. 47, tour 45–46, 4, Place Jussieu, F-75005 PARIS, France

*E-mail address:* lbm@ccr.jussieu.fr

## GINZBURG-LANDAU VORTICES AND QUANTIZATION EFFECTS

H. Brézis

This lecture is concerned with the study in two dimensions of stationary solutions $u_\varepsilon$ of a complex-valued Ginzburg-Landau equation involving a small parameter $\varepsilon$:

$$-\Delta u_\varepsilon = \frac{1}{\varepsilon^2} u_\varepsilon(1 - |u_\varepsilon|^2) \quad \text{in } \Omega \subset \mathbf{R}^2.$$

Such problems are related to questions occurring in physics: e.g. phase transition phenomena in superconductors and superfluids. The parameter $\varepsilon$ has the dimension of a length, which is usually small. Thus it is of great interest to study the asymptotics as $\varepsilon \to 0$. In a joint work with Bethuel and Helein (see our book, Ginzburg-Landau Vortices, Birkhäuser 1994) we prove that $u_* = \lim_{\varepsilon \to 0} u_\varepsilon$ exists. Moreover $u_*$ is smooth except at a finite number of points called defects or vortices in physics. If $u_\varepsilon$ is a minimizer of the corresponding energy, then the number of defects is precisely the Brouwer degree of the boundary condition (as a map from $\partial\Omega$ into $S^1$). The singularities have infinite energy, but after removing the 'core energy' one is led to a concept of finite 'renormalized' energy. The location of these singularities is completely determined by minimizing the renormalized energy among all possible configurations of defects.

A blow-up analysis of the singularities yields the equation

$$-\Delta v = v(1 - |v|^2) \quad \text{on } \mathbf{R}^2,$$

which has surprising quantization properties. In particular, if $v$ is any solution one has

$$\frac{1}{2\pi} \int_{\mathbf{R}^2} (1 - |v|^2)^2 = q^2$$

with $q = 0, 1, 2, \ldots, \infty$.

We will also present numerous open problems.

Département de Mathématiques, Université Pierre et Marie Curie, 4 place Jussieu, F-75252 PARIS CEDEX 05, France and Department of Mathematics, Rutgers University, New Brunswick, NJ 08903, USA

*E-mail address:* brezis@ann.jussieu.fr

## THE HEAT KERNEL LEFSCHETZ FIXED POINT FORMULA FOR THE SPIN$_{\mathbf{C}}$-DIRAC OPERATOR

J. J. Duistermaat

Let $M$ be a compact complex analytic manifold and $L$ a holomorphic complex vector bundle over $M$. Suppose that the tangent bundle of $M$

and $L$ are provided with connections which leave the complex structure, and a Hermitian structure in it, invariant. Let $\gamma$ be an automorphism of $M$, which lifts to an automorphism $\gamma_L$ of $L$, and leaves the Hermitian structures invariant. This implies that the connected components $F$ of the fixed point set $M^\gamma$ are closed complex analytic submanifolds of $M$. The holomorphic Lefschetz fixed point formula of Atiyah and Singer expresses the alternating sum of the traces of the action of $\gamma$, on the cohomology groups $\mathrm{H}^q(M, \mathcal{O}(L))$, as the sum over the $F$' of the integral over $F$ of the $F$-volume part of the differential form

$$\det\left(\frac{\Gamma_F}{1-\mathrm{e}^{\Gamma_F}}\right) \operatorname{trace}\left(\gamma_L \, \mathrm{e}^{\Gamma^L}\right) \det\left(1-\gamma_N \, \mathrm{e}^{-\Gamma_N}\right)^{-1}.$$

Here $\Gamma_F$, $\Gamma_L$, $\Gamma_N$ is the Chern curvature matrix of the connection in $\mathrm{T}F$, $L$ and the normal bundle $N$ of $F$ in $M$, respectively. The determinants and the trace are the complex ones. Because these are conjugacy-invariant polynomials, we get an expression in terms of Chern forms, or Chern classes in the de Rham cohomology. Finally, $\gamma_N$ denotes the action of $\mathrm{T}\gamma$ on $N$.

Patodi and Gilkey later proved that, in the Kähler case, this anwer can be obtained as the short time limit of the alternating sum of the traces of a corresponding heat kernel. This is quite miraculous, because in general the asymptotic expansion contains singular terms, and the constant term is a complicated expression in derivatives of the geometric data, the order of which increases with the dimension of $F$. In the non-Kähler case, one has to replace the Dolbeault operator by the so-called Spin$_{\mathbf{C}}$-Dirac operator. This is a zero order perturbation of the Dolbeault operator, which is carefully chosen such that again the short time limit exists and the limit is equal to the form mentioned above, modulo an exact one. The proof actually works equally well in the somewhat more general framework of almost complex manifolds. A very conceptual proof can be given with the methods in the book on Heat Kernels and Dirac Operators, by Berline, Getzler and Vergne. Because in the literature I could only find sketches of the proof for the Spin$_{\mathbf{C}}$-Dirac operator, I needed quite some time for writing up all the details.

DEPARTMENT OF MATHEMATICS, UNIVERSITY OF UTRECHT, BUDAPESTLAAN 6, 3584 CD UTRECHT, HOLLAND

*E-mail address:* duis@math.ruu.nl

## WAVE TRACE INVARIANTS

V. GUILLEMIN

Let $X$ be a compact manifold without boundary and $H$ a self-adjoint first order elliptic operator on $X$. We will describe the singularities of the wave trace, $\operatorname{tr}\exp(itH)$, at $t = T$, where $T$ is the period of a non-degenerate periodic bicharacteristic, in terms of a "microlocal Birkhoff canonical form" for $H$ about this bicharacteristic. In particular we will discuss an inverse result which says that the wave trace singularities determine the classical Birkhoff canonical form of the Poincare map about this bicharacteristic.

DEPARTMENT OF MATHEMATICS, MIT, CAMBRIDGE, MASSACHUSETTS 02139, USA
*E-mail address:* vwg@math.mit.edu

## A REGULARITY THEORY FOR SINGULAR YANG-MILLS FIELDS

J. RÅDE

In this lecture I will develop a regularity theory for Yang-Mills fields in 4 dimensions with co-dimension two singularities. The Yang-Mills equation in 4 dimensions is conformally invariant; if it holds in say a Euclidean metric then it also holds in a hyperbolic metric with the singularity at infinity. To obtain the required estimates it is natural, and indeed necessary, to split the field into two components, and analyze one in the Euclidean metric and one in the hyperbolic metric.

DEPARTMENT OF MATHEMATICS, UNIVERSITY OF LUND, BOX 118, S-221 00 LUND, SWEDEN
*E-mail address:* rade@math.lu.se

# NEUMANN RESONANCES IN LINEAR ELASTICITY (JOINT WORK WITH P. STEFANOV)

G. VODEV

We study resonances (scattering poles) associated to the elasticity operator in the exterior of an arbitrary obstacle with Neumann boundary conditions. We prove that there exists an infinite sequence of resonances tending rapidly to the real axis, which is due to the existence of Rayleigh waves on the boundary. In the case of a strictly convex obstacle we obtain a large region free of resonances.

IRMAR, UNIVERSITÉ DE RENNES I, CAMPUS DE BEAULIEU, F-35042 RENNES CEDEX, FRANCE

*E-mail address:* Georgi.Vodev@univ-rennes1.fr

## Progress in Nonlinear Differential Equations and Their Applications

***Editor***

Haim Brezis
Département de Mathématiques
Université P. et M. Curie
4, Place Jussieu
75252 Paris Cedex 05
France
*and*
Department of Mathematics
Rutgers University
New Brunswick, NJ 08903
U.S.A.

*Progress in Nonlinear Differential Equations and Their Applications* is a book series that lies at the interface of pure and applied mathematics. Many differential equations are motivated by problems arising in such diversified fields as Mechanics, Physics, Differential Geometry, Engineering, Control Theory, Biology, and Economics. This series is open to both the theoretical and applied aspects, hopefully stimulating a fruitful interaction between the two sides. It will publish monographs, polished notes arising from lectures and seminars, graduate level texts, and proceedings of focused and refereed conferences.

We encourage preparation of manuscripts in some form of TeX for delivery in camera-ready copy, which leads to rapid publication, or in electronic form for interfacing with laser printers or typesetters.

Proposals should be sent directly to the editor or to: Birkhäuser Boston, 675 Massachusetts Avenue, Cambridge, MA 02139

PNLDE1 Partial Differential Equations and the Calculus of Variations, Volume I
Essays in Honor of Ennio De Giorgi
*F. Colombini, A. Marino, L. Modica, and S. Spagnolo, editors*

PNLDE2 Partial Differential Equations and the Calculus of Variations, Volume II
Essays in Honor of Ennio De Giorgi
*F. Colombini, A. Marino, L. Modica, and S. Spagnolo, editors*

PNLDE3 Propagation and Interaction of Singularities in Nonlinear Hyperbolic Problems
*Michael Beals*

PNLDE4 Variational Methods
*Henri Berestycki, Jean-Michel Coron, and Ivar Ekeland, editors*

PNLDE5 Composite Media and Homogenization Theory
*Gianni Dal Maso and Gian Fausto Dell'Antonio, editors*

PNLDE6 Infinite Dimensional Morse Theory and Multiple Solution Problems
*Kung-ching Chang*

PNLDE7 Nonlinear Differential Equations and their Equilibrium States, 3
*N.G. Lloyd, W.M. Ni, L.A. Peletier, J. Serrin, editors*

PNLDE 8 Introduction to Γ–Convergence
*Gianni Dal Maso*

PNLDE 9 Differential Inclusions in Nonsmooth Mechanical Problems: Shocks and Dry Frictions
*Manuel D. P. Monteiro Marques*

PNLDE 10 Periodic Solutions of Singular Lagrangian Systems
*Antonio Ambrosetti and Vittorio Coti Zelati*

PNLDE 11 Nonlinear Waves and Weak Turbulence with Applications in Oceanography and Condensed Matter Physics
*N. Fitzmaurice, D. Gurarie, F. McCaugham, and W. A. Woyczyński, editors*

PNLDE 12 Semester on Dynamical Systems: Euler International Math Institute, St. Petersburg, 1991
*Kuksin, Lazutkin, and Poeschel, editors*

PNLDE 13 Ginzburg-Landau Vortices
*F. Bethuel, H. Brezis, and F. Hélein*

PNLDE 14 Variational Methods for Image Segmentation
*Jean-Michel Morel and Sergio Solomini*

PNLDE 15 Topological Nonlinear Analysis: Degree, Singularity, and Variations
*Michele Matzeu and Alfonso Vignoli, editors*

PNLDE 16 Analytic Semigroups and Optimal Regularity in Parabolic Problems
*A. Lunardi*

PNLDE 17 Blowup for Nonlinear Hyperbolic Equations
*Serge Alinhac*

PNLDE 18 The Heat Kernel Lefschetz Fixed Point Formula for the Spin-c Dirac Operator
*J. Duistermaat*

PNLDE 19 Chaos and Nonlinear Dynamical Systems
*I. Hoveijn*

PNLDE 20 Topics in Geometry: Honoring the Memory of Joseph D'Atri
*Simon Gindikin, editor*

PNLDE 21 Partial Differential Equations and Mathematical Physics: The Danish-Swedish Analysis Seminar, 1995
*Lars Hörmander and Anders Melin, editors*